P9-DND-951

An astonishingly comprehensive survey of nature's fundamental ingredients . . . By combining juicy anecdotes and fun with a wealth of up-to-date reference material, *Nature's Building Blocks* hits the spot.

New York Times

From actinium to zirconium, Emsley marshals the details into a well-organized, user-friendly reference about every little thing that makes up every big thing that makes up the universe as we know it . . . fascinating . . . a lot of fun to flip through during idle moments.

San Francisco Chronicle

This is a very readable and non-technical survey of the building blocks of not just chemistry, but also biology, biochemistry, geology, medicine, metallurgy and nutrition. It contains stories of the elements of war and peace, the history of elements, the elements of murder and medicine. All of the ingredients of a good work of fiction are here. It really is a good bedtime read for all.

Times Higher Education Supplement

. . . general readers will find the book more than useful as a work of reference. When the latest scientific "breakthrough" or "disaster" is linked to this or that element in the news, when the public are advised not to cook in aluminium pans, or, when pregnant to "take magnesium or whatever other element", it will be instructive to turn to Emsley . . .

Times Literary Supplement

This is a fine, amusing and quirky book that will sit as comfortably on an academic's bookshelf as beside the loo, to be browsed and savoured . . .

Nature

John Emsley's colorful account of all the elements in the universe is a succinct history of everything . . . this marvelous reference work is the kind of book people consult in the pursuit of a single fact, but this fact will lead to another and another, drawing the reader in an enjoyable chase . . .

New York Times Book Review

"I had it for Christmas and I can't put it down."

David Bellamy in *New Scientist*

Emsley's design layout and presentation is logical, clear and beautifully written. The introduction itself is both informative and full of unexpected yet valuable information . . .

Chemistry in Britain

This is a splendid publication, to be read at one's leisure, to dip into at random, or to be used as a reference volume.

Physics World

Nature's Building Blocks

An A–Z Guide to the Elements

John Emsley
Department of Chemistry, University of Cambridge, UK

OXFORD
UNIVERSITY PRESS

Great Clarendon Street, Oxford OX2 6DP

Oxford University Press is a department of the University of Oxford.
It furthers the University's objective of excellence in research, scholarship, and education by publishing worldwide in

Oxford New York

Auckland Bangkok Buenos Aires Cape Town Chennai
Dar es Salaam Delhi Hong Kong Istanbul Karachi Kolkata
Kuala Lumpur Madrid Melbourne Mexico City Mumbai Nairobi
São Paulo Shanghai Taipei Tokyo Toronto

Oxford is a registered trade mark of Oxford University Press
in the UK and in certain other countries

Published in the United States
by Oxford University Press Inc., New York

© John Emsley, 2001

The moral rights of the author have been asserted

Database right Oxford University Press (maker)

First published 2001
First issued as an Oxford University Press paperback 2003
Reprinted with corrections 2003

All rights reserved. No part of this publication may be reproduced, stored in a retrieval system, or transmitted, in any form or by any means, without the prior permission in writing of Oxford University Press, or as expressly permitted by law, or under terms agreed with the appropriate reprographics rights organization. Enquiries concerning reproduction outside the scope of the above should be sent to the Rights Department, Oxford University Press, at the address above

You must not circulate this book in any other binding or cover
and you must impose this same condition on any acquirer

British Library Cataloguing in Publication Data

Data available

Library of Congress Cataloging in Publication Data

Data available

ISBN–13: 978–0–19–850340–8

9

Typeset in Plantin
by Footnote Graphics, Warminster, Wilts
Printed in Great Britain by
Clays Ltd, St Ives plc

To T.L.C.

Preface

When I wrote the data book *The Elements*, first published by Oxford University Press in 1989, I wrote it for scientists, and especially for chemists. Its pages consisted of tables of numerical data. I also included a few items of general interest, such as when and where each element was discovered, how it came by its name, whether it had a role in living things and how much there was of it in the rocks of the Earth's crust and in sea water. In the second and third editions of *The Elements*, published in 1991 and 1998, there were sections devoted to the biological and environmental roles, and these included things like human intake, toxicity, chief minerals, industrial production, etc. Even so, it was still a data book for scientists.

Nature's Building Blocks aims to please a much wider audience, and especially the general reader who may have had little scientific training. As a writer of popular science, I am aware of the desire of people to know more about the world around them. Most of us are aware that the world is composed of a limited number of elements, and probably we can name many of them, as well as knowing some of their well publicized uses or misuses. We know that sodium is a component of common salt, that without oxygen we would quickly die, that arsenic is poisonous, that we need calcium for teeth and bones, that some people lack iron, and that radon is a dangerous gas.

Such information is to be found in *Nature's Building Blocks* as well, but this book will tell you much more about the elements, and, I hope, it is presented in a way you will find easy to access and to understand. For each element there is a brief account of its discovery, a longer account of what we use it for, and an even longer account of its importance to living things and in particular to humans. What foods contain a particular element? Does our body store it, or shed it quickly? Is it an element essential for health? If so, why? I have also included a lot of environmental data and economic facts about production, known reserves and mineral deposits. I have added a short table of numerical data for each element, such as melting point, density, etc., because such information gives us clues to why an element behaves as it does.

In writing this book I have tried to bear in mind the kinds of question that I think readers are likely to be seeking answers for when they consult it. For example, is the element a threat to the environment? Are supplies of it in danger of running out? Do living things need it? Is it toxic? What is it used for?

To make this book easy to use I have arranged the elements in alphabetical order, and within each element I have provided information in the same format.

This planet is all we have. If we spoil it, our children will have to pay to clean up the mess we make. Our own body is all that we have. If we abuse it, or in our ignorance feed it badly, then one day there will be a price to pay. It is no longer good enough to say that we can do nothing, that we must leave the wellbeing of our planet to those who claim to be its guardians. Similarly, when it comes to knowing about the minerals that our body needs, we cannot simply leave it to doctors and dieticians to correct things when they go wrong, because our eating habits are faulty.

We owe it to ourselves to find out as much as we can about the world in which we live, although much of the information that we pick up from the media may seem contradictory. Let me give you an example. Each time you see or hear an advertisement praising the healthy benefits of fluoride toothpaste, you may wonder how this can be so, because fluorine is such a deadly and dangerous gas. If you turn to *Nature's Building Blocks*, you should find the answer: fluorine can hurt and it can can heal – it all depends on its chemical formulation. Discover what other wonderful things fluorine is doing in the fight against disease, and what it may do in the future.

What we do, we must do from knowledge not ignorance or prejudice. Only sound science should be our guide, and I have tried to make the science in *Nature's Building Blocks* as reliable as possible. This is a resource book that everyone can understand and I hope that those who read it will also enjoy it.

J.E.
London
April 2001

Acknowledgements

Three people have been of enormous assistance in writing *Nature's Building Blocks*: Michael Rodgers, Norman Greenwood and Walter Saxon. Dr Michael Rodgers originally commissioned the book and he has acted both as a friend and as a sympathetic editor. Professor Norman Greenwood FRS, of the University of Leeds, and co-author of one of the leading textbooks of chemistry, *Chemistry of the Elements*, provided particularly helpful suggestions regarding the chemistry of the elements. Walter Saxon, of Jackson Heights, New York, a physicist by training and a skilful editor of reference works, checked the manuscript, made equally useful suggestions for improving the coverage of the heavier elements, and supplied some most interesting data concerning the use of hydrogen and helium in airships. In some instances, where Walter has provided information that would otherwise have been omitted from the book, I have acknowledged his input. To all of them I owe a deep debt of gratitude.

There are a few others who I should also like to thank for taking an interest in the project, reading appropriate sections and adding titbits of information to them. Oliver Sacks provided more amusing items about some of the elements; and it was he who alerted me to the reason why antimony medication often relieved those for whom it was prescribed, and provided information about the notorious perpetual pills (see p. 34). Bill Catling, an industrial chemist of long standing, gave me information about early chemical pesticides. Professor Iain Campbell provided information about the discovery of selenium's role in enzymes. Hugh Aldersey-Williams tracked down the location of Ytterby, Sweden, which gave its name to the elements erbium, terbium, ytterbium and yttrium. Ian Donaldson of Bishopston sent me useful material on the early history of alum, included in the aluminium section, plus other historical information. Nick Lane, author of *Oxygen*, gave much appreciated data concerning that element. Michael Utidjian of Wayne, New Jersey, has also provided some useful comments about the toxicology of the elements. Professor Peter Craig of De Montfort University commented on the entries for antimony and arsenic. Others who also checked parts of the text were Mark Sephton of the Open University, Lesley Coldham of De Beers, Paul Dargan of Guy's Hospital, Peter Harris of Reading University, and Trevor Watts of King's College Dental School. Tapani Mutanen of Rovaniemi, Finland, has provided much useful geological data. Eric Banks of UMIST gave extra data about fluorine. Thanks also to Bill Haller of Countryside, Illinois, for telling me about the Minesota copper mine in Michigan. Albert Cotton informed me of the further eruption of Kudriavy volcano (p. 360). Mark Stoyer of the Lawrence Livermore National Laboratory, California, provided an account of elements 114, 116, and 118. Jim and Jenny Marshall of the University of North Texas informed me of their investigations into the discoverer of radon. Finally, many thanks to Michael Charney for his invaluable suggestions on a variety of topics, especially those related to health and environmental issues.

Finally, and most important, a heartfelt thanks to my wife Joan for her help and forbearance.

Contents

Introduction

There are many ways of arranging the elements. In *Nature's Building Blocks* they are in alphabetical order, or at least the first one hundred of them are, with those of atomic numbers greater than 100 being collected together as the 'transfermium' elements. (Fermium is the element of atomic number 100.) Why choose an alphabetical order? The answer is simple: it is for the convenience of the general reader. There is a price to pay, however, because this leads to the elements being arranged in random order-chemically, so that any similarities between elements are lost. A full list of the elements is printed at the end of the book. (In case you want to check the shorthand chemical symbol against an element's name, these are given in alphabetical order as well.) If you want to know the *chemical* order of the elements, i.e. in sequence of their atomic numbers, this is given in the so-called periodic table at the back of the book.[1]

There are other ways of listing the elements, but none so amusing as the poem that Tom Lehrer wrote back in the 1950s. He entertained a whole generation of chemists when he sang his poem to the tune of the song 'I Am the Very Model of a Modern Major General' from Gilbert and Sullivan's operetta *The Pirates of Penzance*. It goes as follows:

There's antimony, arsenic, aluminum, selenium,
And hydrogen and oxygen and nitrogen and rhenium,
And nickel, neodymium, neptunium, germanium,
And iron, americium, ruthenium, uranium,
Europium, zirconium, lutetium, vanadium,
And lanthanum and osmium and astatine and radium,
And gold and protactinium and indium and gallium,
And iodine and thorium and thulium and thallium.
There's yttrium, ytterbium, actinium, rubidium,
And boron, gadolinium, niobium, iridium,
And strontium and silicon and silver and samarium,
And bismuth, bromine, lithium, beryllium and barium.
There's holmium and helium and hafnium and erbium,
And phosphorus and francium and fluorine and terbium,
And manganese and mercury, molybdenum, magnesium,
Dysprosium and scandium and cerium and cesium,
And lead, praseodymium and platinum, plutonium,
Palladium, promethium, potassium, polonium,

[1] In textbooks of the elements they are most likely to be arranged in the more scientifically logical order, grouping elements of similar chemistry together. It would also have been possible to arrange *Nature's Building Blocks* in order of atomic numbers, starting with hydrogen (atomic number 1), then helium (2), lithium (3), beryllium (4) and so on. Logical though this would seem, it would still take even a chemist a bit of searching to locate, say tin, in the book because it would come as element number 50, and few chemists carry this information around in their heads.

And tantalum, technetium, titanium, tellurium,
And cadmium and calcium and chromium and curium.
There's sulfur, californium and fermium, berkelium,
And also mendelevium, einsteinium, nobelium,
And argon, krypton, neon, radon, xenon, zinc and rhodium,
And chlorine, cobalt, carbon, copper, tungsten, tin and sodium.
These are the only ones of which the news has come to Harvard,
And there may be many others, but they haven't been discarvard.

©Tom Lehrer. Used by permission.

Lehrer wrote this poem at a time when new elements were being synthesized every year, thanks to advances in atomic and nuclear physics and it includes every element up to nobelium (atomic number 102). I am sure news has reached Harvard of the 10 or so new elements that have been synthesized since then.

There are over a hundred known elements – 115 at the last count. They start at hydrogen, with atomic number 1, and extend to the element with atomic number 118, which is called ununoctium, although elements 113, 115 and 117 have yet to be made. We may know of 115 elements but not all are to be found on Earth. Indeed elements with atomic numbers greater than 83, i.e. those above bismuth, are inherently unstable, and undergo radioactive decay. The rate at which this happens may be so slow for some of the heavy elements, like thorium and uranium, that large quantities of these still exist on Earth. Some other elements, like technetium (atomic number 43) and promethium (atomic number 61), are also unstable and may once have been abundant but have long since decayed away.

We know so much about some elements that whole books have been written about them. We know so little about others that it can all be written on a postcard.

There are several sections under each element and some of these may need a little explanation. The sections are as follows, although not all are given with every element:

- **Name**: how the name is pronounced and where it came from.
- **Cosmic element**: the element's production and occurrence in the stars.
- **Human element**: the amount of the element in the average person and what it does.
- **Food element**: this section is included for elements that are essential nutrients, and a few others that we take in with our food, yet which serve no useful purpose.
- **Medical element**: what happens if we get too much, or not enough, of the element, and how the element has been used by doctors through the ages.
- **Element of history**: who discovered the element, how, when and where.
- **Element of war**: the role of the element in weaponry, explosives, munitions and armour.
- **Economic element**: annual production, where the element comes from, and what it is used for.
- **Environmental element**: the amounts of the element in the Earth's crust, soils, seas and air, and the threats it poses to the environment.
- **Chemical element**: main properties, reactions and isotopes.
- **Element of surprise**: a curious use or feature of the element.

The book ends with a chapter on the periodic table.

NAMES

In *Nature's Building Blocks* I have tried to avoid words and phrases that only a chemist would understand. Although we think of chemistry as a perfectly logical science, there are a few illogicalities in our use of language and symbols. One aspect which seems puzzling is the choice of chemical symbols for the elements. Why is gold Au, and silver Ag? The answer lies in the past: Au comes from the Latin *aurum*, and Ag from the Latin *argentum*. In fact there are only 11 common elements with symbols that cannot be guessed from their names: antimony (Sb), copper (Cu), gold (Au), iron (Fe), lead (Pb), mercury (Hg), potassium (K), silver (Ag), sodium (Na), tin (Sn) and tungsten (W). These long-known elements generally have ancient names which are different in different languages and it seemed best to use a symbol derived from the oldest names, which were generally Latin.

The symbol for other elements uses the initial letter as a capital and the second letter as lower case, unless that symbol has already been used for an element, in which case you take the next letter, and so on. So we have Ca for calcium, Cd for cadmium, Ce for cerium and Cs for caesium. This rule is not always held to, so chlorine is Cl not Ch, magnesium is Mg not Ma, and radon is Rn not Ra (this stands for radium), Rd or Ro. Such inconsistencies are not likely to confuse you. The common elements of organic chemistry have single letters: C for carbon, H for hydrogen, O for oxygen, N for nitrogen, P for phosphorus, S for sulfur and F for fluorine. The same goes for boron (B), iodine (I), uranium (U), vanadium (V) and yttrium (Y).

Finally, a little reminder about the spelling of the names of the elements. Three elements have different spellings of their English names:

- aluminium is the preferred spelling, but in the USA it is spelt aluminum;
- caesium is the preferred spelling, but in the USA it is spelt cesium;
- sulfur is the preferred spelling, but in the British Commonwealth it is often spelt sulphur.

Who says which name is preferred? The answer is the International Union of Pure and Applied Chemistry. This is a worldwide association of chemists from all countries, and it has a committee that meets to decide such weighty issues.

The pronunciation of the name in English is based on the following:

Symbol	as in	Symbol	as in	Symbol	as in	Symbol	as in	Symbol	as in
a	cat	e	set	j	joy	oo	soon	th	thin
ah	part	ee	see	k	kit	ow	cow	u	sun
ai	hair	er	term	l	lid	oy	joy	uh	your
aw	saw	f	fat	m	mat	p	pin	v	vat
ay	day	g	go	n	not	r	red	w	win
b	big	h	hat	ng	sing	s	set	y	yet
ch	chin	i	pit	o	dog	sh	shop	z	zip
d	dig	iy	lie	oh	go	t	top	zh	treasure

COSMIC ELEMENT

The Universe contains roughly 76% hydrogen, 23% helium and 1% for all the other elements. It came into existence around 15 billion years ago when, in the incredible temperature of the Big Bang, protons and neutrons were able to combine to form the nuclei of three elements. A proton by itself is simply hydrogen, and adding a neutron to it gives so-called 'heavy' hydrogen, also known as deuterium or hydrogen-2. The '2' is the mass number of one proton plus one neutron. When two of these nuclei combine they form helium-4, one of the most stable of all nuclei and one whose formation releases a vast amount of energy. It is this energy that continues to fuel stars like the Sun. Other nuclei that formed in the first few minutes of the Big Bang were helium-3 (from hydrogen-2 plus another proton) and lithium-7 (from helium-4 and hydrogen-3) but that was about as far as it went before the Big Bang Universe cooled down to temperatures below which these nucleus-forming reactions could occur.

As the expanding Universe grew, galaxies and stars formed as matter condensed under the pull of gravity. Given enough matter, it could produce a star with enough pressure at its centre to create temperatures of several millions of degrees. When this happens it allows the hydrogen-to-helium reaction to begin again. Indeed, inside such a star the temperature is hot enough to cause other nuclear reactions to occur, such as the combination of helium-4 nuclei to form other elements such as carbon-12, oxygen-16 and neon-20, and these in their turn can combine to form silicon, sulfur, magnesium and iron, with all of these nuclear fusions releasing yet more energy – but with iron the process comes to an end. Iron-56 is a particularly stable nucleus and turning this into heavier elements results in energy being *absorbed* instead of released; consequently normal stars do not form elements heavier than iron. These are only formed when a star explodes and scatters its contents throughout its galaxy, where later they too may condense under gravity to form another solar-type system.

We know this is what happens because the light from stars can be analysed to show the elements they contain. So-called astrochemistry began 200 years ago when, in 1802, William Wollaston examined the spectrum of sunlight and noticed that it contained many dark bands, but he could make nothing of them. In 1814 Joseph Fraunhofer carried out a more careful analysis of the Sun's spectrum and counted some 600 dark lines and these became known as the Fraunhofer lines. Then, in 1864, Sir William Huggins matched some of these with lines in the atomic spectra of elements and rightly deduced that these elements were present in the Sun.[2] So began the study of other stars and these became classified according to their spectra, which indicate the temperature at the surface of the star, the elements that are present, and the star's age.

Thus O-type stars are blue and are the hottest, having surface temperatures above 25 000 degrees, and consist only of hydrogen and helium. B-type stars are blue-white and cooler, having temperatures of between 11 000 and 25 000 degrees. Cooler still are A-type stars, which shine white and have temperatures between 7500 and 11 000 degrees. Sirius and Vega are A-type stars. Next come F-type stars, which are yellow-white

[2] Free atoms absorb and emit light in the visible spectrum (and also in the infrared and ultraviolet). The wavelengths of the absorbed and emitted light correspond to the movement of electrons between orbits, and an atom can easily be identified by the pattern of lines of colour in its visible spectrum.

and have surface temperatures of around 6000–7500 degrees; lines due to metallic elements are noticeable in their spectra.

The Sun is cooler than these, and is a yellow G-type star, which means it has a surface temperature of 5000–6000 degrees, and with metallic lines aplenty in its spectrum. After that come K-type orange stars at 3500–5000 degrees, where metallic lines dominate, and finally M-type red stars with temperatures below 3500 degrees and noticeable bands of titanium oxide in the spectrum. Betelgeuse is one such star. Some of these latter types of star have outer atmospheres enriched with heavier elements and are designated R- and S-type stars.[3] Some K- and M-type stars even show spectral features of carbon compounds, C_2, CN and CH, and are sometimes referred to as carbon or C-type stars.[4] (L and T dwarfs have recently been added.)

The composition of the Sun is roughly 94% hydrogen, 6% helium, and less than 0.2% other elements. The rankings for its 10 most abundant elements are given in the table below. These are ranked on a comparative basis, showing the number of atoms for each million atoms of hydrogen.

The top 10 elements in the Sun and their relative abundance*

Ranking	Element	Proportions
1	Hydrogen	1,000,000
2	Helium	63,000
3	Oxygen	690
4	Carbon	410
5	Nitrogen	87
6	Silicon	45
7	Magnesium	40
8	Neon	37
9	Iron	32
10	Sulfur	16

*Taken from *The Elements*.

HUMAN ELEMENT

At the head of this section for every element is a table giving the amount of an element in the main components of the human body: blood, bone and tissue (tissue generally meaning muscle tissue). Sometimes a single figure is quoted, which means that there is little variation, with the body maintaining an equilibrium, balancing absorption against loss. Sometimes a range of values is given when research has shown that the particular element varies considerably from person to person. There is also a figure for the total

[3] R stands for rapid and S for slow; they refer to the formation of elements by the capture of fast moving or slow moving neutrons.

[4] If this letter code for star types seems somewhat arbitrary, it is for historical reasons. However, the sequence O, B, A, F, G, K, M, R, N, S can be remembered using the mnemonic 'Oh Be A Fine Girl, Kiss Me Right Now Sweetheart!'.

amount in the average person; this is taken to be an adult weighing 70 kilograms.[5] Often the amounts in this table are tiny and given in units of p.p.m. (parts per million) or p.p.b. (parts per billion); if you find such small numbers difficult to imagine then you should consult the section 'Quantities and numbers, large and small' on p. 15.

The human body requires 25 elements if it is to function properly. These are called the 'essential elements' and are listed in the table below as fractions of the total body mass. In the tables for individual elements, the actual weights are given. Those tables also give the amount of all the non-essential elements as well, because we contain traces of all of the elements that exist on Earth.

Essential elements for the human body

Element	Fraction of total body mass
Oxygen	61%
Carbon	23%
Hydrogen	10%
Nitrogen	2.6%
Calcium	1.4%
Phosphorus	1.1%
Sulfur	0.2%
Potassium	0.2%
Sodium	0.14%
Chlorine	0.12%
Magnesium	0.027% (270 p.p.m.)
Silicon	0.026% (260 p.p.m.)
Iron	60 p.p.m.
Fluorine	37 p.p.m.
Zinc	33 p.p.m.
Copper	1 p.p.m.
Manganese	0.2 p.p.m.
Tin	0.2 p.p.m.
Iodine	0.2 p.p.m.
Selenium	0.2 p.p.m.
Nickel	0.2 p.p.m.
Molybdenum	0.1 p.p.m.
Vanadium	0.1 p.p.m.
Chromium	0.03 p.p.m. (30 p.p.b.)
Cobalt	0.02 p.p.m. (20 p.p.b.)

The fact that the human body is more than 60% water (H_2O) explains why oxygen and hydrogen feature so prominently; oxygen also features in bone, which is formed from

[5] This is the same as 154 pounds (USA) or 11 stone (UK).

calcium and phosphate (PO_4^{3-}), and because the skeleton of the average person amount to 9 kilograms this explains why calcium and phosphorus come high up in the list.

The element that is most important is carbon, which is the basis of all the myriad of different molecules that make up the living cell. The amino acids, which account for most body tissue, are made up of carbon, hydrogen, oxygen, nitrogen and sulfur, and again this explains why these elements come high in the list.

Many elements serve no known purpose, but they come with the food we eat, the water we drink and the air we breathe, and our body absorbs them, perhaps mistaking them for more useful elements. As a result we find that the average adult contains significant amounts of aluminium, barium, cadmium, lead and strontium, and trace amounts of many others, including silver, gold and uranium.

Because strontium so closely resembles calcium, we absorb a lot of this element: the average person has about 320 milligrams of it in their body, far more than many of the essential elements. On the other hand, the weight of gold in the average person is only 7 milligrams, and the weight of uranium is a mere 0.07 milligrams (which is 70 micrograms).

Our body tends to retain some of these unwanted intruders and deposits them either in our skeleton or in our liver. In the case of uranium, this has a special propensity to bind to phosphate and, as bone contains a large amount of this, this is where most uranium ends up. The liver, on the other hand, contains proteins that can trap metals like cadmium and gold.

FOOD ELEMENT

Some elements need to be taken in on a daily basis because the body needs a regular supply for its requirements and to replace that which is being continually lost. For these elements there is a section explaining why the body needs them, how much it needs each day, and the foods that are richest in them. There is also a 'Food element' section for some that are not essential but which are present in the diet and which in the past may have been a cause for concern. Indeed the intake of some is still a cause for concern.

MEDICAL ELEMENT

Down the centuries doctors have prescribed all kinds of medicines. Many of these were compounds of elements that we now regard as dangerous. The 'Medical element' section discusses those that were given to treat diseases, and explains why some elements are still used in the fight against disease. These are now used mainly on a diagnostic basis. Some metals, such as titanium, are inserted into the body as implants.

HISTORICAL ELEMENT

This section deals with the discovery of the elements, by whom this was made, where, when and how. For some elements this information is lost in the mists of time, because they were simply there to be found, such as gold and sulfur. Even when we *know* that someone must have been the first to extract a previously unknown metal from its ores, who that person is cannot now be discovered. In the seventeenth century, elements began to be discovered by known individuals, although often these people were not aware that they were discovering chemical elements as such. The concept of chemical elements was then still an alien concept, locked as people were in the Greek idea of there being just four elements (earth, air, fire and water) or the alchemical belief that all metals were composed of the elements salt, sulfur or mercury.

In his book *Traité Élémentaire de Chimie*, published in 1789, the great French chemist Antoine Lavoisier crystallized the concept of a chemical element and listed those he knew (see p. 514). From that time on it was possible to give full credit to those who discovered them, because they knew the nature of what they had found. The last decade of the eighteenth century and the first quarter of the nineteenth century saw many such discoveries, which petered out only in 1925 when rhenium, the last stable element to be discovered, was found. However, this was not the end of the story. The remainder of the twentieth century saw the synthesis of new elements, rather than their discovery in nature, and today new elements can be made using massive machines and at vast expense. Sometimes only a single atom is produced and that lasts for less than a second.

During the course of history, the same element may have been discovered by different chemists at different times, while some prolific chemists discovered several elements during the course of their researches. The Appendix on p. 529 lists the order in which the elements were probably discovered, the person to whom each discovery can be attributed, and where this occurred.

ELEMENT OF WAR

This may appear a strange category to include in a book on the elements, but I felt it merited a separate section rather than have this as a category under the 'Economic element' section, which was where I first put them if they had military applications. Clearly weapons of war can be thought of as 'uses', but they are intrinsically detrimental to human life and I felt they detracted from the positive information that the 'Economic element' section sought to convey.

Sixteen elements have an 'Element of war' section and these are: antimony (a component of Greek fire), arsenic (mustard gas), beryllium (hydrogen bombs), carbon (gunpowder), chlorine (war gas), hydrogen (bomb), iodine (discovered as a result of Napoleon's need for alternative sources of nitrate for gunpowder manufacture), iron (swords), magnesium (incendiary bombs), manganese (armour), nitrogen (explosives), phosphorus (bombs, nerve gases), plutonium (Nagasaki bomb), sulfur (gunpowder), tungsten (armour), uranium (atomic bombs) and vanadium (armour).

ECONOMIC ELEMENT

This section begins with the common minerals, where they are to be found, the extent to which they are extracted, and the reserves that have been estimated to be economically extractable. In many cases the chemical formula of the mineral(s) is given, but some minerals contain such a complex mixture that there is little point in providing the data. I have tried to reduce the formula to the simplest form possible – and without including molecules of water that are often associated with them.

The tonnage of the various elements that are mined every year is not easy to obtain because not all countries file reliable returns, and for this reason the figures given in this section should only be taken as a rough guide. The same goes for estimates of reserves. The production of the 10 most commonly mined metals is given in the table below, which is based on figures for the totals of the capitalist economies.

World mining and production of the most important metals*

Ranking	Element	Thousand tonnes
1	Iron	1 000 000
2	Aluminium	60 000
3	Manganese	18 000
4	Copper	12 000
5	Zinc	11 000
6	Lead	5000
7	Barium	5000
8	Titanium	4300
9	Chromium	3750
10	Fluorine	3500

* Taken from *Trace Elements in Soils and Plants*, 2nd edn, A. Kabata-Pendias and H. Pendias, CRC Press, Boca Raton, Florida (1991).

ENVIRONMENTAL ELEMENT

The table at the head of this section gives the proportion to which an element makes up the Earth's crust, the order in which it comes in terms of ranking, and its concentration in soil, sea water and the atmosphere. Again the quantities range from the very large to the very small; if the units are unfamiliar then please consult the section 'Quantities and numbers, large and small' on p. 15 for guidance. The following table gives the top 10 most abundant elements in the Earth's crust:

The top 10 elements in the Earth's crust and their abundance*

Ranking	Element	Proportion (p.p.m.)
1	Oxygen	466 000
2	Silicon	277 000
3	Aluminium	82 000
4	Iron	41 000
5	Calcium	41 000
6	Sodium	23 000
7	Potassium	21 000
8	Magnesium	21 000
9	Titanium	4400
10	Hydrogen	1400

* Taken from *Macmillan's Chemical and Physical Data*, A.M. James and M.P. Lord, Macmillan, London (1992).

The amounts of elements dissolved in the oceans can be equally vast, and even when the concentration is low the total can still be quite surprising. For example, although an element is present at only 5 p.p.m. in sea water, its total mass amounts to almost 7 trillion tonnes.[6] The lure of such vast amounts has tantalized people down the ages, and the amount of gold in the sea particularly so. At a concentration of only 10 parts per trillion (p.p.t.), even this amounts to more than 13 million tonnes of gold in the sea. But gold is not the rarest metal in the sea. For example, the concentration of holmium, one of the rare-earth metals, is only 0.4 p.p.t., amounting to only 550 000 tonnes in all. Even so, this is five times as much holmium as is estimated to be in land-based reserves in ores.

The 10 most abundant elements in the oceans are given in the following table:

The top 10 elements dissolved in the oceans and their abundance*

Ranking	Element	Proportion (p.p.m.)
1	Chlorine	19 400
2	Sodium	10 800
3	Magnesium	1300
4	Sulfur	904
5	Calcium	411
6	Potassium	392
7	Bromine	67
8	Carbon	28
9	Strontium	8
10	Boron	5

* Taken from *Macmillan's Chemical and Physical Data*, A.M. James and M.P. Lord, Macmillan, London (1992).

The least weighty part of the environment is the atmosphere, although it has a disproportionate role to play in human affairs. The next table lists the top 10 gaseous elements in the atmosphere. There are other gases present in more abundance than some of these, such as carbon dioxide (approx. 350 p.p.m.), methane (1.5 p.p.m.), nitrous oxide (0.5 p.p.m.) and ozone (0.4 p.p.m.) although this last one is also a form of elemental oxygen. In addition there are trace amounts of many more gases. There is also air-borne dust and sea-spray in the atmosphere and these carry traces of most other elements.

[6] The volume of the oceans is 1.37 billion billion (1.37×10^{18}) cubic metres, hence 5 p.p.m., which is 5×10^{-6}, results in a total mass of 6.85×10^{12} tonnes. (A cubic metre of water weighs a tonne.)

The top 10 gaseous elements in the atmosphere and their abundance*

Ranking	Element	Proportion (p.p.m.)**
1	Nitrogen	780 900
2	Oxygen	209 500
3	Argon	9300
4	Neon	18
5	Helium	5.2
6	Krypton	1.14
7	Hydrogen	0.5
8	Xenon	0.086
9	Radon	traces
10	Chlorine	virtually nil, but measurable in the stratosphere

* Data taken from Kaye and Laby's *Table of Physical and Chemical Constants*, 15th edn, Longman, Harlow, (1993).

** The values are for dry air. Normally air also contains water vapour and this can vary considerably from place to place and from day to day, accounting for up to 4%.

Finally, the section on the environment will also concern itself with actual and potential pollution by the element.

CHEMICAL ELEMENT

At the head of this section is a 'data file' giving the most important properties of an element, including its chemical symbol. This is followed by its atomic number, which starts at 1 (for hydrogen) and goes up to 116 (and maybe even more by the time you are reading this). The atomic number is the number of positive particles, or protons, which the element has in its nucleus.

The atomic weight of an element is a physical measurement and this is generally accurately known. For historical reasons these are all *relative* weights compared with the weight of an atom of carbon, which is taken as exactly 12.0000.[7]

Melting and boiling points are reported on the Celsius (C) scale, in which the freezing point of water is 0°C and the boiling point is 100°C. If you want to convert these to absolute temperatures, which is the Kelvin (K) scale, then add 273. If you prefer the Fahrenheit (F) scale, in which the freezing point of water is 32°F and its boiling point 212°F, then multiply the temperature in °C by 1.8 and add 32.

Density is reported in kilograms per litre, which is the same as grams per cubic centimetre.

[7] The actual weight of a carbon atom is 12 grams divided by the number of atoms it contains, which is 6.022×10^{23}, and that comes to 1.91×10^{-23} grams. In other words we are talking of weights in the range of a trillion-trillionth of a gram.

The chemical formula of an element's oxides reflects its valencies or oxidation numbers. The former term refers to the number of bonds a compound forms, whereas the latter term is a way of expressing its electronic state, and is more general when talking about metals. A few elements have zero valency, which means they are reluctant to form any compounds.

The 'Chemical element' section also describes the physical characteristics of the element, the class into which it falls in the periodic table, its stability in air and water, and its reaction with acids and alkalis. The terms used to describe the various groups of elements are as follows:

- The **alkali metals**, which constitute group 1 of the periodic table.
- The **alkaline earth metals**, which constitute group 2.
- The **transition metals**, which are the d-block elements of the periodic table.
- The **platinum-group metals** are a subset of the d-block and are the heavier elements of groups 8, 9 and 10.
- The **coinage metals** (copper, silver and gold) are those of group 11.
- The **typical metals** are the metals of the p-block.
- The **metalloids** are those elements that lie on the border between metals and non-metals in the p-block.
- The **non-metals** are those of the 1s- and p-blocks.
- The **halogens** are the elements of group 17.
- The **noble gases** are the elements of group 18, plus helium.
- The **lanthanides** are the first row of the f-block elements. These, along with the elements of group 3, used to be known as the rare earths, and this term is still in use.
- The **actinides** are the lower row of the f-block.
- The **transuranium elements** are those elements with atomic numbers greater than 92 (uranium).
- The **transfermium** elements are those with atomic numbers greater than 100 (fermium).

The 'Chemical elements' section also lists those isotopes of an element that make up the bulk of the element as it is generally encountered on Earth, noting when these isotopes are radioactive. I should point out that many radioactive isotopes occur in the Earth's crust that come from the fission of uranium atoms, but these are in such tiny amounts that there is no estimate of how much is present.

It comes as a surprise to many people to realize that the Earth and all it contains is radioactive. Each cell of our body contains billions of radioactive atoms and these are disintegrating at the rate of thousands per day. All this is perfectly natural and we should not to be too worried about minute traces of 'man-made' radioactivity that we might encounter when we undergo medical diagnosis.[8]

So what are the naturally radioactive elements? Radioactive decay emits three types of radiation: α-rays, β-rays and γ-rays (alpha, beta and gamma rays). The first of these is a particle consisting of two protons and two neutrons and is, in effect, the nucleus of a helium atom. When such a particle is ejected from an atom, that atom undergoes transformation to an atom with two fewer atomic units and four fewer mass units. Thus when uranium-238 (atomic number 92, mass 238) emits an α-ray, it becomes thorium-234 (atomic number 90, mass 234).

The second type of emission, the β-ray, is a negative electron, so that when a β-ray is emitted the element ejecting it *increases* its atomic number by one and its mass remains the same. For example, thorium-234 (atomic number 90) decays by β-emission to give protoactinium-234 (atomic number 91). Often accompanying α- and β-emission are γ-emissions; these are high-energy rays of electromagnetic radiation which are like X-rays and highly penetrating.

Radioactivity takes several forms and can vary from very weak to very dangerous. The common stable elements carbon and potassium exhibit mild radioactivity due to the presence of radioactive isotopes carbon-14 and potassium-40. Uranium is much more radioactive, and the elements that are made in nuclear reactors, such as plutonium, are highly radioactive.

The radioactive decay of uranium atoms into fission fragments supplies the Earth's crust with a large variety of radioactive isotopes of many other elements, especially those with atomic numbers 30–63. The amounts are tiny and they are not listed as such, nor do they have any effect on the overall isotopic composition of the elements as given in this section.

ELEMENT OF SURPRISE

These tailpieces contain information that could easily have come under one of the other sections, but which I have separated out because it reveals something quite unexpected about an element, such as its role in history, or use in modern life. However, there is one element that is surprising only in that I could find nothing to say that was surprising about it!

THE MOST IMPORTANT ELEMENTS?

I became aware, while writing *Nature's Building Blocks,* that some elements covered several pages of text while others struggled to fill one page – but was this really a reflection of their relative importance? In fact, I believe it is. However, before looking at the ranking of elements in order of lines of text in this book, let us look at the importance of elements in other ways. Of course there are two aspects to be considered: abundance

[8] The main radioactive isotopes in the human body, which come from natural sources, and which are present in the food we eat, are carbon-14, potassium-40, lead-210, polonium-210, radium-226, rubidium-87, thorium-232, and uranium-238. There are also minute traces of cesium-137, plutonium-239, plutonium-240, and strontium-90, which are there as a result of environmental contamination due to above-ground nuclear weapons testing.

and importance. An element can be abundant without being important, but whereas abundance is a scientific property that we can measure, importance is a subjective thing and what may be important to one person or their culture may be of little significance to others on this planet. Gold, silver and platinum spring to mind.

Clearly certain elements are far more important than others. For example, it goes without saying that the five elements that make up DNA (carbon, hydrogen, nitrogen, oxygen and phosphorus) all are equally important. And the same is true for the other essential elements that the human body requires, such as calcium, iron, magnesium, potassium, zinc, etc.

Then, of course, there are metals which may not be important to our metabolism, but nevertheless are essential to a modern lifestyle, such as aluminium. The same is true of uranium, without which a significant portion of the world's electricity needs could not be met.

Before I began writing *Nature's Building Blocks*, I was aware from my own collection of articles and cuttings, which goes back many years, that I have gathered masses of information about elements such as lead and mercury, because they too were important, if only because they have caused so much trouble down the centuries. I was also aware from compiling the data book *The Elements* that from a chemist's point of view some elements were particularly significant, such as chlorine and manganese, where the book's designers had really struggled to fit all the information on the two pages provided for each element.

Judged, then, by the number of words written about them in *Nature's Building Blocks*, the top 10 most important elements would appear to be:

The top 10 elements in *Nature's Building Blocks*

Ranking	Elements	Words
1	Hydrogen	3700
2	Sulfur	3600
3	Oxygen	3600
4	Lead	3500
5	Phosphorus	3400
6	Mercury	3400
7	Nitrogen	3000
8	Iron	2900
9	Zinc	2900
10	Silicon	2800

Somewhat surprisingly, and despite being the most important element in chemistry and the human body, carbon fails to make the list! But it only just fails, and is number 11.

The list comprises six non-metals and four metals; of the latter, the highest two are lead and mercury, both metals with shady pasts because of their damaging health and environmental implications. The other two metals are iron and zinc, both of which have important dietary requirements as well as being industrially important.

Other elements whose entries exceed 2000 words are 27 in number. In alphabetical order these are aluminium, antimony, arsenic, boron, cadmium, calcium, carbon, cerium, chlorine, copper, fluorine, gold, iodine, lithium, magnesium, nickel, platinum, plutonium, potassium, selenium, silver, sodium, thorium, tin, titanium, tungsten and uranium.

There are 37 elements whose entries exceed 1000 words; they are, again in alphabetical order: argon, barium, beryllium, bromine, caesium, chromium, cobalt, dysprosium, erbium, europium, gadolinium, germanium, hafnium, helium, iridium, lanthanum, manganese, molybdenum, neodymium, niobium, palladium, promethium, radium, radon, rhenium, rubidium, samarium, scandium, strontium, tantalum, technetium, tellurium, thallium, vanadium, xenon, yttrium and zirconium.

Of the remaining 44 elements, 16 are merely short-lived curiosities made in nuclear facilities, 14 are longer lived but highly radioactive and also made in nuclear power plants and the remaining 14 are stable but generally pretty rare and little used.

QUANTITIES AND NUMBERS, LARGE AND SMALL

The most common quantities when talking about elements are weights, volumes and proportions. On the everyday scale, Europeans speak of weights in grams and kilograms, volumes in cubic centimetres (or millilitres) and litres, and percentages (or fractions of a per cent). In the USA the common units of weight are ounces and pounds, while those for volumes are pints and gallons.[9] However, *Nature's Building Blocks* is basically a science book, so the quantities used are those of the *Système International*, or SI system, in which the basic unit of weight is the kilogram, and that of volume is the cubic metre. This latter quantity is large (a cubic metre of water weighs a tonne) and more useful units are the litre, which is a thousandth part of a cubic metre, and the cubic centimetre, which is a thousandth part of a litre.

Nature's Building Blocks will deal with both the very small, such as the concentration of a rare element in the sea, and the very large, such as the total amount of an element on Earth. To talk about tiny proportions we will need to speak of parts per million (p.p.m.), parts per billion (p.p.b.) and even parts per trillion (p.p.t.).[10] At the other extreme millions, billions and trillions of tonnes will be used when talking of global quantities. Each of these is a thousand times smaller than the next, so that a million tonnes (10^6 tonnes) is a thousand times smaller than a billion tonnes (10^9 tonnes), which is a thousand times smaller than a trillion tonnes (10^{12} tonnes). Such amounts are harder to visualize than extremely small quantities. 1 p.p.m. is equivalent to 1 milligram (about a grain of sand) in a litre of water; 1 p.p.b. is 1 microgram (about a speck of dust) in a litre; while 1 p.p.t. is 1 nanogram (invisible to the naked eye) dissolved in a litre.

A trillionth of anything is difficult to imagine but one second in time is a trillionth of 30 000 years and the span of a hand (15 centimetres) is a trillionth part of the distance

[9] A pint is just over half a litre, 0.568 of a litre to be exact, and an ounce is roughly 30 grams, 28.35 grams to be exact. A US gallon is 3.79 litres.

[10] One part per trillion is a thousand times smaller than 1 p.p.b., which is a thousand times smaller than 1 p.p.m.

from the Earth to the Sun. When talking about some of the radioactive elements that have been made in only the tiniest of quantities, then the units may be as small as a billionth of a gram, a quantity known as a nanogram (10^{-9} gram), or even smaller, such as a picogram (10^{-12} gram). Analytical chemists can even detect amounts a thousand times smaller than this.

A note of caution needs to be sounded here, because two scales of measurement overlap: p.p.m. and percentages. When talking about small quantities I could say 0.01% or 100 p.p.m., or for larger concentrations talk of 0.1% rather than 1000 p.p.m. Thus there comes a point where it may be more informative to the reader to use one or the other. The context will determine which it is to be, although in some cases I will use both for the sake of clarity; for example, I might say that the concentration of an element in the Earth's crust is 10 000 p.p.m. (1%).

Finally a word of caution. My previous book, *The Elements*, is a compilation of numerical data that are scientifically derived and, one hopes, unlikely ever to change in a significant way. The same can be said for some of the data in *Nature's Building Blocks* such as that in the various tables at the head of each section. Other data in this book are less secure and liable to change with time, such as figures for mining and amounts used for various processes. For this reason I have often rounded figures up or down.

The elements (A–Z)

Pronounced ak-tin-iuhm, the name comes from the Greek word *aktinos*, meaning a ray. French, *actinium*; German, *Actinium*; Italian, *attinio*; Spanish, *actinio*; Portuguese, *actínio*.

HUMAN ELEMENT

Actinium in the human body
None

Actinium is never encountered outside nuclear facilities or research laboratories, and it has no role to play in living things.

ELEMENT OF HISTORY

Actinium was discovered in 1899 by André Debierne at Paris, France, who extracted it from the uranium ore pitchblende in which it occurs naturally in trace amounts. Three years later, the German chemist Friedrich Otto Giesel independently extracted it; unaware it had already been discovered, he named it 'emanium', from the Latin *emanare* meaning to flow out or emanate.

ECONOMIC ELEMENT

Actinium extracted from uranium ores is the isotope actinium-227 which is a β-emitter with a half-life of approximately 22 years. It decays to form a sequence of isotopes, the first of which is thorium-227. The actinium used for research purposes is made by bombarding radium-236 with neutrons, but even the tiny amount that is produced has no economic value beyond research purposes. It is used as a source of neutrons, being about 150 times more active than radium.

ENVIRONMENTAL ELEMENT

Actinium in the environment	
Earth's crust	present but only as a trace
	Actinium is among the 10 least abundant elements.
Sea water	nil
Atmosphere	nil

Actinium occurs naturally in the Earth's crust, being produced as one of the sequence of isotopes that starts with the decay of uranium-235, but there is little of it, because it in its turn decays to thorium. The average lifetime of an actinium atom is only 33 years (the half-life of actinium-227 is 22.6 years). There are no detectable levels of the element in sea water or the atmosphere.

CHEMICAL ELEMENT

Data file	
Chemical symbol	Ac
Atomic number	89
Atomic weight	227.0 (isotope Ac-227)
Melting point	1047°C
Boiling point	approx. 3200°C (*est.*)
Density	10.06 kilograms per litre (10.06 grams per cubic centimetre)
Oxide	Ac_2O_3

Actinium is a soft, silvery-white metal that is little more than a chemical curiosity. However, it is the first member of a group of elements known as the actinides, which have atomic numbers 89 to 102 (nobelium); they are all radioactive metals. Although 99% of actinium-227 decays by β-emission it also emits intense γ-radiation which makes it dangerous to work with. Pure samples of the metal have been obtained only with difficulty. Actinium reacts with water to release hydrogen gas.

ELEMENT OF SURPRISE

All actinium samples glow in the dark because the intense radioactivity of this element excites the air around it.

Aluminium *or* Aluminum (USA)

Pronounced al-yoo-min-iuhm or al-oo-min-um (USA), the name is derived from the ancient name for alum,[11] which was *alumen* (Latin, meaning bitter salt). Aluminum was the original name given to the element by Humphry Davy but others called it aluminium and that became the accepted name in Europe. However, in the USA the preferred name was aluminum and when the American Chemical Society debated on the issue, in 1925, it decided to stick with aluminum.

French, *aluminium*; German, *Aluminium*; Italian, *alluminio*; Spanish, *aluminio*; Portuguese, *alumínio*.

HUMAN ELEMENT

Aluminium in the human body	
Blood	0.4 milligrams per litre
Bone	between 4 and 27 p.p.m.
Tissue	between 1 and 28 p.p.m.
Total amount in body	approx. 60 milligrams

No living species has been discovered to need aluminium as an essential element, but because it is so abundant in soil, all plants absorb it. Grasses can accumulate it, to the extent of its making up more than 1% of their dry weight. Some plants, such as tea bushes, absorb a lot of aluminium and alum is used as a fertilizer in tea plantations.

Most aluminium in food passes through the gut without being absorbed. Absorption is prevented by the presence of silicon, but appears to be encouraged by citric acid since this forms a soluble compound with aluminium. Once aluminium has entered the blood stream, the body finds it very difficult to remove.

Aluminium is no longer regarded as dangerous at the levels normally associated with foods and water supplies.

FOOD ELEMENT

Food crops also take up aluminium; levels are highest in spinach (104 parts per million (p.p.m.), dry weight), oats (82 p.p.m.), lettuce (73 p.p.m.), onions (63 p.p.m.) and

[11] Potassium aluminium sulfate.

potato (45 p.p.m.). The average daily intake of aluminium is about 5 milligrams, but can be double this depending on the diet. Only a small fraction is absorbed, probably less than 10 micrograms per day.

Foods with most aluminium are processed cheese, sponge cake mix, lentils, chickpeas and basmati rice. Cakes and biscuits are often made with the leavening agent sodium aluminium phosphate. One of the richest sources in food is processed cheese which can have almost 700 p.p.m. Cooking in aluminium pans does not greatly increase the amount of aluminium in food except when cooking acidic foods such as rhubarb.

Beverages account for 50% of the average daily intake of aluminium, especially among tea drinkers, with bread and cereal together accounting for 25%.

MEDICAL ELEMENT

As long ago as the first century AD, the Roman army doctor Dioscorides wrote a medical text, *De materia medica*, in which he recommended alum to stop bleeding – which it undoubtedly does – and for various skin conditions, such as eczema, ulcers and dandruff – for which it would be of little benefit. Today it is still available as an over-the-counter treatment in the form of styptic pencils for stopping bleeding from shaving cuts, but that is its only medical use.

In the 1970s doctors discovered that aluminium could be a serious health problem, when so-called dialysis dementia was diagnosed. Patients on kidney machines suffered progressive brain damage resulting in premature deaths. The cause was eventually traced to high levels of aluminium coming from the equipment. The aluminium attached itself to a molecule in the blood, called transferrin, and in this way it gained entry to the brain.

At one time it was thought that aluminium was the primary cause of Alzheimer's disease (senile dementia) because it was reported that the plaques which form in the brain of people with this disease contained high levels of aluminium. However, later analysis by a group at Oxford University, and reported in *Nature* in 1992, showed that there was much less aluminium than had been supposed, and what was present may have been inadvertently introduced with the dyes used to stain the brain tissue to identify the plaques.

ELEMENT OF HISTORY

Analysis of a curious metal ornament removed from the tomb of Chou-Chu, a military leader in third century China, showed that it was 85% aluminium. How it was produced remains a mystery.

The ancient world used aluminium in the form of alum, which is potassium aluminium sulfate, a mixed salt which occurs naturally in Greece, Turkey and Italy where it was used for centuries as a fixing agent for dyeing. It was exported mainly from Turkey via Constantinople, but in the Middle Ages it was discovered that it could also be made from clay and sulfuric acid. However, the discovery of alunite, a potassium aluminium sulfate rock, at Tolfa in territories controlled by the Pope, led to a Papal monopoly of its manufacture in Europe from the 1460s onwards. The industry employed 8000 men and produced 1500 tonnes of alum a year. This state of affairs continued until English alum production started in the early seventeenth century in north Yorkshire where a large deposit of alum shale was discovered. Production there broke the Papal monopoly and the price of alum fell dramatically. Production continued in Yorkshire for more than 250 years.

Alum was used by paper-makers as a preservative, by doctors to stop bleeding and to preserve anatomical preparations, and as a mordant by dyers whereby it fixed natural dyes to fabrics. Despite alum-making being the first true chemical industry, there was little understanding of the nature of the product or the process of its manufacture.

By the end of the eighteenth century, aluminium oxide was known to contain a metal, but it defied attempts to decompose it by conventional means, such as heating with carbon. Humphry Davy, who had used electric current to decompose other oxides to get the metals sodium and potassium, found his new process did not release aluminium.

The first person to produce aluminium, in 1825, was probably Hans Christian Oersted at Copenhagen, Denmark, and he did this by heating aluminium chloride with potassium metal, but the sample he produced was impure. It fell to the German chemist Friedrich Wöhler to perfect the method and, using sodium instead of potassium, he obtained the first pure samples of the new metal. It was expensive to produce by this method since sodium metal was costly, and consequently aluminium was for many years treated as a novelty of the rich and powerful. In the 1860s the Emperor Napoleon III of France impressed visiting heads of state with special cutlery made of the metal. For another 20 years or so aluminium continued to be much admired but its cost of production meant there were few uses made of it.

However, on 23 February 1886, a 21 year old US student, Charles M. Hall, of Oberlin College, Ohio, finally developed a method for extracting aluminium from molten aluminium salts using an electric current. The process had been shown to be chemically feasible 30 years earlier but, at the time, it was not commercially viable. Hall perfected his process in a makeshift laboratory in the family woodshed using a home-made battery, and on 9 July 1886 he was granted a patent for the method. However, his patent came under attack from two directions.

Hall had tried to get a local company interested in his work and for a short time they employed him while he carried out his researches. This allowed them a claim on his invention, as it normally does for anyone who is employed and whose work leads to a patent. Thankfully, the courts did not uphold their claim. The second attack was far more serious: a young Frenchman had applied for a patent on 23 April 1886 for the same process, several weeks before Hall. The inventor was 23 year old Paul-Louis-Toussaint Héroult, another young student who had worked alone on the problem. The US Patent Office accepted Hall's claim to be the first, on the basis of two letters he had written to his brother the day after he had first made aluminium describing what he had done, and which were still in their postmarked envelopes. Nevertheless, litigation continued until finally an agreement was reached which gave Hall the American rights and Héroult the European rights.

Both men had discovered that, in order to reduce aluminium oxide to aluminium electrolytically, it had to be dissolved in molten cryolite, in which it was quite soluble, and that carbon electrodes had to be used. Cryolite is sodium aluminium fluoride,

Na_3AlF_6. It occurs naturally and was discovered at the end of the eighteenth century in Greenland from where it was exported as raw material for making sodium carbonate, aluminium sulfate and aluminium oxide. Cryolite's most remarkable property is that it melts easily on heating to give a liquid which is one of the few substances to readily dissolve aluminium oxide.

As a result of the Hall–Héroult process, the price of aluminium fell until it cost one-thousandth of the price when Napoleon III had purchased his precious spoons.

ECONOMIC ELEMENT

Today aluminium is made by essentially the same process, passing an electric current through the oxide dissolved in molten cryolite, which is now synthesized instead of being mined. The electrodes are made of graphite, the cathode being the one at which aluminium forms and where it collects as molten metal, while the anode removes the oxygen by converting it to carbon dioxide, a process that also produces a lot of heat which helps keep the cell at its operating temperature of 960°C. Anodes last about 3 weeks before needing to be replaced. A typical aluminium plant producing about 120 000 tonnes a year requires more than 200 megawatts of electricity, enough to meet the needs of a small town. Around two-thirds of this is used to heat the cell and a third to convert the oxide to aluminium metal. Aluminium production consumes 5% of the electricity generated in the USA.

Industrial production world-wide of new metal is around 20 million tonnes a year, and a similar amount is recycled. Known reserves of ore are 6 billion tonnes. Important ores are boehmite, $AlO(OH)$, and gibbsite, $Al(OH)_3$, and bauxite, which is a mixture of the two. They were created geologically as a result of the exhaustive weathering of clays which removed all other elements but insoluble aluminium oxide and hydroxide, and this explains the large deposits in Australia, Brazil, Guinea and Jamaica. The main mining areas are Surinam, Jamaica, Ghana, Indonesia and Russia while the main smelting countries are USA, Russia, Canada, Australia, Norway and Brazil.

Aluminium is easy to recycle, as recycled aluminium requires only 5% of the energy needed to extract aluminium from aluminium oxide. Recycling aluminium drink cans is done simply by putting them into a molten mixture of sodium and potassium chlorides which dissolves the impurities in the scrap, leaving a pool of molten aluminium which sinks to the bottom of the furnace and can be drawn off.

Aluminium metal is used in hundreds of ways in the home, industry and transport. The main uses are in window frames, door handles, metal tubing, power cables, boats, car bodies, motor engines, kegs, aircraft parts, cooking foil and drink cans. Aluminium is lightweight, strong and protected from reacting with air and water by a micrometre-thick oxide film which forms rapidly on the surface. This can be deliberately made thicker by anodizing, a process in which the aluminium acts as the anode of an electrolytic cell with sulfuric acid as the electrolyte. Some alloys of aluminium, such as duralumin, are much stronger than aluminium itself. This alloy is mainly aluminium with 4% copper, plus small amounts of magnesium, manganese, iron and silicon. Duralumin can be made even stronger by quenching in water after heating the alloy to 600°C. Curiously, it continues to gain strength for about a week thereafter, a process known as 'ageing'.

Aluminium's ability to conduct electricity makes it ideal for cables, and its equally good reflectance properties find use in insulation, heat-reflecting blankets and solar mirrors.

Aluminium compounds, such as aluminium sulfate, are also manufactured on a large scale for paper treatment and water purification. For over a century, water engineers have been using aluminium sulfate as a flocculating agent. When water is cloudy with silt and bacteria it can be made sparklingly clean by adding a small amount of lime (calcium hydroxide) and aluminium sulfate. This combination precipitates aluminium hydroxide which carries the impurities down with it. Aluminium hydroxide is so insoluble that it leaves behind only 0.05 p.p.m. of dissolved aluminium in the water, well below the 0.2 p.p.m. recommended maximum for drinking water suggested by the World Health Organization. Surprisingly, flocculation can even remove excess aluminium which may be present naturally.

The most recent development in aluminium technology is the production of aluminium foam by adding to the molten metal a compound (a metal hydride) which releases hydrogen gas. The molten aluminium has to be thickened before this is done so that the bubbles of hydrogen gas don't escape, and this is achieved by adding aluminium oxide or silicon carbide fibres. The result is a solid foam which is used to line traffic tunnels because it acts to muffle sounds and is very durable. Aluminium foam is also used in the space shuttle.

ENVIRONMENTAL ELEMENT

Aluminium in the environment	
Earth's crust	82 000 p.p.m. (8.2%)
	Aluminium is the third most abundant element.
Soil	main constituent, varying between 0.5 and 10%
Sea water	approx. 0.5 p.p.b. Waters of the Atlantic Ocean contain 50 times as much aluminium as those of the Pacific.
Atmosphere	traces as dust particles

Aluminium is the most abundant metal in the Earth's crust, where it is present as the ion Al^{3+}. Aluminium contributes greatly to the properties of soil, where it is present mainly as insoluble aluminium hydroxide, with the result that soil water generally has only 0.4 p.p.m. of aluminum. However, if the soil becomes acidic and the pH drops below 4.5, then solubility increases sharply to more than 5 p.p.m. The result is greatly to affect plants and crops by reducing root growth and phosphate uptake.

CHEMICAL ELEMENT

Data file	
Chemical symbol	Al
Atomic number	13
Atomic weight	26.981538
Melting point	661°C
Boiling point	2467°C
Density	2.7 kilograms per litre (2.7 grams per cubic centimetre)
Oxide	Al_2O_3

Pure aluminium is a soft and malleable metal, and a member of group 13 of the periodic table. Aluminium is soluble in concentrated hydrochloric acid and sodium hydroxide solution.

Aluminium has only one naturally occurring isotope, aluminium-27, which is not radioactive.

ELEMENT OF SURPRISE

Several gemstones are made of the clear crystal form of aluminium oxide known as corundum. The presence of traces of other metals creates various colours: cobalt creates blue sapphires, and chromium makes red rubies. Both these are now easy and cheap to manufacture artificially. Topaz is aluminium silicate coloured yellow by traces of iron.

Americium

Pronounced amer-iss-iuhm, it is named after America where it was first made.

French, *américium*; German, *Americium*; Italian, *americio*; Spanish, *americio*; Portuguese, *americio*.

HUMAN ELEMENT

Americium in the human body
None

Although americium is a radioactive element that is widely used in smoke detectors, it poses little threat and is unlikely to get into the human food chain. It has no role to play in living things. Where americium is encountered, in nuclear facilities and research laboratories, stringent precautions have to be taken because it is an α-emitter, making it particularly dangerous if it gets into the body, where it tends to concentrate in the skeleton. When americium compounds are handled in gram amounts its γ-radiation is also a problem.

ELEMENT OF HISTORY

Americium was first made late in 1944 at the University of Chicago, Illinois, USA, by a team which included Glenn T. Seaborg, Ralph A. James, Leon O. Morgan and Albert Ghiorso. It was produced as a result of the bombardment of plutonium with neutrons in a nuclear reactor.[12] It was in fact discovered after curium, the element which follows it in the periodic table.

ECONOMIC ELEMENT

Americium, as the isotope americium-243, is produced in kilogram quantities from plutonium-239, and this is the most stable isotope with a relatively long half-life of 7370 years. Americium-241 is also extracted from nuclear reactors; this has a half life of 430 years. Most americium is used for research purposes, but a little of the americium-241 isotope is employed as a source of radiation for γ-radiography and in smoke detectors.

[12] The overall reaction involved plutonium accepting two neutrons: $^{239}Pu + 2n \rightarrow {}^{241}Am + e$.

A smoke detector contains 150 micrograms of americium oxide, and it relies on americium's α-radiation to ionize the air in a gap between two electrodes, thereby causing a tiny current to flow between them. This flow is monitored by an electronic circuit which sounds an alarm if the current drops below a certain level. This happens when smoke gets between the electrodes; particles of soot adsorb the ions, causing the current to fall. When the smoke is wafted away, the current rises again and the alarm stops.

Although 33 000 americium atoms per second undergo radioactive decay in a smoke detector, none of the α-particles escapes because these cannot penetrate solid matter – a sheet of paper will stop them. Even in air they travel only a few inches before colliding with an oxygen or nitrogen molecule, and as they do, an α-particle grabs electrons and becomes an atom of helium gas.

The US Atomic Energy Commission first offered americium oxide for sale in March 1962, at a price of $1500 per gram, and this is still the price today. A gram will supply enough americium for more than 5000 detectors. Smoke detectors must conform to standards set out by the National Radiological Protection Board but their disposal is not supervised. However, the radiation hazard they pose is insignificant relative to the benefit which they confer by saving lives.

Americium can also be made to produce neutrons, and these can be used in analytical probes. When an α-particle hits a beryllium atom it transmutes this element into carbon, and as it does so it emits a neutron. The stream of neutrons from such americium–beryllium sources is used to test flasks designed to hold radioactive materials, to ensure they are completely radiation proof.

The γ-rays that americium gives off have shorter wavelengths than X-rays and so are more penetrating. They were once used in radiography to determine the mineral content of bones and the fat content of soft tissue, but are now only used to determine the thickness of plate glass and metal sheeting. The transmission of the rays shows how thick the material is.

ENVIRONMENTAL ELEMENT

Americium in the environment	
Earth's crust	may be present but only in trace amounts
Sea water	nil
Atmosphere	nil

Americium probably does occur naturally on Earth, but only in incredibly tiny amounts in uranium minerals where nuclear reactions may occasionally produce an atom. Even if there had been a *billion* tonnes of americium when the Earth was formed, this would have reduced to a single atom in less than a million years. All that which has been available for study has been manufactured.

There are no natural sources of americium, but it is likely to have been present at times in the past when local concentrations of uranium were sufficient to cause nuclear reactions such as that which occurred at Oklo, Gabon, 2 billion years ago (see p. 481).

CHEMICAL ELEMENT

Data file	
Chemical symbol	Am
Atomic number	95
Atomic weight	243.1 (isotope Am-243)
Melting point	994°C
Density	13.7 kilograms per litre (13.7 grams per cubic centimetre)
Oxides	Am_2O_3 and AmO_2

Americium is a silvery, shiny, radioactive metal and one of the actinide group of elements of the periodic table. The metal itself is made from americium fluoride by reaction with molten barium at 1100°C. Americium is attacked by air, steam and acids, but not by alkalis. It is denser than lead. Several americium compounds have been made and these are generally coloured – for example, the chloride is pink.

ELEMENT OF SURPRISE

The discoverers of americium chose an unusual way to announce their discovery: in a children's radio show in the USA called 'Quiz Kids', broadcast on 11 November 1945. The guest scientist on the panel that week was a 33 year old chemist, Glenn T. Seaborg, who had worked on the top-secret atomic weapons programme that had produced two new elements, curium and americium. Americium came to light as part of the Allied project to develop nuclear weapons, so its discovery was kept secret until the end of World War II. Its existence was officially announced a few days after the broadcast, and the following year Seaborg proposed naming it americium, after the continent on which it was first produced.

Pronounced anti-moni, the name comes from the Greek *anti – monos* meaning not – alone. The chemical symbol, Sb, comes from the Latin word *stibium*, which was the name by which antimony sulfide was known in ancient times. Constantine of Africa, who died in AD 1078, first used the word antimony in his *De Gradibus Simplicibus* and is believed to have coined the name, although he was not referring to the element itself.

French, *antimoine*; German, *Antimon*; Italian, *antimonio*; Spanish, *antimonio*; Portuguese, *antimónio*.

HUMAN ELEMENT

Antimony in the human body	
Blood	3.3 p.p.b.
Bone	0.01–0.6 p.p.m.
Tissue	0.01–0.2 p.p.m.
Total amount in body	about 2 milligrams
Dangerous intake	100 milligrams

Antimony has no biological role. Indeed it is quite toxic and has been known to kill, despite being prescribed for hundreds of years as a treatment for all kinds of illnesses. In small doses, antimony stimulates metabolism, but large doses cause liver damage although the most obvious symptom is copious vomiting. Unlike it sister element, arsenic, antimony is not readily excreted from the body. Antimony attaches itself to particular enzymes because of its attraction to sulfur atoms at the enzyme's active site. Once attached it cannot easily be dislodged. Far more deadly is the gas stibine, which is antimony hydride (SbH_3).

Only tiny amounts of this element are present in soils, although when it is there in a soluble form it is taken up by plants. The level in food crops is of the order of parts per trillion and so poses no threat to human health.

MEDICAL ELEMENT

The Roman physician Dioscorides, who lived in the second half of the first century AD, was familiar with antimony sulfide (Sb_2S_3) or *stibi* as it was then known, and in his book *De materia medica* he recommended it for skin complaints and burns. Abulcasis (Kalaf ibn-Abbas al-Zahrawi) who died *c.* 1013 was a distinguished doctor of Moorish Spain; it was probably he who first advocated the use of antimony sulfide as a medicament.

Antimony compounds really came into vogue in the sixteenth century when they were used to treat all kinds of ailments. They were commonly used by vets and doctors as tartar emetic, the antimony salt of tartaric acid. This compound produces instant vomiting, seen as a way of expelling bad 'humours' from the body, and one way to achieve this was to leave some wine standing overnight in a cup made of antimony. By morning it had usually dissolved enough of the metal to produce the desired effect.

Oswald Croll (1580–1609) wrote *Basilica Chymica* which contained 23 recipes that included antimony, and especially 'butter of antimony' (antimony chloride, $SbCl_3$) made by reacting the sulfide with mercury chloride ($HgCl_2$). Antimony became a popular medicament in the late Middle Ages after the famous physician Paracelsus advocated its use.

James's powder, which consisted of calcium phosphate and antimony oxide, was a patented cure introduced in 1747 by Robert James MD (1705–1776). It had a strong diaphoretic action, i.e. it induced sweating. We now know that about 5 milligrams of antimony will produce this effect, whereas it requires about 50 milligrams or more to act as an emetic. The seventeenth and eighteenth centuries were the heyday for antimony treatments, although its effects could sometimes be devastating. Mozart may well have been prematurely killed by it – see box.

THE INEXPLICABLE EARLY DEATH OF MOZART

In October 1791, the 35 year old composer Wolfgang Amadeus Mozart was living in Vienna and fell ill. Slowly he got worse and died on 5 December of that year. He believed he was being poisoned and although his rival, Antonio Salieri, confessed to his murder many years later, when he did so he was suffering from senile dementia and the claim is not taken seriously.

More likely Mozart had been poisoned accidentally by antimony tartrate which was prescribed by his doctors and he was fond of dosing himself with it. On 20 November 1791 he developed a fever, his hands, feet and stomach became swollen and he had sudden attacks of vomiting. Miliary fever, an illness we no longer recognize, was diagnosed and he died of this 2 weeks later. Its symptoms are identical to those of acute antimony poisoning.

Many doctors opposed the use of antimony, knowing how dangerous it could be. Some Victorian doctors even used it to dispose of their unwanted wives and relatives, the symptoms being disguised as general gastric disorders. In this way Drs Palmer (1855), Smethurst (1859) and Pritchard (1865) carried out their murders – and were all eventually hanged. Other murderers were the society hostess Florence Bravo, who poisoned her husband in 1876 but was not brought to trial, and innkeeper George Chapman, who disposed of a few tiresome girl friends and was executed in 1902.

There was a revival of interest in antimony salts by doctors and especially vets in the twentieth century once it was discovered, in 1915, that they were effective against parasitic infections such as bilharziasis and trichinosis. It was given in doses that would kill these organisms yet not poison the host body. The medicine was given as a complex salt known as stibophen.

ELEMENT OF HISTORY

Antimony and its compounds were known to the ancient world. There is a 5000 year old vase in the Louvre which is almost the pure metal. Antimony sulfide (Sb_2S_3) is mentioned in an Egyptian papyrus of the sixteenth century BC, and from that time onwards the black form of this pigment, which occurs naturally as the mineral stibnite, was used as a type of mascara known as *khol*. The most famous user was the temptress Jezebel whose exploits are recorded in the Bible. (When antimony sulfide is made by precipitation it is not black but orange-red in colour and this form was used to make the heads of household matches in the nineteenth century.)

Another pigment known to the Chaldean civilization, which flourished in the sixth and seventh centuries BC, was yellow lead antimonate. This is found in the glaze of the ornamental bricks which Nebuchadnezzar (604–561 BC) used to adorn the walls of his capital, Babylon. This pigment was still being made in the twentieth century (from lead oxide or carbonate and antimony oxide) and became known as Naples yellow.

In 1604 the influential book *The Triumphal Chariot of Antimony* appeared which accurately described many antimony compounds. This book was reputed to have been written by a monk called Basil Valentine in 1400 but was actually written by a German, Johann Thölde.

ELEMENT OF WAR

Greek fire may have contained antimony sulfide in the form of stibnite. This incendiary liquid was fired from the warships of the Byzantine navy and brought terror to those exposed to it because it was impossible to extinguish; it even burned on the surface of water. How it was made has remained a secret to this day and it was a capital offence to reveal it. It was last used in the defence of the capital, Constantinople, in 1453. The most likely composition of Greek fire was crude oil, stibnite and saltpetre, a combination that would be highly flammable and almost impossible to extinguish with water. Once it is ignited antimony sulfide generates a lot of heat.

Antimony sulfide has a different role to play in modern warfare. Because it reflects infrared radiation, in the same way as green vegetation, it is used in camouflage paints.

ECONOMIC ELEMENT

Antimony is an important metal in the world economy. Annual production is around 50 000 tonnes per year, with virgin material coming mainly from China, Russia, Bolivia and South Africa, where deposits of antimony sulfide ores are to be found, and world reserves exceed 5 million tonnes. The chief ores are stibnite and tetrahedrite, the latter of which is mainly a copper ore but it yields antimony as a by-product. In Finland there is a deposit of elemental antimony. In the USA about half the antimony that is used is recovered from lead batteries.

Antimony is not quite a true metal – it is often referred to as a metalloid – and although in its metallic form it resembles lead it is a poor conductor of electricity. Nevertheless it can be alloyed with other metals and this accounts for its major use. The metal has a curious property in that it expands when it solidifies. This property was appreciated in the ancient world where it was used to produce finely cast objects. It really came into its own with the development of printing when an alloy consisting of 60% lead, 10% tin and 30% antimony was cast to give a sharp typeface. When lead-based pewter fell out of favour because of its toxic properties, it was replaced by an alloy of composition 89% tin, 7% antimony, with 2% each of copper and bismuth.

Antimony–lead alloys, apart from being used for storage batteries, are also used in bearings and cable sheathing, although all these uses are declining as lead gives way to other materials.

Antimony oxide, Sb_2O_3, is the form in which antimony is added to plastics to act as a flame retardant, especially in car components, televisions and cot mattresses (see below). This use accounts for about two-thirds of all consumption. Antimony oxide quenches a fire by reacting chemically with burning materials to form a viscous layer that smothers the flame.

Research into antimony still continues, with chemists using it to make semiconductors such as gallium arsenide antimonide.

ENVIRONMENTAL ELEMENT

Antimony in the environment	
Earth's crust	approx. 0.2 p.p.m.
	Antimony is the 63rd most abundant element.
Soil	approx. 1 p.p.m.
Sea water	approx. 0.3 p.p.b.
Atmosphere	insignificant

A little uncombined antimony occurs naturally as granular masses or nodules, generally in silver-bearing lodes, and has been found in Borneo, Sweden, Germany, Portugal, Italy, Canada and the USA.

Although many of its former uses have now been discontinued because of its toxicity, antimony oxide is still added to PVC as a flame retardant. It was this use which led to its being accused in the 1990s of causing cot deaths in babies. The theory was that antimony was converted to volatile stibine (SbH_3) by fungi in the mattress, and that babies were breathing in enough of the vapour to kill them. The fungus, *Scopulariopsis brevicaulis*, was fingered as the likely culprit, and it is indeed capable of reducing antimony to methyl stibine under laboratory conditions. Support seemed to come from the analysis of tissue from cot death victims, which contained higher than expected levels of antimony.

A panel set up to enquire into the affair reported in 1998, saying that there was no scientific evidence to support the theory. The fungus was only found in a tiny number of cot mattresses, and even when it was present there was no evidence of the toxic gas being emitted from them. Moreover, research by Mike Thompson and Ian Thornton showed that the level of antimony in healthy babies is higher than expected at 13 p.p.m., and that this comes from the domestic environment. In some homes they found levels of antimony in house dust to exceed 1800 p.p.m.. The average infant consumes around 100 milligrams of house dust a day, which would explain the raised levels. Where this antimony is coming from is still not clear, but the most likely source is from lead, which often contains traces of antimony, and which is still present in older homes as piping and layers of lead-based paints.

CHEMICAL ELEMENT

Data file	
Chemical symbol	Sb
Atomic number	51
Atomic weight	121.760
Melting point	631°C
Boiling point	1635°C
Density	6.67 kilograms per litre (6.67 grams per cubic centimetre)
Oxides	Sb_2O_3, Sb_2O_5

Antimony is a metalloid element and a member of group 15 of the periodic table of the elements. It can exist in two forms: the metallic form is bright, silvery, hard and brittle; the non-metallic form is a grey powder. Antimony is stable in dry air and is not attacked by dilute acids or alkalis. Its compounds can exist in two different oxidation states: antimony(III) and antimony(V), the former being the more stable.

There are two naturally occurring isotopes of antimony, antimony-121 which accounts for 57% and antimony-123 which accounts for 43%. Neither is radioactive.

ELEMENT OF SURPRISE

In the Middle Ages antimony metal pills were sold as reusable laxatives. Those suffering constipation would swallow one, about the size of a pea, and then wait for its toxic action to aggravate the intestines, causing them to become more active in order to expel the irritant pill. This was eventually retrieved from the expelled excrement and stored for future use! There are even reports of such pills being passed down from generation to generation.

Argon

Pronounced ar-gon, the name is derived from the Greek word *argos* meaning idle.
French, *argon*; German, *Argon*; Italian, *argo*; Spanish, *argón*; Portuguese, *argônio*.

HUMAN ELEMENT

Argon in the human body	
Blood	trace
Total amount in body	very small

Argon has no biological role. Bacteria in the nodules of certain plants, like beans, which have the ability to absorb nitrogen from the air and convert it to ammonia, can also absorb argon, although they are unable to process it further.

FOOD ELEMENT

As yet argon has no role to play in food preservation, although one day liquid argon may be used as a preservative, and argon gas as a non-oxidizing atmosphere inside containers of oxidizable products.

ELEMENT OF HISTORY

Although it is fairly abundant in the Earth's atmosphere (nearly 1%), argon evaded discovery until 1894 when the physicist John Strutt, the third Lord Rayleigh (1842–1919) and the chemist Sir William Ramsay (1852–1916), reported its existence to the annual meeting of the British Association for the Advancement of Science at Oxford. Rather strangely, they then fell silent about their remarkable discovery, declining to publish their results.

The reason for this was so that they could enter a competition for 'some new ... discovery about atmospheric air' which was organized by the Smithsonian Institution in Washington DC. A condition of the competition was that the discovery should not have been previously disclosed before the closing date at the end of 1894. Rayleigh and Ramsay eventually won the $10 000 prize, worth about $150 000 today, in competition with 218 other entrants.

Then, on 31 January 1895, before an audience of more than 800 people in the main lecture theatre of University College London, Professor Ramsay read a paper he and Rayleigh had submitted to the Royal Society reporting the discovery. In 1904 Rayleigh won the Nobel Prize for Physics and Ramsay won the Nobel Prize for Chemistry.

In fact they were not the first to discover argon, although they were the first to identify the gas. Argon was first isolated in 1785 in Clapham, London, by the wealthy eccentric, Henry Cavendish, who had a private laboratory. There he set about investigating the chemistry of the atmosphere. He passed electric sparks through a sample of air and absorbed the gases which formed, but he was puzzled that there remained about 1% of its volume that would not combine chemically. He did not realize that he had stumbled on a new gaseous element, not that he would have been able to identify it using the primitive techniques of his day. For over a century Cavendish's observations were not understood – but they were not forgotten.

The second encounter with argon occurred in around 1882 when H. F. Newall and W. N. Hartley independently recorded some new lines in the spectrum of air photographed at low pressures. They were unable to identify the element that was giving rise to these lines, although they realized there was something, as yet unidentified, that was present in the atmosphere. However, it was not this research which motivated Rayleigh and Ramsay. They discovered argon as a result of trying to solve a puzzling feature of nitrogen gas: why did the density of this depend on how it was obtained?

The nitrogen that was extracted from air had a density of 1.257 grams per litre, whereas that obtained by decomposing ammonia had a density of 1.251 grams per litre. Rayleigh had reported this in the prestigous journal *Nature* in 1892 but no one could offer a convincing reason why this should be so. Either atmospheric nitrogen must be mixed with a heavier gas, or chemically derived nitrogen diluted with a lighter gas.

The Scottish chemist, William Ramsay, wrote to Rayleigh suggesting that each of them should investigate one of the two kinds of nitrogen: he would look for a heavier gas in that from air, while Rayleigh would look for a lighter gas in the chemically derived nitrogen. Consequently Ramsay devised an experiment to remove all the nitrogen from his sample of the gas obtained from air. He repeatedly passed it over heated magnesium, with which nitrogen reacts to form a solid, magnesium nitride. Like Cavendish more than a hundred years earlier, he was left with about 1% of the volume which would not react, but unlike his predecessor he measured the density of the residual gas, and found it was denser than nitrogen (1.784 grams per litre).

More importantly, Ramsay examined its spectrum and there he saw new groups of red and green lines, confirming that he had isolated a hitherto unidentified gas. The great spectroscopist Sir William Crookes carried out a more detailed analysis of the spectrum and recorded nearly 200 lines. When Newall and Hartley re-examined the photographic plates they had taken 12 years earlier, the same lines were present.

ECONOMIC ELEMENT

Argon is an important industrial gas extracted from liquid air. World production exceeds 750 000 tonnes per year. The supply is virtually inexhaustible, with an estimated 66×10^{12} tonnes (= 66 trillion tonnes) circulating the planet. Hundreds of industrial plants around the world extract it from liquid air. A typical plant processes 375 tonnes of air a day, separating it into oxygen, nitrogen and argon, which are shipped out as liquids in tankers holding 20 tonnes per load.

Argon is used as an inert atmosphere in most kinds of electric lighting, such as tungsten filament bulbs and fluorescent lamps (at a pressure of 3 millimetres of mercury). In fluorescent tubes, the argon makes the lamps easier to start. These uses rely on argon's chemical inertness – nothing will induce it to react with any other material, no matter how high the temperature to which it is heated, nor how strong the electrical discharge passed through it. Even when it is ionized by radiation, as it is in some types of Geiger counter, the gas still refuses to react.

Argon is particularly important for the metals industry. Most goes to make steel, where it is blown through the molten metal, along with oxygen. The argon acts to stir the steel while the oxygen reacts with carbon to form carbon dioxide, thereby reducing the level of this. Argon is also used as a 'blanket' gas when air must be excluded to prevent oxidation of hot metals, as in welding aluminium and the production of titanium metal. Aluminium is welded using an electric arc welder which uses direct current to create a spark that melts the welding rod which is surrounded by argon flowing at a rate of 10–20 litres per minute. Atomic energy scientists protect fuel elements with argon during refining and reprocessing.

The alloys used for making high-grade tools start as metal powders, which are produced by directing a jet of liquid argon, at a temperature of –190°C, at a jet of the molten metal. The result is an ultrafine powder with a clean surface.

Some smelters prevent toxic metal dusts escaping to the environment by venting them through an argon plasma torch. In this, argon atoms are electrically charged to reach temperatures of 10 000°C and the toxic dust particles passing through it are turned into a blob of molten scrap.

Some consumer products contain argon, such as sealed double glazing. Argon in the gap between the panes of glass provides better insulation because it is a poorer conductor of heat than ordinary air. Illuminated signs glow blue if they contain argon and bright blue if a little mercury vapour is also present. The most exotic use of argon is in the tyres of luxury cars. Not only does it protect the rubber from attack by oxygen, but it ensures less tyre noise when the car is moving at speed.

Blue argon lasers are used in surgery to correct defects of the eye, to weld arteries and to destroy tumours. They are also used by chemists to probe molecular states which exist for only a trillionth of a second.

ENVIRONMENTAL ELEMENT

Argon in the environment	
Earth's crust	1.2 p.p.m.
	Argon is the 56th most abundant element.
Sea water	0.45 p.p.m.; the residence time on average is 28 000 years
Atmosphere	0.93% by volume

Argon is present in some potassium minerals because of radioactive decay of the isotope potassium-40. The argon swirling around in the Earth's atmosphere has slowly built up over billions of years, coming from the decay of potassium-40 which has a half-life of 1.27 billion years. Of a million potassium atoms, 117 are potassium-40 and of these 12 are able to seize one of the electrons surrounding the nucleus – a decay mode known as electron capture – to give argon-40 (the rest decays by β-ray emission and converts to calcium-40). By measuring the ratio of potassium to argon in a mineral, it is possible to date the sample. This provides geologists with a valuable technique for dating minerals.

Argon-40 accounts for 99.6% of the argon in the atmosphere; the remainder is mainly the lighter isotopes argon-36 (0.34%) and argon-38 (0.06%) which were present when the Earth formed.

The atmosphere of Mars also contains quite a bit of argon, to the extent of 1.6%, and this too is almost all argon-40.

CHEMICAL ELEMENT

Data file	
Chemical symbol	Ar*
Atomic number	18
Atomic weight	39.948
Melting point	−189°C
Boiling point	−186°C
Density of the gas	1.8 grams per litre
Oxides	none

* Originally the symbol for argon was simply 'A', but this was changed to Ar by IUPAC in 1957 to bring it into line with the other noble gases.

Argon is one of a group of elements, called the noble gases, which constitute group 18 of the periodic table of the elements; the others are helium, neon, krypton, xenon, radon and the yet unnamed element 118. Like them, argon is a colourless, odourless gas, but unlike them it is not rare. It used to be thought that argon was inert towards all other elements and chemicals, and until recently it resisted all attempts to make it bond to other elements and preferred to exist as single atoms. There are some chemical compounds which contain argon, the so-called argon clathrates, but in these it is present merely as atoms trapped in hole in the lattice of a larger molecule and is chemically unbound. As soon as the clathrates dissolve in water the argon escapes.

Argon, element 18, originally puzzled chemists because it turned out to be *too* heavy. Its relative atomic weight ought to be smaller than than that of potassium, element 19, which follows it in the periodic table, but argon's relative atomic weight is 39.95, which is greater than that of potassium (39.10). The reason lies in the isotopic distribution of the two: argon is mainly composed of the isotope argon-40, with 18 protons and 22 neutrons, while potassium is predominantly potassium-39, with 19 protons but only 20 neutrons.

Naturally occurring argon is not radioactive.

ELEMENT OF SURPRISE

Argon turned out to be not quite so chemically inert as had always been assumed. A group of chemists at the University of Helsinki, Finland, led by Markku Räsänen, announced in the magazine *Nature* (24 August 2000) that they had made the first ever compound of argon, argon fluorohydride (HArF). This they got by freezing a mixture of argon and hydrogen fluoride on to caesium iodide at –265°C and exposing the mixture to ultraviolet radiation, which caused the gases to react to form HArF. They identified the new compound by looking at its infrared spectrum. At temperatures below –246°C the compound clearly exists, albeit with extremely weak bonds, but on warming it reverts back to argon and hydrogen fluoride.

Pronounced ars-nik, the name probably comes from *arsenikon*, the Greek name for yellow orpiment, a mineral; it has been thought, wrongly, to have derived originally from the word *arsenikos*, meaning male, whereas it is more likely to have come from the Persian *al-zarnik*, meaning orpiment.

French, *arsenic*; German, *Arsen*; Italian, *arsenico*; Spanish, *arsénico*; Portuguese, *arsénic*.

HUMAN ELEMENT

Arsenic in the human body	
Blood	2–9 micrograms per litre of blood
Bone	0.1–1.6 p.p.m.
Tissue	0.1–1.6 p.p.m.
Hair	1 p.p.m.
Total amount in body	av. 7 milligrams but varies between 0.5 and 15 milligrams

Despite its notoriety as a deadly poison, arsenic is an essential trace element for some animals, and maybe even for humans, although the necessary intake may be as low as 0.01 milligrams per day. Chickens and rats that are fed an arsenic-free diet have stunted growth. The reason for this need is not clearly understood but it is believed to be linked to the metabolism of the essential amino acid arginine. Several marine species, such as algae and shrimps, contain organoarsenic compounds in the form of arseno-sugars and arsenobetaine.

A lethal dose of arsenic oxide is generally regarded as more than 100 milligrams. It exerts its toxic effect by binding to sulfur-containing enzymes and blocking their action, but the body can easily rid itself of this element. Even large doses can be survived once arsenic poisoning is diagnosed, because there are effective antidotes. The symptoms of arsenic poisoning are vomiting, colic, diarrhoea and dehydration. With a dangerous dose these symptoms progress further: the skin feels cold and clammy, the victim may go into a coma; heart failure is likely and death occurs within a day or two.

It is not surprising that arsenic was the agent chosen by many poisoners in the nineteenth and early twentieth centuries, when it was readily available as a weedkiller or could easily be extracted from flypapers. Given in small doses over a period of week, the symptoms could easily be mistaken for other diseases; the victim would become progressively weaker and appear to die of a wasting disease.

Earlier ages had also made use of it, especially in Italy and France where there was a flourishing underground trade in so-called 'succession powders' that removed dukes, kings and even popes. Those who perpetrated such crimes usually escaped prosecution since there was no reliable way to detect small amounts of arsenic in the body. This state of affairs changed radically when James Marsh devised his famous test in 1836 which could measure the element in minute amounts.

Sometimes mass poisonings occurred, such as that which afflicted 6000 beer-drinkers in Manchester, England, in 1900, of whom 70 died. The beer contained 15 p.p.m. of arsenic, so that 6 pints of the brew, equivalent to around 3 litres, would provide a dangerously high dose of 45 milligrams. The cause of the contamination was the invert sugar used to brew the beer. This had been made using sulfuric acid manufactured from the sulfur of iron pyrites that contained arsenopyrite.

FOOD ELEMENT

The average daily intake of arsenic in a normal diet can be anything up to 1 milligram, depending on the kinds of food eaten. Some foods contain relatively large amounts, although not at levels that might affect those who eat them. Plaice and oysters have 4 p.p.m., mussels 120 p.p.m. and prawns as much as 175 p.p.m., but the arsenic clearly does them no harm, nor does it affect those who eat them. The arsenic is in the form of arsenobetaine, and although this is readily absorbed from the gut, it is also rapidly excreted in the urine.

MEDICAL ELEMENT

In small doses arsenic stimulates metabolism and boosts the formation of red blood cells, but prolonged exposure causes chronic dermatitis and this has been suspected of eventually causing skin cancer, although the evidence is not compelling. However, breathing arsenic trioxide fumes, which are given off when metal sulfide ores that contain this element are roasted, has certainly been the cause of lung cancer in some workers.

The stimulatory effect of arsenic was exploited by unscrupulous racehorse trainers to improve an animal's chances of winning; if arsenic is detected in the animal's urine, and exceeds 50 micrograms per litre, then it is a sure sign of its having been doped in this way. Another, more legitimate use, of arsenic was to fatten poultry and pigs; the compound used for that purpose was roxarsone, a derivative of arsenic acid.

Arsenic was prescribed down the ages for all kinds of ailments, such as rheumatism, malaria, tuberculosis and diabetes. The ancient physicians, Hippocrates and Pliny, described the use of arsenic sulfides in medicines, but it really only became popular with the introduction of 'Dr Fowler's solution' in the eighteenth century. This was concocted in 1780 by the eponymous doctor, who found it beneficial when treating

[13] The recipe for making Fowler's solution was 10 grams of arsenic trioxide, 7.6 grams of potassium hydrogen carbonate and 30 millilitres of alcohol, made up to one litre with distilled water.

patients with fevers. It is essentially a solution of potassium arsenite plus lavender water, this being added to prevent accidents.[13] A few drops of the medicine were added to a glass of water or taken with wine. In the nineteenth century it was regarded as a popular cure-all, a general tonic and an aphrodisiac – even Charles Dickens used it. It was often prescribed by doctors to aid convalescence. (Vichy water was also reputed to have a tonic effect, which might also have been due to the 2 p.p.m. of arsenate that it contains.)

Misguided as these uses may have been, arsenic really did find a role in scientific medicine in 1909 when Paul Ehrlich discovered a chemical that was capable of curing syphilis. He had undertaken a systematic study of arsenic compounds guided by the belief that he might discover one that was toxic to the syphilis spirochaete yet not toxic enough to harm the patient. On his 606th attempt he finally discovered the one he was looking for (arsphenamine); it quickly became the recognized treatment for this disease, and was named Salvarsan. Eventually it fell out of favour, being superseded by penicillin, although it continued to be used in the treatment of sleeping sickness and dysentery.

Several modern Chinese medicines include arsenic sulfide as an ingredient. More recently, arsenic trioxide (As_2O_3), as the drug trisenox, was approved for use by the US Food and Drugs Administration for treating promyelocytic leukaemia; it acts by stimulating the production of normal blood cells which have become crowded out by cancerous white blood cells.

ELEMENT OF HISTORY

Contact with arsenic goes back more than 5000 years. We know this because hair from the 'Iceman', who was preserved in a glacier in the mountains of the Italian Alps, contained high levels of the element. His exposure to arsenic is thought to indicate that he was a coppersmith by trade, since the smelting of this metal is often from ores that are rich in arsenic.

Arsenic was certainly known to the ancient civilizations – it is mentioned in the Leiden Papyrus as a way of gilding – although not as the element itself.[14] Theophrastus, Aristotle's pupil and successor, recognized two forms of 'arsenic' (the name given to the arsenic sulfide minerals): orpiment (As_2S_3) and realgar (As_4S_4). The ancient Chinese also knew of arsenic: the encyclopaedic work of Pen Ts'ao Kan-Mu mentions it, noting its toxicity and use as a pesticide in millet fields.

Arsenic compounds are mentioned both in *Physica et Mystica*, the sacred text of Greek alchemy, and in the work of Bolus (a pseudonym of Democritus) written about 200 BC. According to the Roman writer Pliny, the Emperor Caligula (AD 12–41) financed a project for making gold from orpiment (As_2S_3) and some was produced but so small a quantity that the project was abandoned.

A more dangerous form of arsenic, called white arsenic (arsenic trioxide), was also known. Another alchemical treatise mentions a stone, probably realgar, which can be made to yield white arsenic by fusion with natron (natural soda) and mercury. It noted

[14] The Leiden Papyrus, named after the University of Leiden in The Netherlands where it is kept, was found at Thebes (South Egypt) in the early 18th century. It was written around the 3rd century AD and consists of recipes for alloys, dyes, etc.

that this material was a 'fiery poison' and capable of forming a 'pure spirit', probably referring to the sublimate of white crystals that grew on the inside walls of the flue of the furnace in which the realgar was heated. It went on to say that if this was mixed with vegetable oil and heated, it yielded another sublimate, probably arsenic metal itself, which when applied to copper gave it a silver colour. However, the discovery of the element arsenic is attributed to Albertus Magnus (1193–1280) who may have made it this way, or perhaps by using charcoal instead of oil.

ELEMENT OF WAR

An arsenic derivative, named lewisite, was used in World War I as a chemical weapon. It acted to disable soldiers by forming terrible blisters on exposed skin and damaging the lungs if the vapour was breathed. Lewisite is a liquid which boils at 170°C but it is volatile enough to provide a deadly vapour. The compound's chemical name is dichloro(2-chlorovinyl)arsine. Its common name came from the American chemist, Lewis, who developed it.

The antidote for lewisite is British anti-lewisite (BAL), which was injected and formed a compound with the chemical agent, thereby removing it from the body. BAL (also known as 2,3-dimercaptopropan-1-ol) is still used to treat people who have been poisoned by arsenic, mercury and other heavy metals. The two sulfhydryl (SH) groups in this molecule attach themselves so strongly to the arsenic that they can wrench it from the proteins and enzymes to which it has become attached in the body, and in this way it can be eliminated.

ECONOMIC ELEMENT

A little uncombined arsenic occurs naturally as microcrystalline masses, found in Siberia, Germany, France, Italy, Romania and the USA. Most arsenic is found in conjunction with sulfur in minerals such as arsenopyrite (AsFeS), realgar, orpiment and enargite (Cu_3AsS_4). None is mined as such because it is produced as a by-product of refining the ores of other metals, such as copper and lead. World production of arsenic, in the form of its oxide, is around 50 000 tonnes per year, far in excess of that required by industry. China is the chief exporting country. Other arsenic-producing countries are Chile and Mexico. World resources of arsenic in copper and lead ores exceed 10 million tons.

One of the few productive uses to which arsenic was put in its early years was as orpiment. This bright yellow pigment (called 'royal yellow') was much favoured by Dutch painters of the seventeenth century. Unfortunately they were not to know that over the subsequent centuries it would slowly oxidize to arsenic oxide and not only would the colour fade but the pigment would come adrift from the canvas.

Other industrial uses followed, such as the addition of arsenic to molten lead which was found to improve the spherical shape of shot. Arsenic compounds are still used in making special types of glass, as a wood preservative (as copper chromium arsenate) and, lately, in the semiconductor gallium arsenide, which has the ability to convert electric current to laser light.

Arsine gas, AsH_3, has become an important dopant gas in the microchip industry, although this requires strict guidelines regarding its use because it is extremely toxic. The dopant is added in tiny amounts so that a few arsenic atoms become incorporated into the microchip and it is those which determine the degree of semiconductivity.

ENVIRONMENTAL ELEMENT

Arsenic in the environment	
Earth's crust	1.5 p.p.m.
	Arsenic is the 53rd most abundant element.
Soils	1–10 p.p.m.
Sea water	1.6 p.p.b.
Atmosphere	trace, but higher near coal-burning industries and power stations

Arsenic in the atmosphere comes from various sources: volcanoes release about 3000 tonnes per year and micro-organisms release volatile methylarsines to the extent of 20 000 tonnes per year, but human activity is responsible for much more than these natural sources: 80 000 tonnes of arsenic per year are released from the burning of fossil fuels.

A much more widespread threat from arsenic afflicts many in developing countries; this has come about inadvertently as a result of improving water supplies. In countries like India, Bangladesh and Thailand, where drinking water was often taken from polluted streams, villagers were encouraged and paid by the United Nations in the 1960s to dig wells and draw water from the subterranean water table which was free of disease pathogens. Tens of thousands of such wells were dug and the water was used mainly for irrigation, enabling several rice crops to be grown each year. It also provided drinking water for the villagers. Unknown to them, it contained high levels of arsenic – up to 4 milligrams per litre in some places in the Indian state of West Bengal – where it had leached from the underlying rocks. (The World Health Organization's recommended maximum level of arsenic in drinking water is 0.01 milligrams per litre, i.e. 0.01 p.p.m.) The result has been to expose around 70 million people to chronic arsenic poisoning over many years, largely because in hot countries people drink several litres of water per day when working in the heat. This has caused widespread arsenicosis, in which the skin erupts in disfiguring, leprosy-like lesions. After many years of such exposure, cancerous growths begin to appear. The Indian government has issued chlorination tablets that will oxidize the arsenic from AsO_3^{3-} to AsO_4^{3-}, which forms an insoluble salt with the iron which is present in the water.

CHEMICAL ELEMENT

Data file	
Chemical symbol	As
Atomic number	33
Atomic weight	74.92160
Melting point	arsenic does not melt – it sublimes at 616°C
Density	see below
Oxides	As_2O_3 and As_2O_5

Arsenic is a metalloid element and a member of group 15 of the periodic table of the elements. It has two forms: grey, or α-arsenic, which is metallic with a density of 5.8 kilograms per litre (5.8 grams per cubic centimetre); and yellow, or α-arsenic, which is a non-metal with a density of 2.0 kilograms per litre. The metallic form is brittle, tarnishes and burns in oxygen, while the non-metallic form is less reactive but will dissolve when heated with strong oxidizing acids and alkalis.

Arsenic consists of a single isotope, arsenic-75, which is not radioactive.

ELEMENT OF SURPRISES

1. Regular ingestion of arsenic can allow the body to tolerate quite large doses, well in excess of that which would kill a normal person. The 'arsenic eaters' of the Styrian Alps in the seventeenth century were reputed to consume doses of 250 milligrams of arsenic trioxide twice a week. They obtained it from the flues of small metal refineries where it had sublimed. They claimed it produced a fresh complexion in women and enabled men, and horses, to work better at high altitudes.

   This supposed use of arsenic was ridiculed as nonsense by scientists in the nineteenth century after it had been publicized by a physician, J. J. von Tschudi, in 1851. However, in 1875 at a meeting of the Association of German Scientists and Physicians in Graz, a peasant consumed 400 milligrams of arsenic trioxide before a large audience and analysis of his urine subsequently confirmed that he was able to tolerate such a dose, which is twice the fatal dose for a normal adult. Arsenic eating on this scale is possible if the body becomes accustomed to it slowly, starting with doses of around 10 milligrams and increasing the amount gradually. Some peasants were reputed to have eaten arsenic regularly for most of their adult life.

   The villagers of San Pedro de Atacama, in an isolated region of modern-day Chile, are also immune to arsenic because their water supply contains 600 p.p.m.; strangely, they too show no signs of arsenic-related diseases.

2. Accidental arsenic poisoning was a threat in the nineteenth century when wallpapers were often printed with arsenical dyes such as Scheele's green and Paris green (also known as emerald green), which are copper arsenates. When walls became damp they could give off deadly methylarsine gas due to the action of moulds growing on them. Breathing the air in such rooms for any length of time, as in a bedroom, was

capable of producing chronic arsenic poisoning. It was finally proved that this was due to arsenic gases in the 1930s.

Such a fate might have befallen Napoleon when he was confined to the damp island of St Helena after the Battle of Waterloo. High levels of arsenic were detected in his hair when this was analysed by neutron activation analysis, showing he had been exposed to the element, although whether this was deliberate, or through taking Dr Fowler's solution, or from his wallpaper, is not known. The last of these possibilities was confirmed when a sample of wallpaper from Longwood House, his home on St Helena, was found in a scrapbook in the 1980s and was analysed; the green pattern on it was an arsenic pigment.[15]

3. Soil contaminated with arsenic can be cleaned up by growing the Chinese ladder fern, *Pteris vittata*, which grows rapidly and absorbs arsenic to the extent of 5% of its dry weight. The plant grows quickly and is easily harvested.

[15] A case of arsenic poisoning occurred in the US Embassy in Rome in the 1950s when a female ambassador, Clare Booth Luce, was affected. The cause was eventually traced to fungi growing on wallpaper coloured with green arsenical pigments. [Michael Utidjian]

Astatine

Pronounced as-ta-teen, the name comes from *astatos*, a Greek word meaning unstable. French, *astate*; German, *Astat*; Italian, *astato*; Spanish, *astato*; Portuguese, *astato*.

HUMAN ELEMENT

Astatine in the human body
None

Astatine has no biological role and is never encountered outside nuclear facilities or research laboratories. It is highly dangerous because of its intense radioactivity.

ELEMENT OF HISTORY

In 1931 a group of chemists led by F. Allison reported that they had detected the previously unknown element below iodine in the periodic table. They decided to call it alabamine, after the US state of Alabama, but their claim was mistaken. The element was first produced at the University of California in 1940 by Dale R. Corson, K. R. Mackenzie and Emilio Segré (an Italian physicist who fled Mussolini's Italy and who had previously synthesised technetium, see p. 424). Although they reported their discovery they were unable to carry on with their research due to World War II and the demands of the Manhattan project, which diverted all such work into the making of atomic weapons. Only in 1947 were they able to name it astatine.

ENVIRONMENTAL ELEMENT

Astatine in the environment	
Earth's crust	minute traces in some minerals
	Astatine is among the 10 least abundant elements.
Sea water	nil
Atmosphere	nil

Some isotopes of astatine, namely astatine-215, astatine-218 and astatine-219, are present in minute amounts in uranium and thorium minerals as part of the natural decay series, but the total amount present in the Earth's crust is estimated to be less than 30 grams. Isotopes 210 and 211 can be produced by bombarding bismuth with high-energy α-particles; the astatine produced is easily separated because it is volatile.

CHEMICAL ELEMENT

Data file	
Chemical symbol	At
Atomic number	85
Atomic weight	210.0 (isotope At-210)
Melting point	approx. 300°C (*est.*)
Boiling point	approx. 340°C (*est.*)
Density	not reported

Astatine is a reactive, radioactive non-metal element which resembles iodine. Like iodine it is one of the halogen group (group 17) of the periodic table of the elements. It has been little researched because all its isotopes have short half-lives. The half-life of astatine-210, the longest-lived isotope, is only 8 hours. All that is known about the element has been estimated from knowing its position in the periodic table below iodine and by studying its chemistry in extremely dilute solutions.

ELEMENT OF SURPRISE

Total world production of astatine to date is estimated to be less than a millionth of a gram, and virtually all of this has now decayed away. A piece of this element has never been seen with the naked eye, and it is unlikely that enough will ever be made to make this possible. Even if it were, the heat generated by astatine's intense radioactivity would immediately result in its complete evaporation.

Barium

Pronounced bair-iuhm, the name comes from the Greek *barys*, meaning heavy, and derives from the heaviness associated with barium minerals.

French, *baryum*; German, *Barium*; Italian, *bario*; Spanish, *bario*; Portuguese, *bário*.

HUMAN ELEMENT

Barium in the human body	
Blood	70 p.p.b.
Bone	3–70 p.p.m.
Tissue	0.1 p.p.m.
Total amount in body	approx. 22 milligrams

Barium has no biological role, or so it is generally believed, although for some species it appears to be essential – see 'Element of surprise' below. As long ago as the 1770s, it was detected in the ash from plants and vegetables, and in 1865 it was found to be present in seaweed and some sea creatures. It is quite an abundant element, so perhaps it is not surprising to find it in food to the extent that we probably take in about 1 milligram per day, and because this element is so like calcium we absorb some of it into our system, but it serves no useful purpose.

The food plants containing most barium are carrots (13 p.p.m. dry weight), onions (12 p.p.m.), lettuce (9 p.p.m.), beans (8 p.p.m.) and cereal grains (6 p.p.m.), while those containing the lowest are fruits such as apples (1 p.p.m.) and oranges (3 p.p.m.). Brazil nuts have been reported with as much as 10 000 p.p.m. (1%) of barium.

Barium can stimulate metabolism to the extent that it will cause the heart to beat erratically (known as ventricular fibrillation). Its soluble salts are dangerously toxic if taken. The symptoms of barium poisoning are vomiting, colic, diarrhoea, tremors and paralysis. Although barium carbonate is insoluble in water, it is toxic because it dissolves in the acid of the stomach, and it has been used as a rat poison.

MEDICAL ELEMENT

The barium of a 'barium meal' is barium sulfate, which is amongst the most insoluble of all salts, so it is safe to ingest and does not react with stomach acid. This form of barium is given to patients suffering gastric and intestinal disorders and its progress through the body can be followed by X-ray scans. Because it is an atom with many electrons, barium absorbs X-rays and thus shows up clearly on X-ray pictures.

ELEMENT OF HISTORY

Barium sulfate occurs naturally as the mineral barite and was first investigated by Vincenzo Casciarolo, a shoemaker who dabbled in alchemy; he found some shiny pebbles near his home in Bologna, Italy, in the early 1600s. Heating these stones to redness bestowed upon them a rather curious property: they shone in the dark. Indeed, provided it was exposed to the sun in the daytime, the 'Bologna stone', as it became known, would then shine for quite some time the following night.

Another Italian, Ulisse Aldrovandi, published an account of the phenomenon and so brought it to the notice of a wider public. In the eighteenth century Bologna stone was investigated by eminent chemists such as Carl Scheele, who analysed it and realized it was the sulfate of an unknown element. Meanwhile, a mineralogist, Dr William Withering (1741–99), had found another curiously heavy mineral in a lead mine in Cumberland, England, which clearly was not a lead ore. He called it *terra ponderosa* and it was later shown to be barium carbonate, and is now called witherite.

Separating the metal itself was beyond eighteenth century methods of chemistry, such as heating the oxide with carbon, but early in the nineteenth century, in 1808, Sir Humphry Davy was able to produce it by the electrolysis of molten barium hydroxide, at the Royal Institution in London.

ECONOMIC ELEMENT

The chief mined ores are barite (barium sulfate), which is also the most common, and witherite (barium carbonate). The main mining areas are the UK, Italy, Czech Republic, USA and Germany, with about 6 million tonnes being produced each year. Reserves are reported to exceed 400 million tonnes.

A rare barium mineral is benitoite, a barium titanium silicate, which is a translucent blue colour and used as a gemstone.

Barium metal is made from barium oxide (BaO) by heating with aluminium. It has few uses beyond being inserted as a 'getter' which scavenges the last traces of gas in a vacuum tube or fluorescent light so as to produce a perfect vacuum. Some is used to deoxidize copper. There are some alloys. The nickel–barium alloy readily emits electrons when heated and is used in electron tubes and for spark plugs. Frary metal is a calcium–barium–lead alloy used for bearings.

Some barium compounds are useful, such as barium sulfate, which in addition to being used in medical diagnostics is also employed on a large scale as a weighting agent in drilling for oil and gas exploration. The sulfate also finds use as coating for photographic paper and as a filler and delustrant for textiles.

The brilliant white pigment known as lithopone is a mixture of barium sulfate and zinc sulfide, and has the advantage of not darkening in the presence of sulfides. Barium hydroxide, $Ba(OH)_2$, is used to manufacture oil and grease additives; in the refining of beet sugar; as a dehairing agent and in the manufacture of special types of glass. Barium nitrate and chlorate give fireworks a green colour.

ENVIRONMENTAL ELEMENT

Barium in the environment	
Earth's crust	500 p.p.m.
	Barium is the 14th most abundant element.
Soils	approx. 500 p.p.m.
Sea water	10 p.p.b. in the Atlantic, 20 p.p.b. in the Pacific
Atmosphere	minute traces

Barium is suprisingly abundant in the Earth's crust, despite the fact that it is a heavy element: it is more abundant than even such common elements as carbon, sulfur and zinc. Because it forms insoluble salts with other common components of the environment, such as carbonate and sulfate, it is not mobile and poses little risk.

The amount of barium in soils ranges from as little as 20 p.p.m. to as high as 2300 p.p.m. (0.23%). Plants take up some barium, the highest concentrations being found in certain trees and shrubs which can have as much as 1% (dry weight).

CHEMICAL ELEMENT

Data file	
Chemical symbol	Ba
Atomic number	56
Atomic weight	137.327
Melting point	729°C
Boiling point	1637°C
Density	3.6 kilograms per litre (3.6 grams per cubic centimetre)
Oxides	BaO, BaO_2

Although its ores are heavy, metallic barium is unexpectedly light, as is shown by its density, which is half that of iron. Barium is a soft, silvery metal of the group known as the alkaline earths, which comprise group 2 of the periodic table of the elements. It rapidly tarnishes in air and reacts with water.

There are seven naturally occurring isotopes of barium; none is radioactive. The most common isotope is barium-138, which comprises 71.7% of the total; the other isotopes are barium-130 (0.1%), barium-132 (0.1%), barium-134 (2.4%), barium-135 (6.6%), barium-136 (7.9%) and barium-137 (11.2%).

ELEMENT OF SURPRISE

Large, single-celled algae known as desmids manage to thrive in such inhospitable places as pools on cold, upland peat bogs which are supplied mainly by rain water and where there are few nutrients. These curious plants, of which *Closterium* is the most common, can be up to a millimetre in length, and have prominent fluid-filled cavities containing tiny crystals of barium sulfate. The barium can be scavenged from water with only parts per billion of the element. At the other extreme, desmids will happily grow in water which contains as much as 35 p.p.m. of this toxic metal. The barium sulfate is not there in place of calcium sulfate (calcium and barium being very similar): desmids will produce barium sulfate crystals from water that may be 10 000 times richer in calcium than barium. What purpose the crystals serve is unclear, but one suggestion is that desmids use them as gravity sensors to orientate themselves. Failure of desmids to grow in the absence of barium shows how important this metal is to them.

Pronounced berk-eel-iuhm, it was named after the town of Berkeley, California, where it was first made.

French, *berkélium*; German, *Berkelium*; Italian, *berkelio*; Spanish, *berkelio*; Portuguese, *berquélio*.

HUMAN ELEMENT

Berkelium in the human body
None

Berkelium has no role to play in living things and is never encountered outside nuclear facilities or research laboratories. It is highly dangerous because it is a powerful source of radiation; the maximum permitted body burden is 0.0004 millicuries but so little is ever made that the chances of encountering even one atom of it are remote.

ELEMENT OF HISTORY

Berkelium was first produced on 19 December 1949, at the University of California at Berkeley, and was made by Stanley G. Thompson, Albert Ghiorso and Glenn T. Seaborg who bombarded 7 milligrams of pure americium-241 with helium ions for several hours in a 60-inch cyclotron.[16] Although americium had been discovered 5 years earlier, it had been necessary to collect and purify enough americium to form a 'target' large enough for the high-energy helium ions to hit and to stand a chance of combining to form a new element of higher atomic number. The americium had been made by exposing plutonium to neutrons inside a high-flux nuclear reactor.

Later that day, the Berkeley team dissolved the target and subjected the solution to a technique known as ion exchange to separate any new element that had been created. By late that evening their hopes had been realized: a new element was definitely present. They had made the isotope berkelium-243, which has a half-life of radioactive

[16] In a cyclotron, charged sub-atomic particles are fed into the centre and then accelerated outwards in a spiral motion until, by the time they reach the perimeter, they have very high energies, making them able to strike and interact with other similarly charged particles, such as the nuclei of atoms.

decay of about 5 hours. They were even able to deduce some of its chemistry, showing that it had two oxidation states just like the element above it in the periodic table.

It took a further 9 years before enough berkelium had been made to see with the naked eye and even this was only a few micrograms. The first chemical compound of berkelium, berkelium dioxide, was made in 1962.

ECONOMIC ELEMENT

No practical use of berkelium has so far emerged. It can be produced in nuclear reactors and in 100 milligram quantities, as the isotope berkelium-249, by bombarding plutonium-239 with neutrons. This isotope has a half-life of radioactive decay of 320 days.

ENVIRONMENTAL ELEMENT

Berkelium in the environment	
Earth's crust	virtually nil
Sea water	nil
Atmosphere	nil

Berkelium does not occur naturally on Earth. The few atoms of berkelium that could have existed before modern times were made in the Oklo nuclear reactor, a natural nuclear reactor that started up around 1.8 billion years ago and lasted on and off for around a billion years – see uranium (p. 481).

CHEMICAL ELEMENT

Data file	
Chemical symbol	Bk
Atomic number	97
Atomic weight	247.1 (isotope berkelium-247)
Melting point	1047°C
Boiling point	not known
Oxides	BkO_2 and Bk_2O_3

Berkelium is a radioactive metallic element that is a member of the actinide group of elements of the periodic table. A sample of the metal itself, which is silver in colour, was made from berkelium-249 isotope by reacting the oxide with molten barium at 1100°C. Its chemistry has been investigated to a limited extent and several compounds of berkelium have been made. Berkelium metal is attacked by oxygen, steam and acids, but not by alkalis.

Many isotopes of berkelium have been made, the longest lived of which is berkelium-247, with a half-life of 1400 years.

Beryllium

Pronounced be-ril-iuhm, the name comes from the Greek, *beryllo*, the word for the mineral beryl. When the element was first discovered, it was given the name glucinium from the Greek word *glykys*, meaning sweet, because its compounds tasted sweet.

French, *béryllium*; German, *Beryllium*; Italian, *berillio*; Spanish, *berilio*; Portuguese, *berílio*.

COSMIC ELEMENT

Only hydrogen, helium and lithium were formed during the Big Bang itself. The next element, beryllium, is relatively rare in the Universe because it is not formed in the nuclear furnaces of stars. It is generated only during supernova explosions, when heavier nuclei disintegrate into subatomic particles. Yet analysis of the spectrum of the star G64-12 shows that it contains far more beryllium than theory predicts, suggesting that there is yet another way in which this light element can be formed in the cosmos.

Radioactive beryllium-10 is produced when cosmic rays in the Earth's upper atmosphere interact with oxygen. It has been detected in Greenland ice cores and marine sediments; the amount in ice cores deposited over the past 200 years increases and decreases in line with the Sun's activity, as shown by the frequency of sun-spots. The amount of this isotope in marine sediments laid down in the last ice age was 25% higher than that in post-glacial deposits, indicating that the Earth's magnetic field was much weaker than it currently is (a weaker magnetic field deflects cosmic rays less well and so leads to more beryllium-10 being formed).

In 1950 the popular science writer Isaac Asimov wrote a prophetic short story called 'Sucker Bait', which was one of a collection entitled *The Martian Way*. It is about a space expedition which went to colonize a seemingly fertile planet but after a few years the colonists began to die of a mysterious disease which made breathing progressively difficult. Investigations revealed that the planet had abundant plant life and the atmosphere was quite breathable. So what was wrong? The explanation was that there were high levels of beryllium in the planet's environment, making it totally unsuitable for human habitation. Could such planets exist? Very probably. In 1992 a group of European astronomers discovered six old stars in our galaxy whose spectral investigation showed they had large quantities of beryllium.

HUMAN ELEMENT

Beryllium in the human body	
Blood	10 p.p.t.
Bone	3 p.p.b.
Tissue	0.1 p.p.b.
Total amount in body	approx. 35 micrograms

Beryllium has no known biological role, and indeed it is a dangerous metal to take into the human body. The tiny amount in the average person does not affect their health.

MEDICAL ELEMENT

If beryllium dust or fumes from the metal are breathed in, they cause chronic inflammation of the lungs and shortage of breath, and this is persistent even though most of the inhaled beryllium is carried by the blood to other sites in the body, generally to the bone where it concentrates. Brief exposure to a lot of beryllium, or long exposure to a little, will bring on this lung condition, which is known as berylliosis. The disease may take up to 5 years to manifest itself. About a third of those with the disease die of it and the rest are permanently disabled. Sometimes lung cancer may result, and the metal has been shown to be carcinogenic for laboratory animals.

When taken through the mouth, beryllium compounds may be less toxic but they are also dangerous and have been known to kill. Beryllium is related to magnesium, which is an essential element of human nutrition, and it can mimic this and displace it from certain key enzymes which then malfunction. The lungs are particularly sensitive.

Workers in industries using beryllium alloys are most at risk, as were those making early types of fluorescent lamps which were coated inside with phosphors containing up to 12% beryllium oxide. A survey in 1949 uncovered 400 cases of beryllium-related lung disease among workers in the industry in the USA (about 5% of those working in the industry), and in 1950 manufacture of this type of lamp was ended.

ELEMENT OF HISTORY

The beautiful minerals beryl and emerald were known to the ancient Egyptians, Jews and Romans; the Roman writer Pliny recognized them as forms of the same mineral. He said that beryl was mined near the Red Sea and that emerald was imported from India (but see 'Element of surprise' below). Beryl was also mined in the desert of Nubia at the time of Cleopatra. The emperor Nero used a large emerald, the better to view gladiatorial fights in the arena.

That these minerals might harbour a previously unknown element was suspected by the eighteenth century French mineralogist Abbé René-Just Haüy, who asked his compatriot Nicholas Louis Vauquelin to analyse them. This he did, and on 15 February 1798 he presented his results to the French Academy, announcing that the minerals contained a new element although he was unable to separate this from its oxide. The metal itself was isolated in 1828 by Friedrich Wöhler at Berlin, Germany, and independently by Antoine-Alexandere-Brutus Bussy at Paris, France, both of whom extracted it from beryllium chloride ($BeCl_2$) by reacting this with potassium.

Beryllium was to play a historic role in advancing our knowledge of atomic theory, since it helped uncover the fundamental particle, the neutron. This was discovered in 1932 by James Chadwick who bombarded a sample of beryllium with α-rays (from radium) and observed that it emitted a previously unknown kind of subatomic particle that had mass but no charge. The combination of radium and beryllium can be used to generate neutrons for research purposes, although a million α-particles only manage to produce 30 neutrons.

ELEMENT OF WAR

Beryllium has a rather unusual property: it does not absorb neutrons and even reflects them, and for this reason it is used in nuclear weapons and in the nuclear energy industry. In a nuclear warhead, which relies on neutron bombardment releasing energy from uranium, a casing of beryllium ensures a higher neutron flux within the bomb.

ECONOMIC ELEMENT

The chief ores of beryllium are beryl and bertrandite, both of which are silicates. Sometimes truly enormous crystals of the mineral turn up: one specimen found in Maine in the USA was over 5 metres in length and weighed almost 20 tonnes. The main areas where beryl and bertrandite are found are Brazil, USA, Argentina, Madagascar, Russia and India. Known reserves are in excess of 400 000 tonnes.

When copper and nickel are alloyed with beryllium they not only become much better at conducting electricity and heat, but also display remarkable elasticity. For this reason their alloys make good springs and were once used in watch springs. The copper one is also used to make spark-proof tools, which are the only ones allowed in sensitive areas such as oil refineries and plants where flammable gases are stored.

Beryllium is transparent to X-rays, which is why it is used as windows for X-ray tubes. Some beryllium oxide goes into specialist ceramics.

Earlier plans to use beryllium metal itself on a large scale in the aerospace industries did not materialize even though its lightness and strength made it seem an ideal metal for such purposes. At one time it was even thought that beryllium powder would be used as a fuel for rockets because of the colossal amount of heat which it releases when it is burnt.

Commercial production of beryllium metal began in 1957 but it never found the market that was once imagined, although its alloys have played an important part in the space shuttle. Less than 500 tonnes of metal are refined each year.

Beryl and chrysoberyl are beryllium aluminium silicates; when these are of gem quality they are known as emeralds and alexandrite (or cats' eyes) respectively. Their beautiful green colour is due to around 2% of chromium; aquamarine is also a form of beryllium aluminium silicate but with much less chromium.

ENVIRONMENTAL ELEMENT

Beryllium in the environment	
Earth's crust	2.6 p.p.m
	Beryllium is the 47th most abundant element.
Soils	6 p.p.m.
Sea water	0.2 p.p.t.
Atmosphere	minute traces – see 'Cosmic element' above

There is some environmental contamination by beryllium from coal-burning industries. Few people, other than industrial workers in some industries, had been exposed to high levels of beryllium until 1990, when there was an explosion at a Russian military plant near the border with China. It was producing and processing beryllium for nuclear warheads. The blast threw a cloud, of an estimated 4 tonnes of beryllium oxide dust, over the nearby town of Ust-Kamenogorsk and 120 000 people were exposed to the chemical. Subsequent monitoring of the inhabitants showed that only about 9% had been affected and even they had only slightly raised levels of beryllium in their bodies.

Beryllium in soil can pass into the plants grown on it, provided it is in a soluble form. Typical levels in plants vary between 1 and 400 p.p.b. – too low to affect the animals that live off them – even in plants that have some ability to concentrate the metal, such as legumes. Too much beryllium in the soil is toxic to plants and stunts foliage.

CHEMICAL ELEMENT

Data file	
Chemical symbol	Be
Atomic number	4
Atomic weight	9.012182
Melting point	1278°C
Boiling point	2970°C
Density	1.8 kilograms per litre (1.8 grams per cubic centimetre)
Oxide	BeO

Beryllium is a silvery-white, lustrous, relatively soft metal of group 2 of the periodic table of the elements, a group also known as the alkaline earth metals. The metal can be obtained by the electrolysis of molten beryllium chloride ($BeCl_2$). It is unaffected by air or water, even at red heat.

Mined beryllium has but a single isotope, beryllium-9, which is not radioactive. Beryllium-10, which is produced by cosmic rays in the upper atmosphere, is radioactive, however, with a half-life of 1.5 million years.

ELEMENT OF SURPRISE

Emeralds have been traded since the time of the Ancient Egyptians and are to be found in early jewellery across Europe, including that of the Celts. Analysis of the oxygen isotopes in the gems enables their source to be identified because the isotope ratio of oxygen-18/oxygen-16 varies from deposit to deposit. The Romans and Celts got their emeralds mainly from Austria, although some came from as far away as Pakistan.

More surprising is the discovery that the Mogul rulers of India got some of their emeralds from South America. Analysis of gems from the treasure of the eighteenth century Nizam of Hyderabad showed they were of Colombian origin; this has been interpreted to mean that there was trade across the Pacific Ocean by Spanish vessels which carried emeralds to their colonies in the Philippines, from where they eventually reached India.

Pronounced biz-muth, the name comes from the German *Bisemutum*, which was derived from *Wismuth* which, in its turn, was a corruption of *Weisse Masse* meaning 'white mass', probably a reference to bismuth oxychloride.

French, *bismuth*; German, *Wismut*; Italian, *bismuto*; Spanish, *bismuto*; Portuguese, *bismuto*.

HUMAN ELEMENT

Bismuth in the human body	
Blood	approx. 16 p.p.b.
Bone	approx. 200 p.p.b.
Tissue	30 p.p.b.
Total amount in body	less than 0.5 milligrams

Bismuth has no biological role to play and appears to be relatively benign for a heavy element. The amount we ingest each day is probably less than 20 micrograms, since the element is not readily taken up by plants and levels in those that have been measured were a tiny 20 p.p.b.

MEDICAL ELEMENT

So called 'bismuth mixture' appeared in the 1780s when doctors began to prescribe bismuth subnitrate, approximate composition $Bi(NO_3)(OH)_2$, as a treatment for gastric disorders, especially peptic ulcers. This compound was superseded by bismuth subcarbonate, approximate composition $Bi_2O_2(CO_3)$, in the 1860s. Today the preferred form is colloidal bismuth subcitrate (CBS; also known as De-Nol). Its mode of action is still unclear but it appears to act on the mucus which lines the stomach and protects it against the acidic gastric juices the body produces to digest its food. Bismuth also deactivates the pepsin enzyme which is particularly aggressive, and can attack the stomach wall.

Nowadays peptic and duodenal ulcers are thought to be due to the bacterium *Helicobacter pylori*, which can survive the extremely acid conditions and the digestive enzymes of the gastric juices. Bismuth as CBS can be used in combination with powerful antibiotics to eliminate these microbes.

Only one person has ever been reported to have died from an overdose of bismuth, and compounds of this element are still widely used to treat stomach upsets and sold as over-the-counter remedies. Nevertheless, people are warned against taking this popular remedy in excess because bismuth can cause liver damage.

ELEMENT OF HISTORY

Bismuth was discovered by an unknown alchemist around AD 1400. By 1450 it was being alloyed with lead to make cast type and by 1480 objects such as decorated caskets were being crafted in the metal. In Germany, artisans working with bismuth formed their own guild. For three centuries or more bismuth was confused with lead, although Georgius Agricola speculated in 1556 that it was distinctly different, as did Caspar Neuman around 1750; proof that it was so came in 1753 from the investigations of Claude-François Geoffroy.

ECONOMIC ELEMENT

Bismuth occurs naturally as the metal itself and is found as crystals in the sulfide ores of nickel, cobalt, silver and tin. Most bismuth occurs as the mineral bismuthinite, which is bismuth sulfide (Bi_2S_3). Bismuth is mainly produced as a by-product from lead and copper smelting, especially in the USA. The chief areas where it is mined are Bolivia, Peru, Japan, Mexico and Canada, but only to the extent of 3000 tonnes per year. There is no reliable estimate of how much bismuth is available to be mined, but it seems unlikely that there will ever be a shortage of this metal.

Bismuth has a relatively low melting point for a metal (271°C) and when alloyed with tin and lead it melts at an even lower temperature. Because of this it was once commonly used for electric fuses, and Wood's metal, an alloy of bismuth with 12% cadmium that melts at 70°C, is still used in automatic fire-sprinkler systems.

Industry makes use of bismuth compounds as catalysts in manufacturing acrylonitrile, the starting material for synthetic fibres and rubbers.

ENVIRONMENTAL ELEMENT

Bismuth in the environment	
Earth's crust	48 p.p.b.
	Bismuth is the 70th most abundant element.
Soils	approx. 0.25 p.p.m.
Sea water	400 p.p.t.
Atmosphere	virtually none

Bismuth appears to pose no environmental threat to any living species.

CHEMICAL ELEMENT

Data file	
Chemical symbol	Bi
Atomic number	83
Atomic weight	208.98038
Melting point	271°C
Boiling point	1560°C
Density	9.7 kilograms per litre (9.7 grams per cubic centimetre)
Oxide	Bi_2O_3

Bismuth is a heavy, silvery metal with a faint pink tinge and is a member of group 15 of the periodic table of the elements. It is not used as a pure metal because it is too brittle. Rather it is used as an alloy with other metals. Like antimony, the metal above it in the periodic table, it expands on solidifying, by about 3%. Bismuth is stable to oxygen and water but dissolves in concentrated nitric acid. All bismuth salts form insoluble compounds when put into water.

There is only one important naturally occurring isotope, bismuth-209, and this is not radioactive. There are traces of other bismuth isotopes in thorium and uranium minerals where they occur as part of the decay chain of those elements. Thus thorium contains bismuth-212, which has a half-life of 60 minutes, and uranium contains bismuth-214 with a half-life of 20 minutes and bismuth-210 with a half-life of 5 days. Bismuth has the distinction of being the heaviest element with a stable nucleus, although it has been suggested that bismuth-209 might be inherently unstable but that its half-life is so long that it has not been detected undergoing radioactive decay.

Bismuth has been used for making transuranium elements by the 'cold fusion' method. In this it acts as a target and is bombarded with heavy ions of lighter elements in the hope that they will fuse together and form the nucleus of a new element. Bohrium, meitnerium and element 111 were discovered in this way. The cold fusion method was developed in 1974 by Yuri Oganessian and Alexander Demin at the Joint Institute for Nuclear Research in Dubna, Russia.

ELEMENT OF SURPRISE

The pearl effect in cosmetics such as nail varnish and lipstick is produced with bismuth oxychloride (BiOCl), a lustrous crystalline powder. This is made from bismuth chloride ($BiCl_3$), which reacts with water to form a hydroxide and this, on heating, gives the oxychloride.

Bohrium

This is element atomic number 107. Along with other elements above 100 it is dealt with under the transfermium elements on p. 457.

Boron

Pronounced bohr-on, the name is derived from the Arabic *buraq*, which was the name for borax. Humphry Davy devised the name by combining borax and carbon since the element came from the former and resembled the latter.

French, *bore*; German, *Bor*; Italian, *boro*; Spanish, *boro*; Portuguese, *boro*.

COSMIC ELEMENT

Boron's very existence is surprising because there appears to be no way in which it can be created within stars.[17] Nor could it have been made at the time of the Big Bang if this event was homogeneous. In other words, an even distribution of protons, neutrons and electrons can only combine to form hydrogen, helium and lithium. However, if the Big Bang had been inhomogeneous then it could have created boron in neutron-rich regions. The existence of boron is taken as indicating that this might have been what really happened. Otherwise the only way in which boron could be formed would be in a supernova explosion or by interstellar cosmic rays shattering the atoms of other nuclei, some of which splinter to give boron. This process is known as spallation.

HUMAN ELEMENT

Boron in the human body	
Blood	0.13 p.p.m.
Bone	approx. 2 p.p.m.
Tissue	2 p.p.m.
Total amount in body	18 milligrams

[17] The same is true for beryllium.

Boron is an essential element for plants so, not surprisingly, we have a regular daily intake of about 2 milligrams and with about 18 milligrams in our body in total. Foods relatively rich in boron are fruits (in particular apples, grapes, peaches, cherries and pears), and vegetables (broccoli, cabbage and onions), all of which have more than 1 milligram of boron per kilogram. These quantities are well below those at which boron can upset the system. It takes 5 grams of boric acid to make a person ill, and 20 grams or more to put their life in danger. Despite this, boron compounds, such as borax ($Na_2B_4O_5(OH)_4{\cdot}8H_2O$) and boric acid ($H_3BO_3$), are still used in topical medicines.

MEDICAL ELEMENT

Boron compounds are being investigated as a treatment for brain tumours, using a process known as boron neutron capture therapy (BNCT). Patients are given a substance known to be attracted to cancer cells but to which a cluster of boron atoms has been attached. When these have been taken into the tumour it is then bombarded with neutrons which are absorbed by the boron atoms. The boron-10 isotope is an excellent absorber of neutrons; in absorbing neutrons, it undergoes immediate nuclear fission to form lithium, and emits deadly α-rays which kill the target cell and those surrounding it. (α-rays cannot penetrate far from the cell in which they were generated, so they do not threaten healthy cells of the body.)

ELEMENT OF HISTORY

Even though there are references to 'borax' in the writings from ancient Babylon, Egypt and Rome, we cannot know if this really was sodium borate or another salt. The only source of borax in the ancient world was Tibet, where it crystallizes from Lake Yamdok Cho, South of Lhasa. From there it was exported to the Near East and Europe right up to the end of the eighteenth century. It is not certain when trade in borax from Tibet began, but by AD 1100, when the Moors were flourishing, their goldsmiths used borax as a flux to make it easier to work with the molten metal.

Borax was mentioned in Geoffrey Chaucer's *Canterbury Tales*, written in the 1380s. In the Prologue, where he introduces his 29 pilgrims, he says of the summoner (a lawyer for the church courts):

> No quicksilver, lead ointments, tartar creams,
> Boracic, no, nor brimstone, so it seems,
> Could make a salve that had the power to bite,
> Clean up or cure his whelks[18] of knobby white.

'Boracic' was the old name for borax and it was clearly being used as a skin ointment (along with other elements mentioned, namely mercury, lead and sulfur). Borax was also part of Queen Elizabeth I's collection of cosmetics 200 years later. For a time the Venetians had the monopoly for its importation into Europe and they exploited it.

[18] A whelk was another name for a pimple.

The characteristic green flame that came to symbolize boron was first reported in 1732 by Geoffroy the Younger. It is produced by treating borax, or any borate, with sulfuric acid to convert the boron to boric acid, and then adding alcohol to dissolve this and setting it alight. This simple test of analytical chemistry was to be used by mineral prospectors and eventually resulted in the discovery of the world's largest deposit of borate minerals, in Death Valley in the Californian desert.

In 1779 a large deposit of borate was discovered in Tuscany, Italy, but it was not exploited until the early 1800s when Italy became the main exporting country for more than 75 years. Slowly the use of boron became more widespread. The US deposits were discovered in 1873 and soon became more profitable to work. As a result the Italian industry slowly declined and disappeared.

In 1808, Louis-Joseph Gay-Lussac and Louis-Jacques Thenard working in Paris, and Sir Humphry Davy in London, independently extracted boron from borate by heating it with potassium metal. The two Frenchmen reported their results on 21 June that year and Davy on the 30 June. Neither party had produced the pure element, which is almost impossible to obtain. (A purer type of boron was isolated in 1892 by Henri Moissan, but it was not until E. Weintraub, of the General Electric Company of America, produced it by sparking a mixture of boron chloride (BCl_3) vapour and hydrogen that it was finally obtained pure, and was found to have very different properties from those samples previously reported.)

ECONOMIC ELEMENT

Boron is only found as borates (boron–oxygen compounds). The main ores are kernite and tincal (borax), which are both sodium borates, and ulexite and colemanite, which are calcium borates. Borates are mined in the USA, Tibet, Chile (ulexite) and Turkey (colemanite), with world production being about 2 million tonnes a year. Known reserves are of the order of 500 million tonnes.

Boron is an important industrial material, although as the pure element it is little used except to make metal borides, or to be added to metals to improve their conductivity (aluminium), or to make them easier to refine (nickel), or to made them flow better (iron). Some borides conduct electricity many times better than the metal itself and they have very high melting points. Metal borides are used for turbine blades, rocket nozzles, high-temperature reaction vessels and inert electrodes.

Boron compounds are important in many industries, such as in making glass and detergents, and in agriculture. The most important compounds are borax (sodium borate), boric oxide and boric acid. Pyrex glass is tough and heat resistant because of the 12–15% boric oxide used in its manufacture. Glass fibre is also borosilicate glass and this finds use in reinforcing plastics and as insulation in buildings. Ceramic glazes for tiles and kitchen equipment now account for a lot of the demand for boric acid.

An early use of borax was to make perborate, the bleaching agent once widely used in household detergents. Boron compounds also came into the average home in the guise of food preservatives, especially for margarine, fish and other foodstuffs, but were eventually outlawed. However, some is still used to preserve caviar.

Boron compounds such as sodium octaborate ($Na_2B_8O_{13}$) are used to fire-proof fabrics, wood and the flakes of shredded paper that are used to simulate snow on film sets. Boric acid is insecticidal, especially for ants and cockroaches which are particularly susceptible to it and unable to detect its presence in baits such as sugar solution or cereal pellets.

Elemental boron held in a polymer matrix is used to regulate nuclear reactors and to provide a safety system by quenching a reactor should it threaten to go critical.

ENVIRONMENTAL ELEMENT

Boron in the environment	
Earth's crust	10 p.p.m.
	Boron is the 38th most abundant element.
Soils	varies widely between 1 and 450 p.p.m., but generally is around 10–30 p.p.m.
Sea water	4 p.p.m.
Atmosphere	minute trace detectable in rain water

Thousands of tonnes of borate are added to fertilizers each year because boron is vital to plants. Plants grown in soil lacking boron are stunted, fail to be pollinated or produce fruit, and are more susceptible to diseases. Boron deficiency in several plant species is common and its lack can stunt the growth of crops such as sugar beet, celery, sunflowers, legumes and apples. Trees appear to need more boron than most other plants. Boron has revitalized the age-old olive groves of Greece and Spain that were giving poor yields of fruit. Only 60 grams of soluble sodium octaborate is needed per tree per year; it not only increases yield but also eliminates pests such as the olive fly and the olive moth.

Plants take up boron in order to mobilize sugars, and it has been shown that the amount of sugar that sugar beet produces goes up in line with the amount of boron in the soil in which it is grown, although if this is too high then a toxic reaction sets in and yields begin to fall. Too much boron can have an adverse effect on some crops, such as kidney beans and lemons, while at the other extreme there are plants that seem to be able to tolerate any amount of boron, such as turnips, cotton and clover.

CHEMICAL ELEMENT

Data file	
Chemical symbol	B
Atomic number	5
Atomic weight	10.811
Melting point	2300°C
Boiling point	3658°C
Density	2.3 kilograms per litre (2.3 grams per cubic centimetre)
Oxide	B_2O_3

Boron is a non-metallic element and the only non-metal of group 13 of the periodic table of the elements. It has several forms, the most common of which is amorphous boron, which is a dark powder, unreactive to oxygen, water, acids and alkalis. It reacts with metals to form borides.

Naturally occurring boron consists of two isotopes, boron-10 (20%) and boron-11 (80%), neither of which is radioactive.

ELEMENT OF SURPRISE

Boron nitride is a remarkable material, similar in strength and sparkle to diamond. It can be fabricated into items with high heat capacity and high electrical resistance that can cope with extremely high temperatures and even molten metals. It enhances the performance of many other materials, including face powder, to which it adds a lustre and a luxurious silky feel (as a microfine powder), as well as being an excellent abrasive for industry.

Boron carbide is likewise both chemically inert and heat resistant and has a hardness approaching that of diamond. It is made by heating boron or boron oxide with carbon at 2500°C, and is manufactured in tonne quantities. It composition is $B_{13}C_2$ but its exact chemical structure has yet to be determined.

Bromine

Pronounced broh-meen, the name comes from the Greek *bromos*, meaning stench. French, *brome*; German, *Brom*; Italian, *bromo*; Spanish, *bromo*; Portuguese, *bromo*.

Bromine is the name of the element, which exists as the molecule Br_2, whereas bromide, as in potassium bromide, is a bromine atom that has acquired a negative electron and is written Br^-. Br_2 is reactive and dangerous, while Br^- is stable and relatively safe.

HUMAN ELEMENT

Bromine in the human body	
Blood	5 p.p.m.
Bone	approx. 7 p.p.m.
Tissue	approx. 7 p.p.m.
Total amount in body	260 milligrams

Although a small amount of bromine is present as bromide in all living things, and some species even produce organobromine compounds, no biological role has been identified for this element in humans. It has been suggested that some bromide is essential to health, but this has not been proved.

Bromide is not without effect, though, and it depresses mental activity, most noticeably the sex drive. Our average dietary intake varies widely depending on the foods we eat; it may be as low as 1 milligram or as high as 20 milligrams.

A few species contain organobromine compounds; for example, some of these compounds are found in the eggs of sea birds, probably originating from the marine diet of these birds. Marine plants can concentrate bromide to levels a thousand times higher that in the surrounding water. Land plants can do the same from soil: in one case of lettuces grown in a greenhouse fumigated with methyl bromide, the level exceeded 0.1% of the weight.

Bromine itself is corrosive and its vapour attacks the eyes and lungs. It is very toxic, with as little as 100 milligrams being a fatal dose, whereas it requires more than 3000 milligrams of bromide to provoke a toxic response and more than 10 times this amount to pose a threat to life.

MEDICAL ELEMENT

In the nineteenth century, bromide salts were prescribed by doctors for 'nervous' complaints and even epilepsy, and they were the Victorian equivalent of tranquilizers. Indeed the word 'bromide' came to mean something that eased tension or was soothing, but too much leads to depression and loss of weight. The use of bromide salts as sedatives and to reduce sexual activity has now been discontinued because bromide is slightly toxic. Some spa waters owed their popularity to their high bromide content.

Today chemists in pharmaceuticals companies use bromine intermediates during the manufacture of medicinal drugs, even though the element does not appear in the final product.

ELEMENT OF HISTORY

Antoine-Jérôme Balard (1802–1876) discovered bromine while investigating residues from brine at Montpellier, France. By passing chlorine gas into them he liberated an orange-red liquid, which he deduced was a new element. He sent an account of it for publication to the French Academy's journal in 1826 and proposed the name *muride*. The Academy's vetting committee did not like his choice of name, which he had derived from murex, the name of the deep purple natural dye. The Academy decided to change the name to bromine to be consistent with the related elements chlorine and iodine, which this new element clearly resembled, and while Balard was at first upset by this, he eventually agreed.

A year earlier, a student at Heidelberg, Carl Löwig (1803–1890), had brought his professor a sample of bromine that he had produced from the waters of a natural spring near his home at Kreuznach. He was asked to produce more of it, and while he was doing so Balard published his results and so became known as its discoverer.

Once bromine had been discovered, its chemistry was quickly investigated since it could be predicted on the basis of bromine's relationship to chlorine and iodine. Equally quickly other chemists discovered it in many mineral water springs around Europe and showed it to be present in sea water, plants and animals.

ECONOMIC ELEMENT

Bromine used to be produced from naturally occurring brines by adding sulfuric acid and bleaching powder. Then Herbert Dow devised a method using electrolysis and began to extract it from bromine-rich brine deposits, which had up to 0.2% bromine, that underlay the US midwest town of Midland. Its extraction from sea water began in 1934 at Kure Beach, North Carolina, again by what had then become the Dow Chemical Co.

World production of bromine is more than 300 000 tonnes per year; the three main producing countries are the USA, Israel and the UK. In this last case it is extracted from sea water at a plant on the coast of Anglesey, Wales. Israel is a major producer because the Dead Sea is particularly rich in bromide, with levels as high as 0.5%.

Known reserves are virtually unlimited since the oceans contain more than a trillion

(10^{12}) tonnes of bromide. Bromine is extracted from sea water, which contains 65 p.p.m., by bubbling chlorine gas through the water which converts the dissolved bromide salts to bromine. This is then removed by blowing air through the water and the expelled bromine gas is collected. The Anglesey plant produces 30 000 tons of bromine a year.

Bromine is used in industry to make organobromo compounds. A major one was dibromomethane, which formerly was used as a fuel additive for leaded petrol. This compound reacted with the lead during the combustion phase to form lead bromide, which then passed out in the exhaust gases instead of forming a deposit of lead metal in the engine itself. Other organobromines are used as insecticides, in fire extinguishers and to make pharmaceuticals. Some of these uses have come under suspicion – for example, see methyl bromide in the 'Environmental element' section below.

Another example of bromine's use is the organobromine compounds used in so-called halon fire extinguishers. Halon-1211 ($CBrClF_2$) and halon-1301 ($CBrF_3$) are the fire extinguishers of choice for tackling a blaze in a confined space or where water might do as much damage as the fire, for example in archives, museums, art galleries, aircraft, engine rooms of ships, military vehicles such as tanks, and large computer facilities. Although these vapours are highly effective at dousing flames, and are non-toxic, the halons are now frowned upon because the bromine they contain adds to the destruction of the ozone layer in the upper atmosphere. Other bromine compounds, such as polybrominated diphenylether (PBDE), are used as flame-retardants for furniture foams, plastic casings and some textiles, but these too are seen as a threat, in this case to humans because they may act as hormone mimics.

ENVIRONMENTAL ELEMENT

Bromine in the environment	
Earth's crust	0.4 p.p.m.
	Bromine is the 62nd most abundant element.
Soils	varies widely with a range of 5–40 p.p.m.; some volcanic soils can contain as much as 500 p.p.m.
Sea water	65 p.p.m.
Atmosphere	a few p.p.t.

Methyl Bromide

This volatile chemical (CH_3Br; boiling point 3.5°C) was found to be an effective soil fumigant in the 1950s and has been widely used since to kill nematodes,[19] insects, bacteria, mites and fungi. By 1990 more than 50 000 tonnes were being manufactured each year for this purpose alone. Methyl bromide is particularly effective because the gas penetrates to areas that other pesticides cannot reach, and it kills pests at all stages in

[19] These are small parasitic worms.

their life cycles. Equally important is that during 40 years of use there been no indication of pest resistance evolving.

The crops which benefited most were tomatoes, celery, lettuce, strawberries, grapes, tobacco and flowers such as carnations and chrysanthemums. Warehouses where crops were stored were easily protected against insects and rodents by its potent vapour, and it did not contaminate the food. Despite all these advantages, in 1992 it was added to the list of compounds that had to be phased out under the Montreal Protocol because of the threat it poses to the ozone layer. It must no longer be manufactured by 2010; indeed the USA plans to phase it out by 2005.

Removing methyl bromide compounds from the atmosphere may not be as easily achieved as it sounds. Nature also produces volatile organobromine compounds in vast quantities from forest fires and marine plankton, and the salt pans of the bromide-rich Dead Sea increase the amount in the atmosphere. Yet Nature may also play a part in removing methyl bromide since the oceans themselves can dissolve it, bacteria in soil can convert it to bromide, the oxygen in the atmosphere can react with it, and the leaves of many plants will absorb it.

CHEMICAL ELEMENT

Data file	
Chemical symbol	Br
Atomic number	35
Atomic weight	79.904
Melting point	−7°C
Boiling point	59°C
Density of liquid bromine	3.1 kilograms per litre (3.1 grams per cubic centimetre)
Oxides	none stable

Bromine, which exists as a molecule containing two Br atoms (Br_2), is a deep red, dense, sharp-smelling liquid that is highly reactive. It is a member of the halogens, the elements of group 17 of the periodic table. Bromine is soluble in organic solvents and in water; when dissolved in water it forms 'bromine water', which is both a powerful oxidizing agent and a convenient source of the element that was formerly used to test for unsaturated organic molecules, i.e. those with carbon–carbon double bonds in their structure. These react rapidly and the bromine water goes colourless.

Naturally occurring bromine consists of two isotopes, bromine-79 (which comprises 51%) and bromine-81 (which accounts for 49%). Neither is radioactive.

ELEMENT OF SURPRISE

The famous purple of the togas worn by the Roman emperors was produced from naturally occurring dye that contained bromine atoms. It was called Tyrian purple and was extracted from the Mediterranean mollusc *Murex brandaris*. However, it was known in the Middle East well before the time of the Romans and it is mentioned in the Bible in the book of the prophet Ezekiel. Verse 27, line 7 describes a mythical luxury yacht with an awning that was 'purple from the isles of Elishah' and line 16 refers to merchants from Syria who traded in 'emeralds, purple, fine linen, coral and rubies'. Both references indicate how highly valued the dye was. The rich colour of Tyrian purple was not to be rivalled until an 18 year old English chemist, William Perkin, discovered the beautiful violet dye mauve, in 1856.

Pronounced cad-miuhm, the word is derived from the Latin *cadmia*, the name for the mineral calamine (in fact calamine is *zinc* carbonate – see 'Element of history' below).

French, *cadmium*; German, *Cadmium*; Italian, *cadmio*; Spanish, *cadmio*; Portuguese, *cádmio*.

HUMAN ELEMENT

Cadmium in the human body	
Blood	5 p.p.t.
Bone	approx. 2 p.p.m.
Tissue	approx. 2 p.p.m., but it tends to accumulate in the kidney and liver where levels can be 10 times higher than this
Total amount in body	increases with age to around 20 milligrams at age 50

Although it has not been proved, there are suspicions that cadmium could be essential for some species, and it does stimulate metabolism. Marine organisms readily absorb it, which is why there is so little cadmium in the surface layers of the oceans. The strongest evidence of its being essential for one species was the discovery, in the year 2000, that the marine diatom *Thalassiosira weissflogii* produces a cadmium-specific enzyme that catalyses the interconversion of carbon dioxide and carbonic acid. Even so, the amount required is minute.

Cadmium is an accumulative poison which is very insidious and is on the United Nations Environmental Programme's list of top 10 hazardous pollutants. Inhaled cadmium oxide fumes are particularly dangerous, and this has sometimes had fatal consequences for workers in industry – see box.

WORKING THEMSELVES TO DEATH

When a team of men, working on the Severn Road Bridge in England in 1966, had to remove part of a construction tower they decided to use an oxyacetylene torch to melt away the steel bolts that held it together. Unfortunately there was no ventilation inside the tower and the job took all day. The next day, when the men reported for work, they all complained of feeling ill, were coughing a lot and had difficulty breathing. Within a few hours one man had to be taken to hospital, where he died a week later; one by one the others were also admitted for treatment, although they survived. And the cause of their illness? The bolts they were removing were coated with cadmium, which had been volatilized by the heat with the result that the men all succumbed to cadmium poisoning.

FOOD ELEMENT

Cadmium can never be totally excluded from the diet; it is present in foods like offal, shellfish and rice. Some fungi have the ability to absorb cadmium such as the mushroom *Amanita muscaria* which can contain 30 p.p.m. cadmium, even when grown on soil with as little as 0.3 p.p.m. cadmium. Other food crops which seem to absorb the metal are lettuce, spinach, cabbage and turnip, while potatoes, maize, beans and peas are among the least absorbing. Smoking adds to the body burden because tobacco plants absorb it.

Plants growing on contaminated land, either near ancient mines or metal-processing industries, or from land fertilized with human sewage (once known as sewage farms) can all have high levels of the metal; for example, cadmium levels in lettuce can reach 70 p.p.m. Even when such land is no longer used to grow food and is turned over to sheep grazing, these animals can accumulate cadmium in their kidneys and livers.

The human daily intake of cadmium may be as low as 10 micrograms or as high as 1000 micrograms, but the average is probably less than 25 micrograms. The World Health Organization (WHO) has set a recommended safe daily intake of 70 micrograms.

The metal mimics the essential element zinc and so gains access to the body. Luckily the human gut, which needs to allow zinc to be absorbed, is able to keep out most of the cadmium, but as much as 10% of it may slip through, with the result that a few micrograms enter the system each day. There the cadmium triggers a defence mechanism by stimulating the production of the enzyme metallothionein, which contains lots of cysteine amino acid units. The sulfur atoms in these attract the cadmium to the extent that as many as seven metal atoms can attach themselves to every enzyme molecule. These then transport the cadmium to the kidneys with the object of getting get rid of it, but cadmium binds so strongly to the enzyme (300 times stronger than zinc binds) that it tends to accumulate there.

The result is that a cadmium atom remains in the human body on average for around 30 years, which is why there is so much concern about the effect of this metal on human health. Because zinc is part of DNA polymerase, which is crucial to the production of sperm, its replacement by cadmium is particularly damaging to the testicles.

MEDICAL ELEMENT

Treatments with agents that could remove cadmium from the system are risky because they would also remove zinc just as easily. The body can go on storing cadmium but eventually there comes a point when the kidneys can no longer cope. If the level exceeds 200 p.p.m. it prevents reabsorption of proteins, glucose and amino acids, and damages the filtering system; this has occasionally led to cases of kidney failure.

Continued exposure to cadmium leads to a disease known in Japan as Itai-Itai,[20] which weakens the bones and joints, making movement painful. It was at Fuchu, 200 miles North-west of Tokyo, where the condition was first recognized and the cause was traced to a combination of high cadmium intake and low levels of vitamin D in the victims. The cadmium came from rice grown on contaminated land and watered from a source that was polluted by waste from a zinc mine. The rice had 10 times as much cadmium as normal rice.

Cadmium causes cancer in special laboratory-bred rats, but not in mice or hamsters. There has been concern that it may cause cancer in humans, but this is debatable. An analysis of cancer cases, in 1988, among those who worked with cadmium in Britain seemed to show that it did cause cancer, and similar findings were also reported in the USA. However, a follow-up study of 7000 people whose work involved cadmium could find no link between cancer and cadmium.

ELEMENT OF HISTORY

The apothecaries of Hanover, Germany, in the early 1800s used to make zinc oxide by heating calamine, also known as cadmia, a naturally occurring form of zinc carbonate. Sometimes the product they ended up with was an unpleasant yellow colour instead of pure white. When the inspector of pharmacies, Professor Friedrich Stromeyer (1776–1835) of Göttingen University, looked into the problem he traced the discoloration to a component of the mineral that he could not identify and which he deduced must be a previously unknown element. This he separated and named cadmium after the name for the mineral. That was in 1817. Meanwhile two other German chemists, Karl Meissner in Halle and Carl Karsten in Berlin, were working on the same problem; they announced their discoveries of the element the following year.

Interest in the new element generated a spate of research and it was soon found to be present in all zinc ores, but as its presence seemed to have no effect of the performance of the zinc compounds that were derived from it there was little incentive to remove it. It was partly removed when zinc metal was smelted because cadmium is more volatile than zinc and was lost to the atmosphere. (The boiling point of cadmium (756°C) is appreciably lower than that of zinc (907°C).)

[20] 'Ouch-ouch!' would be the equivalent in English.

ECONOMIC ELEMENT

The first true cadmium mineral to be discovered came to light when the Bishopton railway tunnel was being bored near Port Glasgow in Scotland in 1841. This mineral, cadmium sulfide (CdS), was named greenockite after Lord Greenock who was in charge of the project. In fact greenockite had been used for more than 2000 years as a yellow pigment and then it had been mined in Greece and Bohemia.

Other cadmium ores are known, such as cadmoselite (cadmium selenide, CdSe) and otavite (cadmium carbonate, $CdCO_3$) but none is mined for this metal. More than enough is produced as a by-product of the smelting of zinc from its ore, sphalerite (ZnS), in which CdS is a significant impurity, making up as much as 3%. Consequently the main mining areas are those associated with zinc. World production is around 14000 tonnes per year; the main producing country is Canada, with the USA, Australia, Mexico, Japan and Peru also being major suppliers.

Cadmium sulfide used as a common pigment is known as cadmium yellow, although it can be anything from yellow, through orange and red, to brown depending on the proportions of sulfur and selenium in the pigment. The bright yellow cadmium sulfide used to be added to paints, artists' colours, rubber, plastics, printing inks and vitreous enamels. Almost all these uses have now been phased out.

Cadmium has been used particularly to electroplate steel where a film of cadmium only 0.05 millimetres thick will provide complete protection against that most corrosive of environments, the sea. The sea water reacts with the cadmium to form an impervious layer of insoluble cadmium chloride. (Galvanizing with zinc is less protective because any zinc chloride which forms is dissolved away from the surface.)

While some uses for cadmium were phased out as dangerous to the environment and human health, others were taking their place, such as nickel–cadmium batteries, which were seen as environmentally friendly because they could be recharged a thousand times or more and still work well. In these lightweight batteries the cathode is nickel hydroxide, the anode is cadmium oxide and the electrolyte is potassium hydroxide.

The Nissan car company has designed an all-electric car which has a range of 250 kilometres (150 miles) that relies on a new style nickel–cadmium battery that can be recharged in only 15 minutes. Such batteries are more efficient for electric vehicles because they are only a third the weight of conventional lead–acid batteries. Cadmium recycling from batteries now recovers several thousand tonnes of the metal a year.

All other roles for cadmium in motor vehicles have been phased out, even that which formerly coated nuts and bolts that were exposed to the corrosive spray that comes from salted roads in winter.

Since the 1960s, when alarms were first sounded about cadmium, industry has found alternatives for most products that reach the consumer. Loss of cadmium to the environment is now negligible, and its continued use is only justified on the grounds that cadmium makes better batteries and more stable plastics. In the aerospace, mining and offshore oil industries it is needed to protect steel.

Cadmium has the ability to absorb neutrons and so is used in the design of nuclear reactors and for the control rods that regulate the self-sustaining chain reaction that provides the energy.

ENVIRONMENTAL ELEMENT

Cadmium in the environment	
Earth's crust	0.1 p.p.m.
	Cadmium is the 65th most abundant element.
Soils	varies widely but on average is 1 p.p.m. but some polluted soils contain as much as 1500 p.p.m.
Sea water	0.1 p.p.b. with only 1 p.p.t. in the surface layers
Atmosphere	tiny: between 1 and 50 nanograms per cubic metre of air, depending on location, industrial cities having the most

About 8000 tonnes of cadmium enter the atmosphere each year, of which 90% is from human activity. Some of this is washed on to soil used to grow crops and feed animals.

We can trace the 30 micrograms of cadmium in a hamburger to the grass the cattle ate and, ultimately, to the soil on which the grass grew. Some of this is 'natural,' some is a result of air-borne contamination, and some comes from fertilizers added to the soil. Phosphate rock from Morocco, which is no longer used in the European Union, has over 50 grams per tonne of cadmium whereas Florida phosphate has much less than a tenth of this amount. Sewage sludge used as a fertilizer can raise cadmium levels in soils, especially if it comes from industrial areas; in some countries millions of tonnes of sewage are added to the land each year.

All types of fertilizer appear to increase the level of cadmium in farmed soils. Experiments at Rothamsted Experimental Station, UK, showed that both phosphate fertilizers and farmyard manure increased cadmium levels. More surprising was the finding that the level in unfertilized plots went up from 0.5 to 0.8 p.p.m. over a period of 140 years. This was attributed to atmospheric deposition.

Geochemical atlases have been drawn up for some countries; these show that the amount of cadmium in most soils rarely exceeds 1 p.p.m. but that there are local 'hot spots' which exceed the 3 p.p.m. limit recommended by the European Union to protect soils that are fertilized with sewage sludge. Some hot spots have levels exceeding 40 p.p.m.

Cadmium contamination comes from three sources: (i) old lead and zinc mines; (ii) zinc smelters which produce cadmium as a by-product; and (iii) natural outcrops, such as the Carboniferous marine black shales. The UK's worst polluted area is around Shipham in Somerset where the spoil from old zinc mines that operated up to about 1850 have led to soil levels of 500 p.p.m., a world record. This came to light in 1979 when children at a local school were trying to grow vegetables and found that the leaves of their vegetables turned yellow.

CHEMICAL ELEMENT

Data file	
Chemical symbol	Cd
Atomic number	48
Atomic weight	112.411
Melting point	321°C
Boiling point	756°C
Density	8.7 kilograms per litre (8.7 grams per cubic centimetre)
Oxide	CdO

Cadmium is a silvery metal and a member of group 12 of the periodic table of the elements. Its surface has a bluish tinge and the metal is soft enough to cut with a knife, but it tarnishes in air. It is soluble in acids but not in alkalis.

There are eight naturally occurring cadmium isotopes of which cadmium-114 is the most abundant, accounting for 29% of atoms, with cadmium-112 coming next at 24%. The other isotopes, in descending order, are cadmium-111 (13%), cadmium-110 (12.5%), cadmium-113 (12%), cadmium-116 (7.5%) and cadmium-106 and cadmium-108 (1% each).

Cadmium was not considered to be radioactive until 1970 when, much to the surprise of scientists, it was shown that cadmium-113 was a α-emitter, although with an incredibly long half-life of 9 million billion years (9×10^{15} years). The consequence of this radioactivity is negligible; in a 10 gram piece of the metal only one atom of cadmium is lost per minute. Put another way, it means that of every million atoms of cadmium that date from the origin of the Universe, 15 billion years ago, only one has since disintegrated – see also pages 63 and 232.

ELEMENT OF SURPRISE

A solar panel that has an efficiency for converting sunlight into electricity in excess of 10% has been constructed with an absorbing surface of crystalline cadmium telluride. (Silicon-based solar panels have efficiencies of 8% or less.) A square metre of such a panel can generate a power output of nearly 100 watts.

Caesium *or* Cesium (USA)

Pronounced seez-iuhm, the name comes from the Latin *caesius*, meaning sky blue, in reference to the colour its compounds impart to a flame.

French, *césium*; German, *Caesium*; Italian, *cesio*; Spanish, *cesio*; Portuguese, *césio*.

HUMAN ELEMENT

Caesium in the human body	
Blood	4 p.p.b.
Bone	10–50 p.p.b.
Tissue	approx. 1 p.p.m.
Total amount in body	approx. 6 milligrams

Caesium has no known biological role, but it may partly replace potassium, an element that it resembles chemically. However, rats fed a diet of caesium in place of potassium die within 2 weeks, so it can be regarded as a toxic element for that species, although caesium chloride is probably no more toxic than sodium chloride. Certainly it would have serious effects if it were to be taken in excess, and rats given it experienced intense irritability and spasms.

The average daily intake of this metal is only 0.03 milligrams and it comes with foods that are richest in potassium. Caesium can enter the food chain and be absorbed in place of potassium. Although a lone atom of caesium is much larger than one of potassium, when their ions are dissolved in water, the molecules of solvent which cluster around them make them appear to be almost the same size. The amount of caesium has been measured for some plants, e.g. up to 3 p.p.b. in vegetables and fruits, and even 0.2 p.p.m. in some tea leaves.

ELEMENT OF HISTORY

Caesium was almost discovered by Carl Plattner (1800–1858) in 1846 when he investigated the caesium mineral pollucite. He found he could only account for 93% of the elements it contained, but then had no more mineral left to continue his analysis. In fact it was later discovered that he mistook most of the caesium for a mixture of sodium and potassium.

Caesium was eventually discovered in 1860 at Heidelberg, Germany, when it was shown to be present in a mineral water from Durkheim. Atomic spectroscopy was the technique used to show that there was a previously unknown element present. The work was done jointly by Robert Bunsen (1811–1899), who is best remembered for developing the Bunsen burner, and Gustav Kirchhoff (1824–1887). They then took about 30 000 litres of the mineral water which they boiled down, and from which they removed the lithium, sodium, potassium, magnesium, calcium and strontium salts. The remaining liquor was sprayed into a Bunsen flame and the light produced was analysed using a spectroscope, showing two blue lines very close together. This had never been seen before, so Bunsen and Kirchhoff immediately realized that they had stumbled on a hitherto unknown element. Knowing it was there, they were eventually able to extract samples of the metal.

ECONOMIC ELEMENT

Few caesium minerals are known. Pollucite ($Cs_4H_4Al_4Si_9O_{27}$) is the main one and even this is relatively rare. As the silicate magmas of the Earth cooled to form granite, these left behind in the still molten rock higher levels of elements like lithium, rubidium and caesium (among others). In some cases the last fractions to solidify were rich in caesium and this mainly crystallized to form pollucite. Caesium is also present in lepidolite (see lithium, p. 237). World production of caesium compounds is a mere 20 tonnes per year and this comes mainly from Bernic Lake (Manitoba, Canada), with a little from Zimbabwe and South-West Africa. Reserves of pollucite at Bernic Lake are estimated to be around 60 000 tonnes. The main company producing caesium compounds is Chemetall at Langelsheim in Germany.

Caesium is released from its ore by heating the mineral with hydrochloric acid, which leaves behind the silica, and then precipitating the caesium as a mixed chloride with another metal such as lead or tin. Caesium metal itself can be obtained by the electrolysis of a concentrated salt solution, but a more general method is to heat caesium chloride with calcium. These react to form caesium and calcium chloride, and the caesium is then distilled off under vacuum.

Caesium is used in industry as a catalyst promoter, boosting the performance of other metal oxides in this capacity. Caesium nitrate is used to make optical glass. Other types of glass can be strengthened by dipping them into molten caesium salts; the caesium ions exchange with sodium ions on the surface and thereby make it more resistant to etching or breakage. Caesium iodide and caesium fluoride absorb X-rays, γ-rays and other atomic particles; as they do they give off light. This so-called scintillation effect is used in medical diagnostics and radiation monitoring.

Caesium metal will scavenge the last traces of gas in the production of vacuum tubes after they have been sealed. The caesium is made inside the tube by a chemical reaction which converts caesium chromate into caesium vapour, which then reacts with any oxygen, nitrogen or other unwanted gases.

There are other uses to which caesium is put. Solutions of its salts are used in centrifuges to aid the separation and purification of DNA for biochemical research. The 'caesium clock' is the standard measure of time, and depends on the energy difference between two states of the caesium-133 atom. The frequency is exactly 9 192 631 770 cycles per second (equivalent to an accuracy of 1 second in 300 000 years) and the standard second is defined with respect to this number. Several atomic clocks are monitored and atomic time signals are broadcast around the world.

In Japan, so-called 'caesium gel' is sold as a skin rejuvenating cream, supposedly working by boosting enzymes.

ENVIRONMENTAL ELEMENT

Caesium in the environment	
Earth's crust	3 p.p.m.
	Caesium is the 46th most abundant element.
Soils	0.1–5 p.p.b.
Sea water	0.3 p.p.b.
Atmosphere	virtually nil

Although caesium is much less abundant than the other alkali metals, it is still more common than elements like arsenic, iodine and uranium. It has achieved little fame for its beneficial uses, but lots of ill-fame for its artificially produced radioactive isotopes, caesium-134 and caesium-137, released during nuclear accidents. This only added to the radioactive caesium which was released all round the world from the testing of nuclear weapons from 1945 to 1963, when most countries agreed to ban above-ground tests.

Uranium fuel rods in nuclear power stations produce caesium-137, which gives off α- and γ-rays as it decays. Not only that, but the half-life of caesium-137 is 30 years, which means that it takes over 200 years to reduce it to 1% of its former level. For this reason an accident at a nuclear power plant can contaminate the environment around for generations, which is why the Chernobyl accident in the Ukraine in 1986 was such an environmental disaster. It released a large amount of radioactive caesium-137 which drifted all over Western Europe, affecting sheep farms as far West as Scotland, Ireland and Wales, over 1500 miles from the accident. There it was washed to earth by heavy rain and taken up by the roots of plants, thus becoming part of the vegetation that the sheep ate.

Even in 1994, half a million sheep on 500 farms in the British Isles were still classified as affected and could not be sold for slaughter until their meat had been checked for contamination. Because caesium is rapidly excreted from the body, an animal reared on contaminated grass can be decontaminated by being grazed on unaffected pasture for a few days before it is slaughtered. This does not solve the problem of affected land and research has been directed to finding ways to treat farms with chemicals that will lock up the caesium and prevent it being absorbed by plants.

Some clay soils do this naturally and minerals like vermiculite are used specifically for this purpose. Another proposal is to use the old dye Prussian blue, a colour used in ink. There are several forms of this chemical, which contains potassium or ammonium ions that can exchange for caesium and thereby trap it and render it inactive. In Germany,

Austria and Norway these Prussian blues are encapsulated into slow-release pills which are given to animals. There they remain in the animal's stomach for several weeks releasing the agent, which binds the caesium so that it passes out of the sheep without being absorbed.

Humans affected by caesium-137 have also been successfully treated with Prussian blue type agents. For most people who are exposed to low levels of radioactive caesium, the simple expediency of taking in more potassium in their diet is sufficient to reduce absorption of the caesium. (Even though caesium-137 poses a threat to the environment, a little of this isotope is used in medicine where its γ-rays are used to treat certain gynaecological cancers.)

CHEMICAL ELEMENT

Data file	
Chemical symbol	Cs
Atomic number	55
Atomic weight	132.90545
Melting point	28°C
Boiling point	679°C
Density	1.9 kilograms per litre (1.9 grams per cubic centimetre)
Oxide	Cs_2O (there is also a superoxide, CsO_2, which is explosive)

Caesium is a soft, shiny, gold-coloured metal and a member of group 1, also known as the alkali metals, of the periodic table of the elements. When it is dropped into water, caesium reacts violently to give off hydrogen gas, which ignites spontaneously. Caesium metal oxidizes rapidly when exposed to the air and can form the dangerous superoxide on its surface. It is generally stored under oil in sealed cans. Caesium hydroxide is highly caustic and can dissolve glass.

Naturally occurring caesium ores consist entirely of the caesium-133 isotope, which is not radioactive. (There are also minute traces of radioactive caesium isotopes in uranium ores, as a result of the nuclei of this metal undergoing spontaneous fission.)

ELEMENT OF SURPRISE

Caesium 'thrusters' are used to steer satellites. Caesium is ionized in a vacuum chamber and then the ions are accelerated through an electric field and ejected through a nozzle, thereby giving a counter-thrust to the satellite. The heavier the ions, the larger the impulse delivered, and this is why caesium is used because it has a high atomic mass. A kilogram of caesium used in this way will propel a space vehicle 140 times further than by using the same weight of any known fuel. See also xenon, p. 490.

Pronounced kal-sium, the name comes from the Latin *calx*, meaning lime.
French, *calcium*; German, *Calcium*; Italian, *calcio*; Spanish, *calcio*; Portuguese, *cálcio*

HUMAN ELEMENT

Calcium in the human body	
Blood	61 p.p.m., with the plasma having 80 p.p.m. and red blood cells only 4 p.p.b.
Bone	3–17%
Tissue	*approx.* 120 p.p.m.
Total amount in body	1.2 kilograms

Calcium is essential to practically all living things, except a few insects and bacteria, and it performs a variety of functions. As calcium carbonate, $CaCO_3$, it provides the skeleton for most marine creatures and the lens in eyes; as calcium phosphate, $Ca_5(PO_4)_3(OH)$, it is found as the bones and teeth of land-based animals; as calcium oxalate, CaC_2O_4, it is abundant in plants where it acts as a calcium store and possibly as a deterrent to being eaten because it is toxic.

Calcium is the most abundant metal in the human body, the reason being that bone contains a great deal, accounting for 1190 grams of the 1200 grams of calcium in an average adult. The remaining 10 grams has five key metabolic functions:

- it regulates activity at membranes, e.g. it cements cells together;
- it is involved in contraction of muscles and in conduction of nerve impulses;
- it keeps the pH of the blood stable and aids clotting;
- it controls cell division;
- it triggers hormone release.

Calcium is ideal for these purposes because of the affinity of its positive ion Ca^{2+} for particular chemical groups such as carboxylates, which are the negative ions of organic acids.

FOOD ELEMENT

Pregnant women and children are encouraged to eat foods rich in calcium, to promote the growth of teeth and bones. Traditionally they have been advised to eat cheese, milk and white bread, but more recently the trend has been to recommend leafy green vegetables as well. People of all ages would be well advised to eat calcium-rich foods, although calcium deficiency is rarely a problem with a normal diet.[21] When it does occur, its most obvious effect is bone disease, but it is not lack of calcium that is generally the primary cause, but some other factor. For example, rickets in young children is due to lack of vitamin D, and the weakening of the bones known as osteoporosis, associated with ageing, is not cured by taking extra calcium (see box).

The ideal calcium intake is probably about 1000 milligrams per day but, during childhood, pregnancy and old age, when extra calcium is needed, we would probably do best to boost this to 1500 milligrams as insurance. In theory, an adult needs around 500–700 milligrams per day and teenage children around 800–1000 milligrams. Sardines, egg yolk, almonds, Cheddar cheese and milk chocolate are particularly good sources of calcium, providing more than 200 milligrams per 100 grams, while cereal and dairy products generally contain more than 100 milligrams per 100 grams. Vegetables, especially cabbage, broccoli, onions and red kidney beans, can also be a useful source. To utilize this calcium, the diet needs to contain plenty of vitamin D, which is provided by foods like fish liver oils, many types of fish, butter, margarine and eggs.

BONE IDLE?

We tend to think of bone as somehow different from the other cells that make up our body. It is more like a mineral than living flesh, but it is just as 'alive' in that it is endlessly changing, being constructed and broken down at millions of sites throughout the skeleton by cells that are called osteoblasts and osteoclasts, respectively. In this way bone carries out an important function of keeping the level of calcium in the blood steady, thereby ensuring that all its other functions are covered.

When our diet does not provide enough calcium for these essential processes, the deficiency is made good from bone – ideally to be replaced later when there is an excess of calcium in our blood. As we get older this replacement process does not completely compensate for the loss, so to check this depletion we need to make sure we get a daily dose of calcium and vitamin D, the vitamin which regulates bone growth. Even so calcium loss cannot be entirely prevented.

MEDICAL ELEMENT

Calcium has featured in medicines down the centuries: calcium carbonate is used as an over-the-counter antacid for indigestion; calcium lactate can be prescribed to treat

[21] People on a high protein diet should be aware that this increases calcium loss via the urine.

calcium deficiency; calcium chloride in small doses acts as a diuretic and it may be included in a saline drip; and calcium sulfate is used to make plaster casts.

Attention has focused on calcium in recent years, as ageing populations suffer weakened skeletons and bones that break easily. Our bone mass, as it is called, reaches a peak when we are about 35 years old, stays moderately stable for several years thereafter, but then begins to decline, with women losing bone faster than men, particularly after the menopause. Then bone loss might be as much as 1% per year. This can be reduced by hormone replacement therapy, as well as by the drugs bisphosphonate and calcitonin. The latter drug supplements the natural calcitonin secreted by the thyroid gland, specifically to prevent bone loss. Fluoride therapy is used in some countries; given at a rate of 20 milligrams a day, it strengthens bone by forming a tougher kind of calcium phosphate known as fluorapatite, $Ca_5(PO_4)_3F$.

Calcium supplementation to the diet may benefit children too, as shown by Conrad Johnston of Indiana University, USA. He took 60 pairs of identical twins, aged 6–14, and gave one of the pair a tablet of calcium every day for 3 years. In the children who were given calcium tablets, bone growth was faster than in the untreated twin.

ELEMENT OF HISTORY

Lime (calcium oxide, CaO) was used by all the ancient civilizations to make mortar for building and is mentioned in several books of the Roman period. Mortar was made by mixing lime with sand and water; for a while it would remain soft and workable, but on leaving it would set hard, and it got harder over time as it absorbed carbon dioxide from the air to form calcium carbonate.

Lime was also known as 'quick lime'; adding water to it converted it to 'slaked lime', which is calcium hydroxide, $Ca(OH)_2$; this was also a useful material because of its acid-neutralizing ability and soil-improving quality.

In 1755 Joseph Black of Edinburgh proved that when limestone was heated to form lime, it gave off carbon dioxide ('fixed air' was his name for this gas). The great chemist Antoine Lavoisier listed lime as an element because it seemed impossible to reduce it to simpler components, but he also suspected that it was the oxide of an unknown element. This was proved in 1808 when Humphry Davy of the Royal Institution in London obtained calcium metal by electrolysis. He had produced sodium and potassium in this way the previous year and confidently thought that he could do the same by mixing lime with potash, heating the mixture till it melted and then passing an electric current through it, but all this produced was potassium. So he tried a mixture of molten lime and mercury oxide; this was a little more successful since it produced an amalgam of calcium and mercury, but not enough to do anything with.

In May 1808 he got a letter from J. J. Berzelius who told of similar experiments he had conducted with Dr M. Pontin, physician to the King of Sweden. They also produced the amalgam, but more of it. This prompted Davy to try again, using a higher ratio of lime to mercury oxide, and this produced enough of the amalgam for him to be able to distill off the mercury leaving the calcium behind, albeit still contaminated with a little mercury. However, it was possible for him to record some of its properties. Perhaps not surprisingly, it was not until the early twentieth century that a commercial method of making calcium metal was perfected.

ECONOMIC ELEMENT

Many minerals are known: anhydrite, which is calcium sulfate ($CaSO_4$); calcite, limestone and aragonite, which are all forms of calcium carbonate ($CaCO_3$); dolomite, which is a mixed calcium–magnesium carbonate ($CaMg(CO_3)_2$); and gypsum, which is hydrated calcium sulfate ($CaSO_4 \cdot 2H_2O$).

The chief mined ores are calcite, dolomite, gypsum (used in cement and plaster) and anhydrite (used to make sulfuric acid). Marble is a type of calcite. World production of calcium metal is only 2000 tonnes per year, but that of lime is around 120 million tonnes. Reserves of limestone are virtually unlimited.

Calcium metal is produced by heating lime with aluminium metal in a vacuum. When the bulk metal is exposed to air, it reacts mainly with nitrogen and a white coating of calcium nitride and oxide forms on the surface. For this reason it also serves as a 'getter' to remove residual amounts of air from vacuum tubes. The oxide/nitride coating serves to protect calcium and keep it stable, and stable enough even to be machined. The metal is used in the manufacture of zirconium, thorium and the rare-earth metals. It is used as an alloying agent for aluminium, beryllium, copper, lead and magnesium alloys.

Lime is used in metallurgy, water treatment, the chemicals industry, for making cement, etc. Most is used in steel making as a flux, where it reacts with the impurities in the molten iron ore to form a viscous slag that can be tapped off. Another major outlet is in sewage and pollution control to reduce acidity.

Gypsum can also be used as a building material in the form of plaster, but first it has to be heated to remove its water molecules. In addition to covering walls and making plasterboard, it is used to protect broken limbs, when it is known as plaster of Paris.[22] Alabaster is a crystalline form of gypsum that is soft and easy to carve, and is used to make sculptures, which, when polished, become semi-translucent, greatly adding to their beauty.

Cement is made by heating a mixture of limestone and clay in a kiln at 1500°C, which drives off carbon dioxide, leaving a mixture of calcium (mainly) and aluminium silicates. These are then pulverized and gypsum is added. When it is mixed with water, cement undergoes a complex series of reactions which have only recently been understood.

[22] Heating gypsum drives off most of its water and it is then ground to a find powder. When water is added to the powder it forms a paste and slowly reverts to gypsum, eventually setting hard, which produces an ideal protective cast for broken limbs.

ENVIRONMENTAL ELEMENT

Calcium in the environment	
Earth's crust	41000 p.p.m. (= 4.1%)
	Calcium is the fifth most abundant element.
Soils	can be up to 5%, but on average is 1–2%
Sea water	400 p.p.m.
Atmosphere	minute traces

Calcium is one of the most abundant metal elements in the Earth's crust, and the third most abundant metal. It is found mainly as sedimentary rocks formed by precipitation or by the accumulation of the shells of marine creatures over millions of years. Vast areas of calcium carbonate, in the crystalline form known as aragonite, are present as coral; it has been estimated that coral reefs cover about 2 million square kilometres of the planet.

There is a so-called 'calcium cycle' which moves vast amounts of the element through the environment, from land to sea, to sediments, and then back to land as continents uplift. It is estimated that around 500 million tonnes of calcium move through the cycle each year.

Calcium carbonate and phosphate are not very soluble but some calcium dissolves in rain water that percolates through soil and rock, producing what is referred to as 'hard' water. (If the water drips into an underground cave it slowly forms icicle-like stalactites and stalagmites.) Such water is thought to be beneficial for drinking since it acts as a source of this essential element. When used for cleaning purposes the hardness interferes with the cleaning agents, such as soap and detergents, in which case water-softening chemicals are needed to tie up the calcium ions and prevent them interfering with the washing process. Limescale, which is calcium carbonate, deposits from hard water when this is heated in hot water systems and kettles; it can be removed, by treating with an acid descaler, or prevented, by incorporating an ion exchange water softener into the inlet supply, which removes the calcium. Limestone (natural calcium carbonate) is used to neutralise acid soils and surface waters.

CHEMICAL ELEMENT

Data file	
Chemical symbol	Ca
Atomic number	20
Atomic weight	40.078
Melting point	839°C
Boiling point	1484°C
Density	1.6 kilograms per litre (1.6 grams per cubic centimetre)
Oxide	CaO

Calcium is a silvery, relatively soft metal and a member of group 2, also known as the alkaline earth metal group, of the periodic table of the elements. Calcium metal is attacked by oxygen and water.

None of the six naturally occurring calcium isotopes, of which calcium-40 accounts for 97%, is radioactive. The others are, in descending order of abundance, calcium-44 which comprises 2%; calcium-42 at 0.65%; calcium-48 at 0.2%; calcium-43 at 0.14%; and calcium-46 at a mere 0.004%. (Calcium-48 appears to be radioactive but with a half-life of 6×10^{18} years.)

ELEMENT OF SURPRISE

Limelight was invented by a Briton, Thomas Drummond, in 1823 in order to help surveyors who wanted a source of light that could be seen over long distances. He had discovered that when a jet of hydrogen, burning in oxygen, played upon a compact block of calcium oxide (lime) then the lime would glow brilliantly white. The light it emitted could be focused by lenses and was visible for more than 60 miles. (The hydrogen was generated from sulfuric acid and zinc, and the oxygen by heating a mixture of manganese dioxide and potassium chlorate. The two gases were stored in bellows-shaped bags.)

Drummond's invention was quickly seized upon by theatre managers and lighthouse keepers, but it was the former who brought it to the attention of the public when it was used to throw a sharp circle of light on to performers on the stage – and in this way calcium chemistry entered the English language in the phrase 'to be in the limelight,' meaning to be the centre of attention.

Pronounced kali-forn-iuhm, the name comes from the university and state of California where the element was first made.

French, *californium*; German, *Californium*; Italian, *californio*; Spanish, *californio*; Portuguese, *califórnio*.

HUMAN ELEMENT

Californium in the human body
None

Californium has no role to play in living things, nor can it ever have had such a role because of its rarity and intense radioactivity. However, it can be encountered outside nuclear facilities and research laboratories because it is used in mineral prospecting and for medical diagnosis and treatment. Special precautions are needed because it is not only a powerful source of radiation, but also a dangerous neutron emitter.

ELEMENT OF HISTORY

Californium was first made in February 1950 at Berkeley, California, by a team consisting of Stanley G. Thompson, Kenneth Street Jr, Albert Ghiorso and Glenn T. Seaborg. They made it by bombarding curium-242, which had first been made in 1944, with helium ions.[23]

This experiment required the making of enough curium to form a large enough 'target' for the high-energy helium ions to stand a chance of hitting the atomic nuclei and combining to form a new element of higher atomic number. Curium is intensely radioactive and it took the team 3 years to collect the few millionths of a gram needed for the test. Their experiment produced around 5000 atoms of californium, but this was enough to identify that it was a new element.

ECONOMIC ELEMENT

Californium is produced in milligram quantities in nuclear reactors as the isotopes californium-249 and californium-252 by the bombardment of plutonium-239 with

[23] The equation which sums up this process is $^{242}Cm + {}^{4}He \rightarrow {}^{245}Cf$ (half-life = 44 minutes) + ^{1}neutron.

neutrons. The former has a half-life of radioactive decay of 350 years, the latter 2.6 years. Californium-252 is a strong neutron emitter (1 microgram releases 170 million neutrons per minute) and is used in portable neutron sources. These are employed in moisture gauges, in core analysis in drilling oil wells, and in on-the-spot activation analysis in gold prospecting. (Bombarding mineral samples with neutrons produces other radioactive elements which are easily identified by the radiation they emit.)

Californium is also used in cancer therapy (see below), an application that was encouraged by the US Nuclear Regulatory Commission who set its price at a relatively low $10 per microgram.

ENVIRONMENTAL ELEMENT

Californium in the environment	
Earth's crust	virtually nil
Sea water	nil
Atmosphere	nil

Californium does not occur naturally on Earth. All that there now is has been synthesized, but this element was produced in the past when several natural nuclear reactors were in operation about 2 billion years ago in Africa – see uranium (p. 481).

CHEMICAL ELEMENT

Data file	
Chemical symbol	Cf
Atomic number	98
Atomic weight	252.1 (californium-252 isotope)
Melting point	not known
Boiling point	not known
Density	not known
Oxide	Cf_2O_3

Californium is a radioactive metal which is a member of the actinide group of the periodic table of the elements. In 1960, 3 micrograms of californium oxychloride (CfOCl) was made, enough to be seen with the naked eye. A sample of the metal itself has yet to be produced because its compounds resist reduction. It is expected to be readily attacked by air, steam and acids, but not by alkalis.

ELEMENT OF SURPRISE

Californium-252 neutron therapy was first investigated at the University of Kentucky in 1972 as a possible treatment for cancers, and when it proved successful, clinical trials began in 1976. It was possible to increase the life span of more than 80% of women with

Pronounced kar-bon, the name is derived from *carbo*, the Latin for charcoal.

French, *carbone*; German, *Kohlenstoff*; Italian, *carbonio*; Spanish, *carbono*; Portuguese, *carbono*.

In 1787 a group of French chemists tried to bring order to the naming of chemicals; they published the *Méthode de Nomenclature Chimique* in which they proposed that this element be called 'carbone' rather than 'charbon', the name for charcoal.

COSMIC ELEMENT

Carbon is one of the major elements to be found in the interstellar void and is given off by dying stars. A particularly carbon-rich star is CW Leonis (known to astronomers as IRC+10216), which is surrounded by a haze of carbon, making it appear very faint. By analysing the radio waves and microwaves emitted by this cloud of cosmic carbon, it has been possible to identify scores of different carbon molecules, including some with a large number of carbon atoms. It was during laboratory experiments to replicate the formation of such molecules that carbon spheres, the fullerenes, were discovered by Robert Curl Jr, Harry Kroto and Richard Smalley, thereby winning them the 1996 Nobel Prize for Chemistry.

HUMAN ELEMENT

Carbon in the human body*	
Blood	varies minute by minute
Bone	0.8%
Tissue	67% (based on dry matter)
Total amount in body	16 kilograms

* This refers to the total weight of carbon in all its myriad forms, and as such has little meaning except on a comparative basis with other elements.

No element is more essential to life than carbon, because only carbon forms strong single bonds to itself that are stable enough to resist chemical attack under ambient conditions. This gives carbon the ability to form long chains and rings of atoms, and these are the structural basis for many compounds that comprise the living cell, of which the most important is DNA.

Carbon itself is completely non-toxic, but some simple compounds can be very toxic, such as the gas carbon monoxide (CO) or cyanide (CN^-). Carbon black, which is like soot, can be an irritant dust but is not in itself dangerous, although soot may harbour carcinogenic materials.

FOOD ELEMENT

Most of what we eat – carbohydrates, fats, proteins and fibre – is made up of compounds of carbon, giving us a total carbon intake of 300 grams a day. Digestion consists of breaking these compounds down into simpler molecules that can be absorbed through the walls of the stomach or intestines. There they are transported by the blood to sites where they are utilized, some to repair or replace damaged cells, some to be turned into body messengers, and some to make the thousands of different chemicals that our body needs. However, most ingested carbon compounds are oxidized to release the energy they contain, and then we breathe out the carbon as carbon dioxide. This joins the other carbon dioxide in the atmosphere, from where it will again be extracted by plants and become part of the carbon cycle of nature (see below).

MEDICAL ELEMENT

The isotope carbon-11 is employed in medical diagnosis using the technique known as positron emission tomography (PET). With a half-life of only 20 minutes, carbon-11 has to be produced on-site from nitrogen atoms by bombarding them with protons in a cyclotron. Once the isotope has been made, it has to be placed quickly within a simple molecule, such as glucose, and then this has to be incorporated into a more complex molecule that will seek out a tumour when it is injected into a patient's bloodstream.

When a carbon-11 nucleus disintegrates it ejects a particle known as a positron, which is effectively a positively charged electron. When this particle meets an ordinary, negatively charged electron the two suffer mutual annihilation, an event which emits two gamma rays that go off in opposite directions at exactly 180° to each other. The body scanner detects these rays, the organ from which they are emited is highlighted, and in this manner a deep-seated cancer can be accurately located.

ELEMENT OF HISTORY

Carbon occurs naturally as coal, graphite and diamond; these were known to the ancient civilizations, but rarely used. More readily available carbon was that produced by fire, such as soot or charcoal. Charcoal was made by heating wood in a limited supply of air, and it became an important trade supplying the industries that manufactured metals, glass and pottery. Charcoal making is still carried on in parts of the world to this day.

It was with the dawn of chemistry that the various forms of carbon were eventually recognised as the same element. Many of the great chemists of the eighteenth century, such as Scheele, Berthollet and Lavoisier, wrestled with the problem. In view of its hardness and chemical inertness it was diamond that posed the greatest difficulty of identification. Guiseppe Averani and Cipriano Targioni of Florence were the first to discover that diamonds could be destroyed by heating. In 1694 they focused sunlight on to a diamond using a large magnifying glass and the gem eventually disappeared.

The experiment was repeated in July 1771 by P.-J. Macquer and Godefroy de Villetaneuse, who observed that the heated diamond burnt completely away leaving no ash. But it was not until 1796 that the English chemist, Smithson Tennant, proved that diamond was entirely a form of carbon and that it burned to form only carbon dioxide.

Others argued that if diamond was carbon then it should be possible to convert other forms of it into the gem. However, early experimenters were unaware of the high temperatures and intense pressures that were needed to achieve this transformation, so we know that early claims to have made diamonds were false. Even the eminent chemist Henri Moissan, who claimed to have done this in 1893 and exhibited a stone 0.7 millimetres in size, was deluded. That diamond had been surreptitiously slipped into his experiment by one of his laboratory assistants.

ELEMENT OF WAR

Gunpowder is a mixture of charcoal, sulfur and saltpetre (potassium nitrate, KNO_3), and was for many centuries used as an explosive in guns and cannon. Today it is no longer used in firearms and weapons, although it is still the basis of many fireworks. A fuller account of gunpowder is given under sulfur (p. 412).

ECONOMIC ELEMENT

Carbon in its various forms is extracted from the Earth's crust on a far greater scale than any other element. Technically all the mining of limestone, dolomite and marble is the mining of carbon as carbonate, its most oxidized form. Similarly, the extraction of fossil reserves – coal, oil and gas – is equivalent to mining carbon, in this case reduced carbon. However, both of these categories are beyond the scope of this book, except to note them.

Less well known is the extraction of a relatively pure form of carbon: graphite. Graphite is mined in Sri Lanka, Madagascar, Russia, South Korea, Mexico, the Czech Republic and Italy. It finds use as brushes in electrical motors, electrodes in fuel cells, furnace linings, extrusion dies, and even the so-called 'lead' of pencils.

Carbon itself is mainly used in its amorphous (non-crystalline) forms: as coke in steel making to reduce iron oxide to iron; as carbon black in printing and as a filler for tyres; and as activated charcoal in sugar refining, water treatment, and respirators and extractor hoods in kitchens. Carbon black is made by burning natural gas in a limited supply of air, whereas coke is made by heating coal in the absence of air. Activated charcoal is produced by heating coconut shells, and this gives a granulated form, ideal for absorbing pollutants from gases and water.

Carbon fibre is produced by the controlled heating of acrylic fibre, and other carbon-rich materials such as pitch, until it chars to a carbonized material which is stronger than steel. This can be woven into fabrics which can absorb poisonous gases and so are used for protective clothing and fire-hoods. Carbon fibre is used to make laminates that combine great strength with low weight and are used in rockets and aircraft, and to reinforce plastics for sports equipment such as skis, fishing rods and racquets.

The other form of pure carbon extracted from the Earth is diamond, and the principal countries where diamonds are mined are Botswana, Russia, South Africa, Canada, Namibia, Angola and Australia. Today most diamonds are manufactured, although most are suitable only for industrial purposes. The first synthetic diamonds were made by the Swedish company ASEA in 1953 using temperatures of 3000°C and pressures of 90 000 atmospheres, but they did not publish their results. The General Electric Company in the USA made diamonds using similar conditions and they announced their success in 1955. They went on to develop the process to the extent of making gem-quality diamonds, although these are very costly to make. Today, Russia is the main producer of synthetic diamonds.

Diamond films can be made by a process known as chemical deposition in which a simple carbon-containing molecule such as methane is decomposed to its atoms; these will crystallize out on to the surface to be protected, such as a razor blade. The trick is to get the film to grow into a perfect array of carbon atoms with the same chemical bonding between them as in diamond. This is achieved by adding hydrogen gas to the methane before it is decomposed. The resulting film of carbon is invisible to the naked eye, but it gives the surface the same quality of strength and hardness as a real diamond.

ENVIRONMENTAL ELEMENT

Carbon in the environment	
Earth's crust	480 p.p.m.
	Carbon is the 15th most abundant element.
Sea water	28 p.p.m.
Atmosphere	350 p.p.m. as carbon dioxide, 1.6 p.p.m. as methane and 0.25 p.p.m. as carbon monoxide (these are global averages and vary widely depending on location)

The carbon of the Earth comes in several forms, as fully oxidized forms of carbonate (CO_3^{2-}) in rocks or hydrogen carbonate (HCO_3^-) in the seas, or as reduced carbon in fossil deposits. There it is present mainly as hydrocarbons as in natural gas (methane; CH_4) or oil which has a carbon:hydrogen ratio of about 1:2, or coal which has a ratio of about 1:1 although this is a particularly complex material. Known reserves of these fossil fuels amount to a vast 1 trillion tonnes of coal, 980 billion barrels of oil and 141 trillion cubic metres of natural gas; known reserves are still increasing as more is being discovered than is being used up. Canada has 2500 billion barrels of oil as tar sands. Humans use about 7 billion tonnes of fossil carbon each year.

Most of the reduced carbon in the Earth's crust is too widely dispersed to be economically exploitable and is known as kerogen, but altogether there is an incredible

375 000 billion tonnes of it, of which only a few per cent will ever be accessible as fossil fuel. Gas hydrates may also hold vast amounts of methane.

Carbon is probably the most important element from an environmental point of view. The Earth's early atmosphere may have contained a lot of carbon dioxide and methane, but once life evolved this began to change.[24] Today, there is very little of these gases and a lot of oxygen instead, thanks chiefly to the action of plants which convert carbon dioxide and water into carbohydrate and oxygen by photosynthesis. The Earth's atmosphere contains an ever-increasing concentration of carbon dioxide and carbon monoxide, from fossil fuel burning, and of methane, from paddy fields and cows. Human contributions to these sources are still a minor component compared with natural sources: most carbon dioxide comes from plants, microbes and animals, while methane is given off by swamps, marshes and termite mounds.

Marine species, especially algae, are thought to have been responsible for deposits of oil and natural gas, while plants on land laid down the layers of rotting vegetation. This is first turned into peat, then lignite, then bituminous coal, and finally anthracite coal which is almost entirely carbon. It takes around 250 million years to complete this geological sequence of events.

Carbon dioxide also has a controlling effect on the acidity of rain water, since this gas dissolves to form carbonic acid which leads to a pH of 4–5.

The carbon cycle

This is the term used to describe the way in which this element moves between various parts of the terrestrial ecosphere (so-called 'reservoirs'); there are large-scale transfers between them. The cycle rules the tempo of life on Earth and turns over 200 billion tonnes of carbon each year. The amounts of carbon in the various 'reservoirs' are as follows:

Atmosphere	724 billion tonnes
Living things on land	2000 billion tonnes
Oceans	39000 billion tonnes (mainly as dissolved carbonate)
Living things in the seas	40 billion tonnes
Earth's crust	100 million billion tonnes as carbonate rock; 375 000 billion tonnes as reduced carbon, which includes coal, oil and gas

Perhaps more important than movement between these reservoirs is the movement which occurs when carbon is incorporated into living things, i.e. the land-based and marine-based biological cycles. The first step in these cycles is the absorption of carbon dioxide by plants, which then provide food for animals. In this way carbon is passed up the various food chains, with each recipient releasing some as carbon dioxide, until most carbon is back where it started. Nevertheless, a significant proportion of this carbon is

[24] Other planets also have a lot of carbon in their atmosphere: on Venus the atmosphere is almost all carbon dioxide, while on Jupiter and Saturn it is methane.

deposited on the land and is eaten by microbes and creatures that inhabit the soil. Similar food chains exist among the species in the oceans, although there some of the carbon ends up as carbonate and settles down to the bottom sediments where it will remain for millions of years.

CHEMICAL ELEMENT

Data file	
Chemical symbol	C
Atomic number	6
Atomic weight	12.0107
Melting point (diamond)	3550°C; sublimes at 4800°C
Density	graphite, 2.3 kilograms per litre (2.3 grams per cubic centimetre); diamond, 3.5 kilograms per litre (3.5 grams per cubic centimetre)
Oxides	CO and CO_2

The chemistry of carbon compounds is referred to as organic chemistry and its study constitutes the bulk of the subject, with more than 20 million different organic molecules having been made and registered.

Carbon comes at the head of group 14 of the periodic table of the elements. It occurs in three chemically distinguishable forms: graphite, diamond and fullerenes (of which buckminsterfullerene (C_{60}) is the best known). Rather oddly, diamond is less stable than graphite, but the rate of conversion is immeasurably slow. More mundane forms of carbon, such as carbon black, are amorphous, i.e. they have no well defined structure.

Diamond consists of carbon atoms bonded to each other in a three-dimensional array, each carbon forming four bonds to neighbouring carbons, and that is what gives it strength, rigidity and hardness. Graphite consists of carbon atoms in sheets, each carbon bonding to three neighbours, with the sheets stacked one on top of the other, in a rather weak array. This is what makes graphite easy to cleave, why it feels slippery and why it is used as a lubricant. It also makes graphite able to conduct electricity in the direction of the sheets. Fullerenes and nanotubes have similar chemical bonding to graphite but instead of the sheets being flat they are rolled up into hollow balls or tiny tubes. They are less stable than graphite and on exposure to the air they are slowly oxidized.

There are two main isotopes of carbon: carbon-12, which makes up 99% of all carbon, and carbon-13, which accounts for the remaining 1%. These are non-radioactive isotopes, but a third form, carbon-14, is radioactive, and there is also a trace of this in the atmosphere (see box).

ELEMENT OF SURPRISE

Diamonds are called 'ice' not because they look like ice but because they feel cold to the touch. The reason is the high thermal conductivity of diamond, which quickly drains

CARBON DATING

The American scientist Willard F. Libby developed a method of dating ancient relics by analysing the organic matter they contained for its radioactive carbon-14 content. He received the Nobel Prize for Chemistry in 1960 for his work, which has proved invaluable not only for archaeologists but for detecting forgeries. Even fake wine has been uncovered this way, by its *lack* of radioactivity.

Carbon-14 is produced by cosmic rays bombarding nitrogen in the upper atmosphere and it is radioactive. About 7 kilograms of carbon-14 are produced per year; this may not sound very much but it is almost a trillion trillion atoms, which is enough to ensure that it becomes incorporated into all living plants. When the plant dies, the amount of carbon-14 in it is fixed and no more is added. If the plant is a tree, and used in building or furniture, or if it is a fibrous plant, and used to make paper or textiles, then these can be dated by the amount of carbon-14 they contain.

The half-life of carbon-14 is 5730 years, which means that objects from even the earliest civilizations can be dated. It also enables forgeries to be recognized and in this way the Round Table of Lancelot, which is in Winchester Cathedral, was shown to be a forgery, and so was the Turin Shroud.

Fake wine and spirits can be easily made from industrial alcohol which is produced by the petrochemical industry as a solvent. Since the carbon in this has been underground for so long that all its carbon-14 has disappeared, it produces a drink whose alcohol also lacks radioactivity. Showing this to be so proves that the drink is phoney – although it will be just as potent as the real thing!

heat to itself, a property that explains why diamond in conjunction with copper is used as a 'heat sink' for electronic microchips.

Not all diamonds are white and glass-like. In fact they can be anything from pale brown ('champagne') to green; orange, blue and pink are rare, but the rarest of all is red, which gets its colour because its lattice is deformed. Only about one diamond in a thousand is naturally coloured and they command high prices. Coloured diamonds can be produced by natural radiation in the Earth, when they can take on a light green colour, or they can be artificially irradiated in a nuclear reactor, when they take on a spectrum of hues.

The town of Nördlingen in Bavaria is built of stone that contains millions of minute diamonds that were formed when a 1 kilometre diameter meteor crashed there about 15 million years ago, creating what is known as the Ries crater. Such was the force of the impact that it turned a local deposit of graphite to diamonds, creating an estimated 72 000 tonnes of them, but all less than 0.2 millimetres across.

Other impact diamonds are found in the huge Popigai crater in Northern Siberia, at Sudbury in Canada, and Lappajärvi in Finland. The impact crater at Yucatan, Mexico, strewed the planet with them – see also p. 203.

Pronounced seer-iuhm, the name comes from the asteroid, Ceres, which had been discovered 2 years prior to the discovery of the element, and which was itself named after Ceres, the Roman goddess of agriculture.

French, *cérium*; German, *Cer*; Italian, *cerio*; Spanish, *cerio*; Portuguese, *cério*.

Cerium is one of 15, chemically similar, elements referred to as the rare-earth elements, which extend from element atomic number 57 (lanthanum) to element atomic number 71 (lutetium). The term 'rare-earth elements' is a misnomer because some are not rare at all. They are more correctly called the lanthanides, although strictly speaking this excludes lutetium. The minerals from which they are extracted, and the properties and uses they have in common, are discussed under lanthanides, on p. 219.

HUMAN ELEMENT

Cerium in the human body	
Blood	approx. 1 p.p.b.
Bone	approx. 3 p.p.m.
Tissue	approx. 0.3 p.p.m.
Total amount in body	40 milligrams

In 1880 Alfonso Cossa reasoned that, because cerium was often found in calcium phosphate ores, then it might also be found in bone, which is mainly calcium phosphate. He analysed bones and found traces of cerium. Cerium can mimic calcium and was also soon found in measurable amounts in barley, beech wood and tobacco plants. This metal is not generally taken up by plant roots so little gets into the main food chain. The small amount that does is not considered a health risk. Like all the lanthanides, cerium is regarded as non-toxic.

As yet, no one has monitored the cerium content of the diet, so it is difficult to judge how much is consumed. It is probably less than a milligram per day, although the metal can be detected in urine. Cerium has no known biological role, but it has been noted that cerium salts stimulate metabolism.

ELEMENT OF HISTORY

Cerium was first identified by the chemist Jöns Jacob Berzelius (1779–1848) and the geologist Wilhelm Hisinger (1766–1852) at Vestmanland, Sweden, in the winter of 1803/4. The German chemist Martin Klaproth (1743–1817) independently discovered it around the same time, and wanted to call it cererium, again linking its name to the newly discovered asteroid.

Although cerium is one of the rare-earth elements, it was discovered independently of them because there are some minerals that are almost exclusively cerium salts. Cerite, a cerium silicate, comes into this category. A lump of this reddish-brown mineral had been found in 1751 by Axel Cronstedt at a mine in Vestmanland. He thought it was an iron ore, albeit strangely heavy, and when tungsten was reported in 1783 it seemed likely that it was an ore of that metal. Wilhelm Hisinger, whose aristocratic father owned the estate on which the mine stood, sent a sample to Carl Scheele, the famous chemist, to analyse it, but he failed to find any tungsten. Hisinger remained curious and, in 1803, he and Berzelius worked on it and proved that it was the ore of a previously unknown element. Shortly thereafter, a black ore from Greenland, called allanite (an aluminium iron silicate) was also found to contain a lot of cerium.

There were several attempts to get a sample of the pure metal itself, but the usual procedures for doing this did not work. Even heating cerium chloride with potassium failed, producing only an impure brown powder which probably contained some metallic cerium. It was not until 1875 that two American chemists, William Hillebrand and Thomas Norton, first obtained a pure specimen of the metal by passing an electric current through the molten chloride.

ECONOMIC ELEMENT

Cerium comes mainly from the major lanthanide ores (see p. 222) but some is obtained from perovskite, a titanium mineral, and allanite, both of which can have enough cerium to make them viable sources.

Cerium oxide, CeO_2, is the most important outlet for cerium. It is produced by heating bastnäsite ore, which oxidizes the cerium to this insoluble oxide. It remains behind when the other metals are dissolved out using hydrochloric acid; the insoluble residue is known as cerium concentrate. Production amounts to about 23 000 tonnes a year, but this is likely to increase greatly in this century as more uses are found for it and existing uses expand. For example, cerium oxide concentrate, as a slurry in water, is used in place of ferric oxide rouge for polishing high-grade glass, giving a better finish more quickly to lenses and television tubes.

During most of the nineteenth century, cerium and its compounds were considered to be of little interest outside the chemistry laboratory, but early in the twentieth century it finally became important in the production of incandescent mantles for gas lighting. Karl Auer discovered that thorium oxide would glow with a brilliant light when heated. Early supplies of town gas (coal gas) relied on traces of benzene to give a bright luminous flame, but once this began to be removed and sold to chemical works as a raw material, an alternative way of boosting the light had to be found.

Auer came up with the gas mantle which did just that. It was made from a closely woven fabric impregnated with a solution of thorium nitrate and then baked to decompose these to the oxides. When the mantle was then heated in a gas flame, it emitted a broad band of light centred on a wavelength of 500 nanometres, although the light was considered rather harsh. This was cured by adding 1% cerium nitrate to the thorium solution; the cerium oxide formed in this way not only produced a more pleasing light but also acted as a catalyst, ensuring more complete combustion of the gas.

Cerium oxalate, when heated, decomposes to cerium oxide. This is the way in which the pure form is made industrially. Cerium oxide (1–2%) protects glass against radiation (X-rays and cathode rays) and will also filter out ultraviolet light. Most damage caused by light is due to ultraviolet radiation of wavelengths below 400 nanometres, but these are strongly absorbed by cerium. Glass for use in nuclear plants, X-ray equipment, television sets and ultraviolet sterilizers always contains some cerium. Adding cerium to the glass for cathode ray tubes prevents the discoloration that comes with time and which is due to the creation of so-called 'colour centres' in the glass caused by prolonged bombardment by high-energy electrons. Optical glass for lenses also includes cerium, along with lanthanum and yttrium, to provide a high refraction index and low colour dispersion.

Cerium oxide is part of the catalyst of catalytic converters used to clean up vehicle exhausts and it enhances the performance of the other metals present, probably by acting to store and release oxygen atoms which convert carbon monoxide to carbon dioxide as well as oxidizing unburnt hydrocarbons. It also catalyses the reduction of nitrogen oxides (NO_x) to nitrogen gas. All new cars are now equipped with converters; the catalyst in these consists of a ceramic or metal substrate, a coating of aluminium and cerium oxides and then a layer of a finely dispersed metal such as platinum or rhodium, which is the active surface. The cerium oxide plays more than just a support role, and when its catalytic effects were realized the amount used was increased. Cerium oxide is also used by the chemical industry as a catalyst for the production of alcohols, phenols and ketones.

One minor use of cerium oxide is in so-called 'self-cleaning' ovens, in which it is incorporated into the walls and there catalyses the oxidation of cooking residues, which are also mainly carbon.

Until recently cadmium red was the preferred choice of red pigment for containers, toys, household wares and crates, and it replaced pigments derived from toxic heavy metals, such as lead and mercury, but cadmium is now considered environmentally undesirable and its pigments are being phased out. The one most likely to replace them is cerium sulfide, Ce_2S_3. This gives a rich red colour, stable up to 350°C, and is completely non-toxic. It is made by heating cerium metal vapour in an atmosphere of sulfur. By adding traces of other rare-earth metals it is possible to produce a range of colours from deep maroons, through brilliant reds, to bright orange.

Cerium is used in many other ways, such as in flat-screen televisions, low-energy light bulbs and magnetic-optic compact discs, and is part of the core material in carbon-arc electrodes for film studio lights and for flood lighting. Cerium is also used in chromium plating, where a little is added to the electrolyte solution of the plating baths and assists in the plating process by preventing the formation of the chromium(III) ions which are difficult to reduce electrolytically to the metal.

There is only a small demand for cerium metal itself, which can be obtained by heating cerium fluoride with calcium, or by passing an electric current through molten cerium oxide. Cerium metal is added to aluminium to improve its corrosion resistance.

ENVIRONMENTAL ELEMENT

Cerium in the environment	
Earth's crust	68 p.p.m.
	Cerium is the 25th most abundant element.
Soils	av. 50 p.p.m., range 2–150 p.p.m.
Sea water	1.5 p.p.t.
Atmosphere	virtually nil

Cerium has three environmentally friendly virtues, as indicated above: the oxide cleans up vehicle exhausts, the sulfide reduces the need to use toxic heavy metals in red pigments, and cerium is an essential part of long-life, low-energy light bulbs. There is also a fourth way in which cerium might one day improve the atmosphere of cities, or wherever diesel engines operate. These emit so-called particulates, which are fine carbon particles only a few micrometres in diameter. A diesel car releases about 330 milligrams of particulates for each mile it travels, whereas a petrol-driven motor emits only 25. Particulates lodge in the lungs and have been blamed for an upsurge of respiratory diseases, such as asthma and bronchitis. One way to reduce particulate emissions is trap them in a ceramic filter, and then burn them off, but this uses more fuel. However, if a little cerium oxide is added to the fuel itself, it will catalyse the burning of the particulates and eliminate them. Only 30 grams of additive are required for 1000 litres of fuel burnt, and during its lifetime a diesel engine would need around 1.5 kilograms of cerium oxide additive. A car could be equipped with a cartridge of this which would be injected into the fuel as required.

CHEMICAL ELEMENT

Data file	
Chemical symbol	Ce
Atomic number	58
Atomic weight	140.116
Melting point	799°C
Boiling point	3426°C
Density	8.2 kilograms per litre (8.2 grams per cubic centimetre)
Oxides	Ce_2O_3 and CeO_2

Cerium is a reactive, grey metal and a member of the lanthanide group of the periodic table of the elements. It tarnishes in air, burns if scratched with a knife, reacts rapidly with water, and dissolves in acids. It is usually stored in light mineral oil to protect it against oxidation. Cerium fluoride and, especially, cerium oxalate are practically insoluble in water.

Natural cerium consists of four isotopes. Cerium-140 comprises 88.5% of the total. Cerium-142, which comprises 11%, is radioactive but with a half-life that is far greater than the age of the Universe, and which may be in excess of 50 million billion years (50×10^{15} years). The minor isotopes are cerium-136 (0.2%) and cerium-138 (0.3%).

Cerium stands out from the other lanthanides in having a stable higher oxidation state, cerium(IV). This can exist in solution as yellow Ce^{4+} ions (the lower oxidation ions, Ce^{3+}, are colourless). Ce^{4+} reagents have played an important part in analytical chemistry as powerful and stable oxidizing solutions.

ELEMENT OF SURPRISE

In 1854, an eminent medic, Professor Sir James Simpson, of the Department of Medicine and Midwifery at the University of Edinburgh, Scotland, reported that cerium nitrate prevented vomiting, especially that associated with morning sickness in early pregnancy. He recommended a dose of 1 grain (65 milligrams) three times a day. He claimed it was also good for treating stomach pains and other gastrointestinal disorders.

Soon others were trying this new medicament and reporting excellent results, while some preferred to prescribe cerium oxalate which, being insoluble, could be given at 10 times this dose. This was particularly recommended for sea-sickness and by the end of the century medical pharmacopoeias were even suggesting doses of a gram at a time. Tests on animals showed that even higher levels were harmless, although some research suggested that extremely high doses caused nausea, vomiting and diarrhoea. Meanwhile some doctors found that cerium oxalate acted as a cough suppressant for people with tuberculosis.

Commercial preparations of cerium salts were on sale as Novonaurin and Cerocol tablets which continued to be manufactured well into the twentieth century. But early in that century it began to fall out of favour and looked likely never to return, until in 1995 a weak solution of cerium nitrate was found to be an effective first treatment for bathing the skin of people suffering extensive third-degree burns. This is now standard procedure in some specialist burns units.

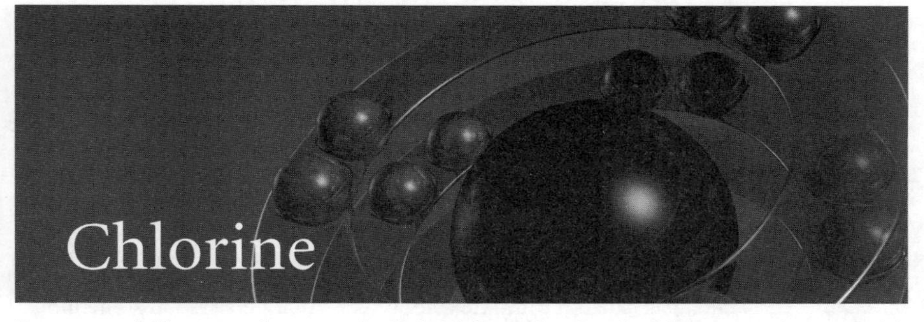

Chlorine

Pronounced klor-een, the name is derived from the Greek *chloros*, meaning greenish yellow.

French, *chlore*; German, *Chlor*; Italian, *cloro*; Spanish, *cloro*; Portuguese, *cloro*.

Chlorine is the name of the element, which exists as the molecule Cl_2, whereas chloride, as in sodium chloride, is a chlorine atom that has acquired a negative electron and is written Cl^-. Chlorine is reactive and dangerous, while chloride is stable and relatively safe.

HUMAN ELEMENT

Chlorine (as chloride) in the human body	
Blood	approx. 0.3%
Bone	900 p.p.m.
Tissue	approx. 0.2–0.5 %
Total amount in body	95 grams

Chloride (Cl^-) is essential to many species, including humans, and is relatively non-toxic. The element itself, chlorine gas (Cl_2), is very toxic. It immediately affects eyes and lungs at a concentration of only 3 p.p.m. in air; at 15 p.p.m. it produces irritation of the throat; at 50 p.p.m. it is dangerous even for a short time; and breathing air with 500 p.p.m. for 5 minutes would almost certainly be fatal – as its use in World War I as a weapon to attack troops showed (see below). Somewhat surprisingly, there is evidence that the white blood cells that defend the body against infection may use chlorine gas to do this; a paper in the *Journal of Clinical Investigation* in 1996 reported that these cells were capable of producing this deadly gas.

Despite hydrochloric acid (HCl) being classed officially as a dangerous chemical, it is produced at comparable concentrations in the human stomach to help break down food and destroy potentially dangerous bacteria.

FOOD ELEMENT

The daily intake of dietary chloride is mostly in the form of common salt (sodium chloride, NaCl) and is on average around 6 grams, which is double the amount we really need. However, it is not the chloride component but the sodium that is seen as potentially dangerous, especially to those prone to high blood pressure and heart disease (see p. 400). Nevertheless food manufacturers have reduced the salt level in many processed foods and many people now try to avoid salt, to such an extent that recently doctors have warned that people who undertake vigorous exercise and/or drink excessive amounts of water need to *increase* their salt intake to stay healthy.

ELEMENT OF HISTORY

Hydrochloric acid was known to alchemists under names such as *acidum salis* or *spiritus salis*, while early chemists called it muriatic acid, a name that persisted in some industries well into the twentieth century. The gaseous element itself, however, was first produced in 1774 by 32 year old Carl Wilhelm Scheele at Uppsala, Sweden, by heating hydrochloric acid with a powdered mineral he knew as brunsten (today it is known as pyrolusite and is native manganese dioxide, MnO_2). A dense, greenish-yellow gas was evolved which Scheele recorded as having a choking smell and which dissolved in water to give an acid solution. Scheele noted that it bleached litmus paper and the leaves and flowers of plants. It attacked almost all metals that were exposed to the gas.

Scheele called it dephlogisticated[25] muriatic acid, and it was known by this name for more than 30 years, until the 29 year old Humphry Davy started to investigate it in 1807 and eventually concluded not only that it was a simple substance, but that it was truly an element. He announced this conclusion to the Royal Society in London in November 1810. Nevertheless, it took another 10 years for some chemists finally to accept that chlorine was an element.

Bleaching with chlorine dissolved in water was first demonstrated by James Watt of Birmingham, England, in 1786. Within a few years this had become the standard method of bleaching both linen and cotton, both of which had hitherto been laid in strips in fields to be bleached by sunlight, a process that took several weeks and depended on the weather. A better bleaching solution was made by dissolving the gas in sodium hydroxide solution to form sodium hypochlorite ($NaOCl$) solution, and this became the general reagent used. It is still commonly used in homes for domestic bleaching and cleaning purposes. Industry preferred to use bleaching powder as a more convenient reagent, not least because it was easier to transport; this was made by absorbing chlorine on to slaked lime, calcium hydroxide, $Ca(OH)_2$.

ELEMENT OF WAR

Chlorine was used as an offensive weapon in World War I in Flanders. It was first deployed on the morning of 22 April 1915 when the German army released the gas from hundreds of cylinders. The prevailing breeze carried the dense gas across no-man's land and into the trenches of the British troops. There it caused chaos, as 5000

[25] Dephlogisticated meant there was no phlogiston present; phlogiston was thought to be a component of matter which was lost on burning. The phlogiston theory was part of chemical theory for most of the 18th century.

men died in agony and 15 000 were disabled by it. The threat was eventually countered by issuing gas masks, after which chlorine gas was little used.

ECONOMIC ELEMENT

Halite (NaCl) is the main mineral mined as salt, but there are other chloride minerals such as carnallite (magnesium potassium chloride) and sylvite (potassium chloride). There are vast salt deposits in the USA, Poland, Russia, Germany, China, India and Australia, with world production exceeding 160 million tonnes annually. Reserves are almost limitless: the sea contains trillions of tonnes of chloride, although most is extracted from deposits of rock salt, and it is from this that chlorine comes.

Chlorine gas is manufactured on a large scale by the electrolysis of brine, which is obtained by pumping water down bore holes to salt beds. The electrolysis produces not only the gas but the equally useful chemical, sodium hydroxide (also known as caustic soda, NaOH) in about the same tonnage. Most chlorine is made in mercury amalgam cells, although these are now being phased out because of the environmental threat this metal poses. Newer electrolysis cells employ either an asbestos diaphragm or an ion-exchange membrane made from a fluorinated polymer, to keep the chlorine and sodium hydroxide separate. Both produce chlorine efficiently but the quality of the hydroxide they produce is not as good as that from a mercury cell.

The primary uses for chlorine are: chemicals for industry (30%); manufacture of poly(vinyl chloride) (PVC) (25%); water purification (20%); solvents (15%); and bleaches (10%), although these figures are only approximate, and are changing all the time.

Chlorine is used to make hundreds of products, directly or indirectly. Directly it is involved in bleaching wood pulp for paper making and in water treatment, although the former is less than it was. Bleach is also used industrially to remove ink from recycled paper. Indirectly chlorine is needed to make the chemicals that are turned into medicinal drugs (85% of pharmaceuticals rely on chlorine in one form or another for their manufacture), silicones and polymers, but most of these no longer contain chlorine by the time they reach market. Some manufactured products do contain chlorine as an essential part of their make-up, such as flame retardants, domestic bleaches, paint-strippers and pesticides, and the hundreds of items made of PVC such as double glazing window frames, garden furniture, vinyl flooring, electrical cable insulation, bottles and blood transfusion bags.

The most important chlorine-containing chemicals are, or have been, sodium hypochlorite, hydrochloric acid, chloroform and carbon tetrachloride, but whereas these last two were once commonly available outside industry, they are now strictly controlled because exposure to them causes liver damage. Other, more benign, organochlorines were substituted as solvents and dry-cleaning agents but these are now disapproved of because, like many volatile organic chemicals (VOCs), they act as greenhouse gases when they leak into the atmosphere.

Chlorine was first used to disinfect tap water at Maidstone, England in 1897 during an outbreak of typhoid; the epidemic was brought under control. Eventually chlorine became the method of purifying drinking water throughout Britain and most of the developed world. Hypochlorite bleach is the usual way in which it is used; this is made by bubbling chlorine up a column down which trickles a solution of the alkali sodium hydroxide, forming NaOCl. This is a strong oxidizing agent, safe enough to be widely used in homes and hospitals, and stable for several months provided it is not exposed to heat, sunlight or metals. Viruses and bacteria are extremely sensitive to it and it quickly destroys them even when very dilute. Because it is persistent, hypochlorite will keep water free of germs at very low concentrations for a long time.

ENVIRONMENTAL ELEMENT

Chlorine in the environment	
Earth's crust	130 p.p.m.
	Chlorine is the 20th most abundant element.
Soils	very variable, ranging from 50 to 2000 p.p.m.
Sea water	1.8%
Atmosphere	mainly as salt spray, and as traces of organochlorine compounds

The amount of chloride in soils varies according to the distance from the sea. The average in topsoils is about 100 p.p.m. Plants contain varying amounts of chloride; it is an essential micronutrient for higher plants where it concentrates in the chloroplasts. Cereals tend to absorb little chloride, having around 10–20 p.p.m. (dry weight), while potatoes can have levels of up to 5000 p.p.m. Growth suffers if the amount of chloride in the soil falls below 2 p.p.m., but this rarely happens. The upper limit of tolerance varies according to the crop, with beans and apple trees suffering if chloride levels exceed 700 p.p.m. while tobacco, tomatoes, cotton and beet can cope with levels above 3000 p.p.m.

To many environmentalists of the last century, chlorine became to be regarded as *the* polluting chemical, and the reason is understandable. In the first half of the twentieth century, many new organochlorine products were marketed, offering great benefits; these included the dry-cleaning fluid carbon tetrachloride, insecticides such as dieldrin and dichlorodiphenyltrichloroethane (DDT), herbicides such as 2,4-D and 2,4,5,-T, engineering oils such as the polychlorinated biphenyls (PCBs), and the refrigerant and aerosol gases known as chlorofluorocarbons (CFCs; see footnote 37 on p. 149). All are now banned or strictly controlled because either they are regarded as potentially damaging to human health, or they are too persistent in the environment and so pose a threat to wildlife. The CFCs in particular are not only greenhouse gases but their chlorine atoms can destroy the protective ozone layer of the upper atmosphere.

Organochlorines contaminate natural waters. The US Environmental Protection Agency and other authorities have set a limit of 100 p.p.b. for chloroform-type chemicals in drinking water. Some organochlorine compounds have caused cancer among those heavily exposed to them in industry, but at the low levels at which they are present in chlorinated water the risk of them causing this disease is unimaginably small.

The best known use for chlorine is water purification, making it safe to drink or swim in. However, such is the chemical reactivity of chlorine that it will react with organic matter in the water to form p.p.b. traces of organochlorine compounds, which some people consider damaging to health and the environment – see below.

For several years organochlorines were thought to be present in the environment solely as the products of the chemical industry, but it is now known that more than 2000 organochlorine compounds occur naturally and that chloromethane (CH_3Cl) is produced by plankton, trees, algae and fungi. Indeed it now seems likely that more than 75% of chlorinated substances at large in the environment come from natural causes, including volcanic eruptions.

Nevertheless, some sites are heavily contaminated by man-made organochlorines; decontamination of these by natural microbes is long and slow, and not always complete. In 1997 a newly discovered bacterium was isolated from sewage that was capable of fully dechlorinating even the solvents tetrachloroethane and trichloroethene (also known as trichlorethylene), which have been widely used as cleaning fluids.

CHEMICAL ELEMENT

Data file	
Chemical symbol	Cl
Atomic number	17
Atomic weight	35.4527
Melting point	−101°C
Boiling point	−34°C
Density of gas	3.2 grams per litre
Oxides	Cl_2O, ClO_2, Cl_2O_7

Chlorine is a greenish-yellow, dense, sharp-smelling gas that is a member of group 17, the halogen group, of the periodic table of the elements. The gas is soluble in water to the extent of 3 litres of gas dissolving in a litre of water at 10°C. Chlorine is extremely reactive and will form compounds with all elements except the noble gases helium, neon, argon and krypton. In some cases elements will form more than one stable chloride.

The chemistry of the compounds of the chloride ion (Cl^-) is equally extensive, and this form of the element is the most stable of all.

Naturally occurring chlorine is composed of two isotopes: chlorine-35, which makes up 76%, and chlorine-37, which makes up the rest. Chlorine-36 is radioactive with a half-life of 300 000 years and is used in research.

ELEMENT OF SURPRISE

The chlorination of drinking water has been common practice for almost a century. Wherever it has been used it has virtually eliminated the water-borne diseases such as typhoid, cholera and meningitis which were once common in overcrowded cites where water supplies were easily contaminated by sewage. Chlorination is cheap and, at low doses, is highly effective at ridding water of disease pathogens.

Despite the benefits of chlorination, there were concerns expressed over the organochlorines that formed from dissolved organic matter when water was chlorinated, because some of them were shown to be carcinogenic when tested on rats, albeit at extremely high doses. People were alarmed when these concerns were widely reported by environmental groups. Those in rich countries who were worried began to drink bottled water that was chlorine-free, and there was a move to use ozone gas to disinfect public swimming pools (although this has still to be supplemented by chlorine because ozone does not persist for long).

Sadly, the public authorities in Peru were persuaded that the health threat from organochlorines outweighed the benefits of chlorination and, in 1991, they discontinued this in many areas. As a result, there was a massive outbreak of cholera with over a million cases being reported throughout the 1990s, of whom 10 000 died. Chlorination was soon reinstated but it took many years to bring the disease under control again. The incident proved, if proof were needed, that human health greatly benefits if chlorine is properly used.

Pronounced kroh-mi-uhm, the name is derived from the Greek *chroma*, meaning colour (chromium's salts are often strikingly coloured). The pigment chrome yellow (lead chromate, $PbCrO_4$) was once popular with artists and designers because of its brilliance.

French, *chrome*; German, *Chrom*; Italian, *cromo*; Spanish, *cromo*; Portuguese, *crómio*.

COSMIC ELEMENT

In 1817 André Laugier detected chromium in the famous Pallas meteorite from Siberia.

HUMAN ELEMENT

Chromium in the human body	
Blood	varies within range 6–100 p.p.b.
Bone	varies within range 100–300 p.p.b.
Tissue	varies within range 25–800 p.p.b.
Total amount in body	1–2 milligrams, but can be as much as 12 milligrams

Chromium is essential to some species, including humans. Like other metals that we require in trace amounts, it can be quite toxic, especially as chromates (CrO_4^{2-}), but not as chromium(III) (Cr^{3+}) compounds.

Those working with chromium compounds are vulnerable to an industrial disease known as chrome ulcers, which was first reported in 1827 among workers in Glasgow, Scotland. It was also common among workers in the chrome-plating, dyeing, French polishing, calico printing and chrome-tanning industries. The symptoms were the sudden appearance of holes in the skin that exposed raw flesh and itched unbearably. Exposure to chromates also leads to stomach ulcers, and these compounds may be carcinogenic.

FOOD ELEMENT

The daily human intake varies according to the diet; it can be as much as 1 milligram, but is more likely to be in the range 15–100 micrograms. Chromium is an essential element because it is needed to help the body utilize glucose, and its presence in RNA may indicate a second role. The organ with the highest level of chromium is the placenta.

It has been shown that animals which lack chromium have an impaired ability to use glucose, suffer mild diabetes and have reduced cholesterol levels. The same might well be true of humans, but chromium supplements are rarely needed, although there are a few cases on record of people suffering chromium deficiency.

It has been observed in studies on Americans that there is a steady fall in chromium levels in the body with age, but what this means is not known. If it were to be shown to be deleterious, then diets could be supplemented with high-chromium foods such as brewer's yeast, molasses, wheat germ and kidney. (It would not be wise to supplement the diet with inorganic chromium salts.)

The foods that contain most chromium, more than 30 micrograms per 100 grams (300 p.p.b.), are oysters, calf's liver, egg yolk, peanuts, grape juice and black pepper. Many commonly eaten foods have some chromium, such as potatoes (18 p.p.b.), beans (9 p.p.b.), carrots (18 p.p.b.) and apples (8 p.p.b.), but such tiny amounts may be all that is needed.

ELEMENT OF HISTORY

Chromium was discovered and isolated in 1798 by Nicholas Louis Vauquelin (1763–1829) in Paris. He was intrigued by a bright red mineral that had been discovered in a Siberian gold mine in 1766 and was referred to as Siberian red lead (it is now known as crocoite, i.e. lead chromate). Vauquelin analysed a sample in 1797 and confirmed that it did contain lead. He precipitated the lead out of a solution that he had made of the mineral, and concentrated what remained. A year later he succeeded in isolating chromium from it. Intrigued by the range of colours that it could produce in solution, he named it chromium. The same year he analysed an emerald and found that its green colour was due to small amount of chromium.[26]

The first commercially exploitable deposits of chromium were in Maryland, Pennsylvania and Virginia in the USA; these supplied all that was needed for 30 years. Then, in 1848, large deposits were found in Turkey. This became the main source of the metal for a long time, and it was mined as the black iron–chromium mineral, chromite ($FeCr_2O_4$).

ECONOMIC ELEMENT

Chromium ores are mined today in South Africa, Zimbabwe, Finland, India, Brazil, Turkey, Albania, Kazakhstan and the Philippines. A total of 14 million tonnes of chromite ore is extracted. Reserves are estimated to be of the order of 1 billion tonnes with unexploited deposits in Greenland, Finland, Russia, Canada and USA. Chromite ore is used to make materials such as refractory bricks, which can withstand the extremely high temperatures encountered in furnaces. World production of chromium metal itself is around 20 000 tonnes per year.

[26] The red of a ruby is also due to this metal.

Chromium can be polished to a high shine and resists oxidation in air. Its main uses are in alloys such as stainless steel (which may contain up to 15%), in chrome plating and in metal ceramics. Chromium plating was once widely used to give steel a polished silvery mirror coating; all that was needed was a 1 micrometre thick layer to achieve this effect, although when used on cars and trucks it needed to be a hundred times thicker than this to resist weathering. It is even possible to chrome-plate plastics. The metal resists corrosion because of a protective oxide layer on the surface.

Chromium is also used as a catalyst in dyeing (as potassium dichromate, $K_2Cr_2O_7$) and leather tanning (as chromium sulfate) and as chromium oxide (Cr_2O_3), which is the most stable green pigment. Today, about 90% of leather is chrome tanned, a process that was introduced in the 1860s, and which is the only way of making leather that will resist hot water. The trouble with chrome tanning is that the effluent it produces generally contains around 5 p.p.m. of chromium; it is for this reason that tanners are searching for alternative tanning agents.

Ruby lasers rely on chromium atoms to release the red light they emit.

ENVIRONMENTAL ELEMENT

Chromium in the environment	
Earth's crust	100 p.p.m.
	Chromium is 21st in order of abundance of the elements.
Soils	varies widely between 1 and 450 p.p.m., with an average of 50 p.p.m.
Sea water	0.24 p.p.b.
Atmosphere	barely detectable

Chromium is not seen as a major environmental pollutant although it has caused problems in rivers taking untreated industrial waste, especially that from tanneries. Soluble chromate in soils gradually turns into insoluble chromium(III) salts and then becomes unavailable to plants. In this way the food chain is protected against excess chromium which cannot pass the barrier between soil and plant roots, so that even chromate-rich sewage from industrial areas does not pose a threat, even though some such sewage has been shown to have chromate levels as high as 4000 p.p.m. (0.4%).

CHEMICAL ELEMENT

Data file	
Chemical symbol	Cr
Atomic number	24
Atomic weight	51.9961
Melting point	1860°C
Boiling point	2672°C
Density	7.2 kilograms per litre (7.2 grams per cubic centimetre)
Oxide	Cr_2O_3

Chromium is a hard silvery metal with a blue tinge, and is a member of group 6 of the periodic table of the elements. It will dissolve in hydrochloric and sulfuric acids, but not in phosphoric acid, due to reaction and formation of a protective layer on the surface.

Naturally occurring chromium consists of four isotopes, of which the most abundant is chromium-52, comprising 84% of the total. The others are chromium-53 (9.5%), chromium-50 (4%) and chromium-54 (2.5%). None is radioactive.

ELEMENT OF SURPRISE

The precious gem alexandrite is the mineral chrysoberyl ($BeAl_2O_4$) with a small amount of chromium which provides its changeable colour. Alexandrite is the birth stone for July and it was named after Tsar Alexander I of Russia who reigned from 1801 to 1825. It was prized not only for the beautiful blue/green colour it exhibited in daylight, but for the way this changed to a deep red when seen at night by artificial light.

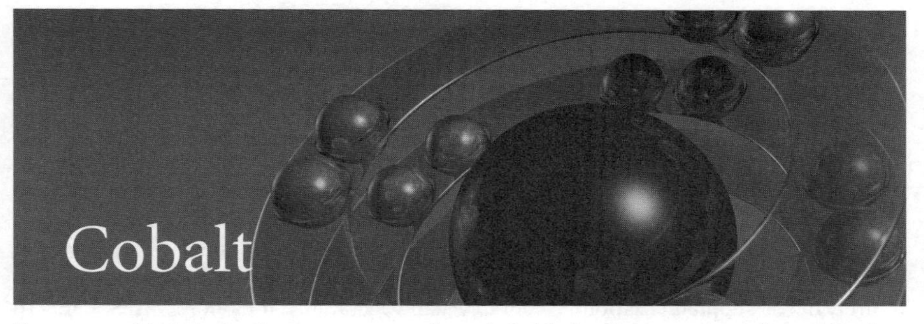

Pronounced koh-bolt, the name comes from *kobald*, the German word for goblin (see 'Element of history' for an explanation of this name). It has been suggested that both words derive from the Greek *cobalos*, meaning a mine.

French, *cobalt*; German, *Kobalt*; Italian, *cobalto*; Spanish, *cobalto*; Portuguese, *cobalto*.

COSMIC ELEMENT

Extraterrestrial cobalt was first detected in a meteorite found at the Cape of Good Hope in 1819.

HUMAN ELEMENT

Cobalt in the human body	
Blood	0.2–20 p.p.b
Bone	1–4 p.p.b.
Tissue	approx. 0.3 p.p.b.
Total amount in body	1–2 milligrams

Cobalt is an essential element for humans because it is at the heart of vitamin B_{12}, in which a cobalt atom is bonded to a methyl group, one of the rare examples of a metal-to-carbon bond in nature. Lack of this vitamin causes pernicious anaemia, a condition in which the body cannot produce enough red blood cells to transport all the oxygen that it needs. Vitamin B_{12} also helps maintain nerve tissue, is essential to the formation of red blood cells and is needed for the release of energy from carbohydrates and fats.

Cobalt compounds are generally thought to be of low toxicity but they may affect the thyroid and damage the heart, and cobalt is a suspected carcinogen. Contact with cobalt salts can lead to dermatitis.

FOOD ELEMENT

It has been said that nothing that grows out of the ground needs vitamin B_{12} but that every creature that walks, flies or swims must have it. Most animal species can make what they require themselves, and this is true for humans to a certain extent, because there are bacteria living in our colon that produce it. However, we do not absorb it very effectively, so we need to supplement it from our diet. This should pose no problem because we require only 1.5 micrograms per day and the average daily intake is more like 5 micrograms.

The total daily intake of cobalt is variable and may be as much as 1 milligram, but almost all will pass through the body unabsorbed, except that in vitamin B_{12}. Foods particularly rich in vitamin B_{12} are clams, sardines, salmon, herring, liver and eggs. The vitamin is depleted in foods exposed to light. Canning also has an adverse effect, although the presence of vitamin C acts to protect vitamin B_{12}. The 1–2 milligrams of vitamin B_{12} we store in our body will last us for a long time, maybe 2 years or more, and it rarely needs supplementation, except in vegan vegetarians, for whom it is essential to take a vitamin supplement. Herbivores, such as cows and sheep, make enough vitamin B_{12} to satisfy their own needs provided they graze on land that is not deficient in cobalt.

ELEMENT OF HISTORY

Among the treasures in the tomb of Pharaoh Tutankhamen, who ruled from 1361 to 1352 BC, was a small glass object that was coloured deep blue with cobalt; it is clear that using a cobalt mineral to achieve this rich colour was not unknown to the ancient world. Pieces of synthetic lapis lazuli[27] dating from 1400 BC have been found at Nippur, the religious centre of the Sumerians, and these owed their colour to cobalt. Cobalt blue was known in China long before this time, however, and was used for pottery glazes, but it was always a rare pigment because cobalt minerals themselves are rare.

Later generations regarded cobalt ores as more of a curse than a colouring agent. When sixteenth century silver miners in Saxony tried to smelt what they believed was a silver ore, they were disappointed to find that toxic fumes of arsenic were given off and no silver was to be won from it. This is not surprising, because they were dealing with an ore now known as smaltite, which is cobalt arsenide ($CoAs_2$). The miners cursed the ore, saying it had been bewitched by goblins and referred to it by the German name for these evil spirits, *kobald* (see also the origin of the name nickel, p. 278).

Others, however, saw cobalt's potential as a deep-blue pigment and it soon began to colour glass and pottery. The material used was called zaffer or smalt, and was made by roasting smaltite with varying amounts of sand, which forms cobalt silicate, or with sand and potash, to form potassium cobalt silicate. The material was then ground to a fine powder. The Venetian glass-makers were particularly fond of using it to make the fine blue glass for which they were famous. Low-grade zaffer was sold as a blue starch for laundry purposes and 'dolly blue' is still used in some parts of the world to make white clothes appear whiter.

Georg Brandt of Stockholm, Sweden, lived from 1694 to 1768 and during his life became one of Europe's most respected chemists. In 1730 he began to take an interest in a dark blue ore from some local copper workings in Vestmanland. He eventually proved that this contained a hitherto unrecognized metal, cobalt, and he gave it the

[27] The blue colour of natural lapis lazuli, which is the aluminosilicate mineral sodalite ($Na_8Cl_2Al_6Si_6O_{24}$), is due to sulfur atoms replacing some of the chlorine.

name by which the same was known in Germany. He published his results in 1739. For many years his claim to have uncovered a new metal was disputed by other chemists who said his 'cobalt' was merely a compound of iron and arsenic, but eventually it was recognized as an element in its own right.

ELEMENT OF WAR

Cobalt steel with highly magnetic properties was first patented in Japan in 1917, and put to use at the start of World War II by Nazi Germany in the making of magnetic mines. These were anchored under water and detected the presence of a ship by the magnetic field it generated, which then triggered the mine to explode. At one time it appeared that magnetic mines would successfully sink all shipping arriving at or leaving the British Isles, but the problem was quickly solved by placing a 'degaussing' coil round ships; this effectively neutralized the ship's magnetic field. Once it was realized that these mines were ineffective, they were no longer employed against shipping; the unused ones were turned into bombs and dropped on British cities instead, becoming the much feared 'land mines' of the Blitz.

ECONOMIC ELEMENT

Cobalt is commercially important. The main mining areas are Congo Democratic Republic, Zambia, Canada, Australia, Russia and New Caledonia. World production is around 30 000 tonnes per year, most coming as a by-product of nickel production. Cobalt is used in alloys for magnets, in ceramics, in catalysts and in paints. Certain types of stainless steel, including that used to make razor blades, contain it. It is also used in alloys for jet engines. One alloy of cobalt, chromium and tungsten, known as stellite, is used for heavy-duty high-temperature cutting tools.

Cobalt can be magnetized, like iron, although it has only two-thirds the magnetic strength of iron. However, it has the advantage of maintaining its magnetism to much higher temperatures: iron loses its magnetism at 770°C whereas cobalt does so at 1130°C. Alnico, an alloy of aluminium, nickel and cobalt, has particularly high magnetic strength; 25% of all cobalt produced goes into making magnets of this kind.

Cobalt blue is an important part of artists' palette of colours and is used by craft workers in porcelain, pottery, stained glass, tiles and enamel jewellery. Its rich blue colour was known as Sèvres blue and Thenard blue. Cobalt was particularly popular with glass bottle makers in the nineteenth century.

The radioactive isotope, cobalt-60, is made by bombarding cobalt-59 with neutrons in a nuclear reactor. It has a half-life of 5.3 years and undergoes β-decay with the emission of intense γ-radiation (with energy of 1.3 million electron volts)[28] which is very penetrating. This isotope is used in medical treatment and also to irradiate food, in order to preserve the food (by destroying the organisms that cause decay) and protect the consumer (by destroying the harmful bacteria that cause disease). (It is a popular misconception that irradiating food with cobalt-60 will cause the food to become

[28] An electron volt is a measure of the energy an electron accumulates as it moves through a potential difference of one volt: 1ev = 1.6×10^{-19} joules.

radioactive, but due to this misunderstanding this method of food preservation has been little used in some countries.) The radiation emitted by cobalt-60 will penetrate solid objects and is used to detect internal cracks in large metal objects, such as containers, pipes and the hulls of ships. Irradiating diamonds with cobalt-60 colours them blue.

ENVIRONMENTAL ELEMENT

Cobalt in the environment	
Earth's crust	20 p.p.m.
	Cobalt is 32nd in order of abundance of the elements.
Soils	8 p.p.m.
Sea water	less than 1 p.p.t.
Atmosphere	virtually nil

Most of the Earth's cobalt is in its core. Cobalt is of relatively low abundance in the Earth's crust and in natural waters, from which it is precipitated as the highly insoluble cobalt sulfide (CoS).

Although the average level of cobalt in soils is 8 p.p.m., there are soils with as little as 0.1 p.p.m. and others with as much as 70 p.p.m. Some soils have so little cobalt that it is necessary to supplement them with the element if they are to be used to grow crops for animal nutrition. However, a lack of cobalt does not retard the growth of plants. Levels of cobalt are high in some crops, such as sweet corn, lettuce and cabbage (up to 100 p.p.m. dry weight) but low in fruits such as oranges and apples (8 p.p.m. dry weight). Cobalt added to soil finds its way into plants; they can also absorb the metal through their leaves if it is applied as a spray. Several types of plant are able to concentrate cobalt, such as legumes, borage, myrtle and violets.

In the marine environment cobalt is needed by blue-green algae (cyanobacteria) and other nitrogen-fixing organisms.

CHEMICAL ELEMENT

Data file	
Chemical symbol	Co
Atomic number	27
Atomic weight	58.933200
Melting point	1495°C
Boiling point	2870°C
Density	8.9 kilograms per litre (8.9 grams per cubic centimetre)
Common oxides	CoO and Co_3O_4

Cobalt is a lustrous, silvery-blue, hard metal; it is ferromagnetic. It is a member of group 9 of the periodic table of the elements. It is stable in air and unaffected by water, but is slowly attacked by dilute acids.

The only naturally occurring isotope, cobalt-59, is not radioactive.

ELEMENT OF SURPRISE

Cobalt was once popular for making so-called 'sympathetic ink', which later became better known as invisible ink. This remains unseen until it is warmed. Exactly who first came across this phenomenon is unclear, but it was used for sending secret messages in the seventeenth century, and even after Jean Hellot revealed its existence around 1700, it continued to be used in espionage. The ink was made by dissolving cobalt ore in *aqua regia*;[29] cobalt chloride crystals were produced from this solution. These were dissolved in water with a little glycerine (glycerol) to give an almost colourless (very pale pink) solution, which was then used to write with. Once this ink had dried another innocent letter in visible ink was then written at right angles across the paper. When the letter reached its destination, the recipient simply heated it to reveal the hidden writing. The heating drives off the water and glycerol molecules surrounding the cobalt, and chloride ions move in to take their place, giving a dark blue molecule.

A similar trick was popular in the nineteenth century with artificial flowers that responded to changes in the weather. White petals were dyed pink by coating them with cobalt chloride solution; they would remain this colour while the weather was wet or the atmosphere humid. When the weather became sunny and dry, they turned violet, and when the day was very dry they turned blue.

[29] *Aqua regia* is made by mixing concentrated nitric and hydrochloric acids in the ratio of one part of the former to three parts of the latter.

Pronounced kop-er, the name is derived from the element's Old English name, *coper*, which came from the Latin *Cyprium aes*, meaning metal from Cyprus. The chemical symbol, Cu, derives from the Latin name. Cyprus had been the main exporter of the metal long before it became part of the Roman Empire. It has even been suggested that the island, which was called *Kupros* by the Greeks, took its name from the metal rather than the other way round.

French, *cuivre*; German, *Kupfer*; Italian, *rame*; Spanish, *cobre*; Portuguese, *cobre*.

HUMAN ELEMENT

Copper in the human body	
Blood	1 p.p.m
Bone	1–25 p.p.m.
Tissue	2–10 p.p.m. (highest in liver)
Total amount in body	70 milligrams

Copper is essential to all species. There are more than 10 copper-dependent enzymes, such as cyctochrome *c* oxidase, which is required by all cells to produce energy. Other important ones are superoxide dismutase, which protects against free radicals, and tyrosinase, which is needed for the synthesis of the pigment melanin. Yet others are involved in making hormones, removing dangerous amines and repairing connective tissue.

Essential though it be, copper can be toxic in excess. One child who ate the sample of copper sulfate in a toy chemistry set died as a result. Normally vomiting soon starts if an excess of a copper compound is swallowed, and this acts to prevent acute toxic effects.

FOOD ELEMENT

There is little danger that our diet does not provide enough copper, because it is abundant in certain foods. It may also come with drinking water if this is from a soft water area where water is supplied through copper pipes. It has even been suggested that we get *too* much copper and that this works against the iron and zinc in our bodies because copper can displace these metals from their active sites in enzymes.[30]

An adult needs to ingest around 1.2 milligrams of copper per day, and breast-feeding women need around 1.5 milligrams although it has been suggested that 2 milligrams per day would be more beneficial for everyone. Volunteers on low-copper diets have increased levels of cholesterol, high blood pressure and an impaired ability to digest glucose.

We take in copper best from meats, where it is present as a copper-protein. The foods with most copper are seafoods – oysters, crab and lobster; among meats it is lamb, duck, pork and beef which have the most, the liver and kidneys of lamb and beef are especially rich in copper. The plant-derived foods with most copper are almonds, walnuts, Brazil nuts, sunflower seeds, mushrooms and bran. The amount of copper in a person's diet can be as much as 6 milligrams a day, depending on how much of these foods are eaten.

MEDICAL ELEMENT

The genetic diseases known as Wilson's disease and Menke's disease are caused by the body's inability to utilize copper properly. In the former condition copper builds up in the brain and in the latter it is lacking in the body because the necessary instruction for making the copper-transporting protein is missing from the genome. The result is a chronic lack of the element, which leads to retarded growth and death in infancy. Even copper histidine therapy, which was introduced in the 1970s, is not particularly effective in prolonging life: most victims die before they are 6 years old.

Copper compounds (such as copper acetate) and even powdered copper metal were used by the Egyptians as long ago as 3000 BC to treat diseases of the eye. In more recent times a lot of research went into potential medicaments based on copper for use as anti-ulcer, anti-inflammatory, anti-convulsant and anti-cancer drugs. Few, if any, were successful. Some copper compounds, however, proved successful at reversing damage of those accidentally exposed to massive doses of ionizing radiation.

ELEMENT OF HISTORY

Copper beads have been excavated in northern Iraq that were more than 10 000 years old and were presumably made from native copper, nuggets of which can sometimes be found on the Earth's surface. (The biggest single mass of uncombined copper, weighing 420 tonnes, was found in 1857 in the Minesota mine on the Keweenaw Peninsula, Michigan, which held the world's largest quantity of native copper and from where the metal has been extracted for several thousand years.) This orange-gold coloured metal is easy to work and can be drawn into fine wire. Smelting copper ores began around 5000 BC and all cultures made use of this metal for weapons, tools and jewellery, the copper generally being extracted from the easily identified green ore malachite, which is basic copper carbonate, $Cu_2CO_3(OH)_2$.

[30] In many enzymes there are metal atoms, and it is to these that their target molecules become attached prior to their conversion to the desired product molecules.

Later it was discovered that copper could be greatly strengthened and be honed to a fine cutting edge by adding one part tin to two parts copper to produce the alloy known as bronze. This was much more suitable for knives and swords. Cultures which advanced to this technology are distinguished as having moved out of the Stone Age to the so-called Bronze Age, which generally started around 3000 BC and lasted to around 1000 BC (these times varied considerably from one ancient civilization to another). How bronze was discovered we do not know, but the peoples of Egypt, Mesopotamia and the Indus valley were among the first to become familiar with it. Soon it spread to other areas and was smelted on a large scale. The prehistoric workings at Mitterberg in the Austrian Alps suggest that during the Bronze Age they yielded around 20 000 tonnes of copper.

Some of the ancient uses of bronze were spectacular. The Colossus of Rhodes was a 35 metre high, bronze statue of the Sun-god, Apollo, erected at the harbour entrance around 280 BC. It fell down during an earthquake about 50 years later and was then sold as scrap. As the Roman Empire expanded, so did the demand for everyday objects made of bronze, such as cutlery, needles, containers, coins, knives, razors, tools, jewellery and musical instruments. Other cultures used bronze for making bells; the art of bell founding can be traced back to 1000 BC in China, but did not become important in the West until around AD 1000.

As it weathers and ages, copper becomes covered with a green layer, or patina, known as verdigris. In ancient Greece and Rome a similar green copper compound was produced as a pigment and a medicament. It was made by hanging strips of copper over vats of vinegar or the lees of wine; the copper became covered with a green layer of basic copper acetate. This method of production survived as a home-based industry in the South of France well into the Middle Ages and beyond.

Copper is an excellent conductor of heat and so was used to make pans, kettles and boilers.[31] A lot of the metal was used in coins. Copper is traditionally known as one of the coinage metals, along with silver and gold, but it is the most common and therefore least valued of this group. The development of printing created a demand for copper plates for engravings, the first of which were printed around 1430. Later, copper sheeting was used to sheath the hulls of wooden ships; this kept them free of barnacles and other marine growth, thereby improving their speed.

Copper is not difficult to extract from its ores, but mineable deposits were not to be found easily. Some, such as the great copper mine at Falun, Sweden, date from the thirteenth century, were the source of great wealth. Fairly early on it was discovered that one way to extract the metal was to roast the sulfide ore then leach out the copper sulfate that was formed, using water. The copper sulfate was then trickled over scrap iron on the surface of which the copper deposited, forming a flaky layer that was easily removed.

By 1860 the Welsh town of Swansea smelted 90% of the world's copper from ores obtained from Cornwall and Anglesey. From there the copper was sent to the industrial city of Birmingham where the copper was fabricated into all kinds of alloys and objects.

[31] By tradition, whisky stills are always made of copper.

ECONOMIC ELEMENT

Today copper is mined in major deposits in Chile, Indonesia, USA, Australia and Canada, which together account for around 80% of the world's copper (with silver and gold as by-products). The main ore is a yellow copper–iron sulfide called chalcopyrite ($CuFeS_2$) from which the copper is extracted first by roasting it to form copper sulfide and iron oxide. This mixture is then heated with limestone and sand to form a molten slag from which the sulfide separates as a lower layer. This is then tapped off, heated and air blown through it, which converts it to copper metal, while the sulfur dioxide that forms is converted to sulfuric acid. The impure copper is first refined by having methane gas blown through it to remove any remaining oxygen and sulfur, before finally being cast into large anodes for electro-refining which gives a product that is 99.99% pure copper.

The better known copper ore, green malachite, is used for polished slabs, tables and columns and is mined in various countries. Other ores are cuprite (copper oxide) and azurite (copper carbonate)

World production of copper amounts to 12 million tonnes a year and exploitable reserves are around 300 million tonnes, which are expected to last for only another 25 years. About 2 million tonnes a year are reclaimed by recycling. Another source of copper is the spoil heaps of mines which can yield yet more copper by using microbes, such as *Thiobacillus ferrooxidans*. These break down iron(II) sulfide and oxidize the iron to Fe^{3+}, which in this oxidation state can oxidize the copper to soluble copper(II) ions which trickle from the spoil heap and on to scrap iron.

In Chile, at Mansa Mina, 20 000 tonnes per year of copper are extracted directly from low-grade ore by heat-loving bacteria such as *Sulfolobus thermophiles* and *Leptospirillium ferrooxidans* which digest the crushed rock in large open tanks kept at 65–85°C.

Most copper is used for electrical equipment (60%); construction, such as roofing and plumbing (20%); industrial machinery, such as heat exchangers (15%); and alloys (5%). There are several long-established copper alloys apart from bronze; one of these is brass (a copper–zinc alloy), which combines hardness with a gold-like lustre. The alloy of copper with tin and zinc, which was strong enough to make guns and cannons, was known as gun metal.[32]

The mixture of copper and nickel, known as cupronickel, was the preferred metal for low-denomination coins called 'coppers' in the UK. In the USA cupronickel is also used for several low-denomination coins where it forms the outer cladding of the coins. The US 5-cent piece, called a nickel because of its colour, is another copper–nickel alloy consisting of 75% copper and 25% nickel. This coin is not clad but has the same composition throughout.

The alloy consisting of 90% copper and 10% nickel finds use because it resists corrosion by sea water, and is important for marine pumps and propellers and in desalination plants which produce drinkable water from sea water. Aluminium bronzes (copper with 7% aluminium) have a pleasing gold colour and polish well; they are used as a decorative substitute for gold.

[32] The copper content of this alloy was 85–90%.

Copper is ideal for electrical wiring because it is easily worked, can be drawn into fine wire and has a high electrical conductivity. Copper appears in many guises in everyday life – in homes, shops, offices and cars – but mostly unseen as wiring, electrical appliances, piping, screws, hinges and locks.

Copper in the form of Bordeaux mixture (a blue gelatinous suspension of copper sulfate and lime in water) was one of the first agrochemical pesticides, developed to control downy mildew on vines. This, and other variants using sodium carbonate or ammonium carbonate, continue to be used as fungicides. Copper in small amounts is added to water to prevent algal growth since it is deadly to these water plants.

ENVIRONMENTAL ELEMENT

Copper in the environment	
Earth's crust	50 p.p.m.
	Copper is the 26th most abundant element.
Soils	approx. 20 p.p.m.
Sea water	0.2 p.p.b.
Atmosphere	small but significant

Copper is rather immobile in soils and concentrates in the top layers where it is tightly bound to inorganic particles and organic matter. The loss of this copper is partly replaced by that deposited from the atmosphere, which in some industrial countries may exceed 200 grams per hectare per year. It is also replenished by the use of copper-based agrochemicals, or even by using municipal sewage. Copper-contaminated soils can have as much as 3500 p.p.m. of copper in industrial regions and 1500 p.p.m. in farmed land.

The copper in plants varies markedly: apples and cabbage have only 0.03 p.p.m., carrots 0.4 p.p.m., tomatoes 0.7 p.p.m. and sweet corn 1 p.p.m (all based on fresh weight). Lettuce normally has 0.1 p.p.m. but when it is grown on contaminated land this can rise 10-fold. Some microorganisms can take in remarkable amounts of copper: *Penicillium* for example can have 20 000 p.p.m. dry weight (i.e. 2%). Exploitation of land in some areas has led to soils becoming copper-deficient and reduced them to marginal status. Such soils have been brought back into production by fertilizing them with copper sulfate, sometimes requiring as much as 23 kilograms of copper sulfate per hectare, although most have required only a fraction of this amount to make them productive again.

Plants absorb copper but most of it tends to remain in their roots. That which is absorbed is linked to protein and plays a key role in photosynthesis, carbohydrate distribution, cell wall metabolism and especially in the production of DNA and RNA. It also helps plants to resist disease. Crops grown on copper-deficient soils, i.e. with less than 2 p.p.m., produce fewer seeds; this condition may be more widespread than imagined because continued harvesting removes about 30 grams per hectare per year of the available nutrient copper.

CHEMICAL ELEMENT

Data file	
Chemical symbol	Cu
Atomic number	29
Atomic weight	63.546
Melting point	1084°C
Boiling point	2567°C
Density	9.0 kilograms per litre (9.0 grams per cubic centimetre)
Oxides	CuO, Cu_2O

Copper is an orange-gold metal that is a member of group 11, also known as the coinage metals, of the periodic table of the elements. It is malleable and ductile, with an electrical conductivity second only to that of silver. Copper is resistant to air and water but attacked by acids.

Copper has two naturally occurring isotopes: copper-63, which comprises 69%, and copper-65, which accounts for the rest. This ratio produces a relative atomic weight of 63.5, a numerical value which puzzled early chemists (see p. 516).

ELEMENT OF SURPRISES

1. In 1847, E. Harless discovered that the blood of the octopus was blue. The same is true of some other creatures, such as snails, spiders and oysters. The reason is that these creatures rely on a blue copper compound to carry oxygen around the body, whereas most other animals rely on the deep red iron compound haemoglobin. In blue-blooded creatures the copper atom is at the centre of a molecule called haemocyanin, which does the job equally well.
2. A gold-coloured US dollar coin was issued in November 1999, made of copper (88.5%) alloyed with zinc (6%), manganese (3.5%) and nickel (2%). Each coin costs 13 cents to make. More than 15 billion of them have been minted, or around 50 per head of population, yet few are in circulation because people in the USA prefer to hoard them.

Pronounced kyuhr-iuhm, curium was named in honour of Pierre and Marie Curie. French, *curium*; German, *Curium*; Italian, *curio*; Spanish, *curio*; Portuguese, *curio*.

HUMAN ELEMENT

Curium in the human body
None

Curium is never encountered outside nuclear facilities or research laboratories and has no role to play in living things. It is dangerous because of its radioactivity, and tests on animals have shown that if it gets into the body it concentrates in the bone marrow where its intense radiation destroys red blood cells. It also collects in the liver, but from there it is rapidly excreted.

ELEMENT OF HISTORY

Curium was first made by the team of Glenn T. Seaborg, Ralph A. James and Albert Ghiorso in the summer of 1944. They bombarded a piece of the newly discovered element plutonium (isotope 239) with α-particles (helium nuclei) having energies of 32 million electron volts, in the cyclotron at Berkeley, California. The irradiated plutonium was then sent to the Metallurgical Laboratory at the University of Chicago where, after much difficulty, a sample of curium was separated and identified.

The discovery was kept secret until the end of World War II and was first revealed by Glenn T. Seaborg when he appeared as the guest scientist on a children's radio show, 'Quiz Kids', on 11 November 1945 (see also under americium). The secrecy surrounding curium had been lifted a few days previously but the new element was not officially announced until a week later.

The first sample of a curium compound to be visible to the naked eye was of the isotope curium-242 of which 30 micrograms was made in 1947 by L. B. Werner and I. Perlman at the University of California, Berkeley. The metal itself was produced in 1951 by heating curium fluoride and barium vapour at 1300°C.

ECONOMIC ELEMENT

Today, curium-242 is produced in nuclear reactors by bombarding plutonium with neutrons. This isotope is used as power source for pacemakers and navigational buoys, and on space missions because it gives off 3 watts of heat energy per gram of metal; being an α-emitter its radiation can easily be shielded against.

ENVIRONMENTAL ELEMENT

Curium in the environment	
Earth's crust	present, but only as trace amounts
Sea water	nil
Atmosphere	nil

Curium probably does occur naturally on Earth, albeit in incredibly small amounts. Concentrated uranium deposits may produce some atoms of it by a sequence of neutron captures and β-decays, which are the same processes that produce atoms of neptunium and plutonium in these minerals. All the curium that is used, however, has been synthesized.

The environment is not entirely free of this element: a little curium was produced in above-ground nuclear weapon tests which were carried out from 1945 to 1963. The amount, however, is negligible and its radiation contributes only a tiny fraction of the Earth's background radiation.

CHEMICAL ELEMENT

Data file	
Chemical symbol	Cm
Atomic number	96
Atomic weight	247.1 (isotope Cm-247)
Melting point	1340°C
Boiling point	not known
Density	13.3 kilograms per litre (13.3 grams per cubic centimetre)
Oxides	Cm_2O_3 and CmO_2

Curium is a radioactive metal that is a member of the actinide group of the periodic table of the elements. Traces may occur in areas of intense radioactivity in the Earth's crust – see page 481. It is silver in colour although it tarnishes even in dry nitrogen, as well as being rapidly attacked by oxygen, steam and acids, but not by alkalis. Altogether there are 14 known isotopes with atomic weights in the range 238–251. The ones made in kilogram quantities are curium-242 (half-life 163 days) and curium-244 (half-life 18 years). The longest-lived isotope is curium-247, with a half-life of 16 million years.

A few compounds of curium are known, such as the fluorides CmF_3 and CmF_4 as well as curium trichloride, tribromide and tri-iodide, all of which are pale yellow in colour. Attempts to study its solution chemistry are hampered because the heat given off, due to the radioactivity of the element, quickly evaporates the water.

ELEMENT OF SURPRISE

Curium is being produced on a relatively large scale, possibly in tonne quantities, by bombarding plutonium with neutrons in nuclear reactors. This is seen as a way of reducing the accumulating world stockpile of plutonium, by turning it into something which is not only relatively short-lived, but much more useful.

Darmstadtium

This is element atomic number 110; it, along with other elements above 100, is dealt with under the transfermium elements on p. 457.

Dubnium

This is element atomic number 105; it, along with other elements above 100, is dealt with under the transfermium elements on p. 457.

Pronounced dis-pro-zee-uhm, the name is derived from the Greek word *dysprositos*, meaning hard to get, for reasons explained below.

French, *dysprosium*; German, *Dysprosium*; Italian, *disprosio*; Spanish, *disprosio*; Portuguese, *disprósio*.

Dysprosium is one of 15, chemically similar, elements referred to as the rare-earth elements, which extend from the element with atomic number 57 (lanthanum) to the element with atomic number 71 (lutetium). The term rare-earth elements is a misnomer because some are not rare at all; they are more correctly called the lanthanides, although strictly speaking this excludes lutetium. The minerals from which they are extracted, and the properties and uses they have in common, are discussed under lanthanides, on p. 219.

HUMAN ELEMENT

Dysprosium in the human body	
Total amount in body	not known, but small

The amount of dysprosium in the average person is tiny and so far no one has monitored diet for dysprosium content. The metal has no biological role and, judging by what is known of other lanthanides, the levels will be highest in bone, with smaller amounts being present in the liver and kidneys. How much we take in every day is difficult to assess, but is probably only a few micrograms. Dysprosium is not taken up by plant roots, so little gets into the main food chain.

Soluble dysprosium salts, such as the chloride and nitrate, are mildly toxic by ingestion, but insoluble salts are non-toxic. Based on the toxicity of dysprosium chloride to mice, it can be calculated that a dose of 500 grams or more would be needed to put a person's life at risk.

ELEMENT OF HISTORY

Dysprosium was discovered by Paul-Émile Lecoq de Boisbaudran in 1886 in Paris. The steps which led to this are part of a sequence of events which began with the discovery of impure yttrium oxide from which erbium and terbium were later to be separated (see p. 137). In 1878–9, erbium was found to be harbouring two other rare-earth oxides: those of holmium and thulium. But this was not the end of the story, because Lecoq de Boisbaudran, working on holmium oxide, separated from it yet another rare-earth oxide, that of dysprosium. His procedure involved dissolving the oxide in acid and then adding ammonia to precipitate the hydroxide. He repeated this sequence 32 times, and followed it with 26 precipitations of the insoluble oxalate salt from solution.

This tedious operation eventually yielded the previously unknown rare-earth dysprosium. Lecoq de Boisbaudran named it thus because it really had been *dysprositos* (difficult to get). He later revealed that the separation had been carried out on the marble slab of his fireplace at home. Perhaps, because of the obstacles in securing the first sample of dysprosium, this was one of the few elements whose discovery was not challenged.

Pure samples of dysprosium were not available until Frank Spedding (1902–1984) and co-workers at Iowa State University developed the technique of ion-exchange chromatography in around 1950. From then on it was possible to separate the rare-earth elements in a reliable and efficient manner.

ECONOMIC ELEMENT

Dysprosium is found in minerals that include all the rare-earth elements. The most important ores are monazite and bastnäsite; although dysprosium is not a major component of either, it is present in extractable amounts. Dysprosium is also found in several other minerals, such as xenotime, fergusonite, gadolinite, euxenite and polycrase. World production is around 100 tonnes per year; the element is available as dysprosium metal and its oxide.

The metal itself is produced by heating dysprosium fluoride and calcium. The metal can be cut with a knife and it can be machined without sparking, but as a pure metal it is useless, because it rapidly corrodes. Some is used in making alloys for the best kinds of permanent magnets, such as the alloy with neodymium, iron and boron. Dysprosium (and its neighbour, holmium) have the highest magnetic strengths of any elements.

The best known use of dysprosium is in halide discharge lamps where dysprosium iodide, DyI_3, is used to get very intense light. The iodide dissociates into dysprosium atoms in the hot centre of the lamp and these absorb energy and re-emit it in the form of visible light in the spectral regions 470–500 nm and 570–600 nm, making it appear almost like white light.

Dysprosium is used in nuclear reactors as a cermet,[33] which is there to absorb neutrons and so 'cools' the chain reaction that is providing the energy. It makes a good cermet because it neither swells nor contracts under prolonged neutron bombardment. Another use in the field of radioactivity is in dosimeters for monitoring exposure to ionizing radiation. When dysprosium-doped crystals of calcium sulfate or calcium fluoride are exposed to damaging radiation, this excites the dysprosium atoms which then luminesce and indicate the degree of exposure to which the dosimeter has been subjected.

Dysprosium oxide is used as a dopant in special ceramics, such as barium titanium oxide ($BaTiO_3$), which are used to produce small capacitors with high capacitance, for electronic applications. Other uses of dysprosium are in erasable optical laser-read discs and in temperature-compensating capacitors.

ENVIRONMENTAL ELEMENT

Dysprosium in the environment	
Earth's crust	6 p.p.m.
	Dysprosium is the 42nd most abundant element.
Soils	approx. 5 p.p.m., range 1–12 p.p.m.
Sea water	1 p.p.t.
Atmosphere	virtually nil

Dysprosium is one of the more abundant lanthanide elements and is more than twice as abundant as tin. It poses no environmental threat to plants or animals.

CHEMICAL ELEMENT

Data file	
Chemical symbol	Dy
Atomic number	66
Atomic weight	162.50
Melting point	1412°C
Boiling point	2560°C
Density	8.6 kilograms per litre (8.6 grams per cubic centimetre)
Oxide	Dy_2O_3

[33] Cermets are a composite material made of ceramic and sintered metal. They are highly resistant to temperature, corrosion and abrasion.

Dysprosium is a bright, silvery metal and a member of the lanthanide group of the periodic table of the elements. It is slowly oxidized by oxygen, reacts with cold water and rapidly dissolves in acids.

There are seven naturally occurring isotopes, of which dysprosium-162 constitutes 26.5%, dysprosium-163 25%, and dysprosium-164 28%. The others are dysprosium-156 (0.06%), dysprosium-158 (0.1%), dysprosium-160 (2.5%) and dysprosium-161 (19%). None of the natural isotopes is radioactive.

ELEMENT OF SURPRISE

A most unusual property is possessed by one type of dysprosium alloy, known as Terfenol. This visibly lengthens and shortens when exposed to a magnetic field. Such so-called magnetostrictive alloys can absorb a lot of energy and research is currently going on to determine how this might be used to work tiny motors, pumps and injection systems.

Einsteinium

Pronounced iyn-stiyn-iuhm, it was named after Albert Einstein (1879–1955) who won the Nobel Prize for Physics in 1921.

French, *einsteinium*; German, *Einsteinium*; Italian, *einsteinio*; Spanish, *einstenio*; Portuguese, *einstenio*.

HUMAN ELEMENT

Einsteinium in the human body
None

Einsteinium has no role to play in living things and is never encountered outside nuclear facilities or research laboratories. It is highly dangerous because of the radiation it emits.

ELEMENT OF HISTORY

A large number of scientists contributed to the discovery of einsteinium; prominent among them were Gregory R. Choppin, Stanley G. Thompson, Albert Ghiorso and Bernard G. Harvey. Einsteinium was discovered in the debris of 'Mike', the code name for the first thermonuclear explosion which took place on Eniwetok, a Pacific atoll, on 1 November 1952.[34]

Fall-out material, gathered on a neighbouring atoll, was sent for analysis to Berkeley, where tonnes of radioactive coral were sifted and examined. A month later the new element was discovered, although it was not announced for another 3 years, because the work was kept secret for security reasons.

In fact, fewer than 200 atoms had been found in the debris, but this was enough to show that the uranium-238, which had been used to provide the heat necessary for triggering a thermonuclear explosion, had been exposed to such a high flux of neutrons that some of its atoms had each captured several of them. These formed an isotope of a new element, einsteinium-253, which has a half-life of 20 days. (Through the process of neutron capture, followed by β^- decay, many elements were formed, including all those with atomic numbers 93–100.)

[34] 'Mike' was the precursor to the development of a thermonuclear weapon, the so-called hydrogen bomb.

By 1961 enough einsteinium had been collected together to be visible to the naked eye, although it amounted to a mere hundredth of a microgram.

ECONOMIC ELEMENT

Milligram quantities of einsteinium can be produced by bombarding plutonium-239 with a high flux of neutrons inside a nuclear reactor for several months. (Such einsteinium is available for purchase in microgram quantities.) The sequence of transformation is one of alternately accepting neutrons and emitting β-particles, and it goes as follows:

$$\text{Pu-239} + 2\text{n} \rightarrow \text{Pu-241} - \beta^- \rightarrow \text{Am-241} + \text{n} \rightarrow \text{Am-242} - \beta^- \rightarrow \text{Cm-242} + 7\text{n} \rightarrow$$
$$\text{Cm-249} - \beta^- \rightarrow \text{Bk-249} + \text{n} \rightarrow \text{Bk-250} - \beta^- \rightarrow \text{Cf-250} + 3\text{n} \rightarrow \text{Cf-253} - \beta^- \rightarrow \text{Es-253}$$

where 'n' is a neutron.

If isotope einsteinium-253 then accepts a further neutron it becomes einsteinium-254, which then emits a β-particle and decays to fermium-254.

There are, as yet, no known applications of einsteinium.

ENVIRONMENTAL ELEMENT

Einsteinium in the environment	
Earth's crust	nil
Sea water	nil
Atmosphere	nil

Einsteinum does not exist naturally on Earth today, but it has occurred in the past in natural reactor deposits – see p. 481.

CHEMICAL ELEMENT

Data file	
Chemical symbol	Es
Atomic number	99
Atomic weight	252.1 (isotope einsteinium-252)
Melting point	about 860°C
Boiling point	not known
Density	not known
Oxide	Es_2O_3

Einsteinium is a metal element and a member of the actinide group of the periodic table of the elements. It is attacked by oxygen, steam and acids, but not by alkalis.

In all there are 17 known isotopes, including isomers, with atomic masses ranging from 243 to 256. Einsteinium-252 is the longest-lived isotope, with a half-life of 470 days.

ELEMENT OF SURPRISE

The first thermonuclear explosion, which produced the first einsteinium, sadly cost the life of US pilot, First Lieutenant Jimmy Robinson. His job was to fly through the cloud formed by the explosion, collecting samples of dust on filter papers. He waited too long before deciding to return to base, ran out of fuel, ditched his plane in the sea and died.

Erbium

Pronounced erb-iuhm, this element is named after Ytterby, Sweden (see 'The village of the four elements' under yttrium, p. 496).

French, *erbium*; German, *Erbium*; Italian, *erbio*; Spanish, *erbio*; Portuguese, *érbio*.

Erbium is one of 15, chemically similar, elements referred to as the rare-earth elements, which extend from the element with atomic number 57 (lanthanum) to the one with atomic number 71 (lutetium). The term rare-earth elements is a misnomer because some are not rare at all; they are more correctly called the lanthanides, although strictly speaking this excludes lutetium. The minerals from which they are extracted, and the properties and uses they have in common, are discussed under lanthanides, on p. 219.

HUMAN ELEMENT

Erbium in the human body	
Total amount in body	not known, but small

The amount of erbium in humans is quite small and the metal has no biological role, but it has been noted that erbium salts stimulate metabolism. It is difficult to separate out the various amounts of the different lanthanides in the human body, but they are present; the levels are highest in bone, with smaller amounts being present in the liver and kidneys.

There is no estimate of the amount of erbium in an average adult, and no one has monitored diet for erbium content, so it is difficult to judge how much we take in, but it is probably only a milligram or so per year. Erbium is not taken up by plant roots so little gets into the main food chain.

Erbium can be mildly toxic by ingestion, but insoluble salts are completely non-toxic.

MEDICAL ELEMENT

The erbium–nickel alloy, of composition Er_3Ni, is very good at absorbing heat and this helps to ensure the economical operation of magnetic resonance imaging body scanners, which work at the temperature of liquid helium, around –270°C.

The isotope erbium-169 (half-life 9½ days) is a high-energy β-particle emitter which is currently being studied for use in radiation therapy treatments.

ELEMENT OF HISTORY

Erbium was discovered in 1843 by Carl Gustav Mosander at Stockholm, Sweden. Erbium, as its oxide, was one of the first rare-earth elements to be discovered, but the picture is clouded because early samples must have contained other rare-earth elements which were isolated from erbium oxide at a later date. In 1843, Mosander extracted two new oxides from the rare-earth oxide yttrium oxide. These were erbium oxide, which was pink; and terbium oxide, which was yellow. He originally named these new oxides the other way round but later workers who confirmed his discovery got them confused and gave them the names by which we know them today.

Mosander's discovery was not the end of the matter; in 1878 the Swiss chemist Jean-Charles Galissard de Marignac, working at the University of Geneva, extracted another element from erbium oxide and called it ytterbium. This too was impure and scandium was extracted from it a year later.

A sample of pure erbium metal was not produced until 1934, when Klemm and Bommer achieved this by reducing purified erbium chloride with potassium vapour.

ECONOMIC ELEMENT

Erbium is found in minerals that include all the rare-earth elements; these are discussed in more detail on p. 222. The main mining areas are China and the USA, and the most important ores are monazite and bastnäsite. Although erbium is not a major component of either ore, it is present in extractable amounts. Better sources of the element are xenotime (mainly yttrium phosphate) and euxenite (a complex ore of many metals). World production of erbium is 500 tonnes per year, mainly in the form of erbium oxide. The metal is produced by heating erbium chloride with calcium under vacuum and is available as ingots, lumps or powder.

Erbium finds little use as a metal because it slowly tarnishes in air and is attacked by water. However, it is more corrosion-resistant than the other lanthanide elements. Some erbium is added to alloys with metals such as vanadium because it lowers their hardness, making them more workable.

Erbium oxide is important in several situations that involve visible and infrared light. For example, it is added to the glass of special safety spectacles for workers, such as welders and glass-blowers, whose eyes are exposed to intense infrared light, because it absorbs these rays. Erbium is incorporated into phosphors that can convert infrared light into visible light giving a green image.

Erbium is used to dope optical fibres at regular intervals to amplify signals; it does this by converting other wavelengths that are pumped down the fibre to the wavelength of the information-carrying signal. Lasers based on erbium have been introduced for medical and dental use; they work at a wavelength of 2.9 micrometres. This wavelength is strongly absorbed by water and is, therefore, ideally suited to deliver energy without causing overheating.

The isotope erbium-167 is highly neutron absorbing and is used in the making of special nuclear fuel rods that have extended life-spans, enabling pressurized water reactors to work in research submersibles for periods of up to 2 years.

ENVIRONMENTAL ELEMENT

Erbium in the environment	
Earth's crust	4 p.p.m.
	Erbium is the 44th most abundant element.
Soils	approx. 1.6. p.p.m.
Sea water	0.8 p.p.t.
Atmosphere	virtually nil

Erbium is one of the more abundant rare-earth elements; it is almost twice as abundant as tin. It poses no environmental threat to plants or animals.

CHEMICAL ELEMENT

Data file	
Chemical symbol	Er
Atomic number	68
Atomic weight	167.26
Melting point	1529°C
Boiling point	2860°C
Density	9.1 kilograms per litre (9.1 grams per cubic centimetre)
Oxide	Er_2O_3

Erbium is a bright silvery metal which is a member of the lanthanide group of the periodic table of the elements. It reacts with oxygen and water, in both cases only very slowly, and it dissolves in acids.

There are six naturally occurring isotopes or erbium, of which erbium-166 comprises 33.5%; erbium-168 27% and erbium-167 23%. The other isotopes are: erbium-162 (0.1%); erbium-164 (1.5%) and erbium-170 (15%). None is radioactive.

ELEMENT OF SURPRISE

About the only way to tint glass pink is to add erbium oxide to it. Up to 5% of the pigment can be added. The result is an attractive pink colour that is used in sunglasses, crystal glassware and jewellery. Ceramics can also be coloured pink this way. Erbium has this ability because it only absorbs light in a narrow band of the spectrum, that of wavelength 530 nanometres; this is in the green part of the spectrum, which is why erbium's salts appear pink. For the same reason when an erbium-impregnated gas mantle glows it emits an emerald-green light.

Europium

Pronounced yoo-roh-pi-uhm, this element is named after Europe.

French, *europium*; German, *Europium*; Italian, *europio*; Spanish, *europio*; Portuguese, *európio*.

Europium is one of 15, chemically similar, elements referred to as the rare-earth elements, which extend from the element with atomic number 57 (lanthanum) to that with atomic number 71 (lutetium). The term rare-earth elements is a misnomer because some are not rare at all; they are more correctly called the lanthanides, although strictly speaking this excludes lutetium. The minerals from which they are extracted, and the properties and uses they have in common, are discussed under lanthanides, on p. 219.

HUMAN ELEMENT

Europium in the human body	
Total amount in body	not known, but small

The amount of europium in humans is quite small. The metal has no biological role. It is difficult to single out europium from among the various lanthanides in the human body, but it is present and is assumed to follow the same metabolic route as calcium and strontium. There is no estimate of the amount of europium in an average adult, and no one has monitored diet for europium content, so it is difficult to judge how much we take in, but it is probably only a milligram or so per year. Europium is not taken up by plant roots, so little gets into the main food chain.

Europium salts could be mildly toxic by ingestion, but insoluble salts are completely non-toxic.

ELEMENT OF HISTORY

The story of europium really begins with that of cerium in 1803. Carl Gustav Mosander separated lanthanum, and an element he called didymium, from cerium in 1839. However, in 1879 Karl Auer showed didymium to be a mixture mainly of praseodymium and neodymium. It also harboured another, rarer, rare-earth element, samarium, separated in the same year by Paul-Émile Lecoq de Boisbaudran. But even samarium was impure, and in 1886 Charles Galissard de Marignac extracted another rare-earth element, gadolinium, from it. Despite all this activity, samples of samarium and gadolinium still had unexplained lines in their spectra, which indicated the presence of yet another element. It fell to Eugène-Anatole Demarçay (1852–1904) in Paris in 1901 to carry out a painstaking sequence of crystallizations of samarium magnesium nitrate, thereby separating another new element, which he named europium.

Had Demarçay and the other investigators known a little more about the chemistry of europium, they might have isolated it much earlier because it has a reduced form, europium(II), or Eu^{2+}, which could have been the key to separating it easily from the other rare-earth elements. The insoluble salt, europium sulfate ($EuSO_4$), will precipitate from solution without bringing down the other rare-earth elements.

ECONOMIC ELEMENT

Europium is found in minerals such as monazite and bastnäsite, which contain all the rare-earth elements; these are discussed in more detail together – see p. 222. Main mining areas are China and the USA. Reserves of europium are estimated to be around 150 000 tonnes. Although it is not a major component, europium is present in extractable amounts from these ores. Europium in a lower oxidation state, Eu^{2+}, can replace calcium and strontium in certain minerals, such as the calcium feldspars (calcium aluminium silicates), which can have unexpectedly high levels of europium.

World production of the pure metal is around 100 tonnes a year, and this is obtained by heating europium oxide with lanthanum metal in a tantalum crucible under vacuum conditions.

Europium phosphors are used in television tubes to give a bright red colour, thanks to europium's emission line in this part of the spectrum.[35] Indeed, it was this property which alerted early pioneers of all-electronic television to the possibility of developing colour. The phosphor commonly used is yttrium oxysulfide doped with europium.

Low-energy light bulbs, which are referred to technically as trichromatic lamps (see p. 224) produce a light which is almost the same as the warmer glow of incandescent light bulbs. For domestic use they are preferred over the cold glare of fluorescent tubes. Europium provides not only the red component of the spectrum, but its lower oxidation state, europium (II), emits in the blue part of the spectrum. Commercial blue phosphors rely on europium-doped materials, such as europium-doped strontium aluminate. For powerful street lighting a little europium is added to mercury vapour lamps to give a more natural light.

Other uses of europium are in thin-film superconductor alloys and in lasers. A europium-doped barium fluoride bromide crystal will absorb X-rays and, when subsequently stimulated by a laser, will emit visible light.

[35] Phosphors are materials that, when bombarded with electrons, emit light, generally of a specific wavelength (colour).

ENVIRONMENTAL ELEMENT

Europium in the environment	
Earth's crust	2 p.p.m.
	Europium is the 50th most abundant element.
Soils	approx. 1.2 p.p.m.
Sea water	0.2 p.p.t.
Atmosphere	virtually nil

Europium is one of the less abundant rare-earth elements; it is almost as abundant as tin. It poses no environmental threat to plants or animals. Europium is present in some plants but only in the 30–130 p.p.b. (dry weight) range; vegetables generally have much less than this, with some having as little as 0.04 p.p.b.

CHEMICAL ELEMENT

Data file	
Chemical symbol	Eu
Atomic number	63
Atomic weight	151.964
Melting point	822°C
Boiling point	1597°C
Density	5.2 kilograms per litre (5.2 grams per cubic centimetre)
Oxide	EuO, Eu_2O_3

Europium is a soft, silvery metal and a member of the lanthanide group of the periodic table of the elements. It is the most reactive of the lanthanide metals, reacting quickly with oxygen and water. It will ignite spontaneously when heated to around 180°C, and it burns in air. Europium is unusual among the rare-earth elements in having a stable lower oxidation state, which it exhibits in the form of the ion Eu^{2+}.

There are two natural isotopes of the element: europium-151, which accounts for 48%, and europium-153, which accounts for 52%. Neither is radioactive.

ELEMENT OF SURPRISE

Europium was found, in much higher concentrations than expected, in samples of rock brought back from the Moon's surface. It was present in feldspars as the reduced form of Eu^{2+}, rather than the more normal Eu^{3+}, suggesting that the minerals of the Moon were formed under highly reducing conditions. This unnaturally high abundance of europium suggests that the Earth and the Moon were not derived from a common source of cosmic material; as yet there is no agreed theory about how the Moon was formed.[36]

[36] The most likely explanation for the origin of the Moon is that a very large object hit the early Earth, sending a lot of it into space around the Earth, where it eventually coalesced to form the Moon. This then began collecting other debris orbiting in space that was gravitationally attracted to it.

Pronounced ferm-iuhm, this element was named after the nuclear physicist Enrico Fermi (1901–1954) who won the Nobel Prize for Physics in 1938, and who, in 1942 at the University of Chicago, initiated the first self-sustaining nuclear reaction to be produced by human endeavour.

French, *fermium*; German, *Fermium*; Italian, *fermio*; Spanish, *fermio*; Portuguese, *fermio*.

HUMAN ELEMENT

Fermium in the human body
None

Fermium has no role to play in living things and is never encountered outside nuclear facilities or research laboratories. It is highly dangerous because of the radiation it emits.

MEDICAL ELEMENT

The isotope fermium-255, an α-particle emitter with a half-life of 20 hours, is currently being researched for possible use in radiation therapy of cancer.

ELEMENT OF HISTORY

A large number of scientists contributed to the discovery of fermium; prominent among them were Gregory R. Choppin, Stanley G. Thompson, Albert Ghiorso and Bernard G. Harvey. It was discovered in the debris of 'Mike', the code name for the first thermonuclear explosion, which took place on a Pacific atoll on 1 November 1952.

Fall-out material, gathered on a neighbouring atoll, was analysed and found to contain two new elements, einsteinium and fermium. There were only about 200 atoms remaining of the fermium which had been formed during the explosion, in which the uranium-238 in a fission bomb was used to provide the heat necessary to trigger a thermonuclear explosion. The uranium-238 had been exposed to such a flux of neutrons that some of its atoms had captured several of them. Through a process of neutron capture, followed by β-decay, many elements were formed, including those of atomic numbers 93–100, among which was fermium-255 with a half-life of 22 hours.

News of the discovery of fermium was kept secret until 1955, by which time a group of physicists at the Nobel Institute in Stockholm had independently made a few atoms of it, in 1953 and 1954, by bombarding uranium-238 with oxygen-16 nuclei. These produced fermium-250 with a half-life of 30 minutes.

ECONOMIC ELEMENT

Fermium-253 can be formed by the bombardment of plutonium-239 with neutrons in nuclear reactors; it has a half-life of 3 days. There are no commercial reasons for fermium to be produced, but it might one day have some use in medicine (see above). Annual world production of fermium probably totals less than a millionth of a gram.

ENVIRONMENTAL ELEMENT

Fermiuim in the environment	
Earth's crust	nil
Sea water	nil
Atmosphere	nil

Fermium does not exist naturally on Earth today but it has occurred in the past, produced in natural reactor deposits (see p. 481).

CHEMICAL ELEMENT

Data file	
Chemical symbol	Fm
Atomic number	100
Atomic weight	257.1 (isotope 257)
Melting point	not known
Boiling point	not known
Density	not known
Oxide	none yet made, but likely to be Fm_2O_3

Fermium is a radioactive element and a member of the actinide group of the periodic table of elements. So far not enough fermium has been made at one time to be visible by the naked eye, but predictions are that it would be a silvery metal susceptible to attack by air, steam and acids.

There are 20 known isotopes, including isomers, which have atomic weights ranging from 243 to 259. The longest lived isotope is fermium-257 with a half-life of 100 days.

ELEMENT OF SURPRISE

Fermium is the element of highest atomic number that can be produced in a nuclear reactor. However, making it is an almost futile task because no sooner does an atom form than it is gone. The problem lies with fermium-257. Though this is the longest-lived isotope, it is also good at capturing neutrons. As soon as an atom of this isotope has formed, it absorbs another neutron to form fermium-258, which has a half-life of only 0.37 milliseconds.

Nevertheless, the little fermium-257 that can be collected – less than a picogram at a time (a million millionth of a gram, i.e. 10^{-12} grams) – has enabled some investigation of its properties to be carried out; this showed that the most stable state of the element would be fermium(III), i.e. Fm^{3+}.

Fluorine

Pronounced flor-een, the name is derived from the Latin *fluere*, meaning to flow, because the common fluoride mineral fluorspar (calcium fluoride, CaF_2) melts when heated in a flame.

French, *fluor*; German, *Fluor*; Italian, *fluoro*; Spanish, *flúor*; Portuguese, *fluor*.

Fluori<u>ne</u> is the name of the element, which exists as the molecule F_2, whereas fluori<u>de</u>, as in sodium fluoride, is a fluorine atom that has acquired a negative electron and is written F^-. Fluorine is highly reactive and dangerous, while fluoride – the form encountered in nature – is stable.

HUMAN ELEMENT

Fluorine (as fluoride) in the human body	
Blood	0.5 p.p.m.
Bone	varies between 2000 and 12 000 p.p.m.
Tissue	approx. 0.05 p.p.m.
Total amount in body	between 3 and 6 grams

As long ago as 1802, fluoride was detected in ivory, bones and teeth. By the mid nineteenth century it was found to be present in blood, sea water, eggs, urine, saliva and hair. Although it appeared to be present in all living things, this did not *prove* that fluorine was an essential element. That had to wait until animal tests showed that if fluoride was excluded from the diet of laboratory animals, they failed to grow properly, or became anaemic and infertile. Today fluoride is regarded as essential for humans, but only in minute doses. Indeed, in larger amounts it is highly dangerous: a quarter of a gram will provoke a toxic response, while 5 grams would be fatal. For this reason fluoride has been used as an effective insecticide for cockroaches and ants. (At a hospital in the USA in 1943, patients were served scrambled eggs to which sodium fluoride had been added instead of salt (sodium chloride): 163 were taken ill and 47 died.)

In humans, most of the fluoride accumulates in the bones and teeth. It strengthens both of these, converting the calcium phosphate of which they are made into fluoroapatite, a harder material. This, in tooth enamel, can better resist the corrosive effect of acids produced by oral bacteria acting on sugar. For this reason it is added to water supplies and toothpastes.

Fluoride has a powerful effect on enzymes, effectively blocking their activity, although a few are stimulated by it. (These enzyme-deactivating properties are made use of in biochemical research and, more practically, in cleaning fermentation vats in breweries and distilleries.)

Fluorine gas is extremely toxic: breathing it at concentrations of 0.1% for only a few minutes will kill. Luckily the gas is so pungent that it acts as its own early warning system, but in any case stringent legal precautions surround its use.

FOOD ELEMENT

The average person takes in between 0.3 and 3 milligrams of fluoride a day in their diet. Fluoridated tap water may provide much of this, but most people get their fluoride from foods such as chicken, pork, eggs, potatoes, butter, cheese and tea (a cup of tea provides 0.4 milligrams). Cod, mackerel, sardines, salmon and sea salt are particularly rich in fluoride because they come from an environment, sea water, which contains 1 p.p.m. of fluoride. Mackerel has 27 p.p.m. (fresh weight). Vegetables generally have much less than this, between 3 and 20 p.p.m. (dry weight).

MEDICAL ELEMENT

Fluoride is prescribed in some countries as a preventative treatment for osteoporosis in the elderly. However, it is the organofluorine drugs that have proved a more potent source of medicaments, like the antibiotic ofloxacin, used world-wide. There are now about a dozen such drugs; some are so potent that a single dose will clear up certain minor infections; for example, ofloxacin can cure gonorrhoea. Fluconazole is the drug of choice for attacking fungal diseases which can threaten the life of transplant patients whose immune system is deliberately suppressed, as well as curing more mundane infections such as vaginal thrush.

These drugs are easily absorbed and penetrate to parts of the body that conventional antibiotics find difficult to reach, for example the bladder. They work by interfering with microbial enzymes, such as that responsible for coiling bacterial DNA, which then prevents the disease pathogens from multiplying.

The isotope fluorine-18 is used in medical diagnosis through the technique known as positron emission tomography (PET). A sample of this short-lived isotope – its half-life is only 110 minutes – is introduced into the body and the X-ray-like radiation, which it emits as it decays, can be scanned to reveal a picture of the body's vital organs. (See carbon, p. 94, for further details.)

ELEMENT OF HISTORY

Georgius Agricola described the use of fluorspar (calcium fluoride, CaF_2) as a flux for molten metals in 1529, pointing out that this mineral would melt and flow when heated in a fire, and that it was added to furnaces for smelting metals because the melt became more fluid and easier to work with. A second benefit was discovered by Heinrick Schwanhard, who found, in 1670, that if fluorspar was dissolved in acid the solution could be used to etch glass. The first glass to be so affected was reputed to be that of his spectacles, but this story may be more apocryphal than gospel.

The early chemists were aware that fluorides contained an unknown element, fluorine, and that its compounds greatly resembled those of chlorine, but they could not isolate it. The French scientist, André Ampère, coined the name fluorine in 1812 but later changed his mind in favour of phthorine, derived from the Greek *phthoros,* meaning destructive. By then, however, the great Humphry Davy was using the name fluorine and his choice prevailed. He, like many of his contemporaries, was unable to isolate the element, and, like several of them, he became ill by trying to isolate it from its acid, hydrofluoric acid. Some chemists, such as the French chemist Jerome Nicklès of Nancy, died from hydrogen fluoride (HF) poisoning.

George Gore, in 1869, passed an electric current through water-free hydrofluoric acid and found that the gas that was liberated at the anode reacted violently with his apparatus. He thought it might be fluorine but it was probably not. It was not until 1886 that Henri Moissan (1852–1907), in Paris, obtained it by the electro-lysis of a mixture of potassium bifluoride (KHF_2) dissolved in anhydrous hydrofluoric acid. He was awarded the 1906 Nobel Prize for Chemistry for this work (and for the development of the electric furnace).

ECONOMIC ELEMENT

Annual world production of the mineral fluorite (fluorspar) is around 4 million tonnes, and there are around 120 million tonnes of mineral reserves. When these are depleted there are vast deposits of this element such as fluorapatite. The main mining areas for fluorite are China, Mexico and Western Europe. Most is used in the smelting and refining of metals, but a lot goes into the manufacture of hydrofluoric acid (400 000 tonnes per year), which in turn is used to make fluorine gas (15 000 tonnes per year). It is used for etching glass, treating metals and making fluorocarbons.

Fluorine gas is still produced by the method devised by Moissan; he worked his apparatus at –23°C to reduce the gas's reactivity, but modern electrolytic cells work at 87°C and are made of Monel metal (nickel alloy) which resists fluorine attack.

The gas is generally made where it is needed. Its main uses are in making uranium hexafluoride (UF_6) for the nuclear industry, sulfur hexafluoride (SF_6) for the electricity supply industry, and chlorine trifluoride (ClF_3) for the chemical industry. UF_6 is needed to separate the nuclear fuel, uranium-235, from its more common isotope, uranium-238. Chlorine trifluoride is used to make UF_6. Sulfur hexafluoride is the inert insulating gas which surrounds high-power electricity transformers.

Fluorine gas can now be handled in bulk and even transported safely as a liquid in containers that are protected by liquid air. It has even been tried as a possible rocket propellant. Diluted with nitrogen, fluorine is much safer to handle; it is used to treat polyethene containers to give them a coating of organofluoro polymer, so making them impermeable to solvents. Vehicle fuel tanks made of such treated plastic are less likely to rupture in a crash and so less likely to cause a fire.

Fluorocarbons

These can be liquids, based on short chains of carbon atoms, or polymers, with long chains of carbon atoms. Both have lots of uses. Fully fluorinated short-chain carbons make excellent solvents, with boiling points ranging from 25 to 250°C, depending on the length of the chain. They are inert, heat-stable, non-flammable, non-toxic and of low viscosity. They are used as refrigerants, lubricants and cleaning fluids for electric circuit boards, and even as blood substitutes because they dissolve oxygen easily.

The fully fluorinated polymer, popularly known as Teflon – more correctly known as poly(tetrafluoroethene) or PTFE for short – is used for cable insulation, as linings for vessels holding corrosive liquids, for hoses and tubing, for fabric roofing, and for plumber's tape for sealing joints in water pipes. Its use in kitchenware is described in 'Element of surprise' (see below). It is also the basis of Goretex, a weatherproof fabric first made in 1969 by Bob Gore, who found a way of expanding PTFE by heating and stretching the polymer to form a membrane which had invisible pores that were small enough to keep water droplets out, but big enough to allow the water molecules of sweat to escape. Goretex is also used for artificial veins and arteries in the treatment of cardiovascular disorders. Scrap from the PTFE industry is re-used by grinding it to a microfine powder and adding it to printer's ink to make it flow more smoothly.

ENVIRONMENTAL ELEMENT

Fluorine in the environment	
Earth's crust	950 p.p.m.
	Fluorine is the 13th most abundant element.
Soils	approx. 330 p.p.m. with a usual range of 150–400 p.p.m. although some soils can have as much as 1000 p.p.m. and heavily contaminated soils have been found with 3500 p.p.m.
Sea water	1.3 p.p.m.
Atmosphere	0.6 p.p.b.; it is present as salt spray and organochlorine compounds. Up to 50 p.p.b. has been recorded in city environments.

It was known for centuries that cattle which grazed where volcanic dust had settled would become ill and lame; the reason was their high fluoride intake. Research in Iceland in 1970 showed that grass affected in this way could have as much as 4300 p.p.m. fluoride (dry weight), although this fell to 30 p.p.m. within 40 days. Other activities can contaminate soil with fluoride, such as close proximity to an aluminium smelter, a coal-fired power plant, brick works or a fertilizer factory – all industries that emit fluoride. Grass grown on such land can have as much as 5400 p.p.m. (dry weight).

Humans can also be affected by too much fluoride and suffer from fluorosis, the first signs of which are mottled teeth. Later, there may be a hardening of the bones, which can lead to a deformed skeleton. In certain parts of India, such as the Punjab, the condition is endemic, especially where villagers drink water from wells with high levels of fluoride, up to 15 p.p.m. About 25 million Indians suffer a mild form of fluorosis and many thousands show skeletal deformities.

Too much fluoride, whether taken in from the soil by roots, or absorbed from the atmosphere by the leaves, retards the growth of plants and reduces crop yields. Those most affected are corn and apricots. Some crops, such as asparagus, beans, cabbage and carrots, are resistant to fluoride.

Volatile chlorofluorocarbons (CFCs),[37] which were once used in aerosols and refrigerators, are environmentally damaging to the ozone layer which protects the Earth from harsh ultraviolet radiation from the Sun, and so they are being phased out of use. Even their replacements, the hydrofluorocarbons (HFCs), which do not have this effect, are frowned upon because they are potent greenhouse gases, although they are much less persistent in the atmosphere than the long-lived CFCs.

CHEMICAL ELEMENT

Data file	
Chemical symbol	F
Atomic number	9
Atomic weight	18.9984032
Melting point	−220°C
Boiling point	−188°C
Density of gas	1.7 grams per litre
Oxides	OF_2 and O_2F_2

Fluorine is a pale yellow gas (F_2) and the first member of group 17, also known as the halogen group, of the periodic table of the elements. It is the most reactive of all the elements. Only the noble gases helium, neon and argon are unaffected by it.

Fluorine in nature has only one isotope, fluorine-19, which is not radioactive.

ELEMENT OF SURPRISE

Contrary to popular belief, it was not space exploration that led to the development of the non-stick frying pan, but the other way round. The non-stick frying pan relies on Teflon, the polymer poly(tetrafluoroethene), which was discovered by Roy Plunkett at the DuPont research laboratories at Deepwater, New Jersey, on 6 April 1938. That morning he opened a cylinder of tetrafluoroethene which was supposed to hold exactly 1000 grams of the gas but only 990 grams came out. The missing 10 grams had turned into a curious white powder which was a plastic with some remarkable properties. It would not dissolve in anything, was not affected by heat, would not burn and was not

[37] The most common ones were CFC-11 (chemical formula CCl_3F), CFC-12 (CCl_2F_2) and CFC-113 (CCl_2FCClF_2).

attacked by hot corrosive acids. More unusual was its ability to remain flexible down to *minus* 240°C, something that no other plastic was capable of doing. It was this property which made it so important for use in space and on the Moon where such temperatures prevail. In addition, it had a slippery feel, because nothing would stick to it. This was to be the secret of its commercial success as a coating for non-stick kitchen utensils.

How does non-stick Teflon stick to an aluminium frying pan? The problem was solved in the 1950s by Louis Hartmann, who treated the metal surface with acid to form microscopic holes, then applied the polymer as an emulsion and baked it at 400°C for a few minutes. The polymer melts, forms a film over the surface and runs into the tiny etched pits, filling them and thereby becoming wedged fast and so gripping the rest of the Teflon film tightly to the surface.

The French company, Tefal, marketed the first non-stick frying pans 10 years before man set foot on the Moon.

Francium

Pronounced fran-see-uhm, the element is named after France.[38]

French, *francium*; German, *Francium*; Italian, *francio*; Spanish, *francio*; Portuguese, *frâncio*.

HUMAN ELEMENT

Francium in the human body
None

Francium has no role to play in living things and is never encountered outside nuclear facilities or research laboratories. It is potentially dangerous because of its intense radioactivity.

When francium is injected into rats it concentrates in the bladder, gut, kidneys, saliva and liver. In those rats with cancer it also concentrates in the tumour; this finding raised hopes that francium might have medical uses in treating this disease. For reasons given below, this was a forlorn hope.

ELEMENT OF HISTORY

Early in the twentieth century it was clear from the improved forms of the periodic table that there must be an element below caesium in group 1, but all who searched for it were to be disappointed. Yet several chemists reported finding it and gave it names such as russium, alcalinium, virginium and moldavium. Their claims were all proved to be false.

The missing element was finally discovered in 1939 by Marguerite Perey at the Curie Institute in Paris. She had purified a sample of actinium free of all its radioactive decay products and yet she observed that the β-ray emission which it was giving out was more intense than it should have been. She knew that this meant there must be another element present, one that had previously escaped discovery, and which she deduced must be the missing element 87. It was.

[38] France has the distinction of having two elements named after it. The name of gallium is derived from the Latin *Gallia*, the Roman name for that country.

ECONOMIC ELEMENT

Francium is an intensely radioactive metal, of which there are minute traces in uranium ores, but that required for research purposes is made either in a nuclear reactor (by bombarding radium with neutrons) or in a cyclotron (by bombarding thorium with protons). Although francium can be made, it is difficult to investigate because the longest-lived atom, isotope-223, has a half-life of only 22 minutes. Not surprisingly, no use has been found for what little francium can be produced, and indeed no weighable quantity of francium has ever been made.

ENVIRONMENTAL ELEMENT

Francium in the environment	
Earth's crust	present, but only in trace amounts
Sea water	nil
Atmosphere	nil

Francium occurs naturally in uranium minerals, but there is only one francium atom per billion billion (10^{18}) atoms of uranium. These few atoms of francium-223 are there as the radioactive decay product of actinium-227, which in turn is part of a uranium-235 decay series. The short half-life of francium-223 means that there is probably less than 30 grams of francium in the Earth's crust at any one time.

CHEMICAL ELEMENT

Data file	
Chemical symbol	Fr
Atomic number	87
Atomic weight	223.0 (isotope Fr-223)
Melting point	approx. 27°C (est.)
Boiling point	approx. 680°C (est.)
Density	not known
Oxide	would be Fr_2O, if it were ever made

Francium is a radioactive metal element and a member of group 1, also known as the alkali metal group, of the periodic table of the elements. It has been studied by radiochemical techniques, which show that its most stable state is the ion Fr^+. It is possible to predict that its chemistry would be like the element above it in the periodic table, caesium.

Thirty isotopes of francium, including isomers, have been reported, ranging in atomic weight from 201 to 229. The longest lived are francium-223 (22 minutes) and francium-212 (20 minutes). Most have half-lives of less than 1 minute.

Gadolinium

Pronounced gad-oh-lin-iuhm, this element was named in honour of Johan Gadolin, who was born in 1760 at Åbo, then a part of Sweden, now Turku, Finland; he died in 1852. Gadolin investigated the first rare-earth mineral, now named gadolinite,[39] which was found at Ytterby, Sweden, in 1780.

French, *gadolinium*; German, *Gadolinium*; Italian, *gadolinio*; Spanish, *gadolinio*; Portuguese, *gadolíneo*.

Gadolinium is one of 15, chemically similar, elements referred to as the rare-earth elements, which extend from the element with atomic number 57 (lanthanum) to the one with atomic number 71 (lutetium). The term rare-earth elements is a misnomer because some are not rare at all; they are more correctly called the lanthanides, although strictly speaking this excludes lutetium. The minerals from which they are extracted, and the properties and uses they have in common, are discussed under lanthanides, on p. 219.

HUMAN ELEMENT

Gadolinium in the human body	
Total amount in body	not known, but small

Although gadolinium has no biological role, it has been noted that gadolinium salts stimulate metabolism. It is difficult to separate out the various amounts of the different lanthanides in the human body, but they are present; the levels are highest in bone, with smaller amounts being present in the liver and kidneys.

No one has monitored diet for gadolinium content, so it is difficult to judge how much we take in, but it is probably only a milligram or so per year. Gadolinium is not taken up by plant roots in any great amount – the amount detected in vegetables is less than 2 p.p.b. (dry weight) – so little gets into the human food chain. Some species, such as lichens, can have as much as 500 p.p.b. (dry weight) of gadolinium.

Soluble gadolinium salts, such as the chloride and nitrate, are mildly toxic by ingestion, but insoluble salts are non-toxic. Gadolinium salts irritate the skin and eyes, and are suspected tumorigens.

[39] Gadolinite is a beryllium iron silicate of formula $Be_2Fe(X)_2Si_2O_{10}$, where X can be a variety of lanthanide elements and/or yttrium.

MEDICAL ELEMENT

Gadolinium's neutron-capturing ability is put to use in the medical diagnostic technique known as neutron radiography. A screen made of gadolinium is used to trap the neutrons, thereby becoming slightly radioactive; the radioactivity is then recorded on film which is in contact with the screen. The extent to which the film darkens depends on the amount of radiation it receives.

Some gadolinium compounds have been used in medical experiments because this element can be tracked in the human body through its magnetic properties, and gadolinium compounds are used as contrast agents in magnetic resonance imaging for medical diagnosis.

ELEMENT OF HISTORY

Gadolinium was discovered in 1880 by Charles Galissard de Marignac at Geneva, Switzerland. He had long suspected that the didymium oxide that Mosander had claimed to be the oxide of a new element was in fact not a pure substance. His suspicions were confirmed when Marc Delafontaine and Paul-Emile Lecoq de Boisbaudran in Paris reported that the spectral lines of the didymium oxide varied according to the source from which it came. Indeed, in 1879 they had already separated another rare-earth oxide from a sample of didymium oxide which had been extracted from the mineral samarskite,[40] found in the Urals in Russia. This they named samarium oxide.

The following year, 1880, Marignac extracted another new rare-earth oxide from didymium oxide which he tentatively referred to as alpha-yttrium. This was also obtained by Boisbaudran in 1886 and it was he who suggested the name gadolinium, to which Marignac agreed.

ECONOMIC ELEMENT

Gadolinium is found in minerals that include all the rare-earth elements; these are discussed in more detail together (see p. 222). The main mining areas are China, USA, Brazil, India, Sri Lanka and Australia, with reserves estimated to be well in excess of a million tonnes. The most important ores are monazite and bastnäsite. Although gadolinium is not a major constituent of either ore, it is present in extractable amounts. World production of pure gadolinium is about 400 tonnes per year. Gadolinium is separated from its ores by ion-exchange techniques.

Gadolinium metal can be produced by heating gadolinium fluoride and calcium. It is unique in two respects – it has the greatest neutron-capturing ability of any known element, and it is magnetic like iron, but it loses its magnetism when its temperature goes up to 20°C. It has been suggested that this could be used in so-called magnetic refrigeration, because as a magnetic field is applied the metal's temperature drops slightly.

[40] Samarskite is a mixed oxide ore of several uncommon metals, including yttrium, cerium, uranium, niobium and tantalum.

Metallic gadolinium is rarely used as the metal itself, but its alloys are needed to make magnets and electronic components, such as the recording heads for video recorders. The alloys are also used for magnetic data storage discs, which can be permanent or erasable. These consist of ultra-thin layers of various metals deposited on glass or plastic of which gadolinium is a key component, along with other rare-earth metals, such as terbium and dysprosium, plus the more common magnetic metals, iron and cobalt.

Gadolinium's affinity for neutrons is used to regulate neutron activity in the reactor core of nuclear power plants. It is uniquely suited for this purpose, having two isotopes that have superb neutron-absorbing capabilities: gadolinium-155 and gadolinium-157. The latter is four times more absorbing than the former and more than 300 times more absorbing than boron, the other element that is used for this purpose. Consequently around 5% of gadolinium oxide is intimately mixed with the uranium-235 oxide pellets used in the reactors; this makes for more efficient use of the nuclear fuel.

Gadolinium gallium garnets (GGG) have microwave and electro-optical applications.

ENVIRONMENTAL ELEMENT

Gadolinium in the environment	
Earth's crust	8 p.p.m.
	Gadolinium is the 41st most abundant element.
Soils	approx. 6 p.p.m., range 1–15 p.p.m.
Sea water	1.2 p.p.t.
Atmosphere	virtually nil

Gadolinium is one of the more abundant rare-earth elements; it is four times as common as tin. It poses no environmental threat to plants or animals.

CHEMICAL ELEMENT

Data file	
Chemical symbol	Gd
Atomic number	64
Atomic weight	157.25
Melting point	1313°C
Boiling point	3270°C
Density	7.90 kilograms per litre (7.90 grams per cubic centimetre)
Oxide	Gd_2O_3

Gadolinium is a soft, shiny, silvery metal and a member of the lanthanide group of the periodic table of the elements. It corrodes in air due to the formation of an oxide layer which flakes off exposing more fresh metal. Gadolinium reacts slowly with water and dissolves in acids.

There are seven naturally occurring isotopes, of which gadolinium-158 is the most abundant, accounting for 25%, followed by gadolinium-160 (22%) and gadolinium-156 (20.4%). The others are gadolinium-157 (15.6%), gadolinium-155 (15%), gadolinium-154 (2%) and the least abundant, gadolinium-152 (0.2%), which is radioactive, but only very weakly so. This has a half-life of 110 trillion years (1.1×10^{14} years), which is 10 000 times the age of the Universe.

ELEMENT OF SURPRISE

Unlike the other rare-earth elements, which are used to produce the vivid colours in television screens, gadolinium is colourless because it does not absorb light in the visible region of the spectrum (its compounds also appear white for this reason). Because of this property it is sometimes used as a support for other rare-earth phosphors that luminesce, in the knowledge that it will not 'quench' their intense colours.

Pronounced gal-iuhm, the name is derived from *Gallia*, the Latin name for France, but see below.

French, *gallium*; German, *Gallium*; Italian, *gallio*; Spanish, *galio*; Portuguese, *gálio*.

HUMAN ELEMENT

Gallium in the human body	
Blood	less than 0.1 p.p.m.
Bone	not known
Tissue	approx. 1 p.p.b.
Total amount in body	less than 1 milligram

Gallium has no biological role but is known to stimulate metabolism. It presents few health risks because its salts generally have low toxicity, although rats fed 10 milligrams of gallium per day in their diet display some toxic symptoms.

Plants show a slight selective uptake of gallium as witnessed by their having more gallium than the soil in which they grow. Land plants have 0.01–5 p.p.m. (dry weight) with edible vegetables having 0.03–2 p.p.m. (again, dry weight).

ELEMENT OF HISTORY

Gallium was discovered in 1875 by Paul-Émile Lecoq de Boisbaudran at Paris, France, but its existence had already been predicted 6 years earlier by the Russian chemist Dimitri Mendeleyev. He had devised his periodic table of the elements in 1869 and this showed that there was an unfilled place below aluminium. He forecast some of the properties of the missing element, such as its atomic weight being around 68 and its density 5.9 grams per cubic centimetre; these were proved to be correct (see p. 521).

Boisbaudran, working in his private laboratory, first detected gallium in August 1875. When he examined the spectrum of the zinc he obtained from a sample of zinc blende ore (ZnS) from the Pyrenees, he saw lines he had never seen before, which told him that a previously unknown element was present. By November of that year, he had isolated a sample of it and even purified the metal by electrolysis of a potassium hydroxide solution in which it was soluble. He identified it as the missing element that should be like aluminium. On 6 December 1875 he presented a sample to the French Academy of Sciences and christened it gallium.[41]

MEDICAL ELEMENT

Although there are no drugs, as yet, that contain gallium, in 1998 one of its compounds was shown to be effective in attacking strains of malaria that have become resistant to conventional anti-malarial drugs, such as chloroquine.

The radioactive isotope gallium-67, which has a half-life of 78 hours, is used to locate and treat melanomas because it concentrates in these cancerous tissues.

ECONOMIC ELEMENT

Several ores, such as the aluminium ore bauxite, contain small amounts of gallium, and coal may have a relatively high gallium content. The flue dust from a coal-burning plant may have as much as 1.5% of the element. No ore is mined specifically for gallium and all that is produced is recovered as a by-product of zinc and copper refining. Annual world production is only 30 tonnes.

Gallium has semiconductor properties, especially as gallium arsenide (GaAs). This can convert electricity to light and is used in light-emitting diodes (LEDs) for electronic displays and watches. However, it is its semiconductivity which has brought gallium to prominence because it generates less heat than silicon and is therefore more suitable for use in supercomputers as well as the more mundane mobile phone. The crystals are grown layer-by-layer using a process known as molecular beam epitaxy. In this process, a mixture of volatile compounds – trimethyl gallium, $Ga(CH_3)_3$, and arsine, AsH_3 – is heated to disintegrate them to atoms, which then deposit in a regular array on the surface of a wafer-thin crystal substrate to give crystals of pure GaAs. Tiny amounts (p.p.m.) of other atoms, such as tin or zinc, are included as dopants to adjust the semiconductivity to that required.

Chemists and physicists are also intrigued by gallium arsenide because it can trap free electrons on the surface of its crystals, and confine them to a small area. Normally electrons are the mobile component of matter, and in this 'frozen' state they are referred to as quantum dots.

[41] The choice of name, gallium, flattered the French Academy, because it appeared to be derived from Gallia, the Roman name for France, but it has been suggested that Boisbaudran named the element after himself! His middle name, Lecoq, is like '*le coq*', the French for 'the cock', and the Latin name for that is *gallus*.

ENVIRONMENTAL ELEMENT

Gallium in the environment	
Earth's crust	18 p.p.m.
	Gallium is the 34th most abundant element.
Soils	approx. 28 p.p.m., range 1–70 p.p.m.
Sea water	30 p.p.t.
Atmosphere	virtually nil

Gallium is more abundant than lead but much less accessible because it has not been selectively concentrated into minerals by any geological process, so it tends to be widely dispersed. It poses no environmental threat to plants or animals.

CHEMICAL ELEMENT

Data file	
Chemical symbol	Ga
Atomic number	31
Atomic weight	69.723
Melting point	30°C
Boiling point	2403°C
Density	solid: 5.9 kilograms per litre (5.9 grams per cubic centimetre); liquid: 6.1 kilograms per litre (6.1 grams per cubic centimetre)
Oxide	Ga_2O_3

Gallium is a silvery-white metal and a member of group 13 of the periodic table of the elements. It is soft enough to be cut with a knife. It is stable in air and water; but it reacts with and dissolves in acids and alkalis.

Natural gallium consists of two isotopes: gallium-69, which comprises 60%, and gallium-71, 40%. Neither is radioactive.

ELEMENT OF SURPRISE

Gallium will melt when held in the hand and on so doing it also shrinks in volume, albeit by only 3%, which is contrary to the way other metals behave. (The semi-metals antimony and bismuth behave likewise.) Gallium remains liquid over a wider range of temperatures (2373°C) than any other known substance and at one time was used in high-temperature thermometers.

Pronounced jer-may-niuhm, the name is derived from *Germania*, the Latin name for Germany.

French, *germanium*; German, *Germanium*; Italian, *germanio*; Spanish, *germanio*; Portuguese, *germânio*.

HUMAN ELEMENT

Germanium in the human body	
Blood	approx. 0.4 p.p.m.
Bone	data not available
Tissue	0.1 p.p.m.
Total amount in body	5 mg

There is no biological role for germanium, although it acts to stimulate metabolism. Its salts generally have low toxicity for mammals, but are deadly to certain bacteria although no practical application of this has resulted.

Some plants, including those eaten as food by humans, can absorb germanium from the soil in which they grow, probably taking it in as the oxide. Average levels of the element of around 0.05 p.p.m. (fresh weight) have been measured in grains and vegetables, but not all plants contain it.

FOOD ELEMENT

The estimated daily intake is around 1 milligram, and there have been claims that germanium could be beneficial to health, although this has never been proved scientifically. Those claims began in 1980 when Kazuhiko Asai wrote a book called *Miracle Cure: Organic Germanium*, published in Japan, in which he claimed that all kinds of health benefits would result from taking various germanium medicaments, which he referred to as organic germanium.[42] The principal compounds were GE-123, Asai's name for a complex oxide also known as carboxyethylgermanium sesquioxide, and sanugerman,

[42] 'Organic' was used here in its chemical sense, i.e. indicating that these compounds were compounds of carbon with germanium.

which was germanium citrate lactate. Asai's theory was given wider publicity by Sandra Goodman in her book *Germanium: the Health and Life Enhancer*, published in 1988.

A high intake of germanium was supposed to improve the immune system, boost the body's oxygen supply, make a person feel more alive and destroy damaging free radicals. In addition it was said to protect the user against radiation. Expensive organic germanium food supplements appeared on the shelves of alternative health stores, but the craze was quickly brought to an end in the UK in 1989 when the Government's Department of Health warned against these supplements, noting that they had no nutritional or medical value and that taking them constituted a risk to health, rather than a benefit.

However, in the USA germanium is still considered to be a helpful micro-nutrient. Health advocates there suggest consuming foods that contain it, in order to get any benefits it might produce. Such foods are garlic, ginseng and some micro-algae that have been revered and touted for their supposed health-giving benefits for thousands of years. Eating such foods would not be harmful.

ELEMENT OF HISTORY

Germanium was discovered by Clemens A. Winkler (1838–1902) at Freiberg, Germany, in 1886, 15 years after its existence had been predicted by Mendeleyev, who saw that there must be an element between silicon and tin in group IV of his periodic table. He forecast some of its properties, predicting that its atomic weight would be about 71 and its density about 5.5 grams per cubic centimetre. Seven years earlier, a London chemist, John Newlands, had suggested that silicon and tin were part of a 'triad', as found with other elements, such as phosphorus/arsenic/antimony and sulfur/selenium/tellurium.[43] If this were so, he said, then there should be another element linking silicon and tin.

In September 1885 a miner working in the deep level (460 metres underground) of the Himmelsfürst silver mine, at St Michaelis near Freiberg, came across an unusual ore. He gave it to the mine manager who sent it to Albin Weisbach at the nearby Mining Academy. Weisbach certified that it was a new mineral, gave it the name argyrodite and passed it to his colleague Winkler to analyse; Winkler found it to be 75% silver and 18% sulfur, but the remaining 7% of its weight could not be accounted for.

Winkler spent the next few months working to separate and identify the unknown component; by February 1886, he believed that it was a new metal-like element and he had produced a sample by heating its sulfide in hydrogen gas. After he had reported his findings, there began a debate as to which of the elements missing from Mendeleyev's periodic table it might be. Mendeleyev favoured a metal in the zinc/cadmium group, while Winkler thought it might be in group V and a neighbour of antimony.

As germanium's properties were revealed, it became clear that it was the missing element between silicon and tin, because its chemical nature was much as Mendeleyev had predicted for the missing element of group IV. Winkler named the element after his native land, and announced that argyrodite, from which it had come, had the formula Ag_8GeS_6.

[43] Triads are explained in more detail on p. 515.

ECONOMIC ELEMENT

Germanium ores are rare. The least rare, germanite, is a copper–iron–germanium sulfide with 8% of the element, but even this is not mined. Germanium is widely distributed in the ores of other metals, such as zinc, and that which is required for manufacturing purposes is recovered as a by-product from the flue-dusts of zinc smelters. World production is about 80 tonnes per year. The metal is first separated by treatment with chlorine, which converts the germanium to germanium tetrachloride, a volatile liquid which boils at 86°C.

Germanium is a semiconductor and, doped with other elements such as arsenic or gallium, it was used on a large scale for many years. Today it has been superseded in electronic devices and most germanium, as the oxide, now finds its way into special glass for wide-angle camera lenses (on account of its high refractive index) and for infrared devices (since it will transmit these rays). A little germanium is used in alloys. In 1993, it was reported that doping silicon chips with germanium made them work almost as well as gallium arsenide, but little has come from this discovery, as yet.

ENVIRONMENTAL ELEMENT

Germanium in the environment	
Earth's crust	2 p.p.m.
	Germanium is the 52nd most abundant element.
Soils	approx. 1 p.p.m., range 0.5–2 p.p.m.
Sea water	0.5 p.p.t., with more in the Pacific than the Atlantic
Atmosphere	virtually nil

Germanium is less abundant than either tin or lead, which are the heavier component metals of group 14, and it is less easily accessed because geological processes have concentrated only small amounts of it into minerals, so that it tends to be widely dispersed. It poses no environmental threat to plants or animals.

Organogermanium compounds have been detected in sea water, suggesting that the element might have an effect on some micro-organisms, but these have not been identified.

CHEMICAL ELEMENT

Data file	
Chemical symbol	Ge
Atomic number	32
Atomic weight	72.61
Melting point	938°C
Boiling point	2830°C
Density	5.3 kilograms per litre (5.3 grams per cubic centimetre)
Oxide	GeO_2

Ultrapure germanium is a silvery-white, brittle, semi-metallic element that comes in group 14 of the periodic table of the elements. It is stable in air and water, and is unaffected by alkalis and acids, except nitric acid.

There are five naturally occurring isotopes, of which the most abundant is germanium-74 at 36.5%, followed by germanium-72 at 27%, germanium-70 at 20.5%, and both germanium-73 and -76 at 8%. None is radioactive.

ELEMENT OF SURPRISE

Germanium was the first element to be used in transistors. Work on it began in 1942 as part of the US war effort. What had attracted attention to germanium was its potential as a rectifier of electrical current, because it had the ability to conduct electricity in one direction but not in the reverse direction. The problem the US Government faced was that the two known sources of germanium were held by Germany. The problem was solved by extracting the element from the waste of the Eagle–Picher Industries' zinc smelter in Oklahoma, where an ore with a high germanium content had been processed for many years. A method of extracting it from this waste was developed and provided all the germanium that was required. By 1948, the first so-called transistor radios were on sale to the public, with germanium transistors instead of the conventional diode valves.

Gold

Pronounced gowld, the name is the Anglo-Saxon word *gold*, which may have come from the word *geolo* meaning yellow; the chemical symbol for the element, Au, derives from the Latin *aurum*, meaning 'glow of sunrise'.

French, *or*; German, *Gold*; Italian, *oro*; Spanish, *oro*; Portuguese, *ouro*.

HUMAN ELEMENT

Gold in the human body	
Blood	1–40 p.p.b.
Bone	16 p.p.b.
Tissue	0.4 p.p.b. in liver
Total amount in body	less than 0.2 milligrams

Gold has no biological role. Both the metal and gold salts generally have low toxicity because they are poorly absorbed by the body.

MEDICAL ELEMENT

Gold is now part of the medical armoury against rheumatoid arthritis, and treatment with it even merits a special name, chrysotherapy, from the Greek word for gold, *chrysos*. It is prescribed when treatment with non-steroid anti-inflammatory drugs is failing to give relief. Gold can take up to 10 weeks to work, but there is a limit to how long it can be given, and even if a patient's symptoms are alleviated, the doctor will have to phase the treatment out after a few years because of the side effects associated with a build-up of gold in the body, which may amount to as much as 500 milligrams.

Gold therapy began in 1927 when gold sodium thiomalate (known as myocrisin) was used to treat tuberculosis, unsuccessfully as it happened, but a beneficial side-effect was noted: patients with arthritis had this condition held in check. That the effect was real was proved in 1956 when the Empire Rheumatism Council conducted double-blind tests, with some patients reporting an improvement up to a year after treatment ended.

Myocrisin had to be injected, but chemists soon developed an oral drug, auranofin (trade-name Ridaura). After successful clinical trials over several years, this was approved by the US Government's Food and Drugs Administration in 1985. Auranofin has fewer side-effects than myocrisin; its main side-effects are diarrhoea and skin rash. The recommended dose, 6 milligrams per day, delivers about 2 milligrams of gold. Of this about 0.5 milligrams is absorbed into the body.

Because platinum drugs are used to cure cancer, it was hoped that drugs based on other metals, such as gold, might be used for this purpose as well, and while some drugs were highly effective, tests on rabbits and dogs showed that they caused irreversible heart damage and research was stopped. (Nevertheless gold has a role to play in radiotherapy treatment for this disease which involves injections of the radioactive isotope, gold-198, which has a half-life of 2.7 days.)

Dental element

Dentists have been using gold since the time of the Etruscans, who lived in Italy from 1000 to 400 BC, where excavated graves have revealed skulls with false teeth held in place by gold bridges. Today more than 60 tonnes a year of gold is used in dentistry; gold fillings, crowns and teeth are particularly popular in Germany and Japan. Dental gold is alloyed with silver and palladium, plus a trace of zinc (0.2%) to harden it, with the proportion of gold being around 75%.

ELEMENT OF HISTORY

Gold has been known since prehistoric times and was one of the first metals to be worked, mainly because it was to be found as grains and even nuggets gleaming in the beds of streams from which it could easily be panned. The Nile particularly was a fruitful source of such alluvial gold, but by 2000 BC the Egyptians were already beginning to mine gold, as they followed the veins of the metal in rocks back underground. The royal graves of ancient Ur (modern Iraq), one of the earliest of civilizations which flourished 5000 years ago, contained gold jewellery that had been skilfully worked.

Gold was easy to refine and reclaim from its alloys; all that was needed was to heat it in a furnace with salt and brick dust which either volatilized or absorbed the other metals leaving the gold unaffected. The minting of gold coins began around 2650 years ago in the Lydian kingdom (modern Turkey); the people there began by using electrum, a native alloy of gold and silver. The first pure gold coins were minted in Lydia under King Croesus, who ruled from about 561 to 547 BC.[44]

All civilizations have valued gold and developed sophisticated techniques for working with it, including weaving gold thread into garments and casting works of art by the lost-wax technique. The lure of gold drove some to seek ways of making it and the alchemists expended much time fruitlessly searching for the 'Philosopher's Stone', the material that they hoped would transmute base metals like lead into gold.

In late Roman times, the worry was that alchemists might succeed. The emperor Diocletian, who ruled from AD 285 to 300, was afraid that his currency reforms might be undermined by the alchemists, so he ordered the destruction of all alchemical texts, thereby consigning to the flames the vast amount of chemical knowledge that had been acquired over the centuries in ancient Egypt. Even so, it did not stem the alchemists'

[44] He was defeated by the Persian King Cyrus II.

ARCHIMEDES, THE FIRST FORENSIC CHEMIST?

When King Hieron II of Syracuse commissioned a new gold crown, he was pleased with what the goldsmith produced but suspected that the metal had been debased with silver. He asked a local scientist, the famous Archimedes (257–212 BC), whether it would be possible to prove this without damaging the crown. The answer came as Archimedes was taking a bath and realized that the volume of water he displaced was equal to the volume of his body. *Eureka!*

The same would be true for the crown; knowing its volume and its weight, it would then be easy to compare it with the weight of a similar volume of pure gold. Today we would talk in terms of measuring the density, which is weight divided by volume, and say that gold had a density of 19.3 kilograms per litre, whereas that of silver is 10.5 kilograms per litre. Alloying gold with silver would reduce its density significantly. As the king suspected, his crown was not pure gold, and the hapless goldsmith's crime was revealed.

quest, although it explained why they worked under a cloud of secrecy thereafter. In the Middle Ages, alchemy became respectable once again, and by the seventeenth century quite eminent individuals, such as the great scientist, Isaac Newton, and his king, Charles II, were engaging in it.

Not all alchemists' time was wasted. In the thirteenth century they found that gold would dissolve in a mixture of concentrated nitric and hydrochloric acids called *aqua regia* ('king of waters'). When this solution was diluted with oil of rosemary, the gold stayed soluble; this potion, called *aurum potabile* (drinkable gold), was prescribed as a cure-all.

Even in the late nineteenth century alchemists were still at work, including August Strindberg (1849–1912), the great Swedish writer. He devoted a considerable amount of effort to the project and believed he had succeeded in 1894 when he sent samples of his 'gold' to the University of Berlin and published his method in a fringe journal, *L'Hyperchimie.* Like all before him he was deluded: later analysis of his samples showed them to be iron compounds, which can sometimes appear a deep gold colour.

Colloidal gold is precipitated when pure tin is added to gold that is dissolved in *aqua regia*. In about an hour a brilliant purple precipitate forms; this is known as Purple of Cassius and is named after Andreas Cassius of Potsdam, Germany, who first published the method of making it in 1685. The pigment was used to colour glass, to which it gave a ruby tint, or to decorate porcelain, such as that of Meissen and Sevres, on which it retained its original purple colour. It is still used to this day for colouring high-quality tableware.[45]

[45] Throughout the eighteenth and nineteenth centuries, chemists tried to explain Purple of Cassius in chemical terms but it was clearly only a form of gold. The answer came with the work of the Viennese chemist and Nobel Prizewinner, Richard Zsigmondy (1865–1929) who researched the nature of colloids and was thus able to explain how colloidal gold could produce different colours.

ECONOMIC ELEMENT

Gold is not the rarest of metals, nor the most expensive – platinum, rhodium, osmium and iridium are rarer and more costly. How much gold has been produced over thousands of years is impossible to estimate, but it has been calculated that in the past 500 years around 100 000 tonnes have been mined (but even that would only fill a cube with each side measuring 17 metres). Gold can sometimes be found as huge nuggets; the largest one on record weighed 112 kilograms (4000 ounces) and was found in Australia. The famous gold rushes of the nineteenth century were driven by the discovery of easily accessible alluvial gold.

Most gold (80%), however, is mined and comes from gravels and quartz veins or is associated with pyrites deposits. The main mining areas are South Africa, USA, Canada, Australia, Peru, and Russia, with world production being around 2500 tonnes per year, but reserves are estimated to be tens of thousands of tonnes. Even Western Europe annually produces 30 tonnes of gold, mainly as a by-product of zinc and copper refining, but some is mined in France and Scandinavia. Half a million people around the world are employed in gold production, with South Africa being the biggest producer. There, gold is mined at depths of 3000 metres.

Mined gold is extracted using cyanide, which forms a water-soluble compound that can then be reacted with zinc to precipitate the gold. In Brazil, alluvial gold is extracted with mercury, which dissolves the gold, and the mercury is then distilled off.

A new method of extracting gold employs bacteria to liberate the gold from sulfide ores, such as iron pyrites (FeS_2) and arsenopyrite (FeAsS), which bind the metal to such an extent that the cyanide process can only extract half of it at best, and only a quarter in many cases. The heat-loving bacterium *Sulfolobus acidocalderius* releases all the gold by digesting the rock to get at the sulfur which it uses as a source of energy. This digestion takes about 3 days at 50°C. Gold from the Youanmi deposit in Western Australia is extracted in this way.

Gold is used as bullion and in jewellery, glass and electronics. Traditionally the purity of gold has been measured in carats, with pure gold being 24 carats. The word derives from the word *keratia* used by the Byzantine goldsmiths.[46] The other grades of gold are 22 carat, which is 92%, 18 carat (75%), 14 carat (58%) and 9 carat (38%). This last kind tarnishes with age because of the high amount of alloying metals, and is little used. Most gold is alloyed with silver or copper, plus 0.2% zinc to harden it.

Jewellery consumes around 75% of all gold produced. Goldsmithing is still basically a craft industry although the making of some items, such as gold chains, is now done mechanically. Pforzheim in Germany is a major centre for jewellery, with 25 000 people there being employed in the trade.

Gold for jewellery can be given a range of hues depending on the metal with which it is alloyed. There is white gold (90% by weight of gold, 10% nickel), red gold (50% gold, 50% copper), blue gold (46% gold, 54% indium), purple gold (80% gold, 20% aluminium), green gold (73% gold, 27% silver) and even black gold (75% gold, 25% cobalt). Green gold was known to the Lydians as long ago as 860 BC, and they called it

[46] *Keratia* was derived from *keration*, the bean of the carob tree, which was taken as 4 grains in weight.

electrum; it occurred naturally as a native alloy of gold and silver, and was more of a pale lemon colour than green. Red gold was being produced in Peru in AD 1300. The most recent alloy for jewellery is known as '990 gold' and consists of 99% gold, 1% titanium (23.76% carat). This alloy is much stronger than pure gold and is ideal also for coins and medals.

Colloidal gold is added to glass to colour it red or purple, and metallic gold is applied as a thin film on the windows of large buildings to reflect the heat of the Sun's rays, as gold has high infrared reflective properties. For the same reason it is used in space exploration.

In gold electroplating, the metal object to be coated is placed in a bath of gold solution and acts as the cathode, with a platinum–titanium alloy as the anode. Application of direct current then deposits the gold on to the object. Electroplating can produce a film micrometres thick and is used in the electronics industry to coat electrical connectors and printed circuit boards, to protect their copper components and improve their solderability. Thin coatings of gold are used as lubricants in aerospace industries for environments where there is high temperature, high vacuum or high radiation, as in space applications. Gold plating is also applied to the gears of watches, to artificial limb joints and to cheap jewellery.

Gold has the highest malleability[47] of any element, and this manifests itself in the way the metal can be beaten into paper-thin films, which have been used in ornamental architecture for millennia, and are now also used to coat space satellites. A mere gram of gold – the size of a grain of rice – can be beaten to a gold film covering 1 square metre.

ENVIRONMENTAL ELEMENT

Gold in the environment	
Earth's crust	1 p.p.b.
	Gold is the 73rd most abundant element.
Soil	approx. 1 p.p.b., range 0.05–8 p.p.b.
Sea water	10 p.p.t.
Atmosphere	virtually nil

Gold occurs mainly as the metal, occasionally as crystals, but more generally as grains, sheets and flakes in other rocks. There are some gold ores, such as sylvanite, which is a combination of gold, silver and tellurium. The continuously erupting volcano, Mount Erebus, in Antarctica, spews forth gold dust and is unique among volcanoes in this respect.

Plants can absorb soluble gold, and those plants that metabolize cyanide absorb most gold, because the cyanide helps solubilize the gold. The horsetail plant was once claimed to have as much as 0.5 p.p.m. gold in its ash, but later tests showed that this was not so; the original analysts had mistaken arsenic for gold. Even so there are some

[47] A metal is described as malleable when it can be easily worked and beaten into a thin film. The word is derived from the Latin *malleus* meaning hammer.

plants that absorb gold from the soil in which they grow, such as the Douglas fir, which can have 20 p.p.b. (dry weight), and honeysuckle, with 15 p.p.b. (dry weight). The Indian mustard plant, *Brassica juncea*, is particularly good at doing this, especially if the soil in which it grows is treated with ammonium thiocyanate, which aids gold solubilization and uptake by this plant's roots. It seems unlikely that this plant could be grown as a gold-gathering crop, although technically it could be done at a profit.

CHEMICAL ELEMENT

Data file	
Chemical symbol	Au
Atomic number	79
Atomic weight	196.96655
Melting point	1064°C
Boiling point	2807°C
Density	19.3 kilograms per litre (19.3 grams per cubic centimetre)
Oxide	Au_2O_3 (unstable)

Gold is a soft metal and a member of group 11 of the periodic table of the elements. It has a characteristic shiny yellow colour as the solid metal but is red or purple as colloidal gold, which consists of tiny particles. Gold is unaffected by air, water, alkalis and acids, except *aqua regia* and selenic acid (H_2SeO_4). Supercritical water, which is water heated under pressure to a temperature about 374°C (see also p. 191), will also dissolve gold.

Gold has only one natural isotope, gold-197, which is not radioactive.

HIDDEN GOLD

It has been said that James Franck, the Nobel Prizewinner, dissolved his gold medal in *aqua regia* in 1943 to prevent it falling into the hands of the Nazis when he fled Denmark, and left it unobtrusively in his laboratory. When he returned in 1945 the bottle of acid was still there, so he reclaimed the gold and a new medal was struck.

The reason for gold's imperviousness to oxidation and corrosion lies in the remarkable grip it has over its outermost electron, the one that is available for chemical bonding. It does this by virtue of the large positive charge of the gold nucleus (+79) which exerts a gravitational effect on its electrons, causing them to move at a speed approaching the speed of light. This increases their mass, making them move closer to the nucleus.

ELEMENT OF SURPRISE

The gold in the sea is worth more than $1 500 000 000 000 000 ($1500 trillion). Many have been tantalized by the idea of claiming some of this bounty, and even some eminent chemists have been led astray by it, including the winner of the 1918 Nobel Prize for Chemistry, Fritz Haber (see nitrogen, p. 291). He thought that it would be possible to extract enough gold from the sea to repay the punitive reparations of 20 billion marks imposed on Germany by the Allies after World War I. Sadly he estimated the concentration in sea water to be 10 p.p.b. which is 1000 times higher than it really is, and his scheme came to grief. Indeed all who have attempted to reclaim gold from the sea have failed to do it economically. Even at the actual concentration of 10 p.p.t., though, the world's oceans contain more than 10 million tonnes of gold.

Hafnium

Pronounced haf-ni-uhm, the name is derived from *Hafnia*, the Latin name for Copenhagen.

French, *hafnium*; German, *Hafnium*; Italian, *afnio*; Spanish, *hafnio*; Portuguese, *háfnio*.

HUMAN ELEMENT

Hafnium in the human body	
Total amount in body	not known, but small

Hafnium has no known biological role, and its salts generally have low toxicity. The element is poorly absorbed by the body and poisoning by hafnium compounds is unheard of.

ELEMENT OF HISTORY

In 1911, Georges Urbain (1872–1938) reported the discovery of a 'missing' element, the one that came below zirconium in group 4 of the periodic table. He called it celtium. However, further research, done in collaboration with a young Englishman, Henry Moseley, eventually showed he was mistaken.

Moseley, born at Weymouth, England, in 1887, was in his early twenties and was already regarded as a scientific genius, following his discovery in 1913 of the fundamental property of the atomic number.[48] This he deduced from the X-ray spectra of elements and as a result he was able to announce with confidence that elements with atomic numbers 43, 61 and 75 still remained to be discovered. The same was true of element 72, but Moseley was originally misled into thinking this was the recently announced celtium. (Indeed he never lived to see the discovery of the missing element because he was killed by a bullet through the head during the ill-fated Dardanelles campaign in World War I. His death, at 27, is regarded as one of the great tragedies of science.)

With celtium discredited, the search was once again on for the missing element. It was finally tracked down in 1923 by the Hungarian chemist George Charles de Hevesy[49] (1889–1966) working with Dirk Coster (1889–1950) at the University of Copenhagen.

[48] This is the number of protons in an atom's nucleus, and it determines what element it is.

[49] De Hevesy won the Nobel Prize for Chemistry in 1943 for his work on isotopes.

The element turned up in a zirconium ore, perhaps not surpisingly on account of their chemical similarity. When de Hevesey and Coster examined a sample of Norwegian zircon, using Moseley's method of X-ray analysis, they observed lines that showed that another element was present. This proved to be hafnium, but it also proved very difficult to separate from the zirconium. Although neither of the discoverers was Danish they named the new element after the city in which it had been found.

Historic specimens of zirconium minerals from museums were examined by Hevesy; some were found to contain as much as 5% hafnium. That it had remained undiscovered for so long is not so surprising because these two elements share many features in common, for reasons explained in the 'Chemical element' section below. These findings necessitated the revision of the atomic weight of zirconium once samples free of hafnium had been produced.

The first pure sample of hafnium was made in 1925 by decomposing hafnium tetraiodide (HfI_4) over a hot tungsten wire.

ECONOMIC ELEMENT

Hafnium ores are rare, but two are known: hafnon, which is hafnium silicate ($HfSiO_4$), and alvite, which is a mixed hafnium–thorium–zirconium silicate, $(Hf,Th,Zr)SiO_4$. Hafnium is obtained commercially as a by-product of zirconium refining because its ores contain around 5% of hafnium. Industrial production of hafnium metal is not much more than 50 tonnes a year. The ore that is extracted for its hafnium content is treated with chlorine gas to form hafnium tetrachloride ($HfCl_4$) which is then reduced by magnesium or sodium to hafnium metal. Known reserves of ore are not recorded, but can be estimated from those of zirconium.

Hafnium, and its alloys, are used for control rods in nuclear reactors and nuclear submarines because hafnium is excellent at absorbing neutrons (being more than 500 times more efficient at this than zirconium) and it has a very high melting point and is corrosion resistant. However, it is expensive: the 50 or so control rods needed for a large nuclear reactor can cost as much as $1 million.

Hafnium is used in high-temperature alloys and ceramics. Hafnium carbide and nitride are some of the most refractory materials known, in other words they will not melt except under the most extreme temperatures; in the case of hafnium nitride, for example, the melting point is 3310°C.

ENVIRONMENTAL ELEMENT

Hafnium in the environment	
Earth's crust	3.3 p.p.m.
	Hafnium is the 45th most abundant element.
Soils	approx. 5 p.p.m., with a range of 2–20 p.p.m.
Sea water	0.007 p.p.t.
Atmosphere	virtually nil

Hafnium was the next to the last of the non-radioactive elements to be discovered – rhenium came 2 years later – but whereas rhenium is extremely rare, hafnium is relatively abundant, being more abundant than tin. It poses no environmental threat to plants or animals. Plants take up small amounts of hafnium from the soil in which they grow; levels of 0.01–0.4 p.p.m. (dry weight) have been recorded.

CHEMICAL ELEMENT

Data file	
Chemical symbol	Hf
Atomic number	72
Atomic weight	178.49
Melting point	2230°C
Boiling point	5200°C
Density	13.3 kilograms per litre (13.3 grams per cubic centimetre)
Oxide	HfO_2

Hafnium is a lustrous, silvery, ductile metal and a member of group 4 of the periodic table of the elements. It resists corrosion due to formation of a tough, impenetrable oxide film on its surface. However, powdered hafnium will burn in air. The metal is unaffected by alkalis and acids, except hydrofluoric acid.

Hafnium is difficult to separate from its group 4 partner, zirconium, because the two elements have atoms that are the same size. Normally the atoms get larger as you go down a group of the periodic table, but for hafnium and zirconium this general rule does not hold. The extra electrons in hafnium occupy an orbit deep inside the atom. With atoms of the two metals being the same size, and sharing the same chemistry, they are almost indistinguishable.

Naturally occurring hafnium consists of six isotopes, of which hafnium-180 accounts for 35% of the total, with hafnium-178 next at 27.5%, then hafnium-177 (18.5%), hafnium-179 (13.5%), hafnium-176 (5%) and finally hafnium-174 (only 0.16%). This last isotope is a weak α-emitter with a half-life of 2 million billion years (2×10^{15} years) yet it is this isotope that gives hafnium its high neutron-capturing ability.

ELEMENT OF SURPRISE

In the late nineteenth century, an American, Edgar Smith, was investigating monazite sand, the mineral that contains the rare-earth elements (see p. 222). Other metals are also present in this ore and one of these was zirconium; Smith removed this by converting it to the sulfate. This was an unwanted by-product but Smith collected it nevertheless, telling his co-workers that he was convinced there was another element concealed within it, and that he would find this when he had the leisure to do it. Smith was right in his belief – there was hafnium present – but he was wrong in thinking he would be the one to discover it.

Hassium

This is element atomic number 108 and along with other elements above 100 it is dealt with under the transfermium elements on p. 457.

Helium

Pronounced hee-lee-uhm, the name comes from the Greek *helios*, meaning the Sun.[50] Helium has the distinction of being discovered on the Sun before it was found on Earth.

French, *hélium*; German, *Helium*; Italian, *elio*; Spanish, *helio*; Portuguese, *hélio*.

COSMIC ELEMENT

Helium was one of the elements produced during the Big Bang, along with hydrogen and lithium, and its presence throughout the Universe supports the theory that such an event occurred. Helium is the second most abundant element in the Universe, making up 23% of its elemental matter (hydrogen is the most abundant, making up 76% of the elemental mass of the Universe), although in number terms there are 8 times as many hydrogen atoms as helium atoms. Together these two elements account for 99% of the *observed* matter in the Universe.[51]

Stars like the Sun produce helium by the fusion of hydrogen nuclei, and in so doing release massive amounts of energy. The fusion of two helium nuclei, however, to form beryllium, is not energetically favourable, but the fusion of three heliums to form carbon does release energy.[52]

[50] Unlike the other noble gases – neon, argon, krypton, xenon and radon – helium's name does not end in '-on'. Although it has been suggested that it should be called 'helion', the name has not been taken up, and the element is still named as if it were a metal, ending in '-ium.'

[51] Recent cosmological research suggests that only 4% of the Universe is elemental matter, 76% is energy (as a quantum field), and 20% is dark matter.

[52] A helium nucleus is also known as an α-particle and this process of forming carbon is referred to as the triple-α reaction.

HUMAN ELEMENT

Helium in the human body	
Blood	trace
Total amount in body	very small

Helium has no biological role. Although it is a harmless gas, it could asphyxiate if it were to exclude oxygen from the lungs.

MEDICAL ELEMENT

Although helium has no medicinal benefits, it is needed for medical diagnosis and about a third of all helium that is gathered is used this way. Liquid helium, at a temperature of −269°C, is required to cool the magnets and so produce the intense magnetic fields for magnetic resonance imaging (MRI). When a human is placed in such a magnetic field and exposed to radio waves, these interact with the hydrogen atoms of water and other molecules in the body and an image is produced, enabling different organs to be identified and possible cancers (for example) to be diagnosed.

ELEMENT OF HISTORY

The technique of atomic spectroscopy (see p. 4) led to the discovery of several elements in the nineteenth century – see e.g. caesium (p. 80), rubidium (p. 364) and thallium (p. 433). It was even possible to identify some elements such as sodium in the light from the Sun, but such was the intensity of this light that faint lines from other elements could not be seen.

However, at the moment of a total eclipse, most of the Sun's light is blocked out, so Pierre J. C. Janssen (1824–1907) reasoned that analysis of the spectrum of the corona surrounding the eclipse might well reveal more information. He travelled to India to study a total eclipse forecast for 18 August 1868. Luckily for him the sky was clear, so he could make his measurements. When he observed a line in the corona spectrum at a wavelength of 587.49 nanometres, he knew that a hitherto unknown element was present.

On 20 October of that same year, the British astronomer Norman Lockyer[53] (1836–1920) and the chemist Edward Frankland (1825–1899) made their observations of the Sun through the smoky skies of an overcast London. They confirmed the same line in the spectrum and agreed that there was a new element which they postulated could exist only on the Sun. Lockyer assumed it would be a metal and so he called it helium. His presumption to have discovered a new element on the Sun, and having the effrontery to name it, were ridiculed by some but in the end the last laugh was to be his.

Unbeknown to scientists at the time, helium existed on Earth but only in tiny amounts in the atmosphere, although more abundantly in uranium minerals. When the US geologist, William Hillebrand dissolved the mineral uraninite (UO_2) in acid, he

[53] Lockyer was a civil servant from Wimbledon, London, who wrote the first book setting down the rules of golf as played at St Andrews in Scotland. He was also the founder of the Science Museum in London.

observed that bubbles of gas escaped from it. He could not identify the gas and did not pursue his observations further. However, when Per Teodor Cleve and Nils Abraham Langer at Uppsala, Sweden, in 1895, made the same observations from a new mineral (which at the time was named cleveite after Cleve, although it later turned out to be uraninite), they were able to collect enough of the gas positively to identify it as the element helium and to measure its atomic weight accurately. Quite independently, that same year Sir William Ramsay and Morris Travers at University College, London, also collected a sample of the new gas which they too identified as helium.

ECONOMIC ELEMENT

Helium is present in some minerals, but the chief source is natural gas which may contain several per cent of helium. Although it is present in the atmosphere, it is currently uneconomical to extract it from air.

Helium is needed for low-temperature instruments, such as the superconducting magnets mentioned above, for providing inert atmospheres for industrial uses such as welding air-sensitive metals, for deep-sea diving, for rocket launches (the valves that release the hydrogen and oxygen for the rockets are controlled by compressed helium), for weather balloons and airships, and for gas lasers. The helium–neon gas laser is used in supermarket checkouts to scan bar codes.

Annual production of helium is in excess of 100 million cubic metres, of which 90 million cubic metres are produced in the USA. Some is also produced in Poland.

Helium is used in balloons and airships and has about the same lifting power as hydrogen gas. A cubic metre of helium gas can lift a load of 1 kilogram. To lift a tonne would require a thousand cubic metres, or a balloon of around 12 metres in diameter.

Helium is used in deep-sea diving. For every 10 metres of depth, the pressure increases by 1 atmosphere, which in turn increases the solubility of air, and particularly nitrogen, in the blood. Nitrogen has a narcotic effect on the brain, whereas helium has no such effect and is easier to breathe at high pressures. The deeper a diver works, the higher the helium-to-oxygen ratio needs to be, to prevent too much oxygen from entering the brain, as this can poison it. At depths below 600 metres even helium cannot always prevent what is known as high-pressure nervous syndrome and then a mixture of hydrogen, helium and oxygen, known as 'hydreliox', is used.

Breathing helium dramatically raises the pitch of the voice because sound travels three times faster in helium than in air, and while the wavelength of the sounds that the human voice box generates does not change, the frequency changes to compensate.

ENVIRONMENTAL ELEMENT

Helium in the environment	
Earth's crust	8 p.p.b.
	Helium is the 71st most abundant element.
Sea water	4 p.p.t.
Atmosphere	5 p.p.m. by volume

The atmosphere contains helium but, unlike most other gases in the atmosphere, helium can escape into space. Even though there are hundreds of million tonnes of the gas, it is not worth extracting from this source at present. Helium is less soluble in water than any other gas, so there is very little in the sea.

Helium is continuously being formed from those radioactive elements in the Earth's crust which decay by emitting an α-particle, a process known as α-decay. An α-particle consists of two protons and two neutrons and as such is the nucleus of a helium-4 atom. An α-particle does not have to travel far before it collects the two electrons it needs to form a helium atom.

CHEMICAL ELEMENT

Data file	
Chemical symbol	He
Atomic number	2
Atomic weight	4.002602
Melting point	–272°C (0.95K) at 26 atmospheres
Boiling point	–269°C (4.2K)
Density of the gas	0.18 grams per litre (0.18 kilograms per cubic metre)
Oxide	none known, nor possible

Helium is one of the so-called noble gases and is assigned to group 18 of the periodic table of the elements. It is a colourless, odourless gas which is unreactive chemically. It is virtually inert towards all other elements and chemicals and so only exists as individual atoms. However, some species have been made, such as VHe^{3+} and $HePtHe^{2+}$ although they are highly unstable.

Helium has two stable isotopes, helium-4 and helium-3, although there is only one atom in a million of the lighter isotope. Neither is radioactive. There are also six other, radioactive, isotopes with mass numbers ranging from 3 to 10.

When the temperature of liquid helium falls below 4K (–269°C), it is at first simply a colourless liquid, albeit extremely cold, but when the temperature drops below 2K something rather strange happens to it. It becomes a liquid with bizarre properties and is known as helium-II. Not only does it expand on cooling, but it is a million times more heat conducting than normal liquid helium and its viscosity is zero, in other words nothing can slow its flow. It will escape from any container that is not totally sealed; it does this by flowing up the sides and out of the top, apparently defying even gravity.

Helium can only be solidified at just above absolute zero (–273°C) by applying a pressure of about 26 atmospheres.

ELEMENT OF SURPRISE

In May 1903 a company drilling for oil near the town of Dexter, in the USA, struck a gas geyser. The pressure of the escaping gas caused a loud howling that could be heard for miles. Before the well was capped, it was decided to celebrate the company's good fortune by igniting what they assumed was natural gas, which was escaping at the rate of a quarter of a million cubic metres a day. Crowds assembled to witness the promised pillar of flame which was to be ignited by a burning bale of hay pushed towards a pipe that had been set up to draw off some of the escaping gas. Unexpectedly it not only failed to ignite the gas, but the burning bale was extinguished.

The Kansas state geologist, Professor Erasmus Howard, collected samples of the escaping gas and took them back to the University of Kansas at Lawrence for analysis. Helped by colleagues Hamily Cady and David McFarland, he found that the gas was 72% by volume nitrogen, 15% methane (not enough to make the gas combustible), 1% hydrogen and 12% of something else which refused to be identified.

To remove all the other components, Howard and his colleagues used activated charcoal, which they knew would absorb all gases except hydrogen, helium and neon, when cooled to liquid air temperatures (–190°C). They then tested the gas that was left spectroscopically and found the yellow line of helium, the gas that had only been reported a few years previously and which was still regarded as a rarity. Far from being rare it was present in vast quantities in the gas fields beneath the Great Plains of North America.

In 1925 the US Government voted to set up a strategic National Helium Reserve at Amarillo in Texas, thinking that the helium would be needed for military blimps and commercial airships of the future.[54] In the 1950s the National Helium Reserve was even expanded when the importance of helium to the space race was apparent – it was required to keep cool the liquid hydrogen and oxygen that powered the Apollo space ships. The helium was collected from the gas well and piped to an underground storage facility, an exhausted gas well. By 1995, a billion cubic metres of the gas had been collected and the project had debts of $1.4 billion. Congress voted to end the scheme in 1996 with the gas being sold off to repay the loans.

[54] In the 1920s and 1930s the US Navy explored rigid-design airships, one of which was a zeppelin obtained as war reparation from Germany, and three of which were built in the USA. The three US-built rigid airships crashed and in the late 1930s the Navy cancelled the programme, although non-rigid airships (blimps) continued in service till 1962 and were used successfully in World War II for anti-submarine reconnaissance. [Walter Saxon.]

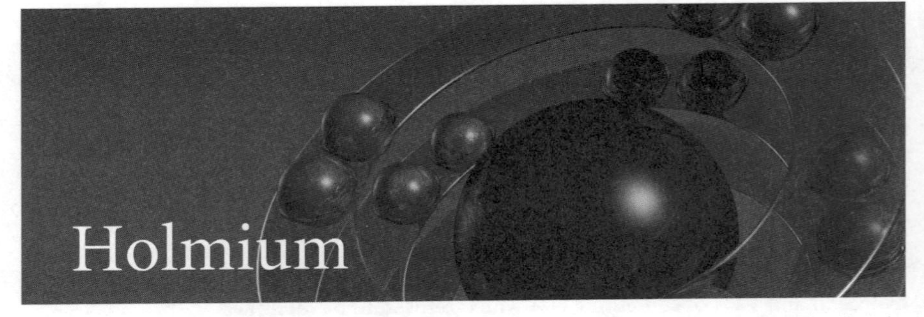

Holmium

Pronounced hol-mi-uhm, the name is derived from *Holmia*, the Latin name for Stockholm.

French, *holmium*; German, *Holmium*; Italian, *olmio*; Spanish, *holmio*; Portuguese, *hólmio*.

Holmium is one of 15, chemically similar, elements referred to as the rare-earth elements, which extend from the element with atomic number 57 (lanthanum) to the one with atomic number 71 (lutetium). The term rare-earth elements is a misnomer because some are not rare at all; they are more correctly called the lanthanides, although strictly speaking this excludes lutetium. The minerals from which they are extracted, and the properties and uses they have in common, are discussed under lanthanides, on p. 219.

HUMAN ELEMENT

Holmium in the human body	
Total amount in body	not known, but small

Holmium has no biological role, but it has been noted that holmium salts stimulate metabolism. It is difficult to identify exactly how much of the various lanthanides are present in the human body, but they are present – see lanthanum (p. 215). However, holmium is likely to be among the least abundant. No one has monitored diet for holmium content, so it is difficult to judge how much we take in, but it is probably only a milligram or so per year. Holmium is not readily taken up by plant roots, so very little gets into the human food chain; vegetables that have been analysed for holmium have been found to have less than 0.1 p.p.b. (dry weight).

ELEMENT OF HISTORY

Holmium was discovered in 1878 by Marc Delafontaine and Louis Soret, at Geneva, Switzerland, and independently by Per Teodor Cleve (1840–1905), who was Professor of Chemistry at Uppsala, Sweden.

The discovery can be traced back to the metal yttrium which, unknown at the time when it was discovered, was contaminated with traces of other lanthanide elements. Later erbium and terbium were separated from it, and then erbium was found to contain ytterbium. When, in 1874, Cleve looked more closely at the residual erbium, left behind after the ytterbium had been removed, he realized it must contain yet other elements because he found that its atomic weight varied depending on how it was made. He separated holmium from erbium in 1878.

However, Delafontaine and Soret had announced a few months earlier that erbium oxide must contain another element, which they referred to as 'element X', because its atomic spectrum gave bands that could not be explained. Eventually, when the spectrum of 'element X' was compared with that of holmium, they were found to be identical.

Even so, we cannot be certain that Cleve had produced a *pure* sample of the new element because within a few years yet another rare-earth, dysprosium, was to be extracted from holmium.

ECONOMIC ELEMENT

Holmium is found in minerals such as monazite and bastnäsite, although it is only a minor component of such ores; for example, holmium is present to the extent of 0.5% in monazite, and it is extracted from those ores that are processed for the yttrium they contain. Main mining areas are China, USA, Brazil, India, Sri Lanka and Australia, with reserves of holmium estimated to be around 400 000 tonnes.

World production of holmium metal itself is only around 10 tonnes per year. The metal is made by heating holmium chloride or holmium fluoride with calcium. It tarnishes easily and is not used as such, but its alloys are used as a magnetic flux concentrator for high magnetic fields because holmium has the highest magnetic strength of any element. The pole pieces of the most powerful magnets are made of holmium.

Holmium is also used in nuclear reactors as a 'burnable poison', i.e. one that burns up while keeping a chain reaction from running out of control.

ENVIRONMENTAL ELEMENT

Holmium in the environment	
Earth's crust	1.4 p.p.m.
	Holmium is the 56th most abundant element.
Soils	1 p.p.m.
Sea water	0.4 p.p.t.
Atmosphere	virtually nil

Holmium is one of the rarer of the rare-earth elements but is, nevertheless, 20 times more abundant than silver. It poses no environmental threat to plants or animals.

CHEMICAL ELEMENT

Data file	
Chemical symbol	Ho
Atomic number	67
Atomic weight	164.93032
Melting point	1474°C
Boiling point	2695°C
Density	8.8 kilograms per litre (8.8 grams per cubic centimetre)
Oxide	Ho_2O_3

Holmium is a bright, soft, silvery metal and one of the lanthanide group of the periodic table of the elements. It is slowly attacked by oxygen and water, and dissolves in acids.

Naturally occurring holmium consists entirely of a single isotope, holmium-165. It is not radioactive.

ELEMENT OF SURPRISE

Holmium-containing lasers operate at a wavelength of 2.08 micrometres, which is eye-safe, and they have been approved for various types of surgery, combining as they do the ability to be transmitted down optical fibres with good tissue-cutting power.

Pronounced hy-dro-jen, the name is derived from the Greek words *hydro* and *genes*, meaning water-forming.

French, *hydrogène*; German, *Wasserstoff*; Italian, *idrogeno*; Spanish, *hidrógeno*; Portuguese, *hidrogênio*.

Hydrogen is the number one element in more ways than one: it was the first element of Creation; it is the first element of the periodic table; it is the most abundant element in the Universe; it is the element that fuels the Sun and stars; it is the lightest of all gases; and it bonds to other elements in its own unique fashion. Much of the role of hydrogen on Earth is linked to water and acids; a separate section on these follows on p. 189.

COSMIC ELEMENT

Hydrogen accounts for 88% of the atoms in the Universe, and was one of the three elements, along with helium and lithium, that were produced in the Big Bang. It continues to be the fuel of the stars. At the centre of a star like the Sun, the temperature is around 13 million degrees and the density about 200 kilograms per litre. Under such conditions hydrogen will begin to 'burn', in other words protons fuse to form helium nuclei releasing vast amounts of energy and radiation as they do so. This radiation takes about a million years to reach the Sun's surface. Our star converts around 600 million tonnes of hydrogen *per second* into helium, and during this process around 5 million tonnes of matter are converted into energy.[55]

The fusing of two hydrogen nuclei first forms deuterium, also known as hydrogen-2, as one proton converts to a neutron and releases a positron (the positive equivalent of an electron). The deuterium then reacts with another hydrogen to form helium-3 and when two of these nuclei collide they form normal helium-4 while ejecting two protons to go through the process again. Because the first step in this process is slow, stars like the Sun have enough fuel to burn for 10 billion years.

On Earth, hydrogen is light enough to reach the upper atmosphere and thence escape into space. Only a planet the size of Jupiter has the gravity to prevent this from happening. A probe in 1995 showed that this planet was composed of 99.8% hydrogen and helium. Indeed such is the pressure at the centre of Jupiter that it is believed that

[55] In accord with Einstein's equation relating energy (E) to mass (m): $E = mc^2$ (c is the speed of light).

the hydrogen there is converted to a metallic form of the element. That such a form of hydrogen exists was proved in 1996 when a layer of liquid hydrogen 0.5 millimetres thick was subjected to a pressure of 2 million atmospheres. Under such conditions the element became able to conduct electricity, which is taken as evidence that it had become metallic.

HUMAN ELEMENT

Hydrogen in the human body	
Blood	whole blood, which is 85% water, is about 9.4% hydrogen
Bone	about 5.2%
Tissue	about 9.3%
Total amount in body	about 7 kilograms

Because water makes up around 65% of the body's weight, and because hydrogen constitutes 11% of the weight of water, then this is going to appear as a key component, but one that varies considerably. Hydrogen is an essential element for life because it is a constituent element of DNA and as such is part of the genetic code. It is a component of almost every molecule in every living cell. The human body requires around 2½ litres of water a day to keep itself healthy: half of this is taken in the form of drinks and the other half as food.

MEDICAL ELEMENT

Hydrogen gas (H_2) is non-toxic and so is its chief compound, water (H_2O), yet both can kill if they exclude air from the lungs. Water can also be fatal if too much is given to people who are badly dehydrated, such as those found dying of thirst. A sudden excess of water upsets the balance of sodium and potassium ions in the heart muscle, resulting in a heart attack.

Blood that is very acidic can pose a threat to health although this is usually a symptom of other conditions. The exchange of sodium ions (Na^+) and hydrogen ions (H^+) in heart tissue plays an important role in coronary heart disease. However, the stomach can tolerate the quite high levels of acid necessary for digestion, although generally such concentrated acids are quite corrosive.

ELEMENT OF HISTORY

Hydrogen gas was known long before its nature was realized. In 1671 Robert Boyle dissolved iron filings in dilute hydrochloric acid and reported that the 'fumes' given off were highly flammable and that when these burned they gave off a little light and a lot of heat 'with more strength than one would easily suspect'.

The person credited with hydrogen's discovery, in 1766, was the eccentric chemist, Henry Cavendish (1731–1810); one of the richest men in England, he lived in Soho in central London, but spent most of his time in his laboratory at his other home in the suburb of Clapham, and it was there that he collected the bubbles that he saw rising from the chemical reaction of iron filings with dilute sulfuric acid. This gas was different from the others that he knew: for example, it was highly flammable and very light. He was also the first person to demonstrate that when hydrogen burned it formed water; this he did in 1781, thereby disproving the Aristotelian theory of four elements of which water was one. As a consequence of Cavendish's observation, the gas was named hydrogen by the great French chemist, Antoine Lavoisier.

Until the twentieth century, hydrogen had little commercial application except in theatres where it was used to produce limelight (see calcium, p. 89) and in airships. In 1898 James Dewar (1842–1923) invented the vacuum flask, and with it he was able to liquefy hydrogen gas for the first time; he produced solid hydrogen the following year.

In December 1931, Harold Urey and colleagues at Columbia University in the USA detected another form of hydrogen in the residue of a large amount of liquid hydrogen which they had allowed to evaporate. This heavier form has a neutron as well as a proton in its nucleus and so has twice the mass of normal hydrogen. They called it deuterium, chemical symbol D. The following year they showed that when water is decomposed to oxygen and hydrogen by electrolysis, deuterium-containing molecules were left behind; this became known as heavy water (D_2O).

Hydrogen provided humans with their first reliable form of air transport. This began in 1783 when the French scientist Jacques Charles invented the hydrogen-filled balloon and demonstrated it a month after the first human-carrying hot-air balloon was successful. Hydrogen balloons were used for sight-seeing flights throughout the nineteenth century.

The first hydrogen-lifted airship was invented in 1852 by Henri Giffard. In the 1890s the German count Ferdinand von Zeppelin promoted the idea of rigid airships which became known as zeppelins; the first of these made its flight in 1900. They were filled with hydrogen gas. They provided the first regularly scheduled air transport service from 1910 with flights to many major cities in Europe. By the outbreak of World War I, in August 1914, they had carried more than 35 000 passengers without a serious incident, demonstrating how safe hydrogen-filled airships could be.

After the war, in 1919, a British airship, the R-34, made the first non-stop transatlantic crossing. In the 1920s and 1930s, zeppelins returned to passenger service. It seemed as if they would be made completely safe when helium from gas wells (see p. 177) was discovered. However, the US Government refused to sell the gas for this purpose, seeing it as a strategic resource.

Despite this, commercial flights continued without incident until the spectacular crash of the Hindenburg on 6 May 1937 as it came into land at an airfield in New Jersey, killing 35 and injuring dozens of others. This was filmed, and broadcast, as it happened. Although a leak of hydrogen was assumed to be responsible for the crash, it was later shown that the aluminized fabric coating was the more likely cause of the fire, and that this has been ignited by static electricity.

ELEMENT OF WAR

Hydrogen balloons were used for military applications, as observation and reconnaissance platforms, soon after they were demonstrated to be feasible. In World War I the Germans used zeppelins to bomb London, often with impunity because they could fly much higher than fighter aircraft, although some were brought down when incendiary bullets ignited them. In that same war the British used hydrogen-filled airships for anti-submarine reconnaissance.

Although it has never been used in war, the most fearful weapon of all is the hydrogen bomb (a thermonuclear explosion). At its heart is an atomic bomb which, when detonated, releases enough neutrons and energy to create a high enough temperature for hydrogen fusion to occur among the surrounding mass of lithium deuteride (LiD) which also contains some tritium (T), the third isotope of hydrogen.[56] When a lithium atom absorbs a neutron, more tritium is formed (in the reaction $Li + n \rightarrow He + T$); this, together with the original tritium, then fuses with the deuterium to release a vast fireball of explosive energy, equivalent to millions of tonnes of TNT, and large enough to obliterate a city of a million people.

Current research is aimed at developing a thermonuclear weapon that would not need a fission explosion (atomic bomb) to set it off. The technique being explored is called inertial confinement fusion (ICF); this uses a powerful focused laser beam to condense hydrogen sufficiently to a density and temperature that can ignite an explosive fusion reaction.

ECONOMIC ELEMENT

World-wide industrial production of hydrogen gas is 30 million tonnes a year (350 billion cubic metres). Known reserves of hydrogen are almost limitless because it can be made from water – and the oceans hold 140 million billion tonnes of the element. (The estimated 130 million tonnes of hydrogen gas in the Earth's atmosphere is too dilute to be reclaimed.) Most hydrogen gas is made in countries with large chemical industries and there is a network of pipelines for moving hydrogen around in North America and Europe.

Currently most hydrogen is produced from natural gas (which is mainly methane, CH_4) by heating it with steam at around 1000°C. It is also produced as a by-product during the manufacture of sodium hydroxide (caustic soda). In theory it could be made by the electrolysis of water, and this has been suggested as a way of using the surplus electricity of nuclear power stations, or where there is plenty of hydroelectricity. It is still not an economical process despite improvements in efficiency such as the electrolysis of steam inside porous electrodes of zirconium oxide.

The action of steam on red-hot coal gives a mixture of hydrogen and carbon monoxide gases ($C + H_2O \rightarrow H_2 + CO$). This so-called synthesis gas makes a good fuel and can be converted into the useful chemical methanol. Alternatively, ethanol, which is

[56] Among the elements, only the isotopes of hydrogen have been given separate names and separate symbols: hydrogen-2 is more commonly called deuterium (D) and hydrogen-3 is tritium (T).

produced on a large scale as a renewable resource, can also be turned into hydrogen and methane by reacting it with water in the presence of a rhodium catalyst, a process that even releases energy.

Sunlight can split water into its component gases, oxygen and hydrogen, a reaction that is catalysed by titanium dioxide doped with platinum metal, but the amount of hydrogen given off is tiny.

There are also anaerobic bacteria that can ferment glucose to acetic acid and as they do they release hydrogen gas. (Hydrogen produced by some bacteria is used by others, known as methanogens, to convert carbon dioxide to methane.)

Hydrogen is needed on a vast scale to make ammonia for fertilizers – two-thirds of the world's hydrogen is used for this. Other uses include the manufacture of cyclohexane for polymers, methanol for fuel, as well as many other intermediates. Some hydrogen is required to turn liquid vegetable oils into solid fats, such as margarine. A little hydrogen is used for special high-temperature welding.

The hydrogen economy

Some scientists see hydrogen gas as the clean fuel of the future – generated from water and returning to water when it is burnt, with almost no pollution. The energy given out by burning hydrogen exceeds that of conventional fuels if we compare them on a weight-for-weight basis: hydrogen releases 25 kilocalories (= 104 kilojoules) per gram compared with 10 kilocalories (= 42 kilojoules) per gram for natural methane gas. However, hydrogen is less dense than methane and it requires three times the volume of gas to provide the same amount of heat. Yet, moving hydrogen through pipelines is a much more efficient way of transporting energy than electric power lines, and hydrogen can easily be stored in subterranean caverns. The difficulty comes with using it for transport when it needs either to be pressurized, stored as a liquid, or absorbed into a special kind of alloy. Such alloys are made of titanium and iron, or magnesium and nickel, and they can soak up hydrogen to such an extent that it is as if they were storing it as liquid hydrogen. They will release the gas as required, but their major drawback is that they tend to lose their efficacy over time.

A tonne of gas occupies a volume of 11 million litres but only 14 000 litres as a liquid, which is how large quantities are transported and stored. The US space programme requires massive amounts of liquid hydrogen and uses road and rail tankers carrying 75 000 litres at a time. One storage tank at Cape Canaveral holds over 3 million litres. A space shuttle requires 500 000 litres of liquid oxygen and 1.5 million litres of liquid hydrogen for take-off, but the dangers are all too apparent as was shown when the *Challenger* space shuttle exploded in a giant fireball on 26 January 1986 during its 25th launch.

Buses running on hydrogen have been demonstrated in some European cities such as Berlin and experimental cars run on hydrogen have been produced by German and Japanese companies. In 1991 such a car travelled 300 kilometres on a 100 litre tankful of liquid hydrogen, held in an insulated aluminium container. Engineers at the Mazda car company have built a hydrogen car which stores its fuel in metal alloys.

There are two ways in which hydrogen gas can be used to drive a car. It can simply be pumped into the engine along with air and sparked, as in a normal engine burning petrol vapour, or it can be turned into electric current in a fuel cell and used to drive an electric motor. Hydrogen cannot compete with hydrocarbon fuels at the moment although it may one day be used to power cars, trains and even aircraft, just as it is now used to launch space vehicles.

Hydrogen fusion

The world could have unlimited power if physicists and engineers could harness the nuclear reactions which fuel the Sun, in other words to turn hydrogen into helium. When a mere 4 grams of hydrogen combines to form a little under 4 grams of helium, 2.5 billion kilojoules of energy is released, equivalent to burning 60 tonnes of oil. But making hydrogen combine to form helium in a controlled way has proved elusive because it requires temperatures of millions of degrees to start the reaction.

ENVIRONMENTAL ELEMENT

Hydrogen in the environment	
Earth's crust	1500 p.p.m. (0.15%)
	Hydrogen is the 10th most abundant element.
Soils	major constituent, as water
Sea water	major constituent, as water, but also contains a little dissolved H_2 gas
Atmosphere	0.5 p.p.m. as H_2, variable proportion as water vapour

The hydrogen of the Earth's crust is chiefly present as water. Hydrogen is also a major component of biomass, constituting about 14% by weight.

CHEMICAL ELEMENT

Data file	
Chemical symbol	H
Atomic number	1
Atomic weight	1.00794
Melting point	−259°C
Boiling point	−253°C
Density	0.090 grams per litre (gas)
Oxides	H_2O (water) and H_2O_2 (hydrogen peroxide)

Molecular hydrogen (H_2) is a colourless, odourless gas, almost insoluble in water. It burns in air, and mixtures with air are explosive. Hydrogen gas can exist in two forms: *ortho* hydrogen and *para* hydrogen, which are distinguished by the spins of the nucleus of the atoms. In *ortho* hydrogen both of the nuclei are spinning in the same direction, while in *para* hydrogen they spin in opposite directions. At room temperature hydrogen is 75% *ortho* and 25% *para*.

There are three naturally occurring isotopes of hydrogen: hydrogen-1, which comprises over 99.9%; hydrogen-2 (deuterium), which accounts for 0.0156%; and hydrogen-3 (tritium), which is present only in trace amounts and is produced naturally in the upper atmosphere by cosmic rays. This last isotope is radioactive with a half-life of 12.5 years.

Hydrogen bonds

Hydrogen bonds influence the structure of all living things. The double helix of DNA is held together by hydrogen bonds. The same is true of the structure of all the protein molecules of our body, from those in the largest organs, such as our muscles and our brain, to the smallest components of cells, such as enzymes. Hydrogen bonds are generally formed between molecules containing oxygen and nitrogen atoms, and occur when a hydrogen attached to one of these atoms in a molecule is attracted to an oxygen or nitrogen atom in another molecule. The two molecules are drawn close together by means of a so-called 'hydrogen bond'.

In living things the most important type of hydrogen bond is between nitrogen and oxygen; these are the ones that hold together the strands of DNA. The genetic code of DNA is represented by the letters A, C, G and T, which stand for the bases adenine, cytosine, guanine and thymine. A and T pair up by forming two hydrogen bonds, while C and G do so by forming three hydrogen bonds. Because these combinations are so well matched, when a strand of DNA separates in order to form a copy of itself, it does so exactly, because each A, C, G and T uses its hydrogen bonds to ensure a match with the right partner.

Although hydrogen bonds are weak compared with normal bonds, having less than a quarter of the energy of these, there are many of them and quantity makes up for quality.

Acids

Some hydrogen-containing molecules can split off one or more of their hydrogen atoms to give positively charged hydrogen ions, H^+. When such molecules are dissolved in water, they behave as acids, because the H^+ ions they produce can add to another water molecule to form the active molecular-ion H_3O^+. The best known acids are sulfuric, nitric and hydrochloric acids, all of which are produced commercially.

The higher the concentration of H_3O^+, the stronger the acid. Acidity can vary so widely that a special scale, known as pH, is needed to measure it. The pH scale is a logarithmic scale in which the concentration of H_3O^+ is expressed in terms of negative powers of 10. For example, in neutral water the concentration of H_3O^+ is very low, only 10^{-7} molecular units per litre, so its pH is 7. The pH scale is an inverse scale as far as acids are concerned, in other words the lower the pH number, the stronger the acid. The range of normal acidities is from 1 (the strongest) to 7 (neutral water); as the scale is logarithmic, this means that an acid of pH 1 has a million times more H_3O^+ than neutral water. Water at pH 7 has equal amounts of H_3O^+ and its counterbalancing ion, OH^-, hence its neutrality.

The pH scale can also be extended even further to cover alkaline conditions from pH 7 to 14, when OH^- predominates and the concentration of H_3O^+ falls by a further factor of a million. The span of pH from 1 to 14 represents a difference in acidity of a million million times. Nevertheless we can come across materials in our everyday life which encompass the full range:

Table 1 The acids and alkalis of everyday experience

pH	Typical conditions	Example
0*	chemical reagent	concentrated sulfuric acid
1	stomach acid and battery acid	dilute hydrochloric acid and sulfuric acid, respectively
2	lemon juice	citric acid
3	vinegar	acetic acid
4	tomato juice	ascorbic acid (vitamin C)
5	beer, rain water	carbonic acid (H_2CO_3)
6	milk	lactic acid
7	blood	neutral conditions
8	sea water	dissolved calcium carbonate
9	bicarbonate solution	sodium hydrogen carbonate in water
10	milk of magnesia	magnesium hydroxide in water
11	household ammonia	ammonia solution in water
12	garden lime	calcium hydroxide solution
13	drain cleaner	sodium hydroxide solution
14	caustic soda	concentrated sodium hydroxide solution

* There are some very strong acids with negative pH values, but these are not met with in everyday life.

ELEMENT OF SURPRISE

There are many surprising things about hydrogen, but none more so than those associated with its most familiar compound, water. Nothing could be simpler than a water molecule, yet nothing is as complex in its behaviour. For example, H_2O, which is a very light molecule, might be expected to be a gas at room temperature, as is hydrogen sulfide (H_2S), which is twice the weight of water; yet water is a liquid. Moreover, when it freezes at 0°C, its solid form, ice, floats instead of sinking, which is contrary to normal behaviour because solids by their nature are denser than liquids. The reason is that each water molecule in ice forms four hydrogen bonds to surrounding water molecules and in so doing arranges itself into a crystal with a very open lattice, making it less dense than liquid water, so it floats.

Without water vapour it has been estimated that the Earth would be much cooler than it is, even too cold to sustain life. Water vapour can make up as much as 4% of the atmosphere and this acts as the principal greenhouse gas, warming the air by more than 30°C and giving an average global temperature of +15°C instead of –20°C. The water of the oceans also acts as a great moderator of global temperature. Indeed, without liquid water on a planet, life would never begin.

Although water boils at 100°C, this is only strictly true at sea level. At the top of Mount Everest it would boil at about 75°C because of the reduced air pressure. If we increase the pressure, we can increase the boiling point up to 374°C, but to do so requires a pressure of 220 times atmospheric pressure. Above this temperature water becomes a so-called supercritical fluid, and behaves both like a gas and a liquid. As such, it will dissolve almost anything, even oils, and when it does so the volume of fluid can suddenly shrink to a half or less. This happens because supercritical water tends to pack tightly around other molecules. More strangely still, organic materials will flame and 'burn' in it. Treatment with supercritical water has been suggested as a way of disposing of sewage sludge, which is converted to a crystal-clear, odourless, germ-free solution by this means. The trouble with supercritical water is that it is capable of corroding almost any metal, even gold.

Heavy water

Heavy water, or D_2O, is used mainly as a moderator in nuclear reactors; by slowing down neutrons it makes it easier for uranium atoms to capture them. Were a person to drink D_2O instead of H_2O, they would not survive long because its chemical nature is sufficiently different from that of H_2O to upset certain key cell processes.

Indium

Pronounced in-di-uhm, the name comes from the Latin *indicum*, meaning violet or indigo, which refers to the brightest line in its atomic spectrum.

French, *indium*; German, *Indium*; Italian, *indio*; Spanish, *indio*; Portuguese, *índio*.

HUMAN ELEMENT

Indium in the human body	
Blood	not known, but low
Bone	not known
Tissue	about 15 p.p.b.
Total amount in body	about 0.4 milligrams

There is no biological role for indium but in small doses its salts stimulate metabolism. However, if more than a few milligrams are consumed, they will provoke a toxic reaction and affect the liver, heart and kidneys.

ELEMENT OF HISTORY

Indium was discovered in 1863 by Ferdinand Reich (1799–1882) and Hieronymus Richter (1824–1898) at the Freiberg School of Mines in Germany. Reich was Professor of Physics and was investigating a sample of the mineral zinc blende (now known as sphalerite, ZnS) which he believed might contain the recently discovered element thallium. He produced a yellow precipitate which he thought was thallium sulfide, but his atomic spectroscope showed lines that were not those of thallium. However, because he was colour-blind he called in his chemistry colleague, Richter, to look at the spectrum, and what was immediately evident to him was a brilliant violet line. It was this, at a wavelength of 451 nanometres, that gave rise to the name indium for the new element.

Reich and Richter realized that indium must be present in the zinc blende, beyond being merely a trace impurity, and they went on to isolate a small sample of it. Together they published a paper announcing the discovery. It was this which eventually led to a rift between the two men, when Reich learned that Richter was laying claim to the discovery when he exhibited samples of the new metal at the Académie des Sciences, Paris, in April 1867.

Searches for minerals of the new element were fruitless until 1876, when indium-rich zinc blendes were discovered in Colorado and at Bergamo, Italy. Indium was detected in the pumice ejected from Mount Krakatoa when it erupted in 1883. Some zinc minerals have sufficient indium to give the characteristic indigo line without the need of chemical concentration.

Although people were interested in the element, little use was found for it, despite the efforts of a Mr Murray who, in 1932, displayed a 3 kilogram ingot of pure indium to the members of the Rotary Club of Utica, New York.

ECONOMIC ELEMENT

Specimens of uncombined indium metal have been found in the Transbaikal region of Russia, but that which is produced for industry comes as the by-product of smelting zinc and lead sulfide ores, some of which can contain 1% indium. (The indium ion, In^{3+}, and the zinc ion, Zn^{2+}, are about the same size, which explains why indium can be accommodated in place of zinc in a zinc sulfide ore.) An indium mineral, indite ($FeIn_2S_4$), has been found in Siberia, but it is rare. World production is around 75 tonnes per year and comes mainly from Canada; reserves of the metal are estimated to exceed 1500 tonnes.

Indium finds little use except in a few low melting alloys such as those used in fire-sprinkler systems in shops and warehouses. One alloy of 24% indium and 76% gallium is even a liquid at room temperature. Some semiconductors and transistors are made of indium–arsenic and indium–antimony, and these have found applications in electronics.

Indium metal will 'glue' itself to glass, and when evaporated and allowed to deposit on glass it produces a mirror as good a quality as that of silver and more resistant to corrosion.

Indium foils are used to assess what is going on inside nuclear reactors. The isotope, indium-115, has a high neutron-capturing ability and as it captures neutrons other radioactive isotopes are produced; these indicate the activity inside the reactor.

ENVIRONMENTAL ELEMENT

Indium in the environment	
Earth's crust	0.1 p.p.m.
	Indium is the 69th most abundant element.
Soils	0.01 p.p.m.
Sea water	0.1 p.p.t.
Atmosphere	virtually nil

Indium is not widely dispersed in the environment and poses no threat to land or marine life. Cultivated soils are reported to be richer in indium than non-cultivated sites; some even have levels as high as 4 p.p.m. This might have a slight inhibitory effect on certain soil micro-organisms such as nitrate-forming bacteria.

CHEMICAL ELEMENT

Data file	
Chemical symbol	In
Atomic number	49
Atomic weight	114.818
Melting point	156°C
Boiling point	2080°C
Density	7.3 kilograms per litre (7.3 grams per cubic centimetre)
Oxide	In_2O_3

Indium is a soft, silvery metal and a member of group 13 of the periodic table of the elements. It is stable in air and in water; but dissolves in acids.

There are two naturally occurring isotopes: indium-113, which accounts for 4%, and indium-115, which accounts for 96%. The more abundant isotope is very weakly radioactive as a β-emitter with a half-life of 600 trillion years (6×10^{14} years).

ELEMENT OF SURPRISE

When a piece of indium metal is bent it lets out a high-pitched shriek.

Pronounced iyo-deen, the name is derived from the Greek *iodes*, meaning violet.

French, *iode*; German, *Jod*; Italian, *iodio*; Spanish, *yodo*; Portuguese, *iodo*.

Iodine is the name of the element, which exists in the form of the molecule I_2 with two iodine atoms joined together. This is the iodine of black crystals, purple vapour and yellow antiseptic solution. When iodine acquires a negative electron, as it is inclined to do, it forms the more stable iodide ion, written I^-. The general term, iodine, is used collectively for all aspects of dietary, medical and environmental iodine, when the actual molecular forms are of secondary importance.

HUMAN ELEMENT

Iodine in the human body	
Blood	0.06 p.p.m.
Bone	0.3 p.p.m.
Tissue	varies widely from 0.05 to 0.7 p.p.m.
Total amount in body	10–20 milligrams, of which most is in the thyroid gland

Iodine is essential to many species, including humans, but not to plants, although these are capable of absorbing it from soil water through their roots, or from the atmosphere via their leaves. The edible plants that take up most iodine are cabbages, onions and mushrooms, all of which can have 10 p.p.m. (dry weight).

Iodine vapour irritates the eyes and lungs. The maximum permissible concentration in industry is 1 milligram per cubic metre. The element itself is toxic: as little as 2 grams could kill. On the other hand, iodides are relatively safe, although even they can cause an adverse reaction if taken in excess.

Most iodine in the body is in the thyroid gland; other organs where it tends to concentrate are the salivary glands, stomach, pituitary and ovaries. In the thyroid gland iodine is converted into two hormones, thyroxine and tri-iodothyronine, which regulate several metabolic functions, most noticeably to control the body's temperature. Iodine is needed for normal growth, development and life-span. Too little iodine in the body results in various deficiency diseases, of which the best known are goitre (which results in a swollen neck caused by an enlarged thyroid gland) and hypothyroidism (when a person becomes listless and feels cold). Too much iodine, or its malfunction in the body, results in hyperthyroidism with its associated restlessness and hyperactivity. The body can recycle a lot of iodine but some is lost each day in the urine. Too much iodide, however, is toxic: experiments with rats showed that it interferes with female fertility.

FOOD ELEMENT

People living in developed societies get the small amount of iodine they require from their diet. Milk is a major source of dietary iodine and the level of iodine in milk has doubled in recent years because cattle feed is supplemented with iodine and because of the use of iodine antiseptics to sterilize udders before milking. Foods richest in iodine are cod, oysters, haddock, shrimp, herring, lobster, sunflower seeds, seaweed and mushrooms. Some foods, such as cassava, maize, bamboo shoots and sweet potatoes, can interfere with iodine uptake, but they only threaten those living in regions where dietary iodine is already in short supply.

Iodine is needed especially during the first 3 months of pregnancy for the development of the baby's nervous system, and mothers who are deficient in the element are likely to produce a baby with cretinism.

The minimum iodine intake needed to prevent goitre is 70 micrograms per day. In the UK the dietary reference value is set at twice this level to provide a measure of safety. A safe upper limit is believed to be 1 milligram per day, above which there is a risk of adversely affecting the thyroid gland.

Iodine dietary deficiency (IDD) was estimated to affect 750 million people in the developing world in 1990 and of these around 10 million were suffering cretinism. Those most at risk were in India and China, where centuries of intensive cultivation had depleted the level of iodine in the soil and in drinking water. The World Health Organization and other agencies, such as UNICEF, ran a successful campaign to have all edible salt iodized by the year 2000 and most of the countries where IDD was endemic have now made this a legal requirement. Salt is iodized at a level of 15 p.p.m. so that a daily intake of around 5 grams of salt would provide the necessary 70 micrograms of iodide.

MEDICAL ELEMENT

In 1820, Dr Jean-François Coindet (1774–1834) introduced iodine into medicine in the form of a solution of iodine and potassium iodide dissolved in alcohol, and he advocated its use for the treatment of goitre. He had noted that earlier treatments for this disease advised taking the ash of seaweed and he pointed out that, since this was a rich source of iodine, this element might well be the active ingredient. He was right, but when patients were given his tincture of iodine as a medicine they suffered from severe stomach pains due to its irritant effects and the treatment did not catch on.

Although iodine solution had failed as a remedy for goitre, it soon became the accepted treatment for open wounds, even though it was not realized at the time that its effectiveness was due to its germ-killing properties. While iodine and organic iodine compounds have been much in vogue as antiseptics and disinfectants, they have generally been replaced by much less painful chemicals.

In 1830, another link between goitre and the level of iodide was noted. Regions where goitre was endemic had noticeably little iodide in the water supply, but this observation was not acted upon. Attempts in the mid-nineteenth century to cure goitre with iodide were also abandoned when patients were adversely affected by too much iodide.

The link with the thyroid gland was uncovered in 1895 by a Dr Bauman who is reputed to have spilled some concentrated nitric acid on a sample of thyroid gland and seen the tell-tale purple iodine fumes rising from the decomposing tissue. However, it was not until 1916 that an American biologist, David Marine, of Ohio, showed that this disfiguring condition could be treated and prevented with iodide provided that it was taken as a dietary supplement; the best way to do this was to add it to common salt. Iodized salt was introduced in the USA in 1930 and in the following two decades goitre was eliminated in that country.

The radioactive isotope iodine-131 has been used in medical diagnosis, particularly for thyroid conditions, although this isotope has accidentally probably caused more cancers that it has ever helped to cure (see below). Iodine-131 was the first radionuclide to be used therapeutically in the form of a drug, becoming available after World War II. Another isotope, iodine-125, is a γ-ray emitter and is used in radiation therapy.

ELEMENT OF WAR

Iodine was discovered as an indirect result of the Napoleonic Wars of the early nineteenth century, during which the British Navy blockaded France in an effort to cut off the supply of saltpetre (potassium nitrate) that was needed to make gunpowder. Saltpetre was imported from the East where its crystals were collected from latrines and cesspits. It is produced by bacteria acting on the nitrogenous components of human waste.

In republican France a cottage industry quickly sprang up to supply the demand for saltpetre. This was produced from heaps of rotting manure and animal offal mixed with humus-rich soil and ashes, these last providing the necessary potassium.

Bernard Courtois, a chemist engaged in saltpetre manufacture on the outskirts of Paris, used seaweed ash as a source of potassium. He boiled this with water and extracted potassium chloride by selective crystallization. One day, in 1811, he added sulfuric acid to the residual liquor and was surprised to see purple fumes rising from the pan. When he repeated the reaction in a retort, the fumes condensed to beautiful crystals with a metallic lustre. Courtois was fascinated and guessed he had discovered a hitherto unknown element. He even carried out a few simple experiments with it, such as adding the iodine to ammonia solution from which he got explosive nitrogen triiodide crystals.[57]

[57] Courtois's business went bankrupt following Napoleon's final defeat at the Battle of Waterloo in 1815, after which the naval blockade was lifted. This allowed cheap imports of saltpetre to flood into Europe from the East, just at a time when the need for gunpowder dramatically declined. Nor did Courtois profit from the growing market for iodine, despite being able to sell it at 6 francs per 10 grams, because he could make so little of it. Sadly, he died penniless in 1838.

ELEMENT OF HISTORY

Courtois gave samples of his new chemical to two chemist friends, Charles-Bernard Desormes and Nicolas Clément, who carried out a more systematic investigation, and on 29 November 1813 they presented it at the Institut Imperial de France. That it was a new element was soon proved by fellow countryman Joseph Gay-Lussac and confirmed by the Englishman Humphry Davy. This famous chemist had been given permission by Napoleon to visit Paris, even though the two countries were at war. Davy was given a sample of iodine and worked on it in his hotel room using his travelling chemistry set. He announced his results on 11 December 1813 to the Institut Imperial, and sent back a report to the Royal Institution in London where it was read at a meeting on 20 January 1814.

The British, not knowing of Gay-Lussac's parallel proof, thought Davy was the first person to show that iodine was an element. A lingering disagreement over priority continued for the next century and right up to 1913 when the French celebrated the centenary of the announcement of iodine's discovery. Authentic samples of Courtois' iodine were exhibited at the event and his part in the discovery was reaffirmed. Even so, Gay-Lussac had been the first to confirm that iodine was a true element and it was he who named it 'iode', from the Greek *iodes*, meaning violet-like.

ECONOMIC ELEMENT

There are some iodine-containing minerals, such as lautarite, $Ca(IO_3)_2$, which is found in Chile, and iodargyrite, AgI, specimens of which are to be found in Colorado, Nevada and New Mexico.

World-wide industrial production of iodine is about 13 000 tonnes per year, mainly in Chile (40%) and Japan (30%) plus smaller amounts in Russia and the USA. Iodine is extracted from natural brines and oil-well brines, which may have up to 100 p.p.m. of the element, or from Chilean nitrate deposits which may contain 5% sodium iodate ($NaIO_3$). Known reserves of easily accessible iodine amount to around 2 million tonnes.

To extract iodine, brines are acidified and then chlorine gas is bubbled through to release the dissolved iodide as iodine. This is made 99.5% pure by converting it first to hydriodic acid (HI) and then back to iodine. The iodate of nitrate ores is extracted using sulfur dioxide to release the iodine, and after further treatment is converted to a solid form known as prill, which consists of tiny iodine spheres that are dust-free and safe to handle. Iodine is not currently being produced from seaweed, the original source. However, this method was employed from the 1820s to the 1950s and was based mainly on kelp. This seaweed contains 0.45% iodide (dry weight) and its ash is 1.5% iodide, permitting 15 kilograms of the element to be produced from a tonne of the ash.

The main outlets for the element are in pharmaceuticals (including iodine-based disinfectants), which account for 25%, animal feeds (15%), printing inks and dyes (15%), industrial catalysts (15%) and photographic chemicals (10%), with a variety of miscellaneous uses accounting for the remaining 20%. For many of these uses the iodine is turned into iodides such as titanium iodide for catalytic use, silver iodide which increases the sensitivity of photographic films, zirconium iodide which is used to prepare the

THE FIRST PHOTOGRAPHS

In 1839 Loius Daguerre published his method of taking photographs, his so-called daguerrotypes. The image was produced on a piece of plated silver which had been exposed to iodine vapour, thereby forming a surface layer of light-sensitive silver iodide. The stronger the light, the more the silver iodide was converted back to silver metal. The iodine formed in this reaction could be washed away, leaving a positive image. In 1851, a wet plate alternative was introduced with the silver iodide film on glass.

pure metal zirconium, and potassium iodide which is used to supplement animal feed. Organic iodine compounds are used in pharmaceuticals and dyes, and to stabilize nylon. One of the first commercial uses of iodine was in photography (see box).

ENVIRONMENTAL ELEMENT

Iodine in the environment	
Earth's crust	0.14 p.p.m.
	Iodine is the 64th most abundant element.
Soils	approx. 3 p.p.m., with between 0.1 and 10 p.p.m. in common soils
Sea water	0.06 p.p.m.
Atmosphere	0.2–60 p.p.b., rain water contains 0.7 micrograms per litre

The iodine cycle in nature

About 400 000 tonnes of iodine escape from the oceans every year as iodide in sea spray or as iodine, hydriodic acid and methyl iodide, produced by marine organisms. Much of this is deposited on land where it may become part of the biocycle. Soil micro-organisms play a significant role, and fungi are known to accumulate iodine, although a lot returns to the oceans via the rivers. More recently it has been shown that rice plants emit methyl iodide, CH_3I, and that this accounts for about 4% of the iodine present in the atmosphere.

In some regions, such as the Baraba Steppe of Russia, the level of iodine in soils can be as high as 300 p.p.m. and soils along the coasts in Japan and Wales can have 150 p.p.m. However, where the soils have been exposed to glaciation and leaching, the amount can be very low, which is why goitre was prevalent in such regions.

Radioactive iodine-131 has been released in large amounts following nuclear accidents such as that in Chernobyl in 1986. This isotope has a half-life of 8 days but is picked up by grazing animals and may enter the human food chain rapidly as milk or in water supplies. To counter its effects, potassium iodide tablets can be issued to the population, especially to children, but this may come too late, as it did in the Chernobyl region, where the incidence of thyroid cancers has increased noticeably. If a large dose of potassium iodide is given immediately exposure is imminent it prevents the radioactive form from being absorbed, thereby limiting the damage it can cause.

CHEMICAL ELEMENT

Data file	
Chemical symbol	I
Atomic number	53
Atomic weight	126.90447
Melting point	114°C
Boiling point	184°C (under pressure)
Density	4.9 kilograms per litre (4.9 grams per cubic centimetre)
Oxide	I_2O_5, which is stable, and I_2O_4 and I_4O_9, which are less stable

Iodine is a black, shiny, non-metallic solid and a member of group 17, also known as the halogen group, of the periodic table of elements. It sublimes easily on heating to give a purple vapour.

In nature iodine occurs as a single, stable isotope, iodine-127, which is not radioactive. Radioactive isotopes occur in the Earth's crust, and there are others on the Earth's surface, most produced from the fission of uranium-238 and plutonium-239.

Iodine dissolves in some solvents, such as carbon tetrachloride, to give a beautiful purple colour, while in other solvents, such as ether and water, it is brown. Its solubility in water is greatly enhanced by the addition of potassium iodide which forms the triiodide ion, I_3^-. Iodine is an important part of chemical analysis, where the dramatic blue colour it becomes in the presence of starch is used in analytical titrations.

As well as gaining an electron to form I^- ions, iodine can lose one to form the positive ion I^+. This form may even be involved in the way in which the organo-iodine compound, thyroxine, is produced in the thyroid gland.

ELEMENT OF SURPRISE

Silver iodide can be used to seed clouds and thereby initiate rainfall. One gram of silver iodide can provide a trillion (10^9) seed crystals around which water vapour will condense into droplets large enough to fall to earth as rain. An aircraft flying through a rain cloud dispersing this chemical as a fine smoke can produce a torrential downpour.

Secret tests carried out by Britain's Ministry of Defence in August 1952 in the West of England resulted in a flash flood that engulfed the seaside resort of Lynmouth, sweeping away houses and sending 31 inhabitants to their deaths. (Induced rain formation was being tested as a strategy for bogging down tanks and troops and thereby hindering their advance.)

Iridium

Pronounced irid-iuhm, the element is named after Iris, the Greek goddess of the rainbow; it was so called because its salts were of such varied colours.

French, *iridium*; German, *Iridium*; Italian, *iridio*; Spanish, *iridio*; Portuguese, *irídio*.

HUMAN ELEMENT

Iridium in the human body	
Blood	not accurately known but very low
Bone	not known
Tissue	about 20 p.p.t.
Total amount in body	not reported

Iridium has no biological role. The metal is clinically inert, but iridium chloride is moderately toxic by ingestion. Most compounds are insoluble and so, if taken, would not be absorbed by the body.

MEDICAL ELEMENT

The radioactive isotope, iridium-192, is a γ-ray emitter and is used in radiation therapy.

ELEMENT OF HISTORY

Iridium and osmium were discovered together, in1803, by Smithson Tennant (1761–1815) in London. He had tried to dissolve some crude platinum in *aqua regia* but found that a black residue was left behind. He was not the first person to have observed this, but earlier reports had suggested the residue was graphite. Tennant was not convinced that it was, and began working on the insoluble material. By treating it alternately with alkalis and acids he was able to separate it into two previously unidentified elements. These he announced to the Royal Institution in London in the spring of 1804, naming one iridium, because its salts were so colourful, and the other osmium, because it had a curious odour (see p. 294).

In July 1813, a chemist, William Allen, recorded in his diary that he had gone with Tennant, Humphry Davy and about 30 other eminent scientists to the home of John

Children, Secretary of the Royal Society, who lived in Tunbridge. There the assembled throng set up a massive battery of electrical cells consisting of copper and zinc plates with a mixture of acids as the electrolyte solution. Such was the current this generated that they were able to melt a sample of Tennant's iridium with it, to great acclaim.

ECONOMIC ELEMENT

Iridium is found as the uncombined element, and also as the iridium–osmium alloys, osmiridium (which has more osmium than iridium) and iridosmine (vice versa), of which osmiridium is the more common. Most iridium, however, is obtained as a by-product of platinum refining and comes from South Africa. Annual world production, expressed in the industrial units of troy ounces, amounts to 100 000 ounces, which is around 3 tonnes. Reserves have not been estimated.

Iridium is the most corrosion-resistant metal known. Traditionally it was the coating of nibs for fountain pens and compass bearings. The standard metre bar, kept in Paris, is made of a platinum–iridium alloy (90% platinum and 10% iridium) but this was superseded as the basic unit of length in 1960 by a line in the atomic spectrum of krypton (see p. 213).

Nowadays demand for iridium comes mainly from the electronics industry (37% ends up this way), the automotive industry where it is used to tip spark plugs (11%) and from the chemical industry (15%) where it is used to coat the electrodes in the chlor-alkali process, and in catalysts. The rest is used in a variety of ways, such as the catalysts for the new types of automobile engine known as the gasoline direct-injection engines which were launched in Japan by Mitsubishi in 1996.

The aerospace industry relies on iridium for certain long-life engine parts, and an iridium–titanium alloy is used in corrosion-resistant pipes for deep-water projects. The new Cativa process for making acetic acid has an iridium catalyst, and single crystal sapphires, which are needed for the lasers used in robot-controlled welding, are grown in iridium crucibles.

ENVIRONMENTAL ELEMENT

Iridium in the environment	
Earth's crust	about 3 p.p.t.
	Iridium is the 84th most abundant element.*
Soils	not recorded
Sea water	not accurately known, but very low
Atmosphere	nil

* Another way of putting this would be to say it is one of the *least* abundant of the metals.

The level of iridium in land plants is below 20 p.p.b. (dry weight) and for all intents and purposes the element today has no impact on the environment or living things, although its presence in certain geological deposits can serve to indicate what might have caused past mass extinctions (see 'Element of surprise').

CHEMICAL ELEMENT

Data file	
Chemical symbol	Ir
Atomic number	77
Atomic weight	192.217
Melting point	2410°C
Boiling point	4130°C
Density	22.6 kilograms per litre (22.6 grams per cubic centimetre)
Oxide	IrO_2

Iridium is a hard, brittle, lustrous, silvery metal of the so-called platinum group of elements and it comes in group 9 of the periodic table. The solid metal has the second highest density known, and for a time it was thought to be equal to, if not slightly ahead, of osmium in this respect. It is unaffected by air, water and acids, including *aqua regia*; the only two substances that the metal will dissolve in are molten sodium cyanide and potassium cyanide.

There are two naturally occurring isotopes: iridium-191, which comprises 37%, and iridium-193, which comprises 63%. Neither is radioactive.

ELEMENT OF SURPRISE

It came as a surprise when geologists discovered iridium in the rocks that were deposited at the boundary between the Cretaceous and Tertiary geological periods.[58] This was the time, around 65 million years ago, when a traumatic event happened on Earth that led to the extinction of most of the species then in existence, most notably the dinosaurs.

One theory is that it was a time of great volcanic activity with flood basalt eruptions in India that covered thousands of square kilometres with hot molten lava, and released vast amounts of the poisonous gas sulfur dioxide into the atmosphere. Another theory blames the extinctions on the impact of a 10 kilometre meteor which struck the region that is now the Yucatan peninsula, off the Gulf of Mexico, forming a crater up to 300 kilometres across, and that dust from this impact caused a darkening of the sky for several seasons, with the result that few creatures were able to survive.

The evidence for the meteor impact having global significance is the amount of iridium in the boundary layer all around the Earth. This metal can only have come from space, and indeed several meteorites have been found to be relatively rich in this rarest

[58] A layer of soot, deposited at the same time, shows that fires ravaged the earth's forests and may have consumed 25% of the planet's biomass.

of metals. Similar meteoric impacts around 36 million years ago left craters at Popigai in northern Russia and Chesapeake Bay in the USA. These also left an imprint of iridium on the planet. However, they did not lead to mass extinctions and it is now thought that, while the Yucatan meteorite had a serious effect, it was not the main reason why the Cretaceous period, and the dinosaurs, came to an end. It is now thought that it might have been one of several contributing factors.

Pronounced iy-on, the name comes from the Anglo-Saxon *iren*, and the chemical symbol, Fe, comes from *ferrum*, the Latin for iron.

French, *fer*; German, *Eisen*; Italian, *ferro*; Spanish, *hierro*; Portuguese, *ferro*.

COSMIC ELEMENT

Iron is the heaviest element that can be made in the nuclear fusion furnace that runs in the interior of a typical star. The nucleus of an iron atom is the most tightly bound of all nuclei, which means that no more energy can be liberated by fusing it with other nuclei. In effect it is the 'ash' of the nuclear burning process; once the core of a star has become mainly iron, that star has run out of its primary source of energy.

HUMAN ELEMENT

Iron in the human body	
Blood	approx. 415 p.p.m. (range 380–450 p.p.m.)
Bone	varies from 3 to 380 p.p.m.
Tissue	approx. 180 p.p.m. (range 20–1400 p.p.m.)
Total amount in body	4 grams

Iron is essential to almost all living things, from micro-organisms to humans, but this does not exclude the possibility of its being poisonous in excess; indeed, an intake of 200 milligrams of an iron(II) compound can provoke a toxic response in a human adult.[59] A continual excess of iron in the body causes liver and kidney damage, and working in an atmosphere of iron dust can lead to a form of pneumoconiosis known as welder's lung.

A more common problem for humans is iron deficiency, which leads to anaemia. Iron has many roles in the body but its best known one is to carry oxygen from the lungs to where it is needed to generate energy, such as to the heart and brain, and then take some of the waste gas, carbon dioxide, back to the lungs where it can be exhaled and replaced by fresh oxygen.

[59] Iron(II) is more toxic than the more common form, iron(III).

MEDICAL ELEMENT

The ancient Greek physician, Galen,[60] suggested taking iron filings as a medicament, recommending iron as a laxative. (Today we know that iron is more likely to have the opposite effect.) In the eighteenth century, following a Dr Sydenham's suggestion that iron had healing powers, pharmacists began to stock it in the form of iron filings coated with sugar, although a more popular way to take the element was to dissolve the filings in an acidic wine and drink that. Today, if iron is recommended, it is given as a soluble iron(II) salt such as iron(II) sulfate.

Iron was found to be present in vegetable ash early in the 1700s, and in 1745 the Italian physician, Vincenzo Menghini, demonstrated its presence in blood. He did this by feeding dogs iron preparations, bleeding them, drying and burning the blood, and then observing that particles from the ash were attracted to the blade of a magnetized knife. Despite this, and the work of others, the medical role of iron was not understood, and it was not until 1832 that a Dr Bland cured a case of anaemia with iron(II) sulfate. Once the role of iron in the blood became clearer, and a deficiency of the metal was associated with anaemia and other symptoms such as listlessness and general fatigue, then iron tonics and pills became popular and were regarded as a way of keeping a person fit and healthy.

As well as being part of haemoglobin, iron plays a key role in various enzymes, such as those involved in the synthesis of DNA, those which scavenge free radicals and protect the body, and those which enable cells to release energy by using glucose. Normal brain function needs iron. There are regions of the brain that are rich in iron, which may explain why iron deficiency in infants and children has been associated with lowered mental development. Surplus iron is stored mainly in the liver in the form of the iron-containing proteins ferritin[61] and haemosiderin; as much as a gram of iron may be held in reserve this way. Bone marrow is rich in iron too, because this is the site where haemoglobin in synthesized.

A person must have a regular intake of iron because we lose a little of this metal every day through the walls of our stomach and intestines. Even so, it is rare for normal people to be lacking iron, even though they may regularly lose blood, such as women when they are menstruating. Those who regularly donate blood should not require iron supplements if they eat a balanced diet, and the same is true of pregnant women, despite a growing foetus requiring a total of about 700 milligrams of iron.

Inside the body, iron, as iron(III), is strongly bound by transferrin, a protein found in serum and other secretions, and it is this which transfers the metal between cells. Transferrin binds iron tightly and, because it does so, it acts as a powerful antibiotic simply by denying this essential metal to any invading bacteria which need iron to multiply. As soon as our body registers a bacterial invasion, it produces more transferrin to mop up any free iron in the blood stream and 'hide' it in the liver. Mother's milk also contains a form of transferrin called lactotransferrin, and egg white contains ovotransferrin, both of which function in the same iron-binding way, and both of which provide antibacterial protection.

[60] He practised medicine in the second half of the second century AD.

[61] Ferritin is a large protein which can contain up to 4500 iron(III) ions.

Nevertheless bacteria have their own molecules, called siderophores, which scavenge free iron and deliver it back to the microbe. Some bacteria go even further: the bacterium that causes gonorrhoea produces a protein that can even wrest iron from transferrin.

Some people retain too much iron and suffer iron overload. Individuals who have the genetic disorder haemochromatosis are afflicted in this way and iron builds up in their pancreas, liver, spleen and heart. Patients who need multiple blood transfusions for inherited anaemic diseases, such as thalassaemia, also accumulate too much and they need to take drugs that can help the body shed its burden of iron.

FOOD ELEMENT

A man needs an average daily intake of 7 milligrams of iron and a woman 11 milligrams; a normal diet will generally provide all that is needed. Of the iron in food only a certain fraction is capable of being absorbed, maybe as little as 25%, but this is enough to replenish that which is lost each day. Iron-rich foods are liver, corned beef, iron-fortified breakfast cereals, red wine, baked beans, peanut butter, raisins, bread, eggs, curry powder, and cakes made with black treacle (molasses). The iron of red meat is particularly easy to absorb and vegetarians need to ensure a higher dietary intake to compensate for the lack of iron from this source. Extra vitamin C helps absorption, while foods rich in phosphate (such as processed cheese) and phytate (such as bran) hinder it.

The average Western diet provides about 20 milligrams of iron per day, but it is estimated that around the world some 500 million people are anaemic through lack of dietary iron. This problem is likely to get worse because the high-yield crops that have been developed do not provide much in the way of dietary iron.

ELEMENT OF HISTORY

No one discovered the element iron as such, although no doubt individuals often came across samples of the metal lying on the ground. We know this because iron objects have been found in Egyptian tombs dating from around 3500 BC. These must have originated from meteors because they contain about 7.5% nickel, which is the typical composition of meteorites known as siderites. These iron artefacts were able to resist the ravages of rusting on account of their nickel content.

The ancient Hittites of Asia Minor, today's Turkey, appear to have been the first to discover the smelting of iron from its ores, around 1500 BC, and this new metal gave them economic and political power, so they kept the process a closely guarded secret – not that it saved them: an invasion of warriors known as the 'Sea People' destroyed their civilization around 1200 BC, after which the iron workers dispersed and took their craft with them. The Iron Age had begun.

Some iron workers ended up in western Iran where they discovered a superior kind of iron: steel. This depends on the carbon content of the metal, although this was not understood, and the quality of the metal depended more on the skill of the blacksmith, who became a respected professional. Iron working was particularly well developed in India, as witnessed by the 7 metre high iron pillar in Delhi. This remarkable object has stood for 1500 years without rusting, although this is due not to some special alloying of the iron but to the climate, which permits the iron to retain enough heat of the day to prevent moisture condensing on it at night.

Thales, the Greek philosopher, reported in 585 BC that pieces of iron ore that came from Magnesia in Lydia, Asia Minor, had the strange power to attract iron filings. They were called magnets after the place from which they came. The ore is magnetite and pieces of it are referred to as lodestone. The ancient Greek potters also used iron as a pigment, having discovered that items fired under oxidizing conditions would be red in colour (due to the oxide Fe_2O_3) while those fired under reducing conditions would be black (due to Fe_3O_4).

Iron-making on a large scale was an essential part of the Industrial Revolution that began in England in the mid-eighteenth century; there was still no recognition that small amounts of carbon would transform iron into steel. The first person to suggest this was R. A. F. de Réaumur who said that steel, wrought iron and cast iron were to be distinguished by the amount of a black combustible material they contained and that this came from the charcoal then used to smelt the metal.

Technological developments in smelting in the eighteenth century, such as the use of coke rather than charcoal, resulted in cheap wrought iron becoming available and so began an era in which it was widely used in transport and architecture. The first all-iron bridge was built in Shropshire, England, in 1778; the town of Ironbridge took its name from this bridge. The first iron ship, *Volcano*, was launched in 1818 and was in service for 50 years on the river Thames, but the biggest boost to the industry was the development of railways from the 1830s onwards.[62]

ELEMENT OF WAR

One undesirable side effect of the Iron Age was the appearance of warriors armed with iron swords against which there was limited defence. These weapons dominated warfare for more than 2000 years. The Roman god of war was Mars and the element associated with him was iron. Although the Romans made extensive use of iron, the author Pliny had mixed feelings about the metal because of its role in warfare: 'With iron we construct houses, cleave rocks and perform many other useful acts of life, but it is also with iron that wars, murders and robberies take place, and not only hand-to-hand but from a distance ... with winged weapons launched from engines.'

Yet this latter role for iron lay well into the future and was linked to the use of gunpowder, which had the power to project iron cannon balls over large distances. In the World Wars of the twentieth century the quantity of iron discharged in the form of bombs and shells reached colossal proportions, with large sections of all the major industrial economies devoted to turning the metal into instruments of death and destruction.

[62] Perhaps the greatest achievement of this era was the Eiffel Tower in Paris, which was built in 1889 and which is now a much revered monument of that city, although its concept and construction were derided by the leading artists and designers of the day.

ECONOMIC ELEMENT

World production of new iron is over 500 million tonnes a year, and recycled iron adds another 300 million tonnes. Economically workable reserves of iron ores exceed 100 billion tonnes. The main ones are magnetite (Fe_3O_4), which is frequently found as black sand along beaches, haematite (Fe_2O_3), goethite ($FeO(OH)$), lepidocrocite ($FeO(OH)$) and siderite ($FeCO_3$), and the main mining areas are China, Brazil, Australia, Russia and Ukraine, with sizeable amounts mined in the USA, Canada, Venezuela, Sweden and India.

Iron is produced by heating its ore with coke (carbon) and limestone (calcium carbonate) in a blast furnace; the result is so-called pig iron. Around 95% of iron is smelted in this way. Pig iron has 3% carbon plus small amounts of sulfur, silicon, manganese and phosphorus; it is not used as such but is the raw material for making steel, which is done by removing the impurities and adjusting the residual amount of carbon. Most steel has around 1.7% carbon and this makes the iron more durable, less brittle and more resistant to corrosion.

Pure forms of iron, of 99.9% or better, are named after the process that produces them, such as ingot iron, electrolytic iron, reduced iron and carbonyl iron, all of which have special uses. Cast iron contains 3–5% carbon and is used for valves, pipes and pumps. It is not as tough as steel but it is cheaper, and can be improved by the addition of small amounts of nickel–magnesium alloy. Then it is known as ductile cast iron, and can be machined.

Iron has more applications than any other metal and it accounts for around 90% of all metal that is refined, even in the twentieth century. From food containers to family cars, from screwdrivers to washing machines, from cargo ships to paper staples, iron has a part to play.

Rusting can be prevented by coating iron with tin to give tinplate, or with zinc to give galvanized iron. However, other alloys of iron offer better resistance to rusting as well as making the metal more suitable for use. Nickel steel is best for bridges, electricity pylons and bicycle chains; tungsten and vanadium steels are used for cutting tools; manganese steel is used for rifle barrels and power shovels; while the ubiquitous stainless steel has high levels of chromium (18%) and nickel (8%). This is known as '18–8 stainless' and finds use in architecture, cutlery and even jewellery.

ENVIRONMENTAL ELEMENT

Iron in the environment	
Earth's crust	41 000 p.p.m. (4.1%)
	Iron is the fourth most abundant element.
Soils	0.5–5%
Sea water	2.5 p.p.b. (greater in the Atlantic than the Pacific)
Atmosphere	traces

Iron is in fact the Earth's most abundant element because the 7000 kilometre diameter core of the planet consists mainly of the molten metal (the very centre may be a solid iron core about 2500 kilometres in diameter). In effect the planet is a large iron sphere, alloyed with nickel. Although iron is an abundant element at the Earth's surface, there are regions of the planet where it is so lacking that it is the limiting factor to life. This is especially true of the surface layers of large stretches of the oceans, which are more devoid of life than any desert on land – but see 'Element of surprise' below.

Soils contain a lot of iron which is easily transformed by organic matter into various types of iron oxide and other compounds. Iron (III) is the common state but can be converted to iron(II) in waterlogged soils and by plant roots, and this facilitates its absorption. Green plants need iron for energy transformation processes and other uses. Fodder plants can have up to 1000 p.p.m. (dry weight) of iron but food plants generally have much less. Beans are particularly rich in the metal, having as much as 9 p.p.m. fresh weight (80 p.p.m. dry weight), while lettuce also has a lot with 5 p.p.m. fresh weight (130 p.p.m. dry weight).[63] At the other extreme apples, carrots, cucumbers, oranges and onions contain relatively little iron.

Some bacteria use particles of iron in the form of magnetite as a magnetic compass to give them a sense of direction. Molluscs use this mineral to form their teeth, although the very strong teeth of limpets are made of goethite.

CHEMICAL ELEMENT

Data file	
Chemical symbol	Fe
Atomic number	26
Atomic weight	55.845
Melting point	1535°C
Boiling point	2750°C
Density	7.9 kilograms per litre (7.9 grams per cubic centimetre)
Oxides	FeO, Fe_2O_3, Fe_3O_4

Iron, when absolutely pure, is lustrous, silvery and soft (workable). The element is a member of group 8 of the periodic table. Iron rusts in damp air, but not in dry air. It dissolves readily in dilute acids.

Natural iron consists of four isotopes: iron-56, which comprises 92%; iron-54 (6%); iron-57 (2%); and iron-58 (a mere 0.3%). None is radioactive.[64]

[63] As is popularly known, spinach contains a lot of iron, but what is less well known is that this is present in a form that the body cannot assimilate. For most of the twentieth century its reputation as a rich source of iron was grossly exaggerated, due mainly to a printing error in an official US Government booklet on the nutrient content of foods.

[64] An isotope of iron, iron-45, is extremely proton-rich and decays by the rare step of losing two protons. This mechanism, long considered theoretically possible, was first detected with nickel-48 in 1999.

ELEMENT OF SURPRISE

The oceans are often thought of as teeming with life, but this is true of only a few regions, and these tend to be over-fished as a consequence. More than 80% of the boundless ocean is empty. In the mid-1980s, John Martin of the Moss Landing Marine Laboratories, in California, put forward the theory that it was the lack of iron in the upper levels of the sea which prevented plankton from growing, and without plankton to feed on, other forms of marine life have no food supply to support them.

Ten years later Martin's idea was tested by a joint USA–UK research team, which used a solution of iron sulfate to fertilize 60 square kilometres of the Pacific Ocean, West of the Galapagos Islands. The results were dramatic. Within a week this barren span of ocean bloomed and turned green with plankton, proving that it was simply lack of this metal that was limiting their growth.

Suddenly, it was realized that the oceans could one day be fertilized, perhaps with ferrous sulfate, which is easily and cheaply made from rusty old iron and sulfuric acid. The seas could bloom again, as they must once have done when the world was young, and when cellular organisms first flourished and learned how to make use of this versatile metal.

Pronounced krip-ton, the name is derived from the Greek *kryptos*, meaning hidden.
French, *krypton*; German, *Krypton*; Italian, *cripto*; Spanish, *kriptón*; Portuguese, *criptônio*.

COSMIC ELEMENT

Krypton gas has been detected in the atmosphere of Mars at 0.03 p.p.m.

HUMAN ELEMENT

Krypton in the human body	
Blood	minute trace
Total amount in average (70 kilogram) adult	very small

Krypton can have no biological role. The gas is harmless although it could asphyxiate if it were to exclude oxygen from the lungs.

ELEMENT OF HISTORY

Having discovered helium, with an atomic weight of 4, and argon, with an atomic weight of 40, William Ramsay (1852–1916), who was based at University College, London, was convinced he had stumbled upon a new group of elements of the periodic table, and so he knew that there must be others waiting to be found. Together with his assistant, Morris William Travers (1872–1961), he began to search for them but to no avail. He thought they might be released from minerals, which is how helium had been found, but none came to light.

Eventually Ramsay and Travers decided to examine the 15 litres of argon gas that they had extracted from air. This had been produced by removing all the oxygen, nitrogen and carbon dioxide by chemical means, which left the unreactive argon behind. They reasoned that perhaps other gases were also present, and by a process of liquefying the argon and allowing it slowly to evaporate they hoped a heavier component might be left behind.

On the afternoon of 30 May 1898, their hopes were realized when they isolated about 25 cubic centimetres (millilitres) of a residual gas. This they immediately tested in a spectrometer, and saw from its atomic spectrum that it was a hitherto unknown element, with orange and green lines that could not be attributed to any other gas. By midnight that day they had measured the density of this gas and this confirmed that they had indeed found a 'new' element, which they appropriately named krypton.

ECONOMIC ELEMENT

Krypton may be one of the rarest gases in the atmosphere, but in total there are more than 15 billion tonnes of this element circulating the planet, of which only about 8 tonnes a year are extracted, via liquid air.

Krypton has some uses. For example, it is used in strip lighting, either on its own, when it produces a soft violet glow, or to modify the colour of other 'neon' lights. Certain high-speed photographic flash lamps and strobe lights use krypton gas because it has an extremely fast response to an electric current. It is also used in lasers.

From 1960 to 1980, the fundamental standard of length was based on the sharp orange line in the spectrum of krypton-86. One metre was defined as exactly 1 650 763.73 times the wavelength of this line.[65]

ENVIRONMENTAL ELEMENT

Krypton in the environment	
Earth's crust	10 p.p.t.
	Krypton is the 83rd most abundant element.
Soils	nil
Sea water	80 p.p.t.
Atmosphere	1 p.p.m.

Krypton has no role to play in any part of the environment because it is almost totally inert chemically and of low solubility in water.

CHEMICAL ELEMENT

Data file	
Chemical symbol	Kr
Atomic number	36
Atomic weight	83.80
Melting point	−157°C
Boiling point	−52°C
Density (of the gas)	3.7 grams per litre (3.7 kilograms per cubic metre)
Oxide	none

[65] The standard changed in 1983 to one based on the speed of light in a vacuum, a metre being the distance light travelled in 1/299 792 458th of a second, as measured by a light beam from a helium–neon laser.

Krypton is a colourless, odourless gas and one of the noble gases that form group 18 of the periodic table of the elements. For many years it appeared to be chemically inert, not reacting with anything but fluorine gas, with which it forms only KrF_2, KrF^+ and $Kr_2F_3^+$, but in 1988 a molecular ion of formula $HCNKrF^+$ was produced. Unfortunately, this compound explodes at temperatures above –50°C. Other compounds are also unstable, unless isolated in a matrix at very low temperatures, but under such conditions compounds with krypton–hydrogen, krypton–carbon and krypton–chlorine bonds have now been made.

Krypton has six naturally occurring isotopes, of which krypton-84 accounts for 57%. The others are krypton-86 (17%), krypton-82 and krypton-83 (12% each), krypton-80 (2%) and krypton-78 (less than 0.4%). None is radioactive.

ELEMENT OF SURPRISE

Radioactive krypton-85 has a half-life of 11 years and is given off by nuclear reactors and nuclear fuel reprocessing plants. It escapes to the atmosphere but is not considered to pose a threat to life because, like non-radioactive krypton, it is inert.

The level of this gas in the atmosphere was carefully monitored during the Cold War years (1950–1990) by the West because it revealed the extent to which the Soviet bloc was producing nuclear material. Knowing how much krypton-85 was coming from reactors in the USA and Europe, it was possible to deduce how much was being emitted by the Eastern powers, and from that to know the amount of material they had available for nuclear arms.

Lanthanum

Pronounced lan-than-um, the name comes from the Greek *lanthanein*, meaning to lie hidden.

French, *lanthane*; German, *Lanthan*; Italian, *lantanio*; Spanish, *lantano*; Portuguese, *lantânio*.

Lanthanum is one of 15, chemically similar, elements referred to as the rare-earth elements, which extend from the element with atomic number 57 (lanthanum) to the one with atomic number 71 (lutetium). The term rare-earth elements is a misnomer because some are not rare at all; they are more correctly called the lanthanides, although strictly speaking this excludes lutetium. The minerals from which they are extracted, and the properties and uses they have in common, are discussed under lanthanides, following this section, on p. 219.

HUMAN ELEMENT

Lanthanum in the human body	
Blood	not reported
Bone	less than 0.08 p.p.m.
Tissue	0.4 p.p.t.
Total amount in body	approx. 1 milligram

The amount of lanthanum in humans is quite small. The metal has no biological role and has almost no impact on living things. There is no estimate of how much lanthanum we take in each day with our diet but it is part of the food chain because plants will absorb the metal, with some vegetables reported to have as much as 0.2% (dry weight) and some plants have been found with 1.5% (dry weight).

When lanthanum chloride solution was injected into mice it ended up in the liver and the bone, whereas when dogs breathed a dust of this salt it was found to collect in the lungs and gut.

Lanthanum salts, such as the chloride, are not regarded as toxic when taken by mouth, and injection of lanthanum chloride into humans was once explored as a possible treatment for cancer. The side-effects of such treatment were somewhat unpleasant and included chills, fever, muscle pains and stomach cramps.

MEDICAL ELEMENT

People suffering from chronic kidney dysfunction generally have too much phosphate in their blood, a disorder known as hyperphosphataemia. (The average diet contains much more phosphate than the body requires.) If this is left untreated it will result in osteodystrophy, a painful bone condition in which there are abnormalities in the structure of the skeleton. Renal dialysis can do little to reduce phosphate levels, and although patients with hyperphosphataemia can be put on a low-phosphate diet, it is almost impossible to prevent the bones from deteriorating.

A new method of treatment is with lanthanum, which binds to phosphate so strongly that it cannot be absorbed from the gut. Lanthanum, in the form of the carbonate, is added to the food of hyperphosphataemic patients; it combines with any phosphate that is released as the food is digested in the stomach and intestines, and so carries it out of the body.

ELEMENT OF HISTORY

Lanthanum was discovered in January 1839 by Carl Gustav Mosander (1797–1858) at the Karolinska Institute, Stockholm, Sweden. Mosander was Professor of Chemistry and Mineralogy, and it was there that he extracted a new 'earth' from cerium nitrate (cerium had been discovered in 1803). He heated cerium nitrate to decompose it to cerium oxide, and then stirred this with dilute nitric acid in which it was supposed to be insoluble. Mosander noticed that some of the substance *did* dissolve, however, so he filtered the solution, evaporated it and obtained crystals. These were not cerium nitrate, but the nitrate of a 'new' element. Mosander told his friend Berzelius about his discovery and it was the latter who suggested the name lanthanum, to which Mosander agreed. Word of the 'new' element quickly spread but Mosander became strangely silent about it and did not even publish an account announcing its discovery.

That same year, Axel Erdmann, a student at the Karolinska Institutet, discovered lanthanum in a previously unknown mineral from Låven Island, Langesundsfjord, Norway. He named this mineral mosandrite, in honour of Mosander himself.

Mosander, meanwhile, was not to be hurried, as most of his time was taken up running a mineral water business. A year passed, and then two, while chemists all over Europe waited and speculated. Finally, Mosander explained the delay, saying that he had extracted yet another 'new' earth from his lanthanum oxide; he called this didymium, from the Greek word *didumos*, meaning twin. Didymium formed a rose-pink oxide; both cerium oxide and lanthanum oxide were contaminated with didymium oxide, which explained why they appeared pink. When pure they were white. It had taken Mosander two long years of patient extraction to separate didymium from his lanthanum. (Didymium's status as an element was not to last, because it too was shown to be a mixture: in 1885 it was separated into the two elements praseodymium and neodymium.)

ECONOMIC ELEMENT

There are no ores which contain only lanthanum as the metal component; it is found in minerals that include all the rare-earth elements – these are discussed in more detail on p. 222. Rare-earth ores contain a lot of lanthanum: monazite has around 25% lanthanum and bastnäsite around 38%. World production of lanthanum oxide is around 12 000 tonnes per year, and currently known reserves of lanthanum are around 6 million tonnes.

Several lanthanum compounds are commercially available, such as the oxide, carbonate, chloride, fluoride, nitrate and sulfate. The metal itself is obtained by the reaction of lanthanum fluoride and calcium metal. There are no commercial uses for pure lanthanum metal, but there are for its alloys. That of lanthanum and nickel, $LaNi_5$, is extremely good at absorbing hydrogen gas; as a powder it can absorb as much as 400 times its own volume. It is being investigated as a storage system for hydrogen, which could be needed if hydrogen is one day to be the fuel for cars and buses (see p. 187).

Lanthanum is used as the core material in carbon-arc electrodes for film and photographic studio lights and for floodlighting. It increases the brightness of the arc lights and gives an emission spectrum almost identical to that of sunlight.

Lanthanum oxide is added to the glass for making lenses because it improves its refractive index. Up to 40% of lanthanum oxide can be incorporated into optical glass, along with yttrium oxide (up to 5%) and gadolinium oxide (up to 10%) to give a high refraction index and low colour dispersion. Lanthanum-containing glass is also much more resistant to attack by alkalis.

Lanthanum salts are included in the zeolite catalysts used in petroleum refining because they stabilize the zeolite at high temperatures.

ENVIRONMENTAL ELEMENT

Lanthanum in the environment	
Earth's crust	32 p.p.m.
	Lanthanum is the 28th most abundant element.
Soils	approx. 26 p.p.m., ranging from 1 to 120 p.p.m.
Sea water	5 p.p.t.
Atmosphere	virtually nil

Lanthanum is one of the more abundant rare-earth elements, being as common as lead and tin together. It poses no environmental threat to plants or animals.

CHEMICAL ELEMENT

Data file	
Chemical symbol	La
Atomic number	57
Atomic weight	138.9055
Melting point	921°C
Boiling point	3460°C
Density	6.1 kilograms per litre (6.1 grams per cubic centimetre)
Oxide	La_2O_3

Lanthanum is a silvery-white metal and is the first member of the lanthanide group of the periodic table of the elements. It is soft enough to be cut with a knife. It tarnishes rapidly in air and burns easily if ignited. It is one of the most reactive of the rare-earth metals; it even reacts with water, releasing hydrogen gas. Lanthanum salts are often very insoluble, such as the oxide, hydroxide, carbonate, fluoride, phosphate and oxalate. The chloride, bromide and nitrate are readily soluble.

Lanthanum consists of two isotopes, the heavier of which, lanthanum-139, comprises 99.9% of the total. The remaining 0.1% is the lighter isotope, lanthanum-138, which is weakly radioactive with a half-life of 100 billion years (10^{11} years).[66]

ELEMENT OF SURPRISE

Although lanthanum forms trivalent ions, i.e. La^{3+}, whereas calcium forms only divalent ones, i.e. Ca^{2+}, the former is used by researchers as an easily monitored substitute for the latter in living things. The reason is that the two ions are virtually the same size, so that receptors that recognize calcium generally accept lanthanum as well.

[66] There are also minute traces of some radioactive lanthanum isotopes, such as lanthanum-140, in uranium ores, where they have been formed by nuclear fission.

The lanthanides

These are the elements with atomic numbers 57–70. Collectively they are named lanthanides after the first member of the series, lanthanum. The older name was rare-earth elements and this is how they were viewed for much of the time that they have been known, but not all are rare. The chemists of the nineteenth century were mystified by the rare earths. Towards the end of the century, William Crookes, the eminent British chemist, summed them up in exasperation as follows:

> The rare earth elements perplex us in our researches, baffle us in our speculations, and haunt us in our very dreams. They stretch like an unknown sea before us, mocking, mystifying and murmuring strange revelations and possibilities.

It was hardly surprising that he viewed the rare earths thus because there were 15 of them which occurred together in the rare-earth minerals, and extracting them was a tedious process. Chemical theory determines that there can be only 14 lanthanides occupying the spaces in the periodic table (see p. 525), although traditionally chemists have always included lutetium as one of the rare-earth elements (lutetium is the next element beyond the lanthanide group).

The story of the discovery of the lanthanides began with that of yttrium, which is not regarded as a rare-earth element but is very similar to them and often found with them. The yttrium mineral gadolinite was to yield the rare-earth elements terbium, erbium, ytterbium, lutetium, holmium, thulium and dysprosium. The other rare-earth elements were all extracted from another mineral, cerite, which yielded first cerium and then lanthanum, neodymium, praseodymium, samarium, gadolinium and europium.

As the table shows, lanthanide elements with even atomic numbers are more abundant than those adjacent to them with odd atomic numbers, a common feature of all the elements. Monazite ores generally have higher concentrations of the heavier lanthanides than bastnasite ores. The cost of the lanthanide metals is in line with their rarity: the most expensive is lutetium, followed by thulium, both commanding prices of several thousand US dollars per kilogram.

What follows is an account of the lanthanides as they impinge on terrestrial and human affairs. In this account no distinction has been made as to the part played by individual elements; they are dealt with as an undifferentiated collection of elements.

Table 2 The abundances of the lanthanides in the Earth's crust and common ores (averages)

Atomic number	Element	Symbol	Abundance (p.p.m.)	Monazite ore	Bastnäsite ore
57	Lanthanum	La	32	20%	33%
58	Cerium	Ce	68	43%	49%
59	Praseodymium	Pr	9.5	4.5%	4.3%
60	Neodymium	Nd	38	16%	12%
61	Promethium	Pm	0	0	0
62	Samarium	Sm	8	2.5%	0.8%
63	Europium	Eu	2	0.1%	0.1%
64	Gadolinium	Gd	8	1.5%	0.2%
65	Terbium	Tb	1	500 p.p.m.	160 p.p.m.
66	Dysprosium	Dy	6	0.6%	300 p.p.m.
67	Holmium	Ho	1.4	50 p.p.m.	50 p.p.m.
68	Erbium	Er	4	0.2%	35 p.p.m.
69	Thulium	Tm	0.5	200 p.p.m.	10 p.p.m.
70	Ytterbium	Yb	3	0.1%	5 p.p.m.
71	Lutetium	Y	0.5	200 p.p.m.	1 p.p.m.

HUMAN ELEMENTS

Lanthanides do not move up the food chain easily, and any amount that does has little toxic effect. When animals have been fed insoluble compounds, such as lanthanide oxides and sulfates, it has proved impossible to register any toxic effect no matter how much was given. Some rats were fed the equivalent of 1 gram per kilogram body weight (which would be equivalent of an average-sized human eating 70 grams) without any effect. Even 10 times this amount had no effect when lanthanum oxide itself was used.

It is difficult to identify the various amounts of the different lanthanides in humans, but they are present; the levels are highest in bone, with much smaller amounts being present in the liver and kidney. In most human organs the levels are measured only in parts per billion although lanthanides have been found at levels of parts per million in the eyes, kidneys and spleen. The amounts in the lens of the eyes increase when cataracts form.

Studies on workers in a smelter who were exposed to lanthanides as part of the job showed twice the level of lanthanides in their body than normal – though that was still not very much. When they moved on to other jobs, and were contacted years later, it was found that the level of lanthanum was unchanged, suggesting that once these metals have been absorbed by the body they stay put.

A lot of research has been carried out into the behaviour of lanthanides in animals, generally by injecting a radioactive isotope and following the way it moves and finding out where it ends up in the animal's body and how long it takes to be excreted. Injections of the lighter lanthanides into rats led to so-called 'rare-earth fatty liver' in which this organ become pale and enlarged with fatty deposits.

When lanthanides are injected directly into the brain they have the same pain-killing ability as opiates such as morphine.

Broken skin is sensitive to lanthanides: exposure results in ulcers which are very slow to heal – terbium will even irritate unbroken skin. Lanthanide dusts irritate the eyes, neodymium especially so. Workers exposed to the fumes and dusts coming from arc lights, in which lanthanides are used, have been known to develop a lung condition called 'rare-earth pneumoconiosis'.

In parts of China, lanthanide salts are added to fish farm tanks to prevent diseases of the scales, gills and intestines caused by microbes. It is even claimed that fish treated in this way grow more quickly. The Chinese also produce a fertilizer with lanthanide salts added, called *Nong-le*, which is said to enhance crop yields of sugar-cane, apples, wheat and rice. Using the fertilizer spreads about 500 grams of lanthanide salts per hectare.

MEDICAL ELEMENTS

Early clinical trials of lanthanide salts for the treatment of tuberculosis, cholera and leprosy were encouraging but eventually led nowhere. There was some concern that these heavy metals might be toxic, but fears were soon allayed when they proved to be harmless. Radioactive isotopes were tried in anti-cancer treatments following the discovery that cancer cells tended to absorb lanthanides, but again this did not prove to be a successful form of treatment. The lanthanides do have some effect on human metabolism, lowering blood pressure and cholesterol levels, preventing blood coagulation and reducing appetite and, in animals, they prevent atherosclerosis. In Hungary a lanthanide cream called Phlogodym is used to reduce inflammation, and cerium nitrate has proved successful as an antiseptic for burn wounds.

ENVIRONMENTAL ELEMENTS

Lanthanides are present in trace amounts in many ores because their tripositive ions, which have the general formula Ln^{3+}, are about the same size as those of calcium, Ca^{2+}, which they can partly replace in the mineral lattice. Consequently when calcium ores (of which there are many) break down under weathering, they release lanthanides into the environment, but these are soon rendered immobile by contact with carbonate and phosphate ions, with which they form insoluble salts. The result is that very little finds its way into living systems.

Lanthanides were detected in plants in the 1870s and the amounts were related to the soil on which they grew. Generally, roots reject lanthanides and the levels in plants are less than 1/10 000th of the level of these metals in the soil in which they are growing. Some plants can concentrate them; for example, hickory trees in the eastern USA have been found to have over 2000 p.p.m. of lanthanides, mainly lanthanum, cerium and neodymium, which are the most abundant of the rare-earth metals. Mosses and ferns also concentrate lanthanides.

Table 3 Lanthanides in cow manure and municipal sewage

Element	Abundance (p.p.m. dry weight)	
	in cow manure	in human sewage
Lanthanum	23.7	12.7
Cerium	55.0	41.9
Praseodymium	10.7	4.3
Neodymium	2.5	2.5
Promethium	0	0
Samarium	5.2	3.5
Europium	0.7	3.7
Gadolinium	1.5	6.8
Terbium	0.3	1.4
Dysprosium	1.0	4.7
Holmium	0.36	0.32
Erbium	0.70	1.16
Thulium	0.14	0.39
Ytterbium	1.76	0.60
Lutetium	0.60	0.12

The amounts of the various lanthanides that become part of the human food chain can be judged from Table 3. While the ratios in cow manure do not correspond exactly with those in human sewage, they are somewhat similar and indicate that the rare-earth elements are part of our dietary intake and may come mainly from dairy products.

Lanthanide minerals are very insoluble in water, which explains why their abundance in natural waters is invariably low.

Lanthanide ores

Lanthanides are to be found in massive rock formations such as basalts, granites, gneisses, shales and silicate rocks, but only in low concentrations (10–300 p.p.m.). None is an exploitable source of the lanthanides. There are over a hundred lanthanide minerals, but most of these are so rare as to be little more than collectors' curios. However, there are a few that are mined specifically as a source of rare-earth metals. These rare-earth minerals were formed deep in the Earth's crust, where they precipitated from superheated solutions or molten rock at very high pressures.

The most important mineral sources are monazite, which is a phosphate ore, and bastnäsite, which is a fluoride carbonate ore. Both contain the full spectrum of lanthanide elements although cerium and lanthanum predominate (see Table 2). Monazite sands have built up over millions of years along the coasts of southern India, western Australia, Brazil, South Africa and Sri Lanka. Unfortunately, monazite also has high levels of the radioactive element thorium. The largest deposits of bastnäsite, which is free of this undesirable impurity, are in China and the USA. Other minerals, such as cerite, loparite and samarskite, are rich in lanthanides but are relatively rare. The main lanthanide-producing countries are China (Inner Mongolia) and USA (Mountain Pass, California).

World reserves of lanthanide metals are estimated to exceed 110 million tonnes. The largest bastnäsite deposit is in China and accounts for 75% of the world's reserves. Other producers are the USA, which has 14%, India 4%, and South Africa and Australia with 2% each. The remaining 3% is to be found in Brazil, Thailand, Sri Lanka and Malaysia. At one time, the open-pit mine at Mountain Pass in south-eastern California supplied most of the world's needs for rare-earth metals.

Today, China produces around 25 000 tonnes of lanthanides per year, the USA slightly less, Russia and Australia around 8000 tonnes and India, Brazil and Malaysia around 2000 tonnes each, making a total world output of around 70 000 tonnes a year. Only a small amount of this is separated into individual elements, most being used as a mixture of lanthanides known as *mischmetall* (see below). In this mixture cerium accounts for about half, lanthanum about a quarter, neodymium around 15%, and the rest together around 10%.

Monazite is processed by treating the ore with alkali, which dissolves the thorium but converts the lanthanides to their hydroxides, which are insoluble and so precipitate. Care has to be taken to remove all traces of the radioactive thorium. Some beach sands are dredged for their titanium and zirconium ores and monazite is also part of these ores. Bastnäsite is a by-product of iron ore mining in China, but the largest deposit of rare earths is found at Bayan Obo in Inner Mongolia, China; this is a mixture of bastnäsite and monazite in the ratio 70:30.

The French chemical company Rhodia (formerly Rhône–Poulenc) is the world's largest supplier of rare-earth elements, processing them at La Rochelle (France), Freeport (Texas, USA) and Pinjawa (Australia). Monazite is processed with sodium hydroxide (caustic soda); its phosphate component is retrieved and also marketed. Bastnäsite is processed by leaching with hydrochloric acid and then heated to remove the carbonate as carbon dioxide.

After these primary digestion processes, the mixture of rare-earth elements has then to be separated. The early chemists tried a slow method of endless crystallizations, each one giving a slightly better separation, but such methods are not suitable for commercial processes. A better method is ion-exchange, in which a solution of the rare-earth elements flows down a column packed with special resin which picks up the metals at different rates. This method, common about 50 years ago, has now been superseded by liquid–liquid extraction, which involves a dissolved compound moving between an oil and a water layer. Liquid–liquid extraction can be very effective because the various metals and salts differ in solubility, but it is a slow process, requiring 60 or more stages.

It is difficult to produce the metals themselves because they have high melting points and are easily oxidized. Two methods are used: chemical reduction and electrolytic reduction. The former consists of heating the metal chloride, or better still the fluoride, with calcium in a tantalum crucible under an atmosphere of argon gas. The electrolytic method involves passing an electric current through a mixture of the metal chloride and sodium chloride in a graphite-lined steel cell, which serves as the cathode, with a graphite rod, which serves as the anode. The lanthanides that are lighter and have lower melting points – cerium, samarium and europium – can be obtained by this method, which is also used for ytterbium. Most lanthanide metals are produced by the calcium process.

The promethium puzzle

One lanthanide is missing on Earth and that is promethium, element number 61, although it was present when the planet formed about 4.57 billion years ago. It disappeared because all its atoms were radioactive and none had a lifetime long enough to survive until now (see promethium, p. 343). (There are still a few atoms of promethium in the Earth's crust, such as promethium-147, produced as a result of nuclear fission of uranium atoms.)

ECONOMIC ELEMENTS

Fluorescent light relies on lanthanide phosphors to convert the emission from a mercury arc, which has wavelength in the ultraviolet part of the spectrum (254 nanometres) into visible light (wavelength 400–700 nanometres). Such lamps are more correctly called trichromatic fluorescent bulbs because they rely on these phosphors to emit bands of light in the blue, green and red parts of the spectrum, i.e. at 450, 550 and 610 nanometres, respectively. The resulting visible radiation is perceived as white light by the eye.

In trichromatic lights, the blue emission band comes from europium, the green from a mixture of lanthanum, cerium and terbium, and the red from a mixture of yttrium and europium. Such light bulbs not only last for 5 years, or more, but an 18 watt bulb will give out as much illumination as a conventional 75 watt incandescent bulb.

Specialized high-intensity lighting using a high-pressure mercury discharge (halogen lamps) also relies on lanthanide elements in the form of metal halides. These dissociate in the arc column, forming free metal atoms which emit visible light in many parts of the spectrum.

An yttrium–cerium phosphor is used also to convert the visible mercury emission at 435 nanometres into more useful light.

Mischmetall[67]

The rare-earth elements need not be separated for some of their uses. Together they are converted into their chlorides and then melted and electrolysed to produce a mixture of metals called *mischmetall*. This consists of about 50% cerium, 25% lanthanum and 25% other rare-earth metals. A form of *mischmetall* was originally patented in 1903 for lighter flints as so-called Auer metal, which was two-thirds *mischmetall* and one-third iron; these are still made for cigarette lighters.

Mischmetall is used in steel-making where it acts to scavenge oxygen and sulfur, thereby improving the physical properties to give high-strength, low-alloy steels. This use is declining, however, as modern steel-making technologies produce 'cleaner' steels, i.e. ones that have lower levels of oxygen and sulfur, but it is still used in China. Iron, aluminium

[67] This is the original name, but in English it is often written as two words and spelt misch metal.

and magnesium become stronger and more workable if they have some *Mischmetall* added to them. Some goes to making cobalt and nickel alloys which are magnetic. *Mischmetall* is also used as a 'getter' in vacuum tubes, i.e. it is ignited to remove unwanted traces of air.

CHEMICAL ASPECTS

All the lanthanides are chemically similar, which is what made them so difficult to separate by the traditional methods available to chemists in the early nineteenth century. The reason for this similarity is that as we go from one element to the next in the series, the additional electron is added not to the outermost orbit of the atom, which would greatly affect its chemical behaviour, but it goes into an orbit deeper in the atom where its effect is muted. The result is that the atoms of the lanthanide elements present the same three outer electrons to the world. Consequently their chemistry is dominated by loss of these to form positively charged ions, Ln^{3+}. This is the form of these elements that is the most stable.

Rather oddly, as the atomic number of the lanthanides increases across the series, and the number of electrons in the atoms increases, the atoms grow smaller, rather than increasing as one would expect. This phenomenon is known as the lanthanide contraction and is due to the extra electrons going into an inner orbit within the atom.

The lanthanides played an important role in the discovery of nuclear fission. When samples of uranium were exposed to neutrons, the scientist in charge of the experiments, Otto Hahn, working at the then Kaiser Wilhelm Institute for Chemistry in Berlin, was puzzled to find that several lanthanides were present in the samples. We now know that these consisted of the radioactive isotopes, lanthanum-140 (half-life 41 hours), cerium-141 (23 days), cerium-143 (4 hours), cerium-144 (284 days), praseodymium-143 (14 days), neodymium-147 (11 days), promethium-147 (2.6 years), samarium-151 (93 years) and europium-154 (16 years). Hahn wrote about his findings to his former colleague Lise Meitner who, being of Jewish descent, had fled to Sweden because the Nazi race laws excluded her from working at the Institute. It was she who realized what Hahn's findings meant: that uranium was undergoing nuclear fission, splitting into lighter elements.

Thankfully the lanthanides in the fall-out from nuclear fission bombs and nuclear accidents are generally short-lived and their insolubility means that they have not posed the same threat as some of the other radioactive elements released into the environment.

Lawrencium

This is element atomic number 103 and, along with other elements above 100, it is dealt with under the transfermium elements on p. 457.

Lead

Pronounced led, the name comes from the Anglo-Saxon word for the metal, *lead*. The chemical symbol, Pb, derives from *plumbum*, the Latin name for the metal.

French, *plomb*; German, *Blei*; Italian, *piombo*; Spanish, *plomo*; Portuguese, *chumbo*.

HUMAN ELEMENT

Lead in the human body	
Blood	0.2 p.p.m.
Bone	3–30 p.p.m.
Tissue	0.2–3 p.p.m.
Total amount in body	120 milligrams

Lead has no biological role. It is a cumulative poison and used to be a common occupational hazard, but is now much more strictly controlled. The symptoms of mild lead poisoning – headaches, stomach pains and constipation – can easily be overlooked or attributed to other causes. A single dose of a lead compound is unlikely to kill; what is more insidious is its unsuspected absorption over a period of time. The lead which penetrates the body's defences tends to remain for a long time, locked away in the bones as lead phosphate.

Today, the level of lead in bones is about twice that found in bones taken from prehistoric burials, but it is many times less than that in bones of people who lived in ancient Rome, the Middle Ages and the early industrial period through to the mid-20th century. In areas where leaded petrol is still used, the dust of city streets is heavily contaminated with it, and it can enter the body via the lungs.

FOOD ELEMENT

A little lead gets into the food chain because all plants contain some lead, albeit a very small amount. Vegetables contain lead only in minute amounts, such as cucumber which has 24 p.p.b. (fresh weight), while sweet corn has 22 p.p.b., cabbage 16 p.p.b., tomatoes 2 p.p.b. and apples a mere 1 p.p.b. However, lettuces grown on soil near a lead processing plant in Canada were found to have up to 3000 p.p.b. (3 p.p.m.); even this level would be unlikely to produce any symptoms of lead poisoning.

Most lead passes through the body without being digested. The average daily intake of about 1 milligram results in only about 0.1 milligram being absorbed. Exposure to lead in the diet depends on several factors but most comes through drinks. Soluble lead can leach from old water pipes, badly glazed pottery and lead crystal decanters.[68] Dining off antique tableware with a lead glaze can also add lead to a meal and it has been found that some dishes can contaminate food with lead to quite high levels. A likely source of contamination for young children can be their habit of nibbling flakes of paint from old woodwork.

Historically, lead was used to sweeten wine, which it did by reacting to form lead acetate (sugar of lead). In the Middle Ages, this often led to mysterious local outbreaks of 'colic' whose symptoms were stomach cramps, constipation, weariness, anaemia, insanity and lingering death. There was Picton colic, Devon colic, and also dry gripes, which afflicted the West Indies and Massachusetts in the seventeenth century. All were caused by lead-contaminated drinks, in these cases of wine, cider and rum, respectively.

In the eighteenth century the English physician George Baker proved that Devon colic was caused by the lead-lined apple presses used in cider making. Adding a little lead to wine, to improve its flavour and keeping quality, was a practice that continued well into the nineteenth century: a piece of lead shot was added to each bottle and this slowly dissolved.

The World Health Organization reduced its recommended upper limit for the level of lead in drinking water from 50 p.p.b. to 10 p.p.b. in 1995, saying that countries should aim to achieve this by 2010. In the UK, however, there are still a few homes in soft water districts with old lead piping where the lead content of drinking water is above the previous limit set by the World Health Organization.

MEDICAL ELEMENT

The effects of lead were known to the Romans. For example, the architect and engineer Marcus Vitruvius, who lived in the first century BC, observed that labourers in lead smelters had pale complexions. The Greek physician Hippocrates (460–377 BC) described a severe attack of colic in a patient who was a lead miner.

Despite the health risks associated with lead, it was used by doctors for around 2000 years to treat illnesses. The practice started with Tiberius Claudius Menecrates, physician to the Emperor Tiberius (ruled from AD 14 to 37), who invented diachylon plasters,

[68] Research in Canada showed that port wine stored in lead crystal decanters contained around 2000 p.p.m. lead (0.2%) after being left for a year, while Scotch whisky had around 1500 p.p.m. (0.15%).

which contained a paste made of lead oxide and olive oil. They were prescribed for skin complaints, although sometimes women sought to procure abortions by eating the paste.

In the eighteenth century Thomas Goulard, a surgeon of Montpellier, France, produced a medicine by boiling lead oxide in vinegar, which reacts to form soluble lead acetate; he recommended this not only for skin diseases, such as abscesses, erysipelas and ulcers, but also for piles and even for cancer, which it clearly could not cure.

Victorian doctors went further and gave lead acetate with opium as a cure for diarrhoea – it was effective because the lead paralysed the gut, while the opium deadened the associated stomach pains. Internal bleeding and gonorrhoea were also treated with lead acetate, and doses as high as a gram a day were given. Lead medicine was likewise seen as a remedy for neuralgia, persistent coughs and 'hysteria'.

LEAD AND GOUT

In the eighteenth and nineteenth centuries, many prominent people suffered from gout. The condition is caused by excess uric acid in the body crystallizing in the joints, making movement very painful. Among those affected were Benjamin Franklin, one of the founding fathers of the USA, British Prime Minister William Pitt, Lord Tennyson the poet, Charles Darwin the biologist, and John Wesley, the co-founder of Methodism.

The popular belief, that gout was a sign of too rich living and drinking port, may have had some foundation. In the twentieth century, it was found that more than a third of those suffering from gout had high levels of lead in their blood. It now seems likely that earlier generations may have exacerbated a propensity to gout by a fondness for port wine, which was invariably 'improved' with lead acetate, and kept in lead crystal decanters.

Andrew Jackson (1767–1845), seventh President of the USA (from 1829 to 1837), was exposed to too much lead, as indicated by samples of his hair from 1815 which were found to have 131 p.p.m. of lead, confirming some historians' belief that he suffered from plumbism, the technical name for chronic lead poisoning. Its source remains a mystery: did it come from medicines or alcoholic drinks? Ludwig van Beethoven's hair has also been found to have 100 times more lead than normal, showing that he was exposed to this metal, which might also explain his erratic behaviour and maybe even his deafness.

Why is lead so dangerous to health? It is absorbed into the blood stream where it deactivates the enzymes that make haemoglobin. This results in the build up of the precursor molecule aminolaevulinic acid (ALA). It is ALA that causes the various symptoms of lead poisoning. It paralyses the gut, hence the stomach cramps and constipation; it results in excess fluid in the brain, causing pressure headaches and loss of sleep; and it affects the reproductive system, leading to infertility and miscarriages. Prolonged exposure to lead eventually results in anaemia, adding the problems associated with that condition.

Metallic lead still finds use in hospitals, but as a protective shield for employees whose job it is to operate diagnostic imaging devices such as X-ray scanners and computerized axial tomography (CAT scanner) equipment.

ELEMENT OF HISTORY

Lead has been mined for more than 6000 years, and was certainly known to the Ancient Egyptians who used white and red lead in pigments and cosmetics as well as casting the metal itself into small figures. Cosmetics found in tombs of the second millennium BC were found to consist of galena (black lead sulfide), cerussite (white basic lead carbonate), laurionite (white basic lead chloride) and phosgenite (brown lead chloride carbonate); since the latter is rare in nature it was deduced that the Egyptians were able to make it chemically from lead oxide and salt.

The proverbial heaviness of lead is mentioned a few times in the Bible, e.g. in Exodus which described how the Red Sea engulfed the Egyptians: 'they sank as lead in the mighty waters.' That passage was written about 900 BC but referred to events that allegedly happened in the thirteen century BC.

The Greeks mined lead on a large scale from 650 to 350 BC when they exploited a large deposit at Laurion near Athens. Although their aim was to extract silver, of which the mine produced 7000 tonnes, it had associated with it more than 2 million tonnes of lead. It was not long before they found uses for this by-product. The ancient Greek writer, Theophrastus (372–286 BC), described how it could be turned into white lead, which was used as a pigment for paint. Lead strips were exposed to vinegar fumes (acetic acid vapour) which formed a white deposit after about 10 days. This was scraped off to expose more fresh metal and the process repeated until all the lead had reacted. The white deposit was then ground to a powder, boiled in water and left to settle to produce white lead.

The Romans were the first civilization to employ lead on a large scale, mining it mainly in Spain and Britain. They used this easily worked metal for water pipes, pewter tableware and, as white lead, in paint, as well as using red lead as colouring for interior decoration.[69] They also used lead to debase their silver coinage.

One of the most dangerous uses the Romans put lead to was for making lead pans in which they boiled down grape juice or sour wine to produce the sweetening agent known as *sapa*, which was essentially lead acetate syrup. This was widely used in cooking. According to one theory, this explained the low birth rate of the Romans; indeed, the population of the Empire stagnated at around 50 million, even though the level of hygiene was high. As the Empire declined so did the need for lead.

Small lead nuggets have been found in pre-Columbian Peru, Yucatan and Guatemala, but the native North American civilizations made less use of the metal.

After the Dark Ages, lead mining began again and, in addition to the traditional uses, new outlets were found for lead, such as pottery glazes, bullets and printing type. A lot was used to weatherproof the roofs of large buildings.

[69] Red lead was made by heating white lead. Another colourful lead pigment is chrome yellow, which is lead chromate.

ECONOMIC ELEMENT

Galena (PbS) is the main lead ore and there are also deposits of cerussite ($PbCO_3$) and anglesite ($PbSO_4$) which are mined. There have also been worked deposits of pyromorphite (a lead chloride phosphate) and boulangerite (lead antimony sulfide), the latter still being mined in France. Galena is mined in Australia, which produces 19% of the world's new lead, followed by the USA (13%), China (12%), Peru (8%) and Canada (6%). Some is also mined in Mexico and West Germany. The ores are crushed, mixed with coke and limestone, and roasted in a furnace. World production of new lead is 6 million tonnes a year, and workable reserves total an estimated 85 million tonnes, which is less than 15 years' supply.

Lead is refined to remove the silver it contains (as much as 1.2 kilograms per tonne). This is done by adding zinc to the molten metal and allowing it to cool slowly until the zinc settles out as a separate layer carrying the silver with it. This is removed and the lead is then heated under vacuum to drive off the remaining zinc, giving a product that is 99.99% pure.

In the twentieth century many of the traditional uses of lead declined or were banned, but some new ones were found such as the insecticide lead arsenate, now no longer used, and tetraethyl lead (TEL) used as an additive to raise the octane level of petrol (gasoline); this additive is now being phased out. Lead is still used for car batteries, to protect underground cables and to shield humans against dangerous radiation. Lead acetate is still the active ingredient in some hair gels for men because it will turn grey hair dark brown.

Around 35% of lead is used to make the electrodes in car and lift-truck storage batteries; the anode is lead, the cathode a paste of lead oxide on a lead alloy grid. The second largest use is in the glass of computer and television screens, where it shields the viewer from radiation; the average monitor contains about 250 grams of lead. (Lead is also used to protect against neutrons, because, unlike other metals, it does not become radioactive.)

Other uses are in sheeting, cables, solders, lead crystal glassware, ammunition, bearings and as weights in sports equipment (for weight-lifting and to balance golf clubs). Lead is very poor at transmitting sound and vibrations, so it is added to plastic sheeting and tiling designed to block out noise.

Lead is still used in architecture as roof cladding and for stained glass windows. It provides protection that will last for centuries, even in industrial and coastal regions, and does not cause discoloration of the surrounding stone or brickwork. The protective layer slowly changes over the centuries from lead oxide to basic lead carbonate, then to lead carbonate, and via lead sulfide to lead sulfate in urban or industrial environments. In all cases the lead compound that is formed is insoluble and, unlike rust, does not flake from the surface.

'Chemical'-grade lead, which contains small amounts of silver and copper, is used in the chemical industry because it resists the action of corrosive liquids.

A lot of lead is recycled, especially that from storage batteries and pipes that are removed from old housing.

ENVIRONMENTAL ELEMENT

Lead in the environment	
Earth's crust	14 p.p.m.
	Lead is the 36th most abundant element.
Soils	approx. 23 p.p.m., range 2–190 p.p.m.
Sea water	2 p.p.t.
Atmosphere	trace

There are two components of the natural environment which ensure that most lead on the planet remains immobile: sulfur and phosphorus. The sulfur (as sulfide, S^{2-}) and phosphorus (as phosphate, PO_4^{3-}), form lead compounds that are extremely insoluble.

Plants will tolerate quite high levels of lead in soil, up to 500 p.p.m., before it affects their growth. At low concentrations it appears even to stimulate plant growth despite its not being an essential element for any form of life. In parts of the UK that have been heavily industrialized for 250 years, the levels of lead in some soils exceed 1500 p.p.m. in urban gardens and more than 20 000 p.p.m. near old lead mines (both dry weight values). The danger of such high levels is mainly to micro-organisms which are a key part of the biological activity of soils. Farmed soil is generally protected by the application of phosphate fertilizers, which forms insoluble lead phosphate, and in non-farmed soils the formation of sulfides by some micro-organisms will also render lead inactive.

Lead in lake sediments has preserved a record of the contamination of the global environment over the ages and so has the snow which has fallen on Greenland. In layers that represent the times before humans began to mine and use lead, there is a little lead (0.5 p.p.t.) present, coming from atmospheric dust. However, during the Roman era the level rose to 2 p.p.t. (it has been estimated that the Roman smelted around 80 000 tonnes per year of the metal).

The lead analysis of a peat bog on a Swiss mountain has allowed William Shotyk of Bern University to construct an atmospheric lead profile going back 14 500 years. He has measured not only the amount of lead in the various layers but the ratio of the two isotopes, lead-206 and lead-207. His results suggest that agriculture started in Europe around 4000 BC with forest clearing and soil tilling, which put soil dust into the atmosphere, carrying lead with it. In about 1000 BC, the isotope ratio changed so that the level of lead-207 increased more than the level of lead-206; this happened when serious lead mining began, probably by the Phoenicians. It increased even more in the nineteenth century when Australian lead ores were mined because these had even more lead-207.

In 1921, Thomas Midgley found that adding TEL to petrol boosted its performance, and by the 1960s all cars were running on leaded petrol. The amount of lead in the snows of the Arctic reflected this and reached a maximum of 300 p.p.t. by the late 1970s, although levels have since declined.

A widely publicized report of 1979 claimed that reduced intelligence and lack of concentration among city children could be attributed to environmental lead pollution. Such children were exposed to traffic fumes from leaded petrol and from leaded paint in older homes. In the USA, inner-city children are automatically given a blood test for lead before starting school. In 1993 more than 400 000 New York State children had their blood lead content measured and about 5% were found to have worryingly high levels. However, official environmental organizations have not been able to support the theory that lead reduces intelligence, and it is now accepted that the dangers were much exaggerated although lead is still regarded as a cause for concern.

In 1997, a study was carried out by the University Hospitals of Lund and Malmö, Sweden, on 38 men who had worked at a lead smelter for 10 years. The conclusion was that, despite having blood lead levels of around 0.4 p.p.m., which is twice the average, they behaved no differently in memory and behavioural tests when compared with a control group whose blood lead levels were low (average of 0.04 p.p.m.).

A report by Columbia University in the USA in 1998 found that sediments in the lake of New York's Central Park showed the highest airborne lead levels in the 1930s, before leaded petrol was introduced, and linked the cause to municipal waste incineration, saying that this was always likely to be the major source of lead pollution in cities.

Another environmental threat, at least to wildlife, were the lead weights used as sinkers by fishermen. These were responsible for the deaths of many swans who scooped up and swallowed lost sinkers as they fed in the mud along the bottom of rivers. These have now been supplanted by non-toxic weights (see tungsten, p. 474).

CHEMICAL ELEMENT

Data file	
Chemical symbol	Pb
Atomic number	82
Atomic weight	207.2
Melting point	334°C
Boiling point	1740°C
Density	11.4 kilograms per litre (11.4 grams per cubic centimetre)
Oxides	PbO, PbO_2, Pb_3O_4 (red lead)

Lead is a soft, weak, ductile, dull silver-grey metal that is a member of group 14 of the periodic table of the elements. It tarnishes in moist air but is stable to oxygen and water. It dissolves in nitric and acetic acids.

Lead has four natural isotopes: lead-208, which accounts for 52.5%, lead 206 (24%), lead-207 (22%) and lead-204 (only 1.5%). None is radioactive, although there have been suggestions that they may be so but with such long half-lives that the decay of a lead nucleus has never been observed. The first three of these are the end-products of the radioactive decay of elements of higher atomic number, namely uranium-238 (which ends up as lead-206), uranium-235 (which becomes lead-207) and thorium (which becomes lead-208). By comparing the ratio of these metals to lead in a rock it is possible to estimate its age. One technique is the well-known dating method of measuring the uranium-to-lead ratio.

ELEMENT OF SURPRISES

1. In May 1845, John Franklin set off to search for the supposed North-West Passage around the top of Canada, which was seen as an alternative route from the Atlantic to the Pacific. His two ships, *Erebus* and *Terror*, were well provisioned, with 5 years' supply of food for the 129 officers and crew, whose quarters were even equipped with central heating. In August that year the ships were seen in Baffin Bay, but then they disappeared. Five years later a second expedition set out to find them, but all they ever discovered were three graves on Beechey Island. These were of crew members John Torrington, John Hartnell and William Brain, who had died in 1846.

   In 1988, researchers at the University of Alberta, Canada, were allowed to exhume and analyse the perfectly preserved remains of the three men. This showed such high levels of lead that it seems they almost certainly died of lead poisoning, probably exacerbated by scurvy. Not only that, the researchers were able to prove, by analysing empty cans found nearby, that the lead came from the solder of the canned food that they ate. The ratio of lead isotopes in the victims was the same as that of the lead solder, and quite different from the ratio of lead isotopes in local Inuit people.

   The first commercial food cannery was that of Messrs Donkin & Hall of Bermondsey, London. It began to supply the Royal Navy with canned meats, vegetables and soups in the 1820s. The cans were filled through a small hole at the top which was then sealed by having a disc soldered over the hole. While these cans preserved food by remaining air-tight – some Donkin & Hall cans survived for more than 100 years – lead slowly leached from the solder into the contents. This, combined with a diet low in vitamin C, is now thought to be the reason why Franklin and all his crew perished.

2. The effect of using white lead in paintings can be seen in many ancient manuscripts. Faces that were originally painted pale pink are now deep black. The reason is that they were exposed to hydrogen sulfide given off by the coal and gas fires that were popular in the nineteenth century. This reacted with the white lead to form black lead sulfide.

Lithium

Pronounced lith-iuhm, the name comes from the Greek *lithos*, meaning stone.
French, *lithium*; German, *Lithium*; Italian, *litio*; Spanish, *litio*; Portuguese, *litio*.

COSMIC ELEMENT

Lithium is rare in the Universe, although it was one of the three elements, along with hydrogen and helium, to be created in the immediate aftermath of the Big Bang. However, young stars have many times the amount of lithium produced at that event, indicating that other processes can produce it, despite the fact that lithium is destroyed in the interior of a star if the temperature exceeds 2.4 million degrees, which in most stars it does. Destruction occurs when a lithium nucleus is struck by a proton and converted to two atoms of helium.

The presence of lithium has been used by astronomers to distinguish brown dwarf stars from red dwarf stars, both of which are much smaller than our Sun. Red dwarfs are hot enough to destroy lithium, so if this metal shows up in the spectrum of light from the star, as indicated by a line at a wavelength of 670.7 nanometres, then it must be a brown dwarf.

In 1992, an orange star was discovered, in the star system called V404 Cygni, whose spectrum revealed it to have enormous amounts of lithium. A possible explanation was the presence of a nearby black hole that was dragging the lithium from the centre of the star to its cooler surface, lithium being heavier than the hydrogen and helium that make up most of a star. Other lithium-rich orange stars have also been found, one of which, Centaurus X-4, orbits a neutron star which may be having the same effect on its partner.

HUMAN ELEMENT

Lithium in the human body	
Blood	4 p.p.b.
Bone	1.3 p.p.m.
Tissue	approx. 24 p.p.b.
Total amount in body	7 milligrams

Lithium is easily absorbed by plants, so the amount found in plants can be a good guide to the lithium status of the soil. No plant species has yet been found for which the element is an essential nutrient, but for some it does appear to affect growth and development if levels reach around 200 p.p.m. While too much lithium can be toxic to some plants, this can easily be overcome by liming the soil, since calcium inhibits uptake of the lighter metal.

The amount of lithium in plants varies widely, with Solanaceae (the family to which potatoes belong) having most – as much as 30 p.p.m. (dry weight) in some cases. Corn grains have around 0.05 p.p.b., oranges 0.2 p.p.m., lettuce 0.3 p.p.m. and cabbage 0.5 p.p.m. (dry weight).

Goats that are deficient in lithium have been observed to put on less weight than those that are given a normal diet. In this respect lithium may be an essential element, at least for one animal, but this has not been proved. The human body appears to have no biological need for lithium, but because this element is widely dispersed in nature we take in some each day. We absorb some of this, although most is excreted.

Lithium is moderately toxic by ingestion but there are wide variations of tolerance. In the 1940s lithium was implicated in the deaths of patients who were given lithium chloride as a salt substitute. Even lithium carbonate, which is used in psychiatry, is prescribed at doses near to the toxic level.

MEDICAL ELEMENT

In the nineteenth century, lithium enjoyed a vogue as a treatment for gout, the painful condition that develops when sharp crystals of uric acid form between the joints, especially of the feet. Uric acid is not very soluble, so that once crystals form they may take a long time to dissolve again. Because the lithium salt of uric acid is very soluble, it was thought that taking lithium-rich spa waters would aid recovery – a theory advanced by a Dr Ure in 1843. This belief held sway even in the twentieth century until, in 1912, a Dr Pfeiffer showed that lithium in fact *slowed down* elimination of uric acid from the body in patients with gout. In any case the lithium in spa waters was too dilute to have any therapeutic effect.

In 1949, an Australian doctor, John Cade, senior medical officer in the Victoria Department of Mental Hygiene, was experimenting with guinea pigs and injecting them with urine taken from manic-depressive patients in the hospital in the hope of proving that their condition was caused by an excess of some body chemical. The animals died. Cade thought this might have been due to excess uric acid in the urine. To test whether this was so, he injected guinea pigs with a solution of the lithium salt of uric acid and noted that soon they became very lethargic, but recovered within a few hours. He then tested the effect the lithium might be having by injecting them with a 0.5% solution of lithium carbonate. To his surprise these normally highly-strung animals became docile, and indeed were so calm that they could be turned on their backs and would lie placidly in that position for several hours.

Oblivious to the dietary dangers of too much lithium, Cade then gave lithium carbonate to his most mentally disturbed patient, who had been admitted to a secure unit 5 years earlier. The man responded so well that within days he was transferred to a normal hospital ward and within 2 months was able to return home and take up his old job. Other manic depressives responded similarly. Cade's treatment was tried by other doctors with equally good results, and within 10 years lithium medication was used throughout Europe; by 1970, it was being used throughout the USA as well.

With the right dose of lithium, generally given as lithium carbonate, a patient can be kept from either of the extremes, mania and depression. The dose is adjusted so that the level of lithium in the plasma is between 3.5 and 8 milligrams per litre. A blood level of 10 milligrams of lithium per litre results in mild lithium poisoning, and when it reaches 15 milligrams per litre there can be side effects, such as confusion and slurred speech. Only if the lithium level in the blood were to exceed 20 milligrams per litre would there be a risk of death.

How lithium works is still not known for certain, but it is now thought that it interferes with an inositol phosphate chemical messenger which is being over-produced. This messenger is made in the brain from glucose, and lithium appears to reduce the amount produced to more normal levels.

ELEMENT OF HISTORY

Lithium was discovered in 1817 by Johan August Arfvedson (1792–1841) at Stockholm, Sweden. The metal itself was first isolated, in 1821, by William T. Brande. The first lithium minerals to be discovered were petalite and spodumene; these were found on the Swedish island of Utö by the Brazilian mineralogist, Jozé Bonifácio de Andrada e Silva, when he toured Europe in the 1790s. The curious nature of these new minerals was noted, in that they were difficult to analyse and that they gave an intense crimson flame when added to a fire; the colour is due to the molecule LiOH.

In 1817, Arfvedson analysed petalite and realized that it contained a previously unknown metal, which he called lithium because it had come from a stone. He announced his discovery in 1818, correctly identifying lithium as a 'new' alkali metal and a lighter version of the element sodium. Soon after that, he found that spodumene and another new mineral, lepidolite, also contained the element.

Although lithium is an alkali metal, like sodium and potassium, Arfvedson had not been able to separate it by electrolysis, and while another chemist, William Brande, did obtain a tiny amount of the metal this way, it was not enough on which to make measurements. Humphry Davy also managed to isolate a small amount, but again was unable to measure its physical properties. It was not until 1855 that the German chemist Robert Bunsen, and the British chemist Augustus Matthiessen (1831–70), independently succeeded in isolating enough of the metal to study it. They achieved this by the electrolysis of molten lithium chloride.

Soon after its discovery, lithium was found in other minerals and even in spa waters at Karlsbad, Marienbad and Vichy. It was detected in sea water by means of a spectroscope in 1859. In the years that followed, this same technique, which allowed lithium to be revealed by the characteristic red line it imparts to the spectrum, also resulted in the metal being found in grapes, seaweed, tobacco, vegetables, milk, blood and human urine. (The level of lithium in the blood of patients on lithium therapy can easily be monitored by measuring the intensity of this red colour.)

ECONOMIC ELEMENT

The main ore is spodumene ($LiAlSi_2O_6$); other common ores are petalite ($LiAlSi_4O_{10}$), lepidolite ($K(Li,Al)_3(Si,Al)_4O_{10}(F,OH)_2$) and amblygonite ($(Li,Na)AlPO_4(F,OH)$). There is a large deposit of spodumene in South Dakota, where a single crystal of the ore was once discovered which weighed over 10 tonnes and was 15 metres long. Lithium is also extracted from the brines of certain lakes, which may have as much as 1 gram of lithium per litre in their waters, such as Searles Lake in California, as well as lakes in Nevada and northern Chile.

The main producer countries are USA, Russia, China, Australia, Zimbabwe and Brazil, but most lithium currently comes from Chilean brines, which are first concentrated by solar evaporation, a process that crystallizes out calcium sulfate, sodium chloride and potassium–magnesium chloride and leaves a solution that has 3% lithium from which lithium carbonate can be obtained by adding sodium carbonate.

World production of lithium ores and brine salts is around 40 000 tonnes a year and reserves are estimated to be around 7 million tonnes, of which over half is in salt lakes. Reported industrial production of new metal is about 7500 tonnes a year; this is produced by the electrolysis of molten lithium chloride and potassium chloride in steel cells at around 450°C. The amounts of lithium produced for nuclear weapons (see Element of Surprise, p. 239) are not published, but may exceed 25 000 tonnes a year in the USA alone.

Lithium is used in several ways. Lithium oxide accounts for about half the production and goes primarily into glass and glass ceramics; about a third of production goes into lithium chemicals, of which lithium carbonate for pharmaceuticals accounts for a significant amount; lithium metal itself accounts for the rest. The metal goes into alloys with aluminium and magnesium; it greatly improves their strength while making them lighter, so much so that the magnesium–lithium alloy is used for protective armour-plating.

Lithium, being the lightest of all metals, produces alloys which are light. But this was not why the first commercial lithium alloy, called *Bahnmetall*, was produced. This alloy of lead, with 0.7% calcium, 0.6% sodium and 0.04% lithium, was developed by the Germans in World War I as a replacement for tin, the importing of which was blockaded; it is still used by the railways there as an anti-friction alloy for bearings.

The most important alloy is with aluminium; this alloy can be used to reduce the weight of commercial aircraft where every kilogram of weight reduction saves over 125 litres of fuel per year. Aluminium–lithium alloys (with 2–3% lithium) not only offer the greatest saving, but the lithium makes the metal stronger and less prone to fatigue. However, the alloy is more brittle and less ductile than aluminium. These drawbacks can be overcome by adding small amounts of other metals like zirconium, copper or magnesium.

An aluminium–lithium alloy with small amounts of silver, magnesium and copper is about 30% stronger than aluminium alloys now used, and is also likely to appear in more down-to-earth applications, such as bicycle frames and high-speed trains, in this century.

Lithium batteries, which operate at 3 volts or more, are used in wristwatches, pocket calculators and camera flashes where compactness and lightness are all-important. They are also implanted to supply the electrical energy for heart pacemakers. They function with lithium as the anode and iodine as the solid electrolyte and have a lifespan of 10 years. This longevity has been extended to lithium batteries which work at the more common 1.5 volts used in toys, personal stereos, cassette recorders, CD players and televisions. The lower voltage was achieved by replacing the usual manganese dioxide cathode with iron disulfide.

Lithium chloride is one of the most water-absorbing solids known. As a result of this property, it is used to dry industrial gases and in air conditioning, such as in submarines. Lithium stearate, made by reacting stearic acid with lithium hydroxide, is an all-purpose high-temperature grease and most greases contain it. It will even work well at temperatures as low as –60°C and has been used for vehicles in the Antarctic.

The ability of lithium salts to act as a flux, in other words to make something flow more easily, is put to use in aluminium smelting, where lithium carbonate is added to the baths of molten salts; this is a major use of lithium. The carbonate is also put as a flux in baths used in dip-soldering and in welding, and it is added to ceramics and glass partly to make melts flow better. Lithium produces glass which is resistant to sudden heating or cooling because of its low thermal expansion coefficients. This type of glass is used for television tubes. Lithium carbonate is also used to treat glass fibres and make them anti-static.

Lithium has a curious relationship with hydrogen. Together they will form lithium hydride, which is produced as a white powder when lithium and hydrogen react at 750°C. In contact with water this will release the hydrogen again. This is an ideal way of storing hydrogen. A kilogram of lithium hydride will release 2800 litres of hydrogen gas when treated with water.

ENVIRONMENTAL ELEMENT

Lithium in the environment	
Earth's crust	20 p.p.m.
	Lithium is the 31st most abundant element.
Soils	approx. 40 p.p.m., range 1–160 p.p.m.
Sea water	0.17 p.p.m.
Atmosphere	traces

Lithium is found in small amounts in nearly all rocks, and in many mineral spring waters. It poses little threat to plants or animals, on land or in the seas.

CHEMICAL ELEMENT

Data file	
Chemical symbol	Li
Atomic number	3
Atomic weight	6.941, but variable (see below)
Melting point	181°C
Boiling point	1347°C
Density	0.53 kilograms per litre (0.53 grams per cubic centimetre)
Oxide	Li_2O

Lithium is a soft, silvery-white, metal that heads group 1, the alkali metals group, of the periodic table of the elements. It reacts vigorously with water. Storing it is a problem. It cannot be kept under oil, as sodium can, because it is so light that it floats, and will even float on such low-density liquids as petrol (gasoline). It is stored by being coated with Vaseline (petroleum jelly).

Somewhat unexpectedly, in view of the behaviour of other elements of group 1, lithium does not react with oxygen unless heated to 100°C, but it will react with nitrogen from the atmosphere to form a red-brown compound, lithium nitride (Li3N).

The metal is made up of two isotopes, lithium-6 (8%) and lithium-7 (92%), neither of which is radioactive. Its atomic weight varies between 6.94 and 6.99 because separation processes can influence the isotopic composition of the element. Lithium, which has been processed to remove some of the lithium-6 for nuclear weapons, has been sold as depleted lithium without its origin being declared.

ELEMENT OF SURPRISES

1. The hydrogen of hydrogen bombs is the compound lithium hydride, in which the lithium is in the form of the enriched lithium-6 isotope and the hydrogen is the hydrogen-2 isotope, also known as deuterium. This 'lithium deuteride' is capable of releasing massive amounts of energy by nuclear fusion. This is achieved by placing it round the core of an atomic bomb, which provides the heat necessary to initiate nuclear fusion reactions as well as providing the necessary neutron flux. When the bomb detonates, it releases neutrons from the fission of its uranium-235, and these are absorbed by the nuclei of lithium-6 which immediately disintegrates to form helium and hydrogen-3. The hydrogen-3 then fuses with the deuterium to form more helium and this releases yet more neutrons. These are absorbed by the casing of the bomb, which is made of uranium-238, and they convert it to plutonium-239 which then, in theory, adds a third explosion, again one of nuclear fission. See also p. 325. The consequence of all this is to release a billion billion joules of energy in a fraction of a second with the explosive force of millions of tonnes of TNT.
2. The nucleus of an atom was assumed to contain only protons and neutrons but atoms of lithium-7 have been produced with a lambda particle in the nucleus. [Walter Saxon].

Pronounced loo-tee-shi-uhm, the name derives from Lutetia, the Romans' name for Paris.

French, *lutétium*; German, *Lutetium*; Italian, *lutezio*; Spanish, *lutecio*; Portuguese, *lutécio*.

Lutetium is the heaviest of 15 chemically similar elements referred to as the rare-earth elements, which extend from the element with atomic number 57 (lanthanum) to the one with atomic number 71 (lutetium). The term rare-earth elements is a misnomer because some are not rare at all; they are more correctly called the lanthanides, although strictly speaking this excludes lutetium. The minerals from which they are extracted, and the properties and uses they have in common, are discussed under lanthanides, on p. 219.

HUMAN ELEMENT

Lutetium in the human body	
Total amount in body	not known, but small

The amount of lutetium in humans is quite small and the metal has no biological role, but it has been noted that lutetium salts stimulate metabolism. It is difficult to separate out the various amounts of the different lanthanides in the human body, but all are present; the levels are highest in bone, with smaller amounts being present in the liver and kidneys.

No one has monitored diet for lutetium content, so it is difficult to judge how much we take in, but it is probably only a few micrograms per year. Only tiny amounts of lutetium are taken up by plant roots, so little gets into the human food chain – some vegetables have as little as 10 p.p.t. (dry weight). Its low impact on humans can be judged by the analysis of sewage which contains less lutetium that any other lanthanide.

Lutetium is mildly toxic by ingestion, but its insoluble salts are non-toxic.

MEDICAL ELEMENT

The radioactive isotope, lutetium-177, which is a high-energy β-particle emitter with a half-life of 6.75 days, is being studied for possible use in radiotherapy.

ELEMENT OF HISTORY

Lutetium was discovered in 1907 by the French chemist, Georges Urbain, at the Sorbonne in Paris.[70] This element was the final link in a sequence of rare-earth discoveries which began with yttrium in 1794. This element was contaminated with traces of other rare earths. First erbium and terbium were extracted from it in 1843, and then erbium yielded holmium in 1878, thulium in 1879 and finally lutetium in 1907. This came as a result of painstaking work by Urbain, who called the element lutecium; the name was later changed to lutetium.

Meanwhile, in 1907, two other chemists were also on the verge of reporting lutetium: the Austrian chemist Karl Auer (1858–1929) working in Germany and Charles James (1880–1928) in the USA. Auer had isolated the element from yttrium oxide; he called it cassiopeium after the constellation Cassiopeia, and this name was widely used in Germany for many years. Meanwhile James, who was Professor of Chemistry at the University of New Hampshire, had also separated lutetium and produced a lot of it. However, he was rather cautious about publishing his findings and so the credit went to Urbain. James, who had been born and educated in England, spent his whole working life on researching the rare-earth elements and devised the method of separating them that was used until the advent of the modern techniques of ion-exchange and liquid–liquid extraction. A sample of pure lutetium metal was not made until 1953.

ECONOMIC ELEMENT

Lutetium is found in minerals that include all the lanthanide elements; these are discussed in more detail on p. 222. The main mining areas are China, USA, Brazil, India, Sri Lanka and Australia, but the amount of lutetium is tiny in all minerals, e.g. monazite contains only 0.003%, and total world reserves of the element are estimated to be around 200 000 tonnes.

World production of lutetium is around 10 tonnes per year, as lutetium oxide. The metal is obtained by heating lutetium fluoride and calcium, but little use is made of it, except for research purposes. One commercial application has been as a pure β-emitter, using lutetium that has being exposed to neutron activation; as such it has been employed in the oil-refining industries.

A tiny amount of lutetium is added as a dopant to gadolinium gallium garnet (GGG; $Gd_3Ga_5O_{12}$) which is used in magnetic bubble memory devices.

[70] Urbain was also a noted sculptor and painter, who lived from 1872 to 1938.

ENVIRONMENTAL ELEMENT

Lutetium in the environment	
Earth's crust	0.5 p.p.m.
	Lutetium is the 60th most abundant element.
Soils	approx. 0.3 p.p.m. (dry weight)
Sea water	0.3 p.p.t.
Atmosphere	virtually nil

Lutetium is one of the rarer of the rare-earth elements but is, nevertheless, seven times more common than silver. It poses no environmental threat to plants or animals.

CHEMICAL ELEMENT

Data file	
Chemical symbol	Lu
Atomic number	71
Atomic weight	174.967
Melting point	1663°C
Boiling point	3400°C
Density	9.8 kilograms per litre (9.8 grams per cubic centimetre)
Oxide	Lu_2O_3

Lutetium is the hardest and densest of the lanthanides, which is to be expected since it is regarded as the last in the series. (In fact it is not truly a member of this group of elements, but is a transition metal in group 3 of the periodic table of the elements.) It is silvery-white and resists corrosion in air.

Naturally occurring lutetium consists of lutetium-175 (97.5%), which is not radioactive, plus 2.5% of isotope lutetium-176, which is a weakly radioactive β-emitter with a half-life of 20 billion years.

ELEMENT OF SURPRISE

Lutetium is the most expensive metal in the world, costing $75 000 per kilogram, well ahead of the highly valued metals, platinum ($12 800 per kilogram) and gold ($12 500 per kilogram).

Pronounced mag-neez-iuhm, the element is named after Magnesia, a district of Eastern Thessaly in Greece, which in turn took its name from the ancient Magnetes tribe. They also colonized parts of western Turkey, which is likewise known as Magnesia.

French, *magnésium*; German, *Magnesium*; Italian, *magnesio*; Spanish, *magnesio*; Portuguese, *magnésio*.

HUMAN ELEMENT

Magnesium in the human body	
Blood	38 p.p.m.
Bone	varies between 700 and 1800 p.p.m.
Tissue	approx. 900 p.p.m.
Total amount in body	25 grams

Magnesium is an essential element for almost all living things, except some insects. It is certainly essential for all green plants because it is at the heart of the chlorophyll molecule, which plants use to capture the energy of the Sun in order to convert carbon dioxide and water into glucose, which then goes into making cellulose, starch and many other molecules. Leaves appear green because magnesium-chlorophyll absorbs the blue and red components of sunlight but not the green. Plants take their magnesium from the soil, and humans get theirs by eating plants, or eating the animals that feed on them.

Magnesium disperses throughout the body, with most (60%) going into the skeleton where it helps maintain bone structure and acts as a store. Magnesium has four main functions: it regulates movement through membranes; it is a part of, i.e. a co-factor for, more than 100 enzymes, including those which release energy from food; it is used for building proteins, which means it is also important in fetal development; and it is involved in the replication of DNA.

There is no evidence that magnesium produces systemic poisoning although persistent over-indulgence in taking magnesium supplements and medicines can lead to muscle weakness, lethargy and confusion. Gross misuse of certain magnesium-based indigestion and laxative cures can be dangerous (see below).

FOOD ELEMENT

Humans take in between 250 and 350 milligrams of magnesium each day and need at least 200 milligrams, but the body deals very efficiently with this element, taking it from food when it can, and recycling what we already have when it cannot. It is possible to suffer from magnesium deficiency as a result of malnutrition, illness or old age but a normal diet usually provides more than enough.

Most foods contain magnesium, but there is none in distilled spirits, soft drinks, sugar or fats. Cooking does not affect magnesium, although it leaches from vegetables into the water in which they are boiled, and they may lose as much as half of their magnesium in this way. Phosphate in foods interferes with magnesium uptake because it forms insoluble magnesium phosphate, which cannot be absorbed. Foods with high levels of magnesium are almonds, Brazil nuts, cashew nuts, soya beans, parsnips, bran, chocolate, cocoa and brewer's yeast, all having more than 200 milligrams per 100 grams.

Some brands of beer contain a lot of magnesium; an example is Webster's Yorkshire Bitter, which may owe some of its unique taste to the high levels of magnesium sulfate in the water used to brew it. Magnesium salts themselves have a bitter taste.

MEDICAL ELEMENT

Magnesium bromide ($MgBr_2$) used to be prescribed as a sedative, but too much magnesium taken at one time acts as a muscle relaxant and mild laxative – witness the effects of taking Epsom salts (magnesium sulfate) for constipation, or 'Milk of Magnesia' (magnesium hydroxide) for indigestion and constipation. Epsom salts were formerly used as a skin medication as well. Using these medicines too often and in too large doses can raise blood magnesium to dangerously high levels; this has even been known to cause death.

Cases of magnesium deficiency do sometimes occur, e.g. through alcoholism. The symptoms of deficiency are similar to those of *delirium tremens*. Lack of magnesium manifests itself as lethargy, irritation, depression and even personality changes, and some medics have suggested that myalgic encephalomyelitis (ME), also known as chronic fatigue syndrome, is a form of magnesium deficiency because it appears to clear up when those who suffer are given injections of magnesium–saline solution. Dietary magnesium supplements are much less effective because the metal is only absorbed slowly through the stomach and the gut, as shown by studies on concentration camp victims after World War II, who took a year to replenish their body's magnesium store.

Transfusions of magnesium salts are reported to reduce the number of deaths among people suffering heart attacks.

ELEMENT OF WAR

In some air raids during World War II, as many as half a million 2 kilogram magnesium incendiary bombs would be scattered over a city in the space of an hour. The result was massive conflagrations; occasionally these would be so widespread that they caused a firestorm to engulf most of the city.

Once magnesium starts to burn, it is impossible to extinguish: it will even burn in the absence of oxygen by combining with the nitrogen of the air (to form magnesium nitride, Mg_3N_2), and it will continue to burn when doused with water which reacts violently with it liberating hydrogen gas. The only way to limit its damage is to cover it with sand.

Magnesium metal is not easily ignited so this had to be done by a thermite reaction at the heart of the bomb. The thermite reaction, occurring between aluminium powder and iron oxide, releases so much energy that it produces molten iron, and more than enough heat to cause the magnesium casing of the bomb to burn fiercely.

It was the need to produce such bombs that led to the development of a method of extracting magnesium from sea water. Processing plants to carry out this extraction were built in the USA and UK during World War II.

ELEMENT OF HISTORY

The summer of 1618 saw England suffering a terrible drought, yet as Henry Wicker walked across Epsom Common he was puzzled to see a pool of water being ignored by nearby cattle. He found that it tasted very bitter. The reason for this became clear when he evaporated some of the water and obtained crystals of magnesium sulfate from it. These became known as Epsom salts. Soon they were being taken as a medicament, reputedly curing all kinds of ailments, but most noticeably constipation. They became popular all over Europe. Most were made not from water from Epsom but from other magnesium-rich brines and even from sea water.

The first person to recognize that magnesium was an element was Joseph Black at Edinburgh, Scotland, in 1755. He showed that magnesia (magnesium oxide) was not the same as lime (calcium oxide), although both were produced by heating their carbonate ores. Another magnesium mineral called meerschaum (magnesium silicate) was reported by Thomas Henry in 1789, who said that it was much used in Turkey to make pipes for smoking tobacco. These became popular in Europe in the nineteenth century.

An impure form of metallic magnesium was first produced in 1792 by Anton Rupprecht who heated magnesia with charcoal; he gave the element the name *austrium* after his native Austria. A tiny sample of the metal was first isolated by Humphry Davy in 1808, by the electrolysis of magnesium oxide. He proposed the name *magnium*, for the simple reason that he said the word 'magnesium' was too much like 'manganese' and would lead to confusion. His suggestion was not taken up either, and the element became known as magnesium, taking its name from the magnesium mineral which came from Magnesia, Greece.

The French scientist, Antoine-Alexandre-Brutus Bussy (1794–1882), made a sizeable amount of metallic magnesium in 1831 by reacting magnesium chloride with potassium. He recorded its properties.

In 1906, Richard Willstätter proved that magnesium was essential to green plants when he burned some purified chlorophyll and obtained about 1.7% of an ash residue that he showed was mainly magnesium oxide. He found that the result was the same, no matter where the chlorophyll had come from.

Even though it was known in the eighteenth century that sea water tasted slightly bitter on account of the magnesium it contained, it was not until 1941 that the Dow Chemical Company at Texas first devised a way of extracting magnesium from this source and produced an ingot of the metal.

ECONOMIC ELEMENT

Many minerals containing magnesium are known; the main ones are dolomite (calcium magnesium carbonate, $CaMg(CO_3)_2$) and magnesite ($MgCO_3$) which are mined to the extent of 10 million tonnes per year, in countries such as China, Turkey, North Korea, Slovakia, Austria, Russia and Greece. Magnesite is heated to convert it to magnesia (MgO), which has several applications: it may be added to fertilizers; it is used as a supplement for cattle feed, especially at the start of the grazing season; and it is put into plastics as a bulking agent. Most goes into making heat-resistant bricks for fireplaces and furnaces.[71] Workable reserves of magnesite exceed 2 billion tonnes as ores. This pales in comparison with the more than million billion tonnes of magnesium dissolved in the oceans, and from which a sizeable proportion of the world's magnesium metal is extracted.

Most magnesium metal is produced by the electrolysis of molten magnesium chloride, although a little is made by the thermal reduction of magnesium oxide with ferrosilicon. This latter method can even extract magnesium from asbestos waste and other calcium–magnesium silicate minerals. World production of the metal is around 400 000 tonnes per year and is increasing year by year; it is expected to exceed 1 million tonnes per year by the year 2010. The metal is produced mainly in the USA (30% of the total), Russia (16%), Canada (10%), Ukraine (10%), Norway (7%) and a few other countries.

Magnesium is alloyed with up to 10% aluminium, plus traces of zinc and manganese to improve its strength, corrosion resistance and welding qualities; this alloy is used for car bodies and aircraft. More than half the magnesium produced ends up in this way. The next major use is in iron and steel production, where it is added to the molten metal to remove sulfur; this accounts for about 20%. The rest ends up in a variety of products, such as lightweight frames for bicycles, car seats and luggage.

Magnesium is increasingly used by car-makers because of its environmental benefits in making vehicles lighter and longer-lasting. Reducing the weight of a car means not only that it uses less fuel but also that it causes less damage in accidents. Magnesium is also used for lawn mowers, power tools, disc drives and cameras, where again lightness is a key factor. At the end of its useful life the magnesium in all these products can be recycled at very little cost.

The chemical industry needs magnesium for various processes, and the dye industries sometime use magnesium sulfate as a mordant. Magnesium hydroxide is being seen as a safer replacement for caustic soda in neutralizing waste acids. It is now added as a fire-retardant to plastics.

[71] The temperature at which magnesium oxide is produced determines its eventual use. Higher temperatures are needed when it is to be made into refractory materials.

ENVIRONMENTAL ELEMENT

Magnesium in the environment	
Earth's crust	23 000 p.p.m. (2.3%)
	Magnesium is the seventh most abundant element.
Soils	approx. 1–2%
Sea water	1200 p.p.m.
Atmosphere	trace

The table above refers only to the magnesium in the Earth's crust; if the mantle is taken into account too then magnesium becomes the third most abundant element, because the mantle is largely composed of olivine and pyroxene, which are magnesium silicates.

Magnesium appears to be environmentally benign in all contexts, although some soils, known as serpentine soils, produced by the weathering of igneous magnesium rocks, can be toxic to plants on account of their high chromium and nickel content.

CHEMICAL ELEMENT

Data file	
Chemical symbol	Mg
Atomic number	12
Atomic weight	24.3050
Melting point	649°C
Boiling point	1090°C
Density	1.74 kilograms per litre (1.74 grams per cubic centimetre)
Oxide	MgO

Magnesium is a silvery white, lustrous, relatively soft metal that is a member of group 2, also known as the alkaline earth group, of the periodic table of the elements. Magnesium ribbon burns in air when ignited and it reacts with hot water. Because it is a relatively electropositive metal,[72] magnesium can be used as a 'sacrificial' electrode to protect iron and steel structures that are exposed to sea water and ground water, such as pipelines, as it corrodes away preferentially.

There are three naturally occurring isotopes of magnesium: magnesium-24, which makes up 79%; magnesium-25, 10%; and magnesium-26, 11%. None is radioactive.

[72] Electropositive refers to the ease with which a metal forms positive ions, in this case Mg^{2+}.

ELEMENT OF SURPRISE

Although magnesium has been used as incendiary bombs and flash bulbs, and is known to burn with an intensely bright light, it is very difficult to ignite the bulk metal. As a result, magnesium tubes and rods can be safely welded. Magnesium was originally introduced for racing bicycles which were the first vehicles to use pure magnesium frames, giving a better combination of strength and lightness than other metals. (A steel frame is nearly five times heavier than a magnesium one, and even an aluminium one is 50% heavier.) Originally frames were welded together but now a complete frame is cast as a single component from molten magnesium which avoids the need to weld joints whilst maximizing lightness and strength.

Pronounced man-gan-eez, the name is derived either from the Latin *magnes*, meaning magnet, because this element's common mineral, pyrolusite, has slight magnetic properties, or from *magnesia nigra*, referring to black magnesia rock, which is manganese dioxide.

French, *manganèse*; German, *Mangan*; Italian, *manganese*; Spanish, *manganeso*; Portuguese, *manganês*.

Manganese can exist in several oxidation states of which the most common are manganese(II), i.e. Mn^{2+}, and manganese(IV), e.g. the oxide MnO_2. Manganese(VII) is best known by its older name of permanganate, MnO_4^-.

HUMAN ELEMENT

Manganese in the human body	
Blood	2–8 p.p.b.
Bone	varies between 0.2 and 100 p.p.m.
Tissue	varies between 0.2 and 2 p.p.m.
Total amount in body	about 12 milligrams

Manganese is an essential element for all species – indeed for some creatures, such as the red ant, it makes up 0.05% of their weight. Some organisms, such as diatoms, molluscs and sponges, accumulate manganese. Fish can have up to 5 p.p.m. and mammals up to 3 p.p.m. in their tissue, although normally they have around 1 p.p.m. In 1931, A.R. Kemmerer and co-workers proved manganese to be an essential requirement of mice and rats, and in 1936 it was shown that the bone disease, perosis, in chickens could be prevented by giving them manganese. Manganese compounds are added to fertilizers and animal feedstuffs because this element may be lacking in certain soils, and animals grazing on such land may suffer manganese deficiency.

Humans also need manganese, although this was only realized in the 1950s, perhaps because the requirement is so modest. It is still not clear exactly for what part of our metabolism it is essential. It is known to be needed for the working of various enzymes, and it has been demonstrated to be involved in glucose metabolism and the operation of vitamin B_1, and it is associated with RNA. Most manganese in the body is in the bones; other organs with above-average amounts are the pituitary and mammary glands, the liver and the pancreas.

The manganese(II) ion, Mn^{2+}, is the normal form in which the element occurs; Mn^{2+} is not poisonous, but the purple-coloured permanganate ion, MnO_4^-, is toxic. While Mn^{2+} may be the most common form of manganese, the biologically active form is manganese(III), Mn^{3+}, which is the least stable form of the element.

Exposure to dust or fumes from manganese is a health hazard. Workers breathing in the fumes from hot manganese suffer 'fume fever', symptoms of which are fatigue, anorexia and impotence. Miners who were affected by this condition also displayed symptoms that came to be known as 'manganese madness', exhibiting involuntary laughing or crying, aggression, delusions and hallucinations. These effects were first noted in France in 1837 and their symptoms were seen to be similar to those of another brain condition, Parkinson's disease. Fortunately, manganese madness is now rare.

FOOD ELEMENT

The daily dietary intake of manganese averages 4 milligrams, with a range of 1–10 milligrams depending on the foods that are eaten. There is never any need for humans to take manganese supplements because we get more than enough of this element from food.

Milk is low in manganese, while liver can have as much as 10 p.p.m., but only a fraction of this – possibly as little as 5% – is absorbed into the body. The foods that provide most people with their supply of this element are cereal products and nuts. Beetroot has one of the highest levels of manganese (36–110 p.p.m. dry weight). Other foods rich in manganese are sunflower seeds, coconuts, peanuts, almonds, Brazil nuts, blueberries, olives, avocados, corn, wheat, bran, rice, oats and tea. The French delicacy, snails, also contain a high level. Tree fruits, such as apples and oranges, have the lowest manganese levels (1–2 p.p.m.).

MEDICAL ELEMENT

In Victorian times, and for a long time thereafter, 'Condy's Disinfecting Liquid' was a staple commodity of pharmacies, despite being highly dangerous if it were drunk. For this reason it was flavoured with lavender oil so it could not be mistaken for anything else. The active ingredient in Condy's solution was permanganate. Despite the dangers, gargling with Condy's solution was recommended for halitosis and sore throats, and for bathing wounds, because it was an effective antiseptic and deodorizing solution. It is no longer available.

ELEMENT OF HISTORY

Manganese was known long before it was isolated as an element. Its common minerals were mined in Germany, Italy and England, and were used commercially by glassmakers for hundreds of years to remove the pale greenish tint of natural glass which is due to traces of iron ions (Fe^{2+}) in the sand from which it is made. The Roman author,

Pliny the Elder, who perished when Pompeii was destroyed in 79 AD, wrote of a black powder that glass-makers used to make their product crystal clear. This was almost certainly pyrolusite (manganese(IV) oxide; MnO_2); it was referred to as *sapo vitri*, glass soap.[73] It was also used as a black pigment by potters.

In 1740, the Berlin glass technologist J. H. Pott produced potassium permanganate ($KMnO_4$), one of the strongest oxidizing agents known to early chemists.

Several eighteenth century chemists tried unsuccessfully to isolate the metal component in pyrolusite. The Swedish chemist and mineralogist Johan Gottlieb Gahn (1745–1818) is generally credited with being the first to succeed, in 1774. However, a student at Vienna, Ignatius Kaim, had already described how he had produced manganese metal, in his dissertation written in 1771.

Manganese dioxide was responsible for the discovery of the element chlorine, because when this compound is added to hydrochloric acid, chlorine gas is given off. This same effect was noted by the Swedish chemist Scheele when he added hydrochloric acid to vegetable ashes; he deduced that these also must contain manganese dioxide. The same happened with the ashes of other plants, so it became recognized that manganese is an essential element for all plants. Manganese was found in ox bones in 1808, in human bones in 1811, and in human blood in 1830.

ECONOMIC ELEMENT

The most common manganese minerals are pyrolusite and rhodochrosite, which is manganese(II) carbonate ($MnCO_3$); others are psilomelane (barium-containing MnO_2); cryptomelane (potassium-containing MnO_2) and manganite (basic manganese oxide, $MnO(OH)$). More than 25 million tonnes of ore are mined every year, representing 5 million tonnes of the metal, and reserves are estimated to exceed 3 billion tonnes. If these were ever to be exhausted then we would have to exploit the manganese nodules on the ocean floor (see 'Element of surprise').

The main mining areas for manganese ores are South Africa, Russia, Ukraine, Georgia, Gabon and Australia. The metal is obtained either by the thermal reduction of the oxide using other metals, such as sodium, magnesium or aluminium, or by the electrolysis of manganese sulfate.

Manganese metal is not used as such because it is too brittle; 95% of mined ore goes into alloys, mainly steel, which contains about 1% manganese to improve its strength, working properties and wear resistance. Manganese steel itself contains around 13% manganese; a patent for this was granted in 1883 to a 24 year old metallurgist, Robert Hadfield, of Sheffield, England. This alloy is extremely strong and is used for railway tracks, earth-moving machinery, safes, helmets, rifle barrels and the bars of prison cells. Other important alloys containing a few per cent of manganese are manganese bronze (mainly copper) and manganese nickel–silver (an alloy of manganese, copper, zinc and nickel).

Commercially, the two most important compounds of manganese are manganese(IV)

[73] If too much was added then it would colour the glass purple. Indeed, some early types of window glass produced in the American colonies, and which have been exposed to bright sunlight for centuries, now have a purple tinge due to the formation of manganese in oxidation state VII; they are highly valued as such. The colour of true amethysts is due to traces of manganese(VII) in otherwise colourless quartz crystals.

oxide (about 180 000 tonnes per year are made) and manganese(II) sulfate (about 120000 tonnes per year). The former was added to the electrolyte paste in older zinc–carbon batteries in order to prevent the build up of hydrogen gas around the carbon electrode as the battery was discharged, thereby reducing the current. Today, manganese(IV) oxide is more likely to be found in rubber, or as a catalyst in industry. Manganese sulfate is needed for the electrochemical manufacture of manganese metal, and for manufacture of the fungicide Maneb (a manganese dithiocarbamate derivative).

Other compounds that find application are manganese(II) oxide (MnO), manganese carbonate ($MnCO_3$) and potassium permanganate. The first goes into fertilizers and ceramics, the second is the starting material for making other manganese compounds, and the third is used to remove organic impurities from waste gases and effluent water.

ENVIRONMENTAL ELEMENT

Manganese in the environment	
Earth's crust	1000 p.p.m. (0.1%)
	Manganese is the 12th most abundant element.
Soils	approx. 440 p.p.m., with a range from 7 to over 9000 p.p.m.
Sea water	10 p.p.b.
Atmosphere	0.01 microgram per cubic metre, but can be 20 times this level in cities

Manganese has caused no known environmental damage. It is one of the most abundant metals in soils, where it occurs as oxides and hydroxides, and it cycles through its various oxidation states, this activity being largely due to microbes. Soil may be deficient in manganese, so that animals grazing on such land may suffer, hence the need to add either manganese salts to fertilizers or manganese supplements to the animals' feedstuffs. On the other hand, in some soils manganese can approach levels that are toxic to plants, especially in acidic or poorly aerated soils. Plants generally tolerate high levels in soils and this is reflected in the variable amounts they contain; grasses, for example, can have between 17 and 300 p.p.m. (dry weight).

CHEMICAL ELEMENT

Data file	
Chemical symbol	Mn
Atomic number	25
Atomic weight	54.938049
Melting point	1244°C
Boiling point	1962°C
Density	7.4 kilograms per litre (7.4 grams per cubic centimetre)
Oxides	MnO and MnO_2

Manganese is a hard, brittle, silvery metal and a member of group 7 of the periodic table of the elements. It is reactive when pure, and as a powder it will burn in oxygen. It reacts with water (it rusts rather like iron) and dissolves in dilute acids.

Natural manganese consists of a single isotope, manganese-55, which is not radioactive.

ELEMENT OF SURPRISE

The most surprising occurrence of manganese is on the ocean floor where there is an estimated trillion (10^{12}) tonnes of manganese-rich nodules scattered over large areas, with the North-East Pacific being particularly rich in them.[74] In 1876 the three-masted sailing ship, *Challenger*, was sent on a scientific expedition to explore the deep oceans. It returned with a number of curious, cone-shaped lumps that had been dredged up from the ocean floor at various sites around the world. These turned out to be mainly made of manganese with smaller amounts of copper, cobalt and nickel. They appear to have formed around sharks' teeth, which are one of the few parts of living things that are capable of surviving the intense pressures at the bottom of the oceans. Whether these ocean nodules will ever be harvested is debatable.

[74] During the Cold War, the USA built a deep-ocean exploration vessel, ostensibly to investigate the possibility of mining these manganese-rich nodules. The event was much publicized, but it was later revealed that the vessel's real purpose was to locate and recover a wrecked Soviet nuclear submarine.

Meitnerium

This is element atomic number 109, and along with other elements above 100 it is dealt with under the transfermium elements on p. 457.

Mendelevium

This is element atomic number 101, and along with other elements above 100 it is dealt with under the transfermium elements on p. 457.

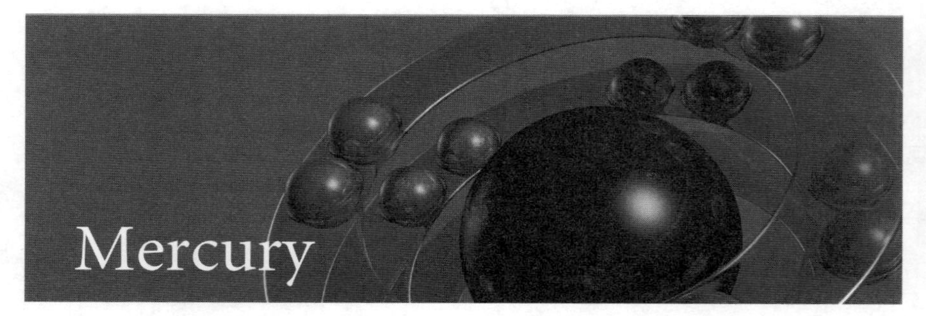

Mercury

Pronounced merk-yoo-ree, this element is named after the planet Mercury. The chemical symbol, Hg, comes from the Latin *hydrargyrum*, meaning liquid silver.[75]

French, *mercure*; German, *Quecksilber*; Italian, *mercurio*; Spanish, *mercurio*; Portuguese, *mercúrio*.

HUMAN ELEMENT

Mercury in the human body	
Blood	8 p.p.b.
Bone	0.5 p.p.m.
Tissue	0.2–0.7 p.p.m.
Total amount in body	6 milligrams

[75] This was the description first given to mercury by Aristotle in about 350 BC. Mercury's common name, quicksilver, derives from the Old English word *cwic*, meaning living, and by implication referred to its being a liquid.

Mercury has no biological role even though it is present in every living thing. It is widespread because it is present in the atmosphere due to its volatility, both as the metal and as the organomercury compounds which are formed by micro-organisms.

Mercury poisoning was once relatively common but is now rare thanks to stringent health and safety regulations and the phasing out of many of its uses. All mercury compounds are toxic, methyl mercury extremely so. This can pass the blood–brain barrier and move across the placenta, with the result that mercury affects the central nervous system and can cause fetal deformities.

Mercury has a particular attraction to sulfur and will attach itself to the sulfur atoms of certain amino acids. When these amino acids are part of an enzyme, that enzyme can be rendered inactive. The Na^+/K^+-ATPase enzyme, which acts as a so-called sodium pump, is essential to the working of the central nervous system; it is particularly sensitive to mercury and this is reflected in the most noticeable symptoms of mercury poisoning, namely the 'shakes' and the mental disturbances. Mercury's toxicity is covered in more detail below. Most mercury in the body is found in the kidneys, followed by the liver, spleen and brain.

FOOD ELEMENT

The human intake of mercury is about 3 micrograms per day for adults and about 1 microgram for babies and young children. Every mouthful of food we eat contains some.

Agricultural soils may hold as much as 0.2 p.p.m. of mercury and this finds its way into plants and food crops, in particular carrots, potatoes and mushrooms; the latter can have as much as 1 p.p.m or more. Grass contains relatively little, around 4 p.p.b., so grazing animals are not contaminated, and meat and dairy products have low levels of mercury. However, ocean fish take up mercury, with tuna and swordfish concentrating it to levels 100 000 times those of the surrounding sea water; this does not render them unsafe to eat, though.

MEDICAL ELEMENT

Despite its toxicity, mercury was much used by doctors down the ages. Calomel (mercury(I) chloride, also called mercurous chloride, Hg_2Cl_2) was prescribed as a laxative and diuretic, and corrosive sublimate (mercury(II) chloride, also called mercuric chloride, $HgCl_2$) was used as a disinfectant. The latter was known in India as early as the twelfth century and was made by heating mercury, salt, brick dust and alum for 3 days in a closed earthenware pot, and then adding water to dissolve out the corrosive sublimate before crystallizing it. Calomel was then made by grinding these crystals with fresh mercury metal. Only this less-soluble form was deemed safe enough to be taken internally.

Metallic mercury dispersed in fat was used as an ointment for skin complaints. This salve first appeared in the thirteenth century; its application was known to induce excess salivation, which we now realize meant that the mercury was being absorbed through the skin and into the bloodstream. Later skin treatments used mercurochrome (also known as merbromin), a complex organic derivative that was developed in the nineteenth century as a household antibacterial and antiseptic agent.

When syphilis became a problem in Europe in the late fifteenth century, the only known cure was corrosive sublimate. Again, intense salivation was noted as a side-effect. The treatment worked because the mercury killed *Treponema pallidum*, the organism that caused the disease, but the 'cure', as it was known, was risky and almost as feared as the disease itself.

In babies, mercury poisoning used to manifest itself as 'pink disease', showing as an unnaturally bright pink coloration of the fingers, toes, cheeks, nose and buttocks. The cause was the use of teething powders containing calomel. In adults, this pink coloration of the skin was seen as a beauty aid and even today there is a Mexican beauty cream, called 'Crema de Belleza Manning', which relies on calomel to achieve the desired effect.

Mercury is still part of traditional Chinese medicine: two preparations, 'antidotal pills' and 'cinnabar sedative pills', contain around 4% of mercury compounds. The former is given at the rate of four pills three times a day for poisonous insect stings, delivering a dose of 0.2 grams of mercury. The latter is given at five pills three times a day as a calmative, delivering as much as a gram.

Mercury poisoning

Mercury is absorbed through the lungs, skin and digestive tract. It varies in toxicity depending on whether it is the metal, a mercury(I) compound, a mercury(II) compound or an organomercury compound. Mercury(I) is much less toxic than mercury(II), because the former is much less soluble than the latter.[76] Examples of mercury poisoning are detailed in the economic and environmental sections which follow.

The physical symptoms of acute mercury poisoning are a severe headache, nausea, vomiting, stomach pains, diarrhoea and a metallic taste in the mouth. After a few days there is excess salivation, swelling of the salivary glands and, after a period of time, loosening of the teeth. Poisoning by smaller amounts over longer periods of time, as occurs with industrial exposure, has a different set of symptoms, most of which are due to effects on the brain. Those affected by chronic mercury poisoning suffer from fatigue, weakness, loss of memory and insomnia. They display psychological symptoms such as irritability, depression and a paranoid belief that other people are persecuting them. They exhibit a tremor of the hands, to the extent that their handwriting becomes spidery. One of the first symptoms is increased salivation.

Liquid mercury is particularly insidious because the metal is slightly volatile and can be absorbed by the lungs, which is why the level in air should not exceed 0.1 milligram per cubic metre. When the British sloop *Triumph* was transporting flasks of mercury from Spain to London in 1810, one of them broke open in a storm, with the result that all 200 members of the crew were eventually affected; three died, as did all the cattle and birds that were on board. In the twentieth century, mercury poisoning was an occupational risk for those detectives whose job it was to search for fingerprints at the

[76] However, cinnabar (HgS) is the least soluble salt of all and a litre of water in contact with this dissolves only 10 nanograms (10^{-8} grams).

MERCURY IN FAMOUS MEN

Hair includes a lot of sulfur-containing amino acids and these attract mercury and provide a permanent record of the level of a person's exposure to the element.

It was often the custom in earlier times to preserve locks of hair from those who had died. When these have been analysed by modern methods, curiously high levels of mercury have been found, leading some to suspect that the person was medically treated with calomel because they had syphilis or that they had taken part in alchemical experiments and thereby breathed in mercury vapour.

The hair of Isaac Newton (1642–1727) contains high levels of mercury, which almost certainly came from his alchemical work because he was known to have been celibate all his life; the mercury in the hair of the Scottish poet Robert Burns (1759–96) suggests he had been treated for venereal disease. Napoleon's hair also shows that he was probably given calomel when he was taken ill on St Helena, but this was not the cause of his death (see arsenic, p. 46).

Mercury in hair does not always imply syphilis or alchemy. One of the saddest deaths of a scientist was that of the Danish astronomer Tycho Brahe (1546–1601) who suffered with prostate trouble. At a royal banquet in Prague he dared not leave the table to relieve himself, with the result that his bladder split and he died of urinary poisoning a few days later. Analysis of a strand of his hair, which had the root intact, showed that the day before he died he was given a mercurial medicine in an effort to save his life.

It has been tempting to assume that when famous people were treated with mercury the reason was syphilis. King Henry VIII of England and Ivan the Terrible of Russia were both dosed this way. Charles II almost certainly died of mercury poisoning but probably not as a result of medical treatment, although he too might well have needed it as a result of his notorious sex life. He was exposed to the metal when doing alchemical experiments which he carried out in a poorly ventilated room in the palace.

scene of a crime. The dusting powder used to be made by grinding chalk and mercury together.

The phrase 'mad as a hatter' owes its derivation to the use of mercury in the hat industry, and described the behaviour of those whose job it was to turn beaver and rabbit fur into felt, the raw material from which hats used to be made. In order to get the short hairs of this type of fur to mat together, the pelts were dipped in a solution of mercury nitrate and then dried. Workers in the industry often suffered from 'hatter's shakes' and 'mercury madness'.

The body rids itself of mercury not only through urine and faeces, but also via the lungs and sweat, as well as locking some away in nails and hair. There are various effective antidotes for treating mercury poisoning once it has been diagnosed.

ELEMENT OF HISTORY

Cinnabar, also known as vermilion (mercury sulfide, HgS), was used as a bright red pigment by the Palaeolithic painters of 30 000 years ago to decorate caves in Spain and France with dramatic pictures of animals and humans. In that respect mercury has been part of human history for a very long time. Whether these artists of the Old Stone Age realized that cinnabar could be made to yield mercury metal by heating it we cannot know.

The oldest known sample of mercury, contained in a coconut-shaped vessel, was found by the German archaeologist Heinrich Schliemann (1822–90) in an ancient Egyptian tomb at Kurna, which dated from around 1600 BC. Mercury was also known in China and India about this time. The Chinese alchemist Ko Hung (281–361 AD) wrote of the wonder of turning bright red cinnabar into silver mercury simply by heating.[77] The Greek and Roman writers such as Aristotle, Pliny the Elder and Vitruvius, knew that it could be obtained this way and that it could be purified by squeezing it through leather. They also knew it was poisonous.

The Almadén cinnabar deposit in Spain has been exploited for 2500 years and has yielded an estimated 200 000 tonnes of mercury, most of which has been lost down the centuries. Pliny reported that more than 4 tonnes of the metal were imported into Rome every year.

In the Americas, cinnabar was mined by the Incas and used as a cosmetic and as war paint. The Spaniards discovered the large deposits at Huancavelica, in 1566, which they mined and smelted to get the mercury they needed to extract gold.[78]

Mercury was thought by the alchemists to be the key to transmuting other metals into gold, but despite centuries of investigation along these lines, little that was useful was achieved. On the other hand, mercury was important to the Scientific Revolution which began in the sixteenth century. It was needed for barometers and thermometers. Mercury was the ideal fluid for these instruments because it did not adhere to glass and so moved freely as the pressure or temperature changed – but see 'Element of surprise'.

Mercury as its oxide, HgO, was also the key to the discovery of oxygen, one of the milestones in the development of chemistry. When Priestley heated this it decomposed into the elements.

ECONOMIC ELEMENT

Native mercury occurs naturally as tiny droplets in cinnabar deposits which are generally associated with volcanic rocks. Cinnabar is the chief ore and is mined in Spain, Russia, Italy, China and Slovenia. Another important mercury ore is livingstonite (mercury antimony sulfide, $HgSb_4S_8$). World production of mercury is around 8000 tonnes per year. Mercury is still traded traditionally in so-called 'flasks', a quantity that has been used since Roman times. A flask containes 34.5 kilograms or 76 pounds imperial

[77] The sulfur is oxidized by the air, forming sulfur dioxide gas, and mercury metal is left behind.

[78] In 1848 the miners of the Californian gold rush used mercury from the New Almaden Mines of California for the same purpose, to dissolve gold and thereby extract it.

of mercury. Mineable reserves of mercury are around 600 000 tonnes, mainly located in Spain (at Almadén), Russia and China.

Industry uses mercury metal as a liquid electrode in the manufacture of chlorine and sodium hydroxide by means of the electrolysis of brine, although it is being phased out in favour of methods which do not require mercury. Some mercury is still used to treat seed corn to make it resistant to fungal disease, a practice that was introduced in the 1920s with the dressing known as Ceresan, which was a 2% solution of ethyl mercury chloride. By the 1960s the procedure had become widespread, with more than 150 proprietary products on the market. Sadly this form of crop protection led to several mass poisonings in developing countries when villagers made bread from the treated grain. In several incidents in northern Iraq, in the early 1970s, more than 5000 peasants went down with mercury poisoning and 280 died. (Crops of wheat, barley, oats and corn that are grown from treated seeds absorb very little of this mercury, but in any case the use of mercury seed dressings has now been curtailed.)

Mercury is still used in some electrical gear, such as switches and rectifiers, which need to be reliable, and for industrial catalysts. Older uses such as in thermometers, felt production, plating, tanning, dyeing and as a de-worming powder have all been superseded by other substances.

Much less mercury is now used in consumer batteries and fluorescent lighting, but it has not been entirely eliminated. In the case of batteries, mercury is now restricted to button cells for hearing aids and other small electronic devices. These batteries consist of an outer zinc casing, which acts as the anode, filled with a paste of zinc hydroxide and mercury oxide surrounding a small steel cathode. The benefit of these batteries is that they provide a voltage of 1.35 volts and maintain this even as they age. In the case of fluorescent lights, a metre-long fluorescent tube now contains only 10 milligrams of mercury as opposed to more than 35 milligrams in previous versions. These lamps can now be disposed of as non-hazardous waste.

Mercury alloy dental fillings were invented by a US dentist, C. V. Black, in 1895. He found that an alloy of 70% silver and 30% tin, when ground to a powder and mixed with mercury, formed a plastic mass that could be pressed into a tooth cavity and would harden in about five to 10 minutes; it expanded as it solidified, thereby exactly filling the hole. The modern version of the alloy powder consists of 60% silver, 27% tin and 13% copper.

Industrial mercury poisoning

Mercury dissolves some metals to give so-called amalgams. This happens with gold and silver. In the past this caused much mercury poisoning. Gold amalgam was used to gild objects made of other metals, such as buttons. When these were heated with the amalgam, the gold would deposit on the surface and the mercury would distill off. A gram of gold would gild 500 buttons. Those engaged in the trade suffered from 'gilder's palsy', and those living nearby were also exposed to mercury. In Birmingham in the early nineteenth century, the metal collected in gutters in the streets surrounding gilding factories. Despite technical improvements for trapping the mercury from the flues of gilding establishments, the process only ceased after 1840, when electroplating took its place. However, Britain's Royal Navy and merchant marine continued with the older form of gilding for a further hundred years or more because it gave a more durable product.

In the early nineteenth century, the gilding of the dome of the cathedral of St Isaac at St Petersburg, Russia, required 100 kilograms of gold to be applied to the copper sheets, with the result that 60 workmen died of mercury poisoning.

Until the middle of the nineteenth century, mirrors were made using silver amalgam. Those engaged in this trade suffered accordingly, until the German chemist Justus von Liebig (1803–1873) showed that silver mirrors could be deposited on glass by chemical means that did not require mercury.

ENVIRONMENTAL ELEMENT

Mercury in the environment	
Earth's crust	50 p.p.b.
	Mercury is the 68th most abundant element.
Soils	0.01–0.5 p.p.b., but contaminated soils can have up to 0.2 p.p.m.
Sea water	40 p.p.t.
Atmosphere	approx. between 2 and 10 nanograms per cubic metre, although in locations over ore deposits it can be as high as 1500 nanograms per cubic metre. Rain water contains between 2 and 5 p.p.t.

Mercury cycles through the various compartments of the environment, primarily because there is an enormous natural input from volcanoes, soil erosion and microbial release of organomercury compounds, together amounting to around 100 000 tonnes a year. On top of this there is a large release – 50 000 tonnes per year – from human activity, mainly from burning coal and oil, and from municipal incinerators where it comes mainly from discarded batteries and fluorescent tubes. Mercury is also emitted by crematoria: it has been estimated that such a place will emit around 5 kilograms of mercury per year from dental fillings, although in some countries such as Sweden the flue gases are filtered to remove it.

Environmental mercury poisoning

One of the worst examples of environmental pollution was the Minamata Bay disaster in Japan in the 1950s. There mercury had been discharged into the bay from a local chemical company at the rate of 100 tonnes per year for around 30 years and had built up in the sediment of the bay to such an extent that fish from the bay had high levels of organomercury compounds in their flesh – some fish contained 0.2% of mercury. These fish were eaten by many living in the area. The result was that more than 10 000 people were crippled with 'Minamata disease', which was basically organomercury poisoning of the central nervous system. The symptoms of the victims were not the normal ones of mercury poisoning, and for a while other toxins were suspected. Eventually it was proved that organomercury compounds were responsible, and that these had been produced by microbes living in the silt of the bay. When a ban on fishing was imposed in 1956, there were no further cases. The factory then introduced a mercury-separating process to clean up its waste waters. However, it took 30 years before some of its victims were properly compensated.

CHEMICAL ELEMENT

Data file	
Chemical symbol	Hg
Atomic number	80
Atomic weight	200.59
Melting point	−39°C
Boiling point	357°C
Density	13.5 kilograms per litre (13.5 grams per cubic centimetre)
Oxide	HgO

Mercury is a liquid, silvery metal and a member of group 12 of the periodic table of the elements. It is stable in air and water, and is unreactive towards acids (except concentrated nitric acid) and alkalis.

There are seven naturally occurring isotopes, of which mercury-202 accounts for 30% and mercury-200 for 23%. The others are: mercury-199 (17%), mercury-201 (13%), mercury-198 (10%), mercury-204 (7%) and, the least abundant, mercury-196 (0.1%). None is radioactive. There are also traces of radioactive mercury isotopes produced by the decay of other heavy radioactive elements.

ELEMENT OF SURPRISE

The alchemists believed that mercury was unique. In their theory it was *the* element, a component of all metals, and so held the key to the transmutation of base metals into gold. It uniquely represented the quintessential property of fluidity. Therefore reports from Siberia, that mercury could freeze solid, and thereby lose its elemental fluidity, were discounted as little more than travellers' tales.

So it came as a shock to two Russian scientists, A. Braun and M. V. Lomonosov, of St Petersburg when, on 26th December 1759, they experimented with snow to see how low a temperature they could achieve. A mixture of ice and salt can produce a temperature drop of many degrees below 0°C and Braun and Lomonosov thought that mixing snow and acids might result in even lower temperatures, and so it did. Suddenly the mercury in their thermometer stopped moving, and appeared to have solidified. Curious to know what had happened, they broke away the glass and found that the mercury in the bulb had become a solid metal ball with the mercury from the tube protruding like a piece of wire, which they could bend, just like other metals. The belief that mercury was unique had been destroyed.

Pronounced mol-ib-den-uhm, the name is derived from the Greek *molybdos*, meaning lead (see 'Element of history' below for an explanation of this name).

French, *molybdène*; German, *Molybdän*; Italian, *molibdeno*; Spanish, *molibdeno*; Portuguese, *molibdênio*.

HUMAN ELEMENT

Molybdenum in the human body	
Blood	1 p.p.b.
Bone	less than 0.7 p.p.m.
Tissue	approx. 20 p.p.b.
Total amount in body	5 milligrams

Molybdenum is essential to all species. As with other trace metals, though, what is essential in tiny amounts can be highly toxic at larger doses. Animal experiments have shown that too much molybdenum causes fetal deformities. The parts of the body with most molybdenum are the bones, skin, liver and kidney.

There are around 20 molybdenum-containing enzymes, used by both plants and animals. The best-known is the nitrogen-fixing nitrogenase, found in the root nodules of legumes, such as beans, which can convert nitrogen of the air into ammonia (see nitrogen, p. 292). Algae also use molybdenum to fix nitrogen, and they have another molybdenum enzyme which helps them get rid of unwanted sulfur by converting it to dimethyl sulfide, which is volatile. It is this gas which attracts sea birds to areas where the sea is rich in nutrients and likely to have lots of fish.

There is a mammalian enzyme, xanthine oxidase, which contains molybdenum, and this produces uric acid, which is how the body excretes unwanted nitrogenous material. If this enzyme is too active it can lead to gout, the painful accumulation of sharp crystals of uric acid in the joints. Modern treatment for this condition targets this enzyme to depress its activity. See also lead (p. 228).

Other enzymes that require molybdenum are aldehyde oxidase, which converts aldehydes into acids, and sulfite oxidase, which detoxifies sulfite (SO_3^{2-}) by oxidizing it to harmless sulfate (SO_4^{2-}). Both of these enzymes are to be found in the liver. Aldehyde oxidase is needed for the metabolism of alcohol, which is converted first to acetaldehyde, by a zinc-containing enzyme, then to acetic acid, by the molybdenum-containing enzyme. Acetic acid is used by cells as a source of energy.

FOOD ELEMENT

The average human takes in about 0.3 milligrams of molybdenum a day. The absolute minimum intake that is needed is not known, although it may be as low as 0.05 milligrams. In any case the intake should not regularly exceed 0.4 milligrams because above this level molybdenum can provoke a toxic response.

Foods which have the most molybdenum are pork, lamb and beef liver with 0.15 milligrams per 100 grams (= 1.5 p.p.m.), while green beans have 0.7 p.p.m and eggs 0.5 p.p.m. Other foods with above average amounts of molybdenum are sunflower seeds, wheat flour, soya beans, lentils, peas and oats. Plant-derived foods can vary between 0.002 and 1 p.p.m. molybdenum (fresh weight), with fruits having the least and legume vegetables, such as beans, having the most. Cereal grains have 0.5 p.p.m. on average, sweet corn has 0.2, potatoes 0.05, tomatoes 0.03 and apples 0.002 p.p.m.

MEDICAL ELEMENT

The artificially produced radioactive isotope, molybdenum-99, with a half-life of 66 hours, is used in hospitals to generate technetium-99, which it does as it decays. This is then given to patients because it collects in various organs of the body and highlights them for diagnosis (see technetium, p. 423).

ELEMENT OF HISTORY

The soft black molybdenum mineral previously called molybdena, and now known as molybdenite (molybdenum sulfide, MoS_2), looks very like graphite. The two were often confused and both were used to make pencils.[79] Even when it was realized that they were different, the former was still mistaken for a lead ore until, in 1778, the Swedish chemist, Karl Scheele, published an analysis of the mineral and showed that it was neither lead nor graphite.

Others investigated it and speculated that it contained a hitherto unknown element, but it proved difficult to reduce it to a metal, mainly because it was impossible to make an intimate mixture of the mineral and carbon by grinding, on account of molybdena's softness. These early chemists also reported that heating the mineral in air gave off sulfur dioxide fumes and left behind an oxide. Treating it with nitric acid formed white crystals of molybdic acid.

Scheele passed the problem over to his friend Peter Jacob Hjelm (1746–1813) who was based at Uppsala, Sweden. Hjelm decided to heat molybdic acid with carbon and ground the two together in linseed oil which formed a paste. This he heated to red heat in a closed crucible, whereupon the heat carbonized the oil and together with the

[79] The 'lead' of a pencil takes its name from the molybdenum mineral which was thought to be a lead-based ore.

charcoal it reduced the oxide to the metal. The newly discovered element was announced in the autumn of 1781.

About this time Scheele discovered a simple and specific test for molybdenum. When molybdenum was oxidized to molybdate (which is the metal in oxidation state VI), it would form an intense blue colour on adding a reducing agent to the solution. Even minute amounts of the metal could be detected this way and the depth of colour was a measure of the concentration. The test was used for almost 200 years, despite the fact that chemists could not identify the agent responsible for the colour. In 1996 the puzzle was solved by a group of German chemists at the University of Bielefeld, who showed it to consist of a cyclical cluster made up of 154 molybdenum atoms interlinked with oxygen atoms.

ELEMENT OF WAR

The British Army were the first to use tanks in warfare in World War I. These were protected by 75 millimetre manganese steel plates, which unfortunately were not strong enough to protect the occupants against a direct hit by a heavy shell. The answer to this problem was armour plating of molybdenum steel. This had been invented 20 years previously by the French company Schneider & Co. Not only did this provide much better protection, but it was much lighter and the thickness of the protective sheeting could be reduced to 25 millimetres, permitting more speed and manoeuvrability.

ECONOMIC ELEMENT

Molybdenite is the chief mineral ore, with wulfenite (lead molybdate, $PbMoO_4$) being less important. Some molybdenum is obtained as a by-product of tungsten and copper production. The main mining areas are the USA, Chile, Canada and Russia, with world production being around 90 000 tonnes per year, and reserves amounting to 12 million tonnes of which 5 million tonnes are in the USA. One of the largest deposits of molybdenite, known as the Climax deposit, was found in Colorado, USA, in 1918.

Molybdenite is processed by roasting, which converts it to molybdenum oxide (MoO_3). This is then reduced to the metal in various ways, such as by converting the oxide into ammonium molybdate followed by treatment with hydrogen to reduce it to the metal. Some molybdenum roasters are equipped to recover rhenium metal as well and these are the only commercial source of that element.

Molybdenum is rarely made as the bulk metal because of its high melting point, and instead is produced and sold as a grey powder. Indeed it could not be cast as a molten metal until 1959 when a special crucible was developed for the purpose. Generally, molybdenum is formed into shape by compressing the powder at very high pressures.

Molybdenum is used in electrical and electronic devices, in engineering, and in glass manufacture where it is used for furnaces because it is not affected by molten glass. Most molybdenum (75%) goes into alloys, such as cast iron and 'moly steel'[80] where just a few per cent of the element can confer desirable properties such as high strength

[80] The famous steel-and-glass *Pyramide du Louvre* in Paris is constructed of it.

up to 2000°C, low expansion on heating (it has the lowest expansion of all the engineering metals), good electrical conductivity, high resistance to corrosion, and resistance to wear even though it is present only as a thin coating. Moly steel is used for automobile and aircraft engine parts, and even for rocket engines.

Other alloys are used for heating elements, support wires for filaments in light bulbs, radiation shields, glass furnace electrodes, and anodes in X-ray equipment. Molybdenum alloys make good tools; the higher the stress these will be subjected to, the higher should be their molybdenum content, with high-speed tools (e.g. those used for drills, gear cutters and saw blades) containing as much as 7%.

Molybdenum disulfide is used as a lubricant and an anti-corrosion additive, because it forms strong, stable films on metal surfaces and will function under extremes of high and low temperatures and high pressures.

Molybdenum powders are used in circuit inks for circuit boards, and in microwave devices and heat sinks for solid-state devices. Power rectifiers rely on molybdenum sheet plate, alloyed with nickel, copper or rhodium, to provide thermal expansion control and heat management, and these are used in the electric motors of trains and in industrial motor power supplies. An increasing amount of molybdenum is used in spray applications where it is blended with binders rich in chromium and nickel and then plasma-sprayed on to piston rings and other moving parts which are exposed to heavy wear.

In the chemical industry a molybdenum catalyst is required for the desulfurization of fossil fuels and the formation of synthetic fibres and rubbers. Some is turned into sodium molybdate, a bright orange pigment used in colouring ceramics and plastics.

ENVIRONMENTAL ELEMENT

Molybdenum in the environment	
Earth's crust	1.5 p.p.m.
	Molybdenum is the 54th most abundant element.
Soils	approx. 2 p.p.m., range 0.1–18 p.p.m.
Sea water	10 p.p.b.
Atmosphere	traces

Molybdenum differs from the other micronutrients in soils in that it is less soluble in acid soils and more soluble in alkaline soils, the result being that its availability to plants is sensitive to pH and drainage conditions. Liming soil is regarded as the better way to mobilize molybdenum, and better than applying it directly because too much molybdenum may then affect the animals that eat the plants. Fodder with more than 10 p.p.m. of molybdenum would put most livestock at risk. However, some plants can have up to 500 p.p.m. of the metal when they grow on alkaline soils.

CHEMICAL ELEMENT

Data file	
Chemical symbol	Mo
Atomic number	42
Atomic weight	95.94
Melting point	2617°C
Boiling point	4612°C
Density	10.2 kilograms per litre (10.2 grams per cubic centimetre)
Oxides	MoO_2, MoO_3, and various others.

Molybdenum is a lustrous, silvery metal and a member of group 6 of the periodic table of the elements. The metal is fairly soft when pure, and is attacked slowly by acids.

There are seven naturally occurring isotopes, of which molybdenum-98 is the most abundant at 24%, while molybdenum-96 accounts for 16.5%, molybdenum-95 for 16%, molybdenum-92 for 15%, molybdenum-97, molybdenum-100 and molybdenum-94 have 9.5%. None is radioactive.[81]

ELEMENT OF SURPRISE

The blades of some Japanese swords of the fourteenth century have been found to contain enough molybdenum for them to be considered as a type of moly steel, with all the strength and corrosion resistance this implies. Yet molybdenum was not recognized as a metal until the eighteenth century, not tested as an alloy for steel until the late nineteenth century, and not widely used until the twentieth century. Whichever Japanese blacksmith chanced upon the benefits of adding molybdenum to iron, he kept the secret to himself and it appears to have died with him.

[81] There are minute traces of radioactive molybdenum isotopes in uranium minerals that have been produced as fragments of nuclear fission.

Neodymium

Pronounced nee-oh-dim-iuhm, the name is derived from the Greek *neos didymos*, meaning new twin.

French, *néodyme*; German, *Neodym*; Italian, *neodimio*; Spanish, *neodimio*; Portuguese, *neodímio*.

Neodymium is one of 15 chemically similar elements referred to as the rare-earth elements, which extend from the element with atomic number 57 (lanthanum) to the one with atomic number 71 (lutetium). The term rare-earth elements is a misnomer because some are not rare at all; they are more correctly called the lanthanides, although strictly speaking this excludes lutetium. The minerals from which they are extracted, and the properties and uses they have in common, are discussed under lanthanides (p. 219).

HUMAN ELEMENT

Neodymium in the human body	
Total amount in body	not known, but small

The amount of neodymium in humans is quite small and, although the metal has no biological role, it can have profound effects on parts of the body; for example, neodymium dust and salts are very irritating to the eyes. However, ingested neodymium salts are regarded as only slightly toxic if they are soluble, and non-toxic if they are insoluble.

It is difficult to separate out the various amounts of the different lanthanides in the human body, but they are present; the levels are highest in bone, with smaller amounts being present in the liver and kidneys. It is impossible to estimate the amount of neodymium in an average adult, and it is difficult to judge how much we take in our diet, but it is probably only a few milligrams a year.

Neodymium is not readily taken up by plant roots, so not much gets into the human food chain. However, there tends to be more of this element in plants than of other lanthanides: vegetables have on average 10 p.p.b. (dry weight) and some plants can have as much as 3000 p.p.b. The level of neodymium in sewage is less than that of most other lanthanides, suggesting that what is taken into the body is preferentially retained.

MEDICAL ELEMENT

Tests on rats showed that an injection of 60 milligrams of neodymium nitrate per kilogram of body weight causes blood clotting to cease altogether, probably by interfering with the calcium ions that are needed for this to occur. It was thought that this anticoagulant property could be useful in treating human conditions, and much work was done trying to put this discovery to use, ultimately to no avail.

ELEMENT OF HISTORY

Neodymium was discovered in Vienna, Austria, in 1885 by Karl Auer (1858–1929), who was also involved in the discovery of lutetium (see p. 241). A sample of the pure metal was first produced in 1925.

Didymium, the element which the Swede, Carl Gustav Mosander, claimed to have separated from cerium in 1839, turned out to be a mixture of lanthanide elements. In 1879, samarium was isolated from didymium, followed a year later by gadolinium. In 1885, Auer obtained neodymium (meaning 'new didymium') and praseodymium (meaning 'green didymium') from didymium, confirming what had already been suspected from its atomic spectrum. This had been studied by Bohuslav Brauner of Prague in 1882 and was shown to vary according to the mineral from which didymium was extracted. Auer, at the time he made his discovery, was a research student of the great German chemist, Robert Bunsen (of Bunsen burner fame). Bunsen had become recognized as the world expert on the 'element' didymium, and he accepted Auer's discovery immediately, whereas other chemists were to remain sceptical for some time.

ECONOMIC ELEMENT

Neodymium is found in minerals that include all the lanthanide elements. The most important ore for neodymium is monazite, but some deposits of bastnäsite are rich in it as well. The main mining areas are China, USA, Brazil, India, Sri Lanka and Australia. Reserves of neodymium are estimated to be around 8 million tonnes.

Little neodymium is produced because it is very expensive and has few applications. World production of neodymium oxide is about 7000 tonnes a year, and the pure metal is obtained by reacting neodymium fluoride with calcium.

Adding a little neodymium metal to magnesium alloys greatly strengthens them, but this use has been overshadowed by a more important alloy, that of neodymium, iron and boron (often referred to as NIB). This was first produced in 1983 and was found to make excellent permanent magnets. Their manufacture now accounts for most of the neodymium that is produced. They are to be found in modern vehicles, being an essential part of starter motors, window lifters, door locks, windscreen wipers and fuel pumps. They are also used in computer data storing, and may well become part of loudspeakers. See also 'Element of surprise' below.

Neodymium oxide is used to tint glass attractive shades of purple. It produces this colour by absorbing light of wavelengths corresponding to yellow and green, so we see the complementary colour, which is purple. It is also used to make other kinds of glass, such as that for the goggles which protect the eyes of welders, and so-called 'tanning' glass which keeps out the heat of infrared rays while letting through ultraviolet rays that cause the skin to darken. Neodymium oxide is also used to dope the glass for power lasers. In the chemicals industry, neodymium oxide and nitrate are used as catalysts in the polymerization of so-called dienes which are used in rubber manufacture.

ENVIRONMENTAL ELEMENT

Neodymium in the environment	
Earth's crust	38 p.p.m.
	Neodymium is the 26th most abundant element.
Soils	approx. 20 p.p.m. (range 4–120 p.p.m.)
Sea water	4 p.p.t.
Atmosphere	virtually nil

Neodymium is the second most abundant of the rare-earth elements (after cerium) and is almost as abundant as copper. It poses no environmental threat to plants or animals.

CHEMICAL ELEMENT

Data file	
Chemical symbol	Nd
Atomic number	60
Atomic weight	144.24
Melting point	1021°C
Boiling point	3070°C
Density	7.0 kilograms per litre (7.0 grams per cubic centimetre)
Oxides	NdO and Nd_2O_3

Neodymium is a bright silvery-white metal, and a member of the lanthanide group of the periodic table of the elements. It quickly tarnishes in air, and so has to be protected either by storing it under oil or casing it in plastic. It reacts slowly with cold water, and rapidly with hot. Neodymium differs from most of the other rare-earth elements in that it has three oxidation states: II, III and IV.

There are seven naturally occurring isotopes of neodymium, of which the most abundant is neodymium-152, accounting for 27%, with neodymnium-144 next at 24%. The latter isotope is also radioactive, albeit weakly so, with a half-life of 2000 trillion years (2×10^{15} years) i.e. more than 100000 times the age of the Universe. The other isotopes are neodymium-146, with 17% of the total, neodymium-143 with 12%; neodymium-145 with 8%; and neodymium-148 and -150, each with 6%. This last isotope is radioactive with a half-life of 1.1×10^{19} years.

ELEMENT OF SURPRISE

Neodymium–iron–boron (NIB) magnets are so powerful that those handling them must wear protective glasses – they fly together with such force that they can shatter and send splinters flying in all directions. At times young people have used these industrial magnets to attach ornaments to their cheeks by putting one of the small magnets on the inside of the mouth. However, the magnet and ornament have then proved impossible to pull apart, sometimes necessitating a visit to a hospital for surgical removal.

In the USA, NIB magnets have been used to check for counterfeit currency because they are powerful enough to detect the magnetic particles in the ink used to print real currency.

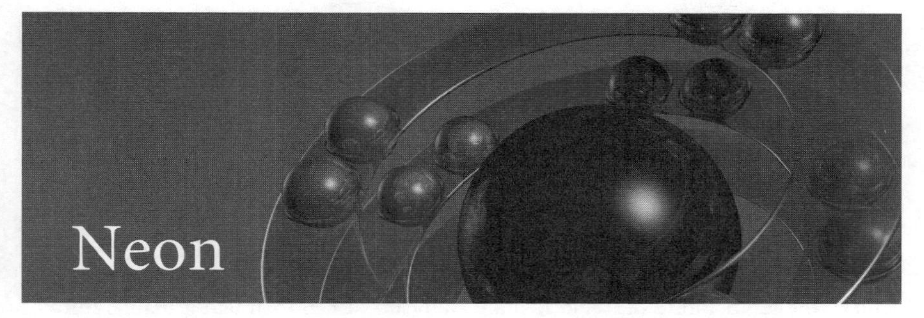

Pronounced nee-on, the name is derived from the Greek *neos*, meaning new.
French, *néon*; German, *Neon*; Italian, *neo*; Spanish, *neón*; Portuguese, *neônio*.

HUMAN ELEMENT

Neon in the human body	
Blood	minute trace
Total amount in body	very small

Neon is a harmless gas which can have no biological role on account if its inability to react with anything.

ELEMENT OF HISTORY

In 1898 at University College, London, William Ramsay and Morris Travers isolated krypton gas by evaporating liquid argon. They had been expecting to find a lighter gas which would fit a niche above argon in the periodic table of the elements. In June of that year they modified and repeated their experiment and allowed solid argon, surrounded by liquid air, to evaporate slowly under reduced pressure. They collected the gas that came off first. This time they were successful and when they put a sample of the newly discovered gas into their atomic spectrometer, the effect startled them by its brilliance. As Travers wrote, 'the blaze of crimson light from the tube told its own story, and it was a sight to dwell upon and never forget. It was worth the struggle of the previous two years...'.

Ramsay's son, Willie, first suggested a name for the new gas, novum, from the Latin word *novus*, meaning new, and while his father liked the idea, he preferred to base it on the Greek word *neos*, so the gas became known as neon.

ECONOMIC ELEMENT

Neon is extracted from liquid air by fractional distillation. It comes off as a gas mixed with helium, traces of which are removed from it by being absorbed by activated charcoal. Commercial needs for neon are met by only a few tonnes of the gas per year, although more could be produced because there are 65 billion tonnes of neon in the atmosphere.

Neon is used mainly in so-called neon signs, although only the red ones are pure neon. These, in particular, are renowned for their intensity of colour, which is why they are used as beacon lights, especially during fog. The first neon sign was made by Georges Claude in 1910, and they were soon adopted by the advertizing industry, appearing in all major cities within a few years. They are quite robust, operating without attention for up to 20 years.

There are other uses for neon, such as in diving equipment, lasers, high-voltage switching gear and as a refrigerant for very low temperatures. While liquid neon cannot achieve the temperature of liquid helium (–269°C), it can maintain a temperature of –246°C and is a much more efficient very-low-temperature refrigerant.

ENVIRONMENTAL ELEMENT

Neon in the environment	
Earth's crust	70 p.p.t.
	Neon is the 82nd most abundant element – in other words it is extremely rare.
Sea water	0.2 p.p.m.
Atmosphere	18 p.p.m. by volume

Neon poses no threat to the environment, and indeed can have no impact at all because it is chemically unreactive and forms no compounds. It is not only to be found in the atmosphere. In 1909, Armand Gautier collected gas bubbling up from fumaroles near Vesuvius, and from hot springs near Naples, and showed that these contained neon.

CHEMICAL ELEMENT

Data file	
Chemical symbol	Ne
Atomic number	10
Atomic weight	20.1797
Melting point	–249°C
Boiling point	–246°C
Density	0.9 grams per litre
Oxide	none

Neon is a colourless, odourless gas and is one of the group of elements known as the noble gases (the others are helium, argon, krypton, xenon, radon and element-118) which make up group 18 of the periodic table of the elements. It is unreactive towards all known chemicals, including fluorine which will react with some of the other noble gases.

There are three naturally occurring isotopes, of which neon-20 is the most abundant at 90.5%; neon-22 makes up 9.2%, and neon-21 accounts for the remaining 0.3%. None is radioactive.

ELEMENT OF SURPRISE

J. Norman Collie (1859–1942) also laid claim to the discovery of neon, and is reputed to have said, 'If anyone happens to write an obituary for me I want two things said: I first discovered neon and I took the first X-ray photograph.' His assertion to have first produced neon is rather suspect, but Collie was a skilled research chemist based at University College, London, where he did collaborate with Ramsay. Indeed he may have been among the first to *observe* neon's red glow. However, a book about his life, *The Snows of Yesteryear: J. Norman Collie*, written by William Taylor in 1973, says that Collie stumbled across neon independently while investigating the atomic spectrum of hydrogen at low pressures. If this were so, then it was unsubstantiated, and it is difficult to reconcile this with the well documented evidence that neon was first isolated by Ramsay and Travers. Yet Collie was a highly respected academic, and eventually became Professor of Organic Chemistry at University College.

Such was Collie's fame as an investigator, and pioneer mountaineer, that he is reputed to have been the character on whom Conan Doyle based his famous fictional detective Sherlock Holmes, who was also a competent chemist.

Pronounced nep-tyoon-iuhm, this element was named after the planet Neptune, because uranium had been named after Uranus. Just as neptunium comes after uranium in the periodic table of the elements, so Neptune comes after Uranus in the Solar system.

French, *neptunium*; German, *Neptunium*; Italian, *nettunio*; Spanish, *neptunio*; Portuguese, *neptúnio*.

HUMAN ELEMENT

Neptunium in the human body
None

Neptunium has no role to play in living things and is never encountered outside nuclear facilities or research laboratories. Research on animals showed that it was not absorbed through the digestive tract, and when injected into the body it tended to accumulate in the bones, from where it was only slowly released.

ELEMENT OF HISTORY

Attempts were made in the 1930s, by teams headed by Enrico Fermi in Italy, Otto Hahn in Germany and Irene Curie in France, to produce elements beyond uranium, by bombarding this metal with subatomic particles. They found atoms with unusual radioactivities which they took to be evidence that they had succeeded, and indeed Fermi and his co-workers even named elements 93 and 94, calling them ausenium and hesperium, respectively. In fact Fermi and his co-workers were looking at uranium fission fragments.[82]

Neptunium was discovered by Edwin M. McMillan (1907–91), with the help of Philip H. Abelson (1913–), at Berkeley, California, USA. McMillan had bombarded a uranium target with slow neutrons in the Berkeley cyclotron and noted that some unusual β-rays were then emitted from it, showing that new isotope was present. It was left to Abelson, who spent a week-long visit there in May 1940, to prove they came from

[82] Fermi's experiments took place in 1934, so this was actually the discovery of fission, yet it is not acknowledged as such because it was not the object of the experiment, nor did the researchers initially realize it had occurred. The officially recognized discovery of fission took place in 1938. [Walter Saxon.]

a new element. He was the right person to do this, having worked in the radiation laboratory at Berkeley for his PhD, which he had been awarded in 1939. Within 3 days of being given a sample of material by McMillan, he had proved that it was a new element and within the week they had sent off a paper to the journal *Physical Review* announcing their discovery; their paper was published later that year.

McMillan had made neptunium from uranium-238 by converting this to uranium-239 with neutron addition after which it underwent loss of a negative electron from the nucleus (β-emission), thereby increasing the nuclear charge by one so that element 92 became element 93. McMillan named the new element neptunium. This isotope, neptunium-239, had a half-life of 2.4 days. A cloak of secrecy then descended on the work because of national security, although work on neptunium continued during World War II. In 1944 the first pure compound, the oxide (NpO_2), was made using a few micrograms of the element that had been extracted from a nuclear reactor.

McMillan shared the 1951 Nobel Prize for Chemistry with Glen T. Seaborg for their discoveries of transuranium elements.[83]

ECONOMIC ELEMENT

Today neptunium-237 is extracted in kilogram quantities from the spent uranium fuel rods of nuclear reactors. This isotope is long-lived, gives off only weak radioactivity (in the form of α-radiation) and can be handled in an ordinary laboratory given the necessary safety measures. Neptunium has been used in neutron detectors.

ENVIRONMENTAL ELEMENT

Neptunium in the environment	
Earth's crust	present but only in trace amounts
Sea water	nil
Atmosphere	nil

Neptunium occurs naturally on Earth, being present in minute quantities in uranium ores, where it is formed when a neutron, emitted by a uranium atom undergoing fission, is captured by another uranium atom. In this fashion uranium-238 becomes uranium-239, which then decays by β-emission, creating an atom of neptunium-239.

[83] Abelson became best known as the editor of *Science*, a post he held from 1961 to 1984.

CHEMICAL ELEMENT

Data file	
Chemical symbol	Np
Atomic number	93
Atomic weight	237.0 (isotope neptunium-237)
Melting point	640°C
Boiling point	3902°C
Density	20.3 kilograms per litre (20.3 grams per cubic centimetre)
Oxides	NpO, NpO_2, Np_2O_5 and NpO_3

The metal itself, which is silver in colour, can be made from neptunium fluoride (NpF_3) by reaction with molten lithium or barium at 1200°C. Neptunium is a member of the actinide group of the periodic table of the elements. Chemically it is very reactive and is attacked by oxygen, steam and acids, but not by alkalis. It can exist in many oxidation states from neptunium(II) to neptunium(VII) and in this respect it is like uranium.

The longest-lived isotopes of neptunium are neptunium-237, with a half-life of 2.14 million years, and neptunium-236, with a half-life of 155 000 years. All the other 19 known isotopes, with mass numbers ranging from 228 to 242, have half-lives of a year or less.

ELEMENT OF SURPRISE

Although it has a half-life of 2.14 million years, this does not mean that any of the neptunium that was around when the Earth was formed, 4.57 billion years ago, could still be around today. Even it there had been a trillion tonnes of neptunium (10^{12} tonnes), and assuming this was composed of the longest-lived isotope, it would all have undergone radioactive decay within 240 million years.[84]

[84] A trillion tonnes of neptunium contains 2.5×10^{33} atoms. After 2.14 million years this would be reduced by a half to 1.25×10^{33}, after 4.28 million years it would halve again to 6.25×10^{32}, and so on, until after 111 half-lives, which amount to 237 million years, it would be down to a single atom.

Pronounced nik-el, the name is a shortened form of the German *kupfernickel*, meaning either devil's copper or St Nicholas's copper.

French, *nickel*; German, *Nickel*; Italian, *nichel*; Spanish, *niquel*; Portuguese, *niquel*.

HUMAN ELEMENT

Nickel in the human body	
Blood	levels vary between 10 and 50 p.p.b.
Bone	not more than 0.7 p.p.m.
Tissue	1–2 p.p.m.
Total amount in body	15 milligrams

Nickel has been proved essential to some species and is linked to growth, but its exact metabolism is not clearly understood. The metal presents some health hazards to humans, and these are of three kinds: contact with nickel solutions can cause dermatitis; breathing in nickel dust can cause lung and nasal cancer; and inhaling nickel carbonyl gas, even in tiny amounts, can kill.

The second and third threats are mining and industrial hazards respectively, but the first hazard is one that affects ordinary people, some of whom are very sensitive to nickel. Even the nickel in stainless steel is able to cause dermatitis, which begins as an irritating itch known as 'nickel itch'. It starts when stainless steel wrist watches, spectacle frames, garment fasteners, jewellery, etc. come in contact with the skin. This is exacerbated by sweat, the acids of which dissolve a little of the nickel.

Nickel is thought to cause cancer by substituting for zinc and magnesium atoms in DNA polymerase. The nickel ion is slightly different from zinc and magnesium and so affects the behaviour of this enzyme, perhaps causing it to bind the wrong nucleotide, resulting in the formation of a rogue sequence of DNA and a cancerous cell.

FOOD ELEMENT

Nickel may be an essential element, yet the human requirement could be as little as 5 micrograms a day, although the daily intake is estimated to be around 150 micrograms. Nickel occurs in some beans where it is an essential component of some enzymes, such as jackbean urease, each molecule of which contains 12 nickel atoms. Another relatively rich source of nickel is tea, which has 7.6 milligrams per kilogram of dried leaves. Other plants generally have less than half this amount.

ELEMENT OF HISTORY

Meteorites contain both iron and nickel, and in earlier ages they were used as a superior form of iron, being worked into tools and swords which were not unlike the stainless steel of today. Because the metal did not rust, it was regarded by the natives of Peru as a kind of silver. A zinc–nickel alloy called *pai-t'ung* (white copper) was in use in China as long ago as 200 BC and exported to the Middle East, from where some even reached Europe. The ore from which it was made came from Yunnan in southern China, where there are nickel sulfide deposits.

A reddish-brown ore called *kupfernickel* was known to German copper miners and they gave it this nickname which meant devil's (or St Nicholas's) copper since it could be used for nothing except colouring glass green. Early mineralogists tried in vain to extract copper from it. Then, in 1751, Alex Fredrik Cronstedt (1722–65), working at Stockholm, investigated a new mineral – now called nickeline (NiAs) – found in a cobalt mine at Los, Hälsingland, Sweden. He thought it might contain copper but he could not extract any of this metal from it. What he did eventually get from this (and from *kupfernickel*) was a previously unknown metal which he christened nickel in 1754. The reason for the 4 year delay was that many chemists would not accept his claim that it was a 'new' metal and believed it was an alloy of cobalt, arsenic, iron and copper. This seemed likely as these elements were present in the 'new' metal as trace contaminants. It was not until 1775 that absolutely pure nickel was produced by Torbern Bergman and this ended the arguments about its elemental nature.

There was little demand for nickel before 1844 until the development of silver plating, when nickel was found to be the most desirable base metal. Thereafter electroplated nickel–silver (EPNS) cutlery became popular.

A lot of the nickel that is mined on Earth may have arrived here in the form of gigantic meteorites. One of these hit the Sudbury region of Ontario, Canada, hundreds of millions of years ago and the deposit there is estimated to contain 200 million tonnes of nickel, although most of this may have welled up from the Earth's mantle. This deposit was discovered in 1856 by construction workers who were building the Pacific Railway and was regarded as a potential source of copper to begin with. By 1905, it had become the world's largest source of nickel once a method of extracting this metal and refining it by electrolysis had been developed. To access this deposit, 17 000 holes were drilled into the rock and each charged with dynamite – the preparations took more than a year – and then exploded together releasing 4 million tonnes of rock for mining.

The extraction and manufacture of pure nickel from its ores was greatly improved in 1888 when the industrialist Ludwig Mond and his assistant, Carl Langer, investigated the problem of leaky valves made of nickel through which carbon monoxide gas was passing. They discovered that the gas was reacting with the nickel to form volatile nickel carbonyl, $Ni(CO)_4$, which boils at 43°C. From this they developed a process, the Mond nickel process, in which impure nickel is treated with carbon monoxide gas; the nickel carbonyl that forms is then passed into a chamber where pellets of nickel are kept at a temperature high enough to release the nickel, which deposits on their surface, so they grow in size. (The carbon monoxide is reused.) The process is potentially very hazardous but was brought into commercial production despite several cases of poisoning, including some deaths, among those who operated the process.

ECONOMIC ELEMENT

Nickel has a particular affinity for sulfur, and sulfide ores are an important source of the metal. Millerite (NiS) is one example but this ore is rare. Most ores from which nickel is extracted are iron–nickel sulfides, such as pentlandite, $(Fe,Ni)_9S_8$. Nickel was first extracted from an ore called nickeline (nickel arsenide, NiAs) but this is now rare. The metal is mined in Russia, Australia, New Caledonia, Cuba, Canada and South Africa. (The Ontario deposit mentioned above still supplies about 30% of the world demand for nickel.) Annual production exceeds 500000 tonnes and easily workable reserves will last at least 150 years.

About half the nickel produced each year ends up alloyed with steel to give stainless steel, whose composition can vary but is typically iron with around 18% chromium and 8% nickel. Stainless steel was invented simultaneously in Germany and England early in the twentieth century. Although stainless steel is generally a shiny silvery metal, its surface can be almost any colour, depending on how it is treated.

Nickel is easy to work and can be drawn into wire. It resists corrosion even at high temperatures and for this reason it is used in gas turbines and rocket engines. Other important uses of nickel are in metal plating and industrial catalysts, and for coinage. The 5 cent US coin, the so-called 'nickel', is copper alloyed with 25% nickel, and was first produced in 1865. Nickel coins were introduced in Belgium in 1860, and today many countries still use coins containing nickel.

Most nickel ends up in alloys, some of which have remarkable properties:

Invar (short for invariable) is 64% iron and 36% nickel; it got its name because it has the curious ability of not expanding when heated. Invar was discovered by Charles Guillaume of the Bureau of Weights and Measures in Paris who was seeking a metal to use for making copies of the standard metre. It was also ideal for use in measuring tapes and for parts within clocks and chronometers.

Nichrome is a nickel–chromium alloy with between 11 and 22% chromium, and small amounts of silicon, manganese and iron. It does not oxidize, even at red heat, and is used in appliances such as toasters and electric ovens.

Monel is an alloy of nickel and copper (e.g. 70% nickel, 30% copper, with traces of iron, manganese and silicon) which is not only hard but will resist corrosion by sea water, so that it is ideal for propeller shafts in boats and in desalination plants. It is also unaffected by acids, and so it is used in the chemical and food processing industries.

Platinite, which is 46% nickel and 54% iron, has the same expansion on heating as ordinary glass and is used as the lead-in wire of conventional electric light bulbs.

INCO-276 is made of 57% nickel with 16% each of chromium and molybdenum, plus a few per cent of other metals. This alloy resists corrosion by hydrogen sulfide gas, which normally attacks stainless steel, and it is used for boring wells that go deep into the Earth's crust where this gas is often encountered.

MM002 is made of 60% nickel with around 10% each of cobalt, tungsten and chromium plus small amounts of several other metals. This alloy is used for the new generation of turbine blades which are grown as single crystals in special moulds.

Finely powdered nickel is used as a catalyst in the hydrogenation of oils to fats. In the 1890s, two French chemists, Sabatier and Senderens, showed that it was possible to convert edible oils into fats, thereby making margarine, a cheap substitute for butter that was equally nutritious, albeit less tasty.[85] Nickel powder is also added to paints that coat the housing of sensitive electronic instruments because it shields them from stray radiation. Rechargeable nickel–cadmium batteries are made from sintered nickel powder and can be recharged and used more than a thousand times.

ENVIRONMENTAL ELEMENT

Nickel in the environment	
Earth's crust	approx. 80 p.p.m.
	Nickel is the 23rd most abundant element in the Earth's crust, but see below.
Soils	approx. 50 p.p.m.
Sea water	0.5 p.p.b.
Atmosphere	insignificant

Most nickel on Earth is inaccessible because it is locked away in the planet's iron–nickel molten core, which is 10% nickel. Indeed, a lot of the nickel that is found naturally in soils and in the seas has been deposited on the Earth from the 'rain' of space dust and meteorites that the planet scoops up each year in its journey round the Sun. The total amount of nickel dissolved in the sea has been calculated to be around 8 billion tonnes (8×10^{12} kilograms). Organic matter has a strong ability to absorb the metal which is why coal and oil contain measurable amounts.

The nickel content in soil can be as low as 0.2 p.p.m. and as high as 450 p.p.m. in some clay and loamy soils. The average is around 20 p.p.m.. Heavily fertilized soil may have its nickel content raised since phosphate fertilizers contain traces of the metal and farmland near oil- and coal-burning industries can also be affected. Sewage sludge from industrial areas can also increase the level of nickel and in some soils fertilized this way the level of nickel is more than 800 p.p.m. In parts of Canada near metal-processing industry nickel levels in soil as high as 26 000 p.p.m. have been recorded.

Plants can tolerate a certain amount of nickel in the soil in which they grow, but some crops fail to grow on heavily contaminated soil. Beans and maize are very sensitive to nickel in the soil: watering them with a 40 p.p.m. solution of nickel will kill them. Clover will not grow on soil that has 80 p.p.m. of nickel, but oats will cope with such levels quite happily and absorb some of the nickel while doing so. Liming such soil, which raises the pH, will reduce nickel absorption quite dramatically. Some plants will

[85] Margarine is equally nutritious in terms of providing energy, but it contains forms of fat that some dietitians regard as harmful, i.e. the so-called *trans* fats..

tolerate extremely high levels of nickel; the flower *Hybanthus floribundus* (shrub violet) will accumulate as much as 6000 p.p.m., which is 0.6% (dry weight), in its leaves and stems, suggesting that this might one day be used to decontaminate polluted soil while acting as a source of reclaimable nickel.

CHEMICAL ELEMENT

Data file	
Chemical symbol	Ni
Atomic number	28
Atomic weight	58.6934
Melting point	1453°C
Boiling point	2732°C
Density	8.9 kilograms per litre (8.9 grams per cubic centimetre)
Oxide	NiO

Nickel is a silvery, lustrous, malleable, ductile metal and a member of group 10 of the periodic table of the elements. It resists oxidation, but is soluble in acids (except concentrated nitric acid), yet unaffected by alkalis. The bulk metal is ferromagnetic.

There are five naturally occurring isotopes of nickel, the most abundant of which is nickel-58, which accounts for 68%. The others are nickel-60, with 26% of the total, nickel-63 (4%) and nickel-61 and -64 (1% each). None of the naturally occurring isotopes is radioactive.[86]

ELEMENT OF SURPRISES

1. Perhaps the most remarkable alloy of nickel is the one it forms with aluminium, called nickel aluminide, which has the exact chemical composition Ni_3Al and as such has an ordered array of atoms in its structure. It was first developed at the US Government's Oak Ridge National Laboratory at Tennessee, but it was found to be too brittle, so research on it was abandoned until it was discovered that adding a fraction of a per cent of boron remedied this failing and created a material of superior strength.

   Nickel aluminide may one day even become part of the engine of the family motor car. Before then it will be incorporated into rockets, high-performance jet engines and heat-exchangers. What makes nickel aluminide so special is its unique property at high temperatures. It is six times stronger than stainless steel, and it gets stronger

[86] The isotope nickel-48, first discovered in 1999, is the most proton-rich isotope known, having 28 protons to 20 neutrons. Both these numbers are so-called 'magic numbers' for protons and neutrons that indicate particularly stable nuclear structures. This nucleus decays by the extremely rare radioactive decay mode of two-proton emission, a mechanism long postulated but rarely detected. [Walter Saxon]

the hotter it gets. At 800°C it is twice as strong as it is at room temperature. The greater the temperature at which you run an engine, the more efficient it becomes, which is why the search is on for materials that will allow engines to run at red-heat temperatures above 1000°C.

2. Another curious alloy of nickel is nitinol (55% nickel, 45% titanium, which corresponds to the chemical formula NiTi), which was developed in the USA in the 1960s. The name is an acronym of Nickel Titanium Naval Ordnance Laboratory. This alloy has the ability to 'remember' a previous shape that it had. Spectacle frames made from nitinol can be bent and twisted into remarkable shapes and, when released, will jump back to their original shape.
3. A nickel compact disc has been developed that will preserve 1.5 million pages of data for around 1000 years without degrading, and it will also resist temperatures of 800°C. The disc, called HD-Rosetta, stores information in an analogue format which means that even in 1000 years it will still be possible to extract its information, albeit using a microscope.

Niobium

Pronounced niy-o-bi-uhm, this element was named after Niobe of Greek mythology who was the daughter of king Tantalus. The name was chosen because of niobium's chemical similarity to the element tantalum. Niobium was originally called columbium after the mineral from which it was first obtained and this name continued to be used for well over 150 years until, in 1950, the International Union of Pure and Applied Chemistry decided that niobium, the alternative name suggested in 1844 – see below – should be the official one. The name columbium is still used in some areas, such as engineering.

French, *niobium*; German, *Niob*; Italian, *niobio*; Spanish, *niobio*; Portuguese, *nióbio*.

HUMAN ELEMENT

Niobium in the human body	
Blood	5 p.p.b.
Bone	70 p.p.b.
Tissue	140 p.p.b.
Hair	2 p.p.m.
Total amount in body	1.5 milligrams

Niobium has no known biological role although we take in enough for it to be measurable in our body. Niobium and its compounds may be toxic (niobium dust causes eye and skin irritation) but there are no reports of humans being poisoned by it, although in Russia there is a limit of 10 p.p.b. for niobium in drinking water.

Apart from measuring its concentration, no research on niobium in humans has been undertaken, but some animals were given radioactive niobium-95 to see how this element acted. These tests showed that at most 5% of the ingested niobium was absorbed into the body, and in most animals only 1% was absorbed. That which did pass through the gut wall tended to accumulate mainly in the bone, but also in the liver and kidneys. Niobium was excreted via the urine.

ELEMENT OF HISTORY

Niobium was discovered in 1801 by Charles Hatchett (1765–1847) in London. When examining some minerals in the British Museum collection, he was intrigued by a small, dark, heavy specimen, labelled columbite, that had been part of the collection of Hans Sloane, who had bequeathed it to the museum many years before and which had been sent to him by one John Winthrop, of Connecticut, early in the eighteenth century.

Hatchett heated the mineral with potassium carbonate and then dissolved the product in water; when the solution was neutralized with acid, the oxide precipitated. The filtered solution contained iron and manganese, which Hatchett also confirmed were present, but it was the precipitate on which he focused his interest and which he deduced must contain a previously unknown metal, although he failed to extract this in a pure form.

Hatchett announced his discovery to the Royal Society of London in November 1801, naming the element columbium, the poetic name for America. For many years it was disputed whether or not columbium was another form of tantalum, discovered the following year. The metals are chemically similar, occur together in nature and are difficult to separate. In 1844 the German chemist Heinrich Rose proved that the mineral columbite contained both elements and he renamed columbium 'niobium' in honour of Niobe, the goddess of grief.[87]

Hatchett died before a sample of the pure metal was produced. That was achieved only in 1864 by C. S. Blomstrand, who reduced niobium chloride to niobium by heating it with hydrogen gas.

ECONOMIC ELEMENT

Niobium was mined chiefly as columbite, $(Fe,Mn)(Nb,Ta)_2O_6$ (which is also known as niobite), and this can vary in terms of the proportions of the four metals that are present. Another mined mineral is pyrochlore, $(Na,Ca)_2Nb_2O_6(OH,F)$, and this is now the most important. Some niobium is obtained as a by-product of tin-mining. Niobium and tantalum are separated by chemical means, such as solvent extraction.

The main mining areas are Brazil, which produces more than 85% of the world's niobium, Zaire, Russia, Nigeria and Canada. World production is around 25 000 tonnes per year. The amount of unmined reserves is not reported but there are extensive deposits of pyrochlore.

Niobium is added in small amounts to improve stainless steel, especially if it is to be welded, and welding rods also contain niobium. Small amounts of niobium impart greater strength to other metals, especially those that are exposed to low temperatures. The alloy with zirconium is particularly resistant to chemical attack, especially to attack by corrosive alkalis, and even to molten lithium and molten sodium. Because niobium has low neutron-capture it was the metal of choice for tubing carrying molten lithium and sodium which were used in fast-breeder nuclear reactors.

Niobium coated with platinum-group metals is used as industrial anodes, while niobium-coated copper anodes are used to give cathodic protection to structures in contact with harsh chloride-containing solutions and these can last for up to 20 years.[88]

[87] All her 12 children were murdered.

[88] Cathodic protection, also known as sacrificial protection, prevents corrosion such as rusting by being more susceptible to chemical attack, and undergoing oxidation preferentially.

The alloys of niobium with tin or titanium are superconducting, albeit only below –250°C, yet despite this, they revolutionized the technique of magnetic resonance which relies on very strong magnetic fields. This is used by research chemists to analyse compounds and by hospital doctors for full-body diagnostic imaging. The above-mentioned alloys are noted for their superconductivity even when carrying high current density, whereas the metal oxide superconducting materials, which can operate at much higher temperatures, tend to lose their superconductivity when exposed to high electrical current.

Niobium alloys are used in surgical implants because they do not react with human tissue.

ENVIRONMENTAL ELEMENT

Niobium in the environment	
Earth's crust	20 p.p.m.
	Niobium is the 33rd most abundant element.
Soils	approx. 24 p.p.m., maximum 300 p.p.m.
Sea water	0.9 p.p.t.
Atmosphere	virtually nil

Plants generally show only traces of niobium and many have none at all, although some mosses and lichens can contain 0.45 p.p.m. However, plants growing near niobium deposits can accumulate the metal to levels above 1 p.p.m.

CHEMICAL ELEMENT

Data file	
Chemical symbol	Nb
Atomic number	41
Atomic weight	92.90638
Melting point	2468°C
Boiling point	4742°C
Density	8.6 kilograms per litre (8.6 grams per cubic centimetre)
Oxides	Nb_2O_5 (also NbO and NbO_2)

Niobium is a shiny, steel-grey metal, which is soft when pure. It is a member of group 5 of the periodic table of the elements. Niobium resists corrosion because an oxide film forms on its surface. The metal is inert to acids, even to *aqua regia* at room temperature, but it is attacked by hot, concentrated acids, and especially by alkalis and oxidizing agents. There is only one naturally occurring isotope, niobium-93, which is not radioactive.[89]

[89] There are minute traces of other isotopes in uranium ores where they are produced as a result of nuclear fission.

ELEMENT OF SURPRISE

Jewellery made from niobium shimmers with iridescent colours. The metal is protected by a thin film of oxide on its surface and, while this is normally transparent, if it is thickened it will reflect light with an interference pattern that appears multicoloured. The interference comes because the waves of light reflected from the oxide surface are not in phase with those reflected from the underlying metal surface. The degree of interference depends on the thickness of the oxide layer and the angle at which it is viewed. The oxide layer is artificially thickened by the process of anodizing in which the item of jewellery acts as the anode in an electrolytic cell (with ammonium sulfate as the electrolyte solution). Depending on the length of time the object is exposed to the current, almost any colour can be produced, but the niobium blues and purples are particularly attractive. By masking parts of the object, it is possible to produce items with varied colours and patterns.

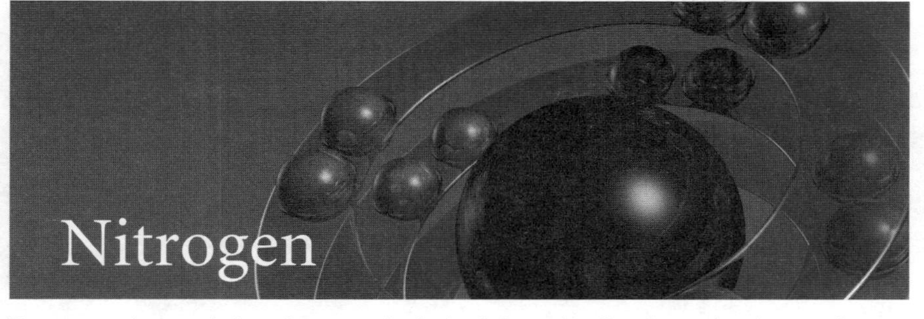

Nitrogen

Pronounced ny-troh-jen, the name is derived from the Greek words *nitron* and *genes*, meaning nitre forming. Nitre was the old name for potassium nitrate (KNO_3) which was more commonly known as saltpetre.

French, *azote*; German, *Stickstoff*; Italian, *azoto*; Spanish, *nitrógeno*; Portuguese, *nitrogênio*. The German word derives from *ersticken* and *stoff*, meaning 'suffocating substance'. Nitrogen was also known as 'choke-damp' in England until a Dr Chaptal suggested the name nitrogen following Cavendish's production of the gas from nitre.

Other words linked with nitrogen are those with the prefix 'azo-' (e.g. azo-dye), azide and ammonia. The first two come from the French word *azote*, meaning 'without life'; the French name for the element was chosen by Lavoisier. Ammonia, the name for the gas NH_3,[90] is derived from Ammon, a god of ancient Egypt, because it was from near his temple that ammonium chloride was to be obtained; this was called *sal ammoniac*, meaning salt of Ammon.

HUMAN ELEMENT

Nitrogen in the human body	
Blood	34 000 p.p.m. (3.4%)
Bone	43 000 p.p.m. (4.3%)
Tissue	72 000 p.p.m. (7.2%)
Total amount in body	1.8 kilograms

Although nitrogen gas (N_2) accounts for 78% of the air we breathe,[91] if it rises much above this level we would die of asphyxiation. Yet nitrogen is an essential element for life, because it is a constituent of DNA and, as such, is part of the genetic code. It is a component of many other biologically important molecules such as haem (part of haemoglobin) and acetylcholine (the neurotransmitter that passes messages from one nerve-ending to another). Nitrogen is also a key part of the many amino acids that form enzymes and other proteins.

[90] The words 'amine' and 'amino acid' refer to derivatives of the NH_2– group, and come from a condensation of 'ammonia' with '-ine', the latter being the archaic word-ending for the names of chemical groupings.

[91] The atmosphere of Mars contains only 2.6% nitrogen.

Amino acids, which are so called because they contain both an amino group (NH_2) and an acid group (CO_2H), can polymerize to long chains of protein by reacting the amino group of one molecule with the acid group of another. These chains are then attracted to one another by hydrogen bonds (see p. 189) and it is these that enable the protein to adopt and hold a particular shape.

Nitrogen is non-toxic as nitrogen gas because it is very unreactive, but some nitrogen derivatives are highly toxic, such as the gases nitrogen dioxide (NO_2), hydrogen cyanide (prussic acid, HCN) and ammonia (NH_3). On the other hand, nitrous oxide (N_2O) induces mild hysteria and eventually anesthesia and was formerly known as laughing gas. It is used as a propellant for cans of whipped cream on account of its solubility in fats.

MEDICAL ELEMENT

Liquid nitrogen is used to freeze blood or to preserve genetic material, such as eggs and sperm, or for performing so-called cryogenic surgery when unwanted tissue can be excised simply by freezing to very low temperatures.

From a metabolic point of view, nitric oxide (NO) is much more important. In 1987, Salvador Moncada discovered that this was a vital body messenger for relaxing muscles, and today we know that it is involved in the cardiovascular system, the immune system, the central nervous system and the peripheral nervous system. The enzyme that produces nitric oxide, called nitric oxide synthase, is abundant in the brain.

Although nitric oxide is relatively short-lived, it can diffuse through membranes to carry out its functions. In 1991, a team headed by K.-E. Andersson of Lund University Hospital, Sweden, showed that nitric oxide activates an erection by relaxing the muscle that controls the flow of blood into the penis. The drug Viagra works by releasing nitric oxide to produce the same effect.

Individuals cannot be treated directly with nitric oxide gas, so drugs that release nitric oxide slowly are given instead. Some older medical treatments did this indirectly; for example, the effect of breathing amyl nitrite vapour was quickly to reduce blood pressure, and this was noted as long ago as 1867, and became the standard treatment for angina. In World War I, doctors noticed that ammunition workers packing shells with nitroglycerine had very low blood pressures and this also led to the use of this compound as a vasodilator (i.e. to dilate blood vessels).

Sometimes, so much nitric oxide may be formed in the body that it threatens life. A leading cause of death among patients in intensive care is septic shock. As the body generates nitric oxide to fight the infection, the blood pressure may become dangerously low. Inhibitors that block the nitric oxide synthase enzymes are needed to prevent this happening.

The naturally occurring isotopes, nitrogen-14 and nitrogen-15, are not radioactive, but nitrogen-13 is, and is used in the modern technique known as positron emission tomography, or PET scanning. This isotope has a half-life of only 10 minutes. When it decays it emits a positron, which is the positive equivalent of the negative electron (its anti-particle). When nitrogen-13 is administered to a patient, the released positrons are immediately annihilated by electrons; this process emits gamma rays. These are monitored to produce a three-dimensional image of the body; PET scanning is particularly useful for examining the brain.

ELEMENT OF WAR

Conventional explosives rely on nitrogen compounds. Gunpowder is a mixture of charcoal (carbon), sulfur and potassium nitrate, and was first discovered in China. Potassium nitrate crystals can be found growing on the walls of cellars and latrines where they form as a result of bacterial action on nitrogen-rich waste; this was the only source of saltpetre for hundreds of years.

The next useful explosive to be discovered, in 1846, was nitroglycerine, which was made by reacting glycerine (today known as glycerol) with nitric and sulfuric acids. It was ideal for blasting in quarries and for tunnelling, but was notoriously temperamental. Then, in 1867, the Swedish chemist Alfred Nobel (1833–96) found that by absorbing it on kieselguhr it became safe to handle and would only explode when it was exposed to a smaller explosion from a detonator. Thus dynamite was invented and Nobel became very rich, eventually using his wealth to set up and fund the Nobel Prizes.

The third nitrogen compound to be used as an explosive was TNT, short for trinitrotoluene, discovered in 1863.[92] Even ammonium nitrate can be used as an explosive,[93] but today there are much more sophisticated ones, in that they are safer to handle, such as HMX (chemical name cyclotetramethylenetetranitramine) and RDX (chemical name cyclotrimethylenetrinitramine) which are widely used by armed forces for munitions.

What makes all such compounds able suddenly to release a lot of energy is the eagerness of the nitrogens they contain to revert violently to nitrogen gas with the evolution of large amounts of heat, which also causes the cloud of released gas to expand even more violently.

ELEMENT OF HISTORY

Sal ammoniac intrigued the alchemists who were fascinated by its volatility. Robert Boyle rightly said in 1661 that it was composed of ammonia (which he called volatile alkali) and hydrochloric acid (which he called muratic acid). It was manufactured in Egypt by heating a mixture of camel and cow dung with salt and urine. Later it was made by heating the antlers of deer ('hartshorn') and other nitrogen-rich natural protein materials such as wool and silk. *Sal ammoniac* was popularly used as smelling salts.

In the 1760s, Henry Cavendish and Joseph Priestley each carried out experiments to remove the oxygen from air, noting that this decreased the volume by about a fifth and left behind 'air' in which a lighted candle was immediately extinguished and in which a

[92] The largest explosion of this occurred accidentally on 6 December 1917 at Halifax, Nova Scotia, Canada, when a ship carrying 2750 tonnes collided with another vessel and thereby destroyed most of the town, killing 1600 and injuring 9000.

[93] On 15 April 1947, a ship in Texas City harbour caught fire and its load of 5000 tonnes of ammonium nitrate exploded, killing 552 people and injuring 3000. Every building within a mile of the ship was demolished.

mouse would soon die. Both men speculated in letters to each other on what this 'air' might be, but neither concluded that it was a substance in its own right, nor an element. Stephen Hales (1677–1761), a clergyman and scientist, had previously carried out similar research but again did not come to the obvious conclusion. So it was left to a young student, Daniel Rutherford (1749–1819), to propose that air consisted mainly of nitrogen gas. He proposed this in his doctorate thesis in September 1772 in Edinburgh, and only a few months before Priestley published a paper on it.[94]

ECONOMIC ELEMENT

Nitrate minerals are known, such as nitratite (Chile saltpetre or sodium nitrate, $NaNO_3$), nitrobarite (barium nitrate, $Ba(NO_3)_2$) and nitrocalcite (calcium nitrate hydrate, $Ca(NO_3)_2{\cdot}4H_2O$), but they are uncommon.

There are more than 4 trillion tonnes (4×10^{12} tonnes) of nitrogen in the Earth's atmosphere and this is the source of nitrogen used in industry, with around 45 million tonnes a year worldwide being extracted. After air has been filtered through molecular sieves,[95] to remove carbon dioxide and water vapour, it is cooled to liquefaction temperature and separated into its component parts by fractional distillation, a process in which the nitrogen boils off first while the oxygen remains behind. Highly purified nitrogen obtained in this way can have less than 10 p.p.m. of other gases. Ultrapure nitrogen, used in the electronics industries, can have as little as 10 p.p.b.

Other methods for producing nitrogen do not require liquefaction, and are employed to generate the gas on site, such as for oil wells. Pressure-swing adsorption uses carbon molecular sieves to absorb oxygen, water vapour and carbon dioxide, and can yield 99% pure nitrogen, while another process uses a semi-permeable membrane that allows the other gases in the air to pass through but prevents nitrogen from escaping.

Nitrogen is primarily used in three different ways:

(1) As an inert atmosphere: here its job is simply to exclude oxygen from materials and systems which are sensitive to it. For example, apples will keep in cold storage for up to 2 years if the container is filled with nitrogen gas. Nitrogen gas is pumped into old oil wells to force out more oil, whereas ordinary air would oxidize some of the oil, producing unwanted by-products.

(2) As a low-temperature coolant: liquid nitrogen boils at −196°C, a temperature that renders most things solid and prevents any kind of chemical interaction from occurring, thereby preserving all manner of sensitive biological material.

(3) As a feedstock for the chemical industry: there are hundreds of chemical compounds that start life as nitrogen gas. These end up in the food we eat, the houses we live in, the clothes we wear, the cars we drive, the objects we use at home, and even the medicines we take.

The sequence of converting the nitrogen of the air to these products is to convert it first to ammonia, then to nitric acid and later to more sophisticated materials. One of

[94] Daniel Rutherford was the uncle of Walter Scott, author of *Ivanhoe* and other historical novels.

[95] These are zeolite materials (types of silicate) which have tiny pores that can trap specific kinds of molecule.

the largest bulk products is ammonium nitrate, 80% of which goes into fertilizers. Crops must have nitrogen. The fertilizer ammonium nitrate (NH_4NO_3) is the best way of applying it. With it, crop yields have more than doubled in the past 20 years, and wheat harvests of 10 tonnes per hectare are possible.

In 1905, the German chemist Fritz Haber (1868–1934) showed that it was possible to make nitrogen and hydrogen gases react to form ammonia by heating them to 500°C, in the presence of a catalyst, and under a pressure of 1000 atmospheres. Although the yield of ammonia was only a few per cent, it was possible to circulate the unreacted gases through the reaction chamber again. Today, thanks to improvements in the catalysts, yields of 20% can be produced at pressures of 150 atmospheres. A newer process makes ammonia from natural gas and air and operates at 380°C and 80 atmospheres – it requires only half the energy to make the same amount of ammonia as the Haber process.[96]

For his achievement, Haber was awarded the Nobel Prize for Chemistry in 1918. The Haber process also increased the supply of nitric acid because ammonia was the feedstock of another process discovered by the German chemist Wilhelm Ostwald (1853–1932), who had found that ammonia could be oxidized to nitric acid (HNO_3) by heating it with oxygen in the presence of a platinum catalyst.[97]

An alternative method of making nitric oxide was to pass an electric spark through air and simulate the reaction that occurs when lightning passes through air: the reaction of oxygen and nitrogen to form nitric oxide and then nitrogen dioxide. This latter gas dissolves in water to form nitric acid. This process is only economically viable when there is a plentiful supply of cheap electricity, such as hydroelectricity.

Other products of the chemical industry that come indirectly from ammonia and nitric acid are plastics such as polyurethane, fibres such as nylon, and the brilliantly coloured azo-dyes. Other nitrogen compounds produced industrially are hydrazine (N_2H_4), which is used as a rocket propellant and an anti-oxidant, and hydroxylamine (NH_2OH), a mild reducing agent which is used in photography and dyeing.

ENVIRONMENTAL ELEMENT

Nitrogen in the environment	
Earth's crust	25 p.p.m.
	Nitrogen is the 30th most abundant element.
Soils	approx. 5 p.p.m.
Sea water	only 0.1 p.p.b at the surface, but up to 0.5 p.p.m. in the deep ocean
Atmosphere	78% in dry air

The nitrogen of the atmosphere came as a result of the out-gassing of the molten Earth. Even today some escapes this way when volcanoes erupt. Most of the planet's

[96] The overall reaction for the process is $7CH_4 + 10H_2O + 8N_2 + 2O_2 \rightarrow 7CO_2 + 16NH_3$. The process is carried out in several stages.

[97] Ostwald also received a Nobel Prize for his work, in 1909.

nitrogen is present as this gas, although there are some deposits of nitrates that formed by the evaporation of natural waters in which they built up.

There are other nitrogen-containing gases in the atmosphere, of which the most important is nitrous oxide, which makes up 0.3 p.p.m. Ten million tonnes of this are released as a by-product of microbial action each year, as well as from human activity. Other nitrogen-containing gases are the oxides nitric oxide (NO) and nitrogen dioxide (NO_2), often referred to as NO_x, which make up 0.05 p.p.b. Although 100 million tonnes of these enter the atmosphere each year, much from the exhausts of vehicles, they are quickly washed out of the atmosphere in rain. NO_x is particularly unpleasant in sunny climates, where these molecules react with hydrocarbons in the atmosphere, such as traces of unburnt fuel, to produce photochemical smog over cities.

Although nitrogen is so important to life, its most stable form – nitrogen gas – is unreactive. All life depends on this being converted to ammonia or nitrogen oxides, a process known as nitrogen fixation. On a global scale, the fixing of nitrogen to ammonia is accomplished to the extent of around 50% by biological processes, 10% by fertilizer manufacture, 10% by lightning, and 30% by the internal combustion engine and fossil fuel burning.

Nitrogen fixation

Bacteria in the soil and algae in the oceans are the primary 'fixers' of nitrogen. The cyanobacteria in soil often form nodules on certain plant roots, providing that plant with ammonia in return for carbohydrate. Species that can do this are clover, legumes such as beans, soya beans and alfalfa. Nitrogen fixation is carried out by the nitrogenase enzyme system which consists of two large proteins; the smaller one contains four iron atoms, while the larger has 12 iron atoms and two molybdenums. Both proteins are needed to fix nitrogen. The nitrogen sticks to the molybdenum where it picks up six hydrogen atoms to form two ammonia molecules.

The conversion of ammonia to nitrate releases energy and some bacteria make use of this. The process is known as nitrification, and is part of the nitrogen cycle (see below). The average nitrogen molecule in the atmosphere only gets trapped into the nitrogen cycle of Nature once every 100 million years, but once there it may go round that cycle many times before it is returned to the atmosphere again.

The nitrogen cycle

The nitrogen cycle is an essential part of the living planet. When nitrogen is 'fixed', it gets incorporated into living things from which it may emerge as excreta from animals or from micro-organisms that decompose dead matter. As a result of all this activity there is a significant amount of nitrogen in the soil in the form of amino acids, humus, ammonium salts, nitrates and nitrites; these last two can be converted back to nitrogen gas or nitrous oxide and returned to the atmosphere.

Nitrate is part of the natural system, but for many years it was suspected of causing stomach cancer because it can be reduced to nitrite (NO_2^-) and thence to nitrosamines, which are carcinogenic. Although this theory is now discredited, it led to the passing of laws limiting the amount of nitrate in drinking water. Nitrate, however, occurs naturally in many vegetables, such as lettuce and spinach, and is produced by microbes in the human gut, with the result that only a small part of the nitrate in the body comes from drinking water.

CHEMICAL ELEMENT

Data file	
Chemical symbol	N
Atomic number	7
Atomic weight	14.00674
Melting point	−210°C
Boiling point	−196°C
Density	1.25 grams per litre
Oxides	N_2O, NO and NO_2

Nitrogen exists as N_2, a colourless, odourless gas which is generally unreactive at normal temperatures. The element is the first member of group 15 of the periodic table of the elements. It has an extensive chemistry, and its compounds exist in a wide range of oxidation states from –III of ammonia to the +V of nitrate.

There are two naturally occurring isotopes of nitrogen: nitrogen-14, which accounts for 99.6%, and nitrogen-15, which represents only 0.4%. Neither is radioactive.

ELEMENT OF SURPRISES

1. Not all explosives are used to destroy and kill: sodium azide (NaN_3) is used to save lives. This is the explosive in airbags that protect passengers in cars when they crash. There is around 200 grams of it in each bag. If the impact velocity is above 10 miles per hour, this will register in a sensor in the electronic control unit which inflates the airbag within 25 thousandths of a second.[98] The explosion is ignited by an electrical impulse which generates a temperature of 300°C. Sodium azide is a white crystalline powder consisting of positive sodium ions and negative azide ions, which are three nitrogen atoms bonded together in a line. The stability of the azide group depends on the metal it is associated with. For example, lead azide is easily exploded by shock and was once used as a detonator, but sodium azide is safe until detonated.
2. In 1999, yet another curious nitrogen compound was announced by the Edwards Airforce Base Laboratory in California. This was the N_5^+ ion, consisting of five nitrogen atoms arranged in a V-shape. What surprised the chemical community was that the ion appeared to be quite stable, instead of decomposing to simpler, more stable structures such as N_2.
3. Atomic nitrogen, in other words single atoms of the element, has recently been produced at the Carnegie Institution of Washington, but to produce it required pressures in excess of 150 gigapascals, which is 1.5 million times atmospheric pressure. Atomic nitrogen appears to be semiconducting.

[98] Five times faster than the blink of an eye.

Nobelium

This is the element with atomic number 102 and along with other elements above 100 is dealt with under the transfermium elements on p. 457.

Osmium

Pronounced oz-mi-uhm, the name is derived from the Greek word *osme*, meaning smell, because the metal's surface gives off volatile osmium tetroxide (OsO_4) which has an irritating, pungent odour.

French, *osmium*; German, *Osmium*; Italian, *osmio*; Spanish, *osmio*; Portuguese, *osmio*.

HUMAN ELEMENT

Osmium in the human body	
Total amount in body	not known but low

Osmium has no biological role. While the metal itself is not toxic, its volatile oxide is: even breathing air containing only 0.11 micrograms per cubic metre of osmium tetroxide is enough to irritate the lungs, eyes and skin of some people, and workers should not be exposed to concentrations of greater than 2 micrograms per cubic metre. The vapour is also known to cause intense headaches.

ELEMENT OF HISTORY

When, in 1803, the English chemist Smithson Tennant (1761–1815) dissolved crude platinum in dilute *aqua regia*, he observed that not all the metal went into solution. Earlier experimenters had assumed that the powder which remained behind was just graphite, but Tennant's tests showed it was not, and he began to investigate it further. By a combination of acid and alkali treatments he eventually separated the powder into two 'new' metal elements, which he named iridium and osmium, christening the latter on account of the strong odour it gave off.

ECONOMIC ELEMENT

Osmium can exist as the uncombined metal and as alloys with iridium known as either iridosmine or osmiridium. Iridosmine is a rare mineral found in Russia and in North and South America. Most osmium is obtained as a by-product of nickel refining, extracted from the Canadian nickel ores of Ontario, but less than 100 kilograms are produced each year. There is little demand for the metal, which is difficult to fabricate, but osmium powder can be worked by sintering at 2000°C and employing high pressures.

The metal is used in a few alloys and in industry as a catalyst. At one time it was to be encountered in the nibs of high-quality fountain pens, compass needles, long-life gramophone needles and clock bearings, where its extreme hardness and corrosion resistance were appreciated. It was used as the filament for a successful electric light bulb devised by Karl Auer in 1897.[99]

ENVIRONMENTAL ELEMENT

Osmium in the environment	
Earth's crust	approx. 0.1 p.p.b.
	Osmium is the 81st most abundant element, in other words it is exceedingly rare.
Soils	not recorded
Sea water	not known, but very low
Atmosphere	nil

Studies show that little or no environmental threat is posed by osmium, even near processing plants.

CHEMICAL ELEMENT

Data file	
Chemical symbol	Os
Atomic number	76
Atomic weight	190.23
Melting point	3054°C
Boiling point	5027°C
Density	22.6 kilograms per litre (22.6 grams per cubic centimetre)
Oxides	OsO_4 and OsO_2

[99] The filament was very fragile and would break if the bulbs were turned upside down! Osmium was replaced by tantalum in 1905 and that, in its turn, was replaced by tungsten, which is still used today.

Osmium is a lustrous, silvery metal, one of the so-called platinum group of metals and a member of group 8 of the periodic table of the elements. It is the densest metal known, although only by the narrowest of margins: its density at 20°C is 22.588 kilograms per litre, while that of iridium is 22.562 kilograms per litre. Osmium is unaffected by water and acids, but dissolves in molten alkalis. Osmium powder reacts slowly with the oxygen of the air and gives off detectable amounts of osmium tetroxide vapour.

There are seven naturally occurring isotopes of which osmium-192 is the most abundant at 41%, with osmium-190 next at 26.5%. The others are osmium-189 (16%), osmium-188 (13.5%), osmium-187 (1.5%), osmium-186 (1.5%) – this is weakly radioactive with a half-life of 2 million billion years (2×10^{15} years) – and osmium-184 (0.02%).

ELEMENT OF SURPRISE

At one time osmium tetroxide, which is a liquid that boils at 130°C, was employed by forensic chemists to highlight fingerprints because the vapour reacts to form a black deposit of the lower oxide, OsO_2, when it comes in contact with the minute traces of oils left by a finger. For the same reason it is also used to stain substances on microscope slides.

Oxygen

Pronounced oksi-jen, from the Greek *oxy genes*, meaning acid-forming.

French, *oxygène*; German, *Sauerstoff*;[100] Italian, *ossigeno*; Spanish, *oxígeno*; Portuguese, *oxigénio*.

COSMIC ELEMENT

Oxygen is the third most abundant element in the Universe, after hydrogen and helium. In stars like the Sun its formation is part of the fuel cycle that provides their energy. When a star has burned up its hydrogen to form helium, and its helium to form carbon, then carbon-burning begins, converting this element to oxygen. The amount of oxygen produced depends on the mass of the star: the more massive a star is, the higher its internal temperature and the more oxygen is formed.

Oxygen-burning will also occur, provided the temperature is a billion degrees. The Sun is not massive enough for this to happen, but in a sufficiently massive star further elements are formed from the fusion of two oxygen nuclei to produce the nuclei of the elements silicon, phosphorus and sulfur.

In the solar system, the Earth is unusual in having a high level of oxygen gas (O_2) in its atmosphere, a sure sign of life because it is formed during the photosynthesis of carbon dioxide and water by plants. The only other planets in the solar system with any oxygen are Mars, whose atmosphere contains 0.15%, and Venus, which has even less. Their oxygen is produced by ultraviolet radiation acting on other molecules.

HUMAN ELEMENT

Oxygen in the human body	
Blood	constituent of water
Bone	28%
Tissue	16%
Total amount in body	approx. 43 kilograms (60% of the total weight) but it varies quite a lot because the oxygen is present mainly as water

[100] Meaning acid-stuff.

Oxygen is essential for all forms of life since it is a constituent of DNA and of almost all other biologically important compounds. It is even more dramatically essential, in that animals must have a minute-by-minute supply of the gas in order to survive. Oxygen in the lungs is picked up by the iron atom at the centre of haemoglobin in the blood, and thereby transported to where it is needed. Because of haemoglobin, a litre of blood will dissolve 200 cubic centimetres of oxygen, much more than will dissolve in the same volume of water.[101]

The haemoglobin passes its oxygen to an enzyme, monooxygenase, which also has an iron atom at its active site. This enzyme is the catalyst for various oxidation processes in the body, including synthesizing new molecules and detoxifying others. The process of using oxygen goes through many stages; intermediate among them is its transformation to two undesirable compounds – the highly dangerous superoxide radical, which is an oxygen molecule that has picked up a single electron, i.e. O_2^-, and the slightly less dangerous hydrogen peroxide, H_2O_2. Both can damage living cells and so there are special enzymes to neutralize them. Superoxide dismutase converts the superoxide to hydrogen peroxide, then glutathionine peroxidase, and similar enzymes, reduce this to water. Superoxide may be dangerous, but some cells in the body purposely make it in order to destroy invading foreign cells.

The proportion of oxygen in the air is mid-way between two extremes that would make life on Earth impossible for humans: if it were below 17% we could not breathe, and if it were above 25% all organic material would be highly flammable. The three astronauts who were destined for the first manned *Apollo* flight were burned alive in minutes in their spacecraft on 27 January 1967, when fire started in the oxygen-enriched air of the cabin.

Generally it is too little oxygen which is a threat to life. It was this which brought the Biosphere project in Arizona to a premature end in the mid-1990s. Eight people had been sealed into this glass-walled ecosystem to see if it was possible for humans to sustain life on the Moon or Mars. Within a month they were gasping for breath as the oxygen in the air fell to 17%. Somehow 30 tonnes of the gas had disappeared (it was thought to have reacted with iron in the soil).

Oxygen gas makes up a fifth of the atmosphere, amounting to more than a million billion tonnes. It is fairly soluble in water, which makes life in rivers, lakes and oceans possible. When we breathe in oxygen it reacts with energy stores in our bodies to provide the warmth and muscular action that keep us alive. We return it to the atmosphere as carbon dioxide, where it is captured by plants and, with the help of sunlight, is turned into carbohydrates. This process, known as photosynthesis, releases oxygen back into the atmosphere.

Oxygen has the capacity to oxidize other substances. With organic material this can be rapid and intense, as in a fire, or slow and gentle, as in a living being, but in both cases oxygen is oxidizing carbon compounds to carbon dioxide. Oxygen also has the ability to kill by oxidizing organisms such as certain bacteria, and to convert metals such as iron to useless rust.

[101] Not all species use haemoglobin as the oxygen carrier – spiders and lobsters use a copper-containing version, which is why their blood is blue; sea worms use an entirely different protein called haemerythrin.

MEDICAL ELEMENT

It is well known that our brain must have oxygen to function and that, if denied oxygen, it will begin to die within minutes, but it is not generally realized that too much oxygen will also poison it. The threat is not appreciated by many sports divers, who need to be warned against breathing pure oxygen below 10 metres since this can lead to convulsions. Several divers have drowned as a result of this. It is better to use 'nitrox', which is air with a boosted oxygen content.

Premature babies used to be kept in oxygen-rich incubators until it was realized that some became blind because of it. Nevertheless, oxygen masks are a key part of medical treatment, supplying oxygen-rich air – about 30% oxygen – to patients whose lungs or heart are impaired in some way.

ELEMENT OF HISTORY

In 1624, the first ever submarine was demonstrated to King James I of Great Britain. It was manned by 12 rowers whose oars protruded through sealed ports and it travelled under water for 2 hours. On board was its Dutch inventor, Cornelius Drebbel, plus a few passengers, one of whom later reported that when the air in the submarine had been consumed it was refreshed from a container. It seems likely that the gas in the container must have been oxygen, and yet this gas was not to be 'discovered' for another 150 years. However, in Drebbel's time it *was* known that when saltpetre (potassium nitrate, KNO_3) was gently heated it gave off a gas, although this was not recognized as oxygen. Drebbel may have produced oxygen in this way.

What made it difficult for early investigators to understand the nature of air was the belief that it was an element, so the idea of its being a mixture of gases was quite alien to the established way of thinking. Yet, in the Middle Ages, Leonardo da Vinci (1452–1519) had noted that air must contain something essential to life and that when this had been used up then neither a candle flame nor a living animal could survive.

In the seventeenth century, scientists like the Englishman, Robert Hooke, the Russian, Mikhail Lomonosov, the Dane, Ole Borch, and the Frenchman, Pierre Bayen, all produced oxygen, but none of them identified it as such. The credit for discovering oxygen is now shared by three other chemists: the Englishman Joseph Priestley (1733–1804), the Swede Carl Wilhelm Scheele (1742–86) and the great French chemist, Antoine Lavoisier (1743–94).

Priestley was the first to publish an account of oxygen, having made it in 1774 by focusing sunlight on to mercuric oxide (HgO) confined in a glass tube, and collecting over water the gas that was evolved. He noted that a candle burned more brightly in the new 'air' and that a mouse lived longer, and appeared more lively, when breathing it. He also tried some on himself, and wrote, 'The feeling of it to my lungs was not sensibly different from that of common air, but I fancied that my breast felt peculiarly light and easy for some time afterwards.'[102]

[102] Aerosol cans of oxygen, and other ways of getting a 'shot' of this gas, can be bought, and although these are promoted as ways of instantly boosting the body's energy, they have little effect. In theory they may resupply the blood with oxygen after strenuous energy has depleted it, but the effect is minimal.

Unknown to Priestley, Scheele had produced oxygen 2 years earlier by heating mercuric oxide, as well as various nitrates, and had written an account of his discovery in a manuscript which he sent to his publisher in 1775, but which did not appear until 1777. It is clear from his preserved note books that not only had he identified the gas as a previously unknown substance, but he had even sent a letter to Lavoisier, posted on 30 September 1774, describing what he had done; a copy of the letter was found among Scheele's papers after his death.

Because Lavoisier never acknowledged this letter, it has always been assumed that it went astray. That autumn, Lavoisier claimed to have made oxygen and that he had done this independently of Priestley. However, we know that Priestley had visited him in October and also told him how he had made oxygen, even though he had not yet published details of his discovery.

What Lavoisier did was, perhaps, more important than claiming priority for the discovery. He realized the importance of oxygen as the key to a different way of looking at chemistry, and this sounded the death knell for the concept of 'phlogiston', which had been the basis of chemical theory for a hundred years. In that theory, when things burned they lost phlogiston to the air and, if they burned in a confined space, the air could become saturated with phlogiston and so the flame went out.[103] Not so, said Lavoisier, in fact just the opposite happened: burning took oxygen *from* the air.

Subsequently Lavoisier proposed that the newly discovered gas be called *oxygène*, meaning 'acid-forming', because he fancied it was an essential component of all acids. The English were opposed to the name, but it entered the language anyway, thanks to a poem called 'Oxygen', which was an ode in praise of the gas. It had been written by Erasmus Darwin[104] and published in 1791 in his popular book *The Botanic Garden.*

The story of the three claimants to oxygen's discovery is told in a play by Carl Djerassi and Roald Hoffman, called 'Oxygen', in which a panel of experts has the job of deciding to award a posthumous Nobel Prize to one of the three protagonists.[105]

ECONOMIC ELEMENT

All oxygen used industrially comes from the air, with 100 million tonnes being extracted every year. Oxygen is produced in one of two ways. The preferred method for large-scale production is from liquefied air, from which nitrogen distils off preferentially, leaving liquid oxygen behind. Oxygen can be shipped as a liquid in specially insulated tankers (1 litre of liquid oxygen is equivalent to 840 litres of the gas). Oxygen is also stored and transported in cylinders of the compressed gas, for use in metal cutting and welding, or for medical purposes.

Another method of producing oxygen is to pass a stream of clean, dry air through a bed of zeolite molecular sieves which absorb the nitrogen, giving a stream of gas that is 90–93% oxygen; this can be pumped down a pipeline to where it is needed. When the

[103] Priestley had even named the newly discovered gas 'dephlogisticated air', on the assumption that it was capable of absorbing a maximum amount of phlogiston which is why things burned longer in it.

[104] The grandfather of Charles Darwin.

[105] This was first performed at the Tricycle Theatre, Kilburn, London in February 2000.

zeolite is saturated with nitrogen, it is regenerated by reducing the pressure, which allows the nitrogen to escape. While this is going on, the flow of air is directed through a second bed of refreshed zeolite so that the process will deliver a continuous supply of oxygen.

Newer methods that give almost pure oxygen are being developed; these use a ceramic membrane based on zirconium oxide through which oxygen gas will preferentially diffuse, forced through it either by high pressure or by an electric current.

Commercially produced oxygen gas is destined mainly for steel-making (55% ends up this way) and the chemical industry (25%), but significant amounts are used in hospitals, water treatment, rocket launches and metal cutting.

Injecting oxygen through a special high-pressure lance into molten iron removes the chief impurities of sulfur and carbon, converting the iron to steel, and as it does this it raises the temperature of the metal to around 1700°C. Cutting through even the toughest steel can be done quickly with an oxygen–acetylene torch whose heat from a burning mixture of the two gases is sufficient to raise the temperature to 3100°C, high enough to melt the steel.

In the chemical industry oxygen is reacted with ethylene to form ethylene oxide which is then turned into materials ranging from antifreeze to polyester bottles and fabrics.

The water in rivers and lakes needs to have a regular supply of oxygen, for when this gets depleted the water will no longer support fish and other aquatic species. Polluted water is assessed by its biochemical oxygen demand (BOD), the amount needed to restore its ecological balance. When neither the plants that live in water, nor the slow absorption of oxygen from the atmosphere, can provide enough oxygen, the BOD can be met by bubbling oxygen gas through the water; this is done to rejuvenate polluted rivers and to treat sewage.

An emergency supply of oxygen automatically becomes available for the passengers in an aircraft when the pressure drops suddenly. This oxygen is stored not as oxygen gas but as the chemical sodium chlorate. In an emergency, the oxygen masks above each seat are activated by a small detonation which mixes the chlorate with iron filings and the two chemicals react to release the gas.

ENVIRONMENTAL ELEMENT

Oxygen in the environment	
Earth's crust	47%
	Oxygen is the most abundant element.
Soils	present as minerals and soil water
Sea water	constituent element of water
Atmosphere	21% in dry air

The crust of the Earth is composed mainly of silicon–oxygen minerals, and many other elements are there as their oxides. Nearly half the known elements are to be found in one oxide form or another, some as oxide minerals such as quartz (silicon oxide), rutile (titanium oxide) and haematite (iron oxide), and others as more complex oxides such as calcite (calcium carbonate), fluoroapatite (calcium fluoride phosphate) and

OZONE (O_3)

This gas consists of three oxygen atoms joined together in a V-shape. It was first identified at the University of Basel in Switzerland by the German chemist, Christian Friedrich Schönbein, in 1840, who based its name on the Greek word *ozon*, meaning to smell.[106] At first he thought he had discovered a hitherto unknown element similar to chlorine, but eventually he proved that it was simply another form of oxygen.

Ozone plays two roles in the Earth's atmosphere: in the troposphere, the lower part of the atmosphere, it is a pollutant; in the stratosphere, the region between 20 and 40 kilometres above the surface, it acts as a shield, protecting the planet from the damaging ultraviolet rays from the Sun.

There is a natural low level of ozone in the air we breathe, about 0.02 p.p.m., but in summer this can increase to 0.1 p.p.m., or more, as a result of sunlight acting on nitrogen dioxide from car exhausts. The legal limit for the occupational exposure is 0.1 p.p.m. because ozone damages the lungs. Some plants are also susceptible to the gas, and even though they do not show visible signs of stress, their growth is reduced in proportion to the level of ozone in the air.

In the stratosphere, the ozone layer is a vital barrier which absorbs ultraviolet light, both in its formation from oxygen, and in its decomposition back to oxygen. Ultraviolet light of wavelengths less than 240 nanometres reacts with oxygen (O_2) and cleaves this into two atoms. Each of these can then combine with another oxygen molecule to form O_3. This in its turn absorbs other ultraviolet rays of wavelengths 230–290 nm, which decompose it back to $O_2 + O$. Without the ozone layer, dangerous radiation capable of harming living cells would penetrate to the Earth's surface.

There are other components of the stratosphere that destroy ozone, such as chlorine atoms, and it was the use of chlorofluorocarbon (CFC) gases in aerosols, plastic foams and refrigerators that posed the greatest threat because these were long-lived in the atmosphere and were a rich source of chlorine atoms. These gases have been replaced with non-chlorine alternatives wherever possible.

Ozone is also an industrial resource, used for purifying drinking water (especially bottled water) and public swimming pools. Ozone is a more potent disinfectant than the commonly used chlorine gas, but the protection it provides is not as long-lasting as that of chlorine, which is why the latter is better for purifying mains water supplies. Ozone is not a stable form of oxygen and reverts back to ordinary oxygen. Because ozone is damaging to the lungs it may be necessary to remove it from air or water; this process can be speeded up by passing the air or water over charcoal which catalyses the decomposition.

Ozone can be condensed into a blue liquid or frozen into a violet-black solid; both are dangerously explosive.

There are two ways of making ozone industrially. The usual method is to pass air through concentric glass tubes with metallized surfaces across which is applied a discharge of 15 kilovolts and 50 hertz. The gas emerging from such treatment contains 2% ozone. Another method, used when only low concentrations of ozone are needed for sterilizing or disinfecting, is to expose air to ultraviolet light.

[106] This Greek word also gave its name to the element osmium, and for the same reason.

kaolinite (an aluminosilicate). The composition of the continental crust is dominated by relatively few complex metal oxides, with silicon oxides and silicates accounting for a massive 62%, aluminium oxides for 16%, calcium and iron oxides for 6% each, and magnesium, sodium and potassium oxides for 3% each.

The oxygen in the Earth's atmosphere comes from the photosynthesis of plants, and has built up over eons of time as they utilised the abundant supply of carbon dioxide in the early atmosphere and released oxygen. Photosynthesis began on the Earth about 3 billion years ago, but the oxygen then produced was mainly used up in oxidizing iron(II) to iron(III). This continued for about a billion years and then the level in the atmosphere began slowly to rise until it reached around 1% about 2 billion years ago, when blue-green algae (cyanobacteria) developed, and then about 500 million years ago it rose relatively rapidly to around 20% when the first land plants started to appear. It has stayed steady at this level since then, although there is some evidence that it may have been higher than this at certain times in the past.

Although plants and animals carry out the reverse process of converting oxygen to carbon dioxide, the average oxygen atom now only takes part in this cycle once every 3000 years. Even the burning of 7 billion tonnes of fossil fuels a year makes almost no impression on the amount of oxygen in the atmosphere. Human activity, however, can have profound effects on other gases in the atmosphere, and in particular the other form of oxygen, ozone – see box.

CHEMICAL ELEMENT

Data file	
Chemical symbol	O
Atomic number	8
Atomic weight	15.9994
Melting point	−218°C
Boiling point	−183°C
Density	1.43 grams per litre

Oxygen is a non-metallic element and a member of group 16 of the periodic table of the elements. It exists as a colourless, odourless, two-atom gas which condenses to a pale blue liquid with magnetic properties. (This was first observed in 1891 by James Dewar, who was able to make liquid oxygen in large enough quantities to study its properties.[107]) The reason for this strange magnetism is the unique chemical bonding between the two oxygen atoms, in which two of the 6 bonding electrons of O_2 refuse to pair up as in normal chemical bonds, and these unpaired electrons generate a magnetic field.

Oxygen is reactive and will form oxides with all other elements except helium, neon, argon and krypton. It is moderately soluble in water (30 cubic centimetres will dissolve per litre of water) at 20°C.

[107] Oxygen was first liquefied in 1877 by the Frenchman Louis Cailletet, although he produced only a few drops.

There are three naturally occurring isotopes of oxygen: oxygen-16 accounts for 99.76%, oxygen-17 for a mere 0.04%, and oxygen-18 for 0.2%. None is radioactive.

Because oxygen-18 is 12% heavier than oxygen-16, it can influence the behaviour of water – see 'Element of surprise'.

ELEMENT OF SURPRISE

The ratio of oxygen-18 to oxygen-16 in the world's oceans has varied slightly over geological time and this has left an imprint on parts of the environment, providing evidence of past climates. When the world is in a cooler part of the global temperature cycle, water molecules containing oxygen-16 evaporate more easily than their heavier oxygen-18-containing counterparts and so the snow that falls at the polar ice-caps, and becomes locked up as ice, is very slightly richer in the former, and the water that is left behind in the oceans is very slightly richer in the latter. Marine creatures therefore lay down shells that have more oxygen-18 than expected and these are preserved in sediments. Analysing the oxygen-18:oxygen-16 ratio in such deposits reveals the cycle of global cooling and warming that has characterized the past half million years with its five ice ages.

The oxygen gas of the atmosphere is also slightly enriched in oxygen-18, although plants release oxygen with the same ratio as that in the oceans. The reason is that oxygen-breathing life-forms consume the oxygen-16 at a slightly greater rate.

The ratio of the two isotopes in the remains of more recent skeletons has been used to show the regions from which people originally came. The nearer a person is to the Equator, the more oxygen-18 there is in their enamel of their teeth. For example, this has enabled the remains of people buried in England during the Roman period to be identified as having come from southern parts of the Roman Empire.

Pronounced pal-ayd-iuhm, this element was name after the asteroid Pallas, which was itself named after the Greek goddess of wisdom, Pallas.

French, *palladium*; German, *Palladium*; Italian, *palladio*; Spanish, *paladio*; Portuguese, *paládio*.

HUMAN ELEMENT

Palladium in the human body	
Total amount in body	not known, but low

Palladium has no known biological role although it has been detected in mammalian tissue, albeit at levels of only 2 p.p.b. It is regarded as of low toxicity, being poorly absorbed by the body when ingested. Palladium(II) chloride ($PdCl_2$) was formerly prescribed as a treatment for tuberculosis and was given in doses of 65 milligrams per day without apparent ill effects – or benefits. Palladium at higher intakes could be poisonous. Tests on rodents have shown it to be carcinogenic, but there is no evidence that it has this effect on humans.

ELEMENT OF HISTORY

The palladium story is part of a wider story that resulted in the discovery of four platinum-group metals – palladium, rhodium, osmium and iridium – by two London chemists, William Hyde Wollaston (1766–1828) and his close friend and collaborator, Smithson Tennant (1761–1815).

As early as 1700, miners in Brazil were aware of *ouro podre*, 'worthless gold', which was in fact a naturally-occurring alloy of palladium and gold. However, it was not from this source that palladium was first extracted, but from platinum. The person to do this, in 1803, was Wollaston, in rather mysterious circumstances.

Wollaston, one of 14 children of a Norfolk clergyman, was educated at Cambridge University, where he got a medical degree, and was a practising doctor for a few years. He also had an interest in chemistry. When he became partially sighted in 1800, he decided to devote more of his time to chemistry. He formed a working partnership with Tennant in 1800 to engage in various chemical enterprises, one of which was to investigate the production of platinum. They kept their partnership, and the work they were doing, secret. To begin with, their business struggled, yet it was in these early years that they discovered four platinum-group metals. Eventually their company became profitable, but it appears to have ceased trading around 1820.

Like others before them, who had dissolved platinum in *aqua regia*, Wollaston and Tennant knew that not all of the platinum went into solution and that it left behind a black residue. Tennant decided to investigate this residue and from it he eventually isolated osmium and iridium. Meanwhile Wollaston concentrated on the solution of platinum from which he extracted palladium and rhodium. First Wollaston reclaimed the platinum from solution by neutralizing the acid with sodium hydroxide and then adding ammonium chloride. This precipitated the platinum as ammonium hexachloroplatinate, free of all other metal impurities, and heating this recovered the platinum.

Then by adding mercury(II) cyanide drop by drop to the platinum-free solution he obtained another, yellow, precipitate, which turned out to be palladium cyanide. He filtered this off and heated it strongly, whereupon it decomposed to the metal.[108] His note book for 1802 shows that he thought first of naming it ceresium, after the asteroid, Ceres, which had been first observed in 1801,[109] but decided instead to honour the asteroid Pallas, which had been discovered in March 1802, around the time that palladium when first isolated, and so it became palladium.

Wollaston estimated that there was about 0.5% palladium in the original platinum, but having separated this impurity and proved to his own satisfaction that it was a 'new' metal, he was faced with a quandary. If he announced the discovery, the venture would become public knowledge, and the two partners did not want this to happen. Yet the chemist in him could not ignore what he had found, so he took most of his palladium to a couple who specialized in minerals, a Mr and Mrs Forster, whose shop was in Soho, London, and there in April 1803 they advertised it for sale. Handbills referred to it as the 'new silver' and listed its properties. It was on sale at a shilling a grain (65 milligram) which was six times the price of gold. Wollaston had left more than 1300 grains (about 80 grams) with them. The editor of the *Journal of Natural Philosophy, Chemistry and the Arts*, one William Nicholson, also published details of the newly discovered metal, again without revealing its discoverer.

A noted chemist, Richard Chenevix (1774–1803), was sufficiently intrigued to purchase some of the 'new silver'. He investigated its properties and declared at a meeting of the Royal Society that it was nothing less than an alloy of mercury and platinum. Yet, when other chemists tried to make it from these ingredients they invariably failed. Wollaston then issued a challenge, via Nicholson's journal, offering £20 (more than £2000 or US$3000 in today's money) to anyone who could produce palladium from other metals.

Naturally the prize remained unclaimed, and in February 1805 Wollaston eventually revealed himself as the discoverer of palladium. In July of that year he read a paper to the Royal Society giving a full account of the metal and its properties. He admitted his

[108] The remaining solution still contained rhodium, which Wollaston obtained as rose-red crystals of rhodium chloride.

[109] That asteroid gave its name to cerium.

secrecy, saying that he hoped to 'take advantage of [palladium], as I have a right to do'. In fact there was little money to be made – even the Forsters had sold only a quarter of what Wollaston had given them, and eventually he reclaimed most of his palladium from them and deposited it with the Royal Society.

The only use to which palladium was put in its early years was for the scales of astronomical and navigational instruments because it appeared resistant to corrosion. Later in the nineteenth century an alloy with 20% silver was introduced into dentistry.

Brazilian gold, which was imported into London in the early nineteenth century for making coins, sometimes gave coins that were much paler than normal and were suspected of being counterfeit. They were, in fact, made of the *ouro podre* mentioned above, which consists of 86% gold, 10% palladium and 4% silver. There was little use for the palladium that had to be extracted from this gold, despite attempts to popularize it as untarnishable silver and showing that it was ideal for medals, some of which were cast to commemorate important events and presented to royalty.

ECONOMIC ELEMENT

Specimens of uncombined palladium are found in Brazil, and there are some minerals rich in palladium, such as stibiopalladinite (palladium antimonide, Pd_5Sb_2) and braggite (a mixed palladium–platinum–nickel sulfide) but most palladium is extracted as a by-product from nickel refining in Russia and Canada, while copper and zinc refining in South Africa, North America, Russia and Zimbabwe also provides some. World production is around 300 tonnes[110] and reserves are estimated to be 24 000 tonnes.

Most palladium (60%) goes into catalytic convertors for cars, 20% finds its way into the electronics and communications industry, 10% is used to make dental fillings and the remaining 10% ends up in the chemical industry and as jewellery. The metal is quite malleable but hammering it increases its strength and hardness. It resists corrosion except in sulfur-contaminated environments.

In 1990, most catalytic converters relied on platinum to reduce emissions from car exhausts but, while this metal is still important, palladium is now the main ingredient because this is even more efficient at removing unburnt and partially burnt hydrocarbons from the fuel. Moreover, it copes better with higher temperatures and was instrumental in solving the problem of the high levels of unburnt fuel emitted when an engine starts from cold.

The recovery of palladium from spent catalysts is increasing, but at present it amounts to only a few tonnes per year. This is bound to increase, however, and recycling may one day be the main source of the metal.

Palladium finds its way into electrical appliances such as widescreen televisions, computers and mobile phones, in the form of tiny multi-layer ceramic capacitors, of which more than 400 billion are made each year; the average digital appliance has 150 or more of them. These capacitors consist of layers of palladium sandwiched between insulating layers of ceramic, sometimes with as many as 50 layers interleaved. (These palladium

[110] The trade measures output in troy ounces, of which this represents around 9 million ounces (1 troy ounce = 31 grams).

capacitors are now facing competition from nickel-based versions.) About 10 tonnes of palladium is reclaimed from old equipment each year.

Palladium is used in dental alloys (with silver), mainly in Japan, although there is now a trend to go back to using gold, as the price of palladium has increased following demand from car makers.

The chemical industry uses palladium-based catalysts in the production of nitric acid, which goes mainly into fertilizers, and for terephthalic acid, which is the raw material from which polyester is produced. This polymer is used as fibre for clothing or padding, or to make transparent plastic bottles for soft drinks. Palladium is also used in other ways: to coat electrical contacts in switches; for non-magnetic springs in clocks and watches; for special mirrors; and in jewellery where its alloy with gold is known as white gold.

ENVIRONMENTAL ELEMENT

Palladium in the environment	
Earth's crust	approx. 0.6 p.p.b.
	Palladium is the 76th most abundant element, in other words it is very rare.
Soils	0.5–30 p.p.b.
Sea water	0.04 p.p.t.
Atmosphere	nil

Palladium has little environmental impact. It is present at low levels in some soils, and the leaves of trees have been found to contain 0.4 p.p.m. (fresh weight). Some plants, such as the water hyacinth, are killed by low levels of palladium salts but most plants tolerate it, although tests indicate that their growth is affected at levels above 3 p.p.m.

CHEMICAL ELEMENT

Data file	
Chemical symbol	Pd
Atomic number	46
Atomic weight	106.42
Melting point	1552°C
Boiling point	3140°C
Density	12.0 kilograms per litre (12.0 grams per cubic centimetre)
Oxide	usual oxide is PdO, but there is a higher oxide, PdO_2

Palladium is a lustrous, silvery-white, malleable and ductile metal. It is one of the so-called platinum metals and a member of group 10 of the periodic table of the elements.

It is the least dense of these metals and has the lowest melting point. It resists corrosion, but dissolves in oxidizing acids and in molten alkalis. Like gold, palladium can be beaten into leaves as thin as 1 micrometre or less.

There are six naturally occurring isotopes of palladium of which the most abundant is palladium-106 (27.5%), followed closely by palladium-108 (26.5%). Then come palladium-105 at 22%, palladium-110 at 12% and palladium-104 at 11%; the least abundant is palladium-102 at 1%. None is radioactive.

ELEMENT OF SURPRISE

Palladium metal has the unusual ability to absorb hydrogen gas (H_2), to the extent of more than 900 times its own volume of the gas at room temperature. When a hydrogen molecule encounters the surface of the metal, it breaks into its component atoms which then rapidly penetrate inwards. Hydrogen gas will even diffuse through the metal, to emerge and recombine as H_2, making this a useful means of separating and purifying hydrogen.

In March 1989, Martin Fleischmann and Stanley Pons held a press conference to announce that they had achieved 'cold fusion', by which they meant they had found a way of generating energy from a nuclear reaction under ordinary laboratory conditions. What they had done was to pass a current through an electrochemical cell containing lithium deuteroxide (LiOD). This liberated deuterium, the heavier isotope of hydrogen, at the palladium cathode. Like hydrogen, deuterium can diffuse into the metal, and Fleischmann and Pons believed that there it underwent nuclear fusion to form helium, with an accompanying release of energy in the form of heat.

While the announcement created a flurry of interest at the time, carefully controlled experiments by others could not verify their claims and eventually they were discredited. Clearly another source of heat was being tapped, probably from a chemical process, although what this was remains unclear.

Phosphorus

Pronounced fos-for-us, the name is derived from the Greek *phosphoros*, meaning bringer of light, and was also the name given to the Morning Star (Venus).

French, *phosphore*; German, *Phosphor*; Italian, *fosforo*; Spanish, *fósforo*; Portuguese, *fósforo*.

HUMAN ELEMENT

Phosphorus in the human body	
Blood	345 p.p.m.
Bone	approx. 70 000 p.p.m. (7%)
Tissue	3000–8000 p.p.m. (0.3–0.8%)
Total amount in body	780 grams

In the form in which it is best known, i.e. white phosphorus, this element is dangerously flammable and a deadly poison – indeed as little as 100 milligrams may be a fatal dose for a human. In the natural world it is never encountered as such, only as phosphate, which consists of a phosphorus atom bonded to four oxygen atoms. This can exist as the negatively charged phosphate ion (PO_4^{3-}) – which is how it occurs in minerals – or as organophosphates in which there are organic molecules attached to one, two or three of the oxygen atoms. Of the phosphate in blood, 92% is organophosphate and only 8% is present as the simple phosphate ion. Blood levels of the latter are around 30 p.p.m. in adults, but double this in children, showing how important this element is for their growth and development. The organ richest in phosphate is brain tissue.

Phosphate is essential to all living cells. Indeed it is a component of DNA itself, albeit only a small part. Most phosphate in animals is in the form of calcium phosphate, which makes up bone. The rest is present as organophosphates, and in particular the energy molecule adenosine triphosphate (called ATP for short), which acts as a source of chemical energy, the messenger molecule guanosine monophosphate (GMP) and the phospholipids of cell membranes. Phosphate is also involved in buffering the fluid between cells and in transporting calcium.

ATP is used by the body on a truly remarkable scale: more than 1 kilogram per hour of ATP is produced, used and recycled. ATP is generated by the breakdown of glucose.

DNA, ATP and most organophosphates are negatively charged and as such they must be balanced by a positive charge, the preferred one being the magnesium ion, Mg^{2+}.

FOOD ELEMENT

Phosphate is a dietary requirement, the recommended intake being 800 milligrams per day, but it is not something that a person need worry about. A normal diet provides between 1000 and 2000 milligrams per day, depending on the extent to which phosphate-rich foods are consumed. These include tuna, salmon, sardines, liver, turkey, chicken, eggs and cheese, all of which have more than 200 milligrams per 100 grams.

The amount of phosphorus that is naturally present in food varies considerably but can be as high as 370 milligrams per 100 grams in liver, or be can very low, as it is in vegetable oils. Lean meat generally contains 180 milligrams per 100 grams. Eggs are also a good source of phosphorus, with 220 milligrams per 100 grams, but cheese can be one of the best, with 500 milligrams per hundred grams, partly because milk has a high level of phosphorus and partly because phosphates are added to certain types of processed cheese.[111] Other foods processed with phosphates are sausages and cooked meats.

The skeleton represents a massive reserve of phosphate that the body can draw on, and it continues to add and subtract from this store throughout life. Bone may appear to be inert but it is constantly being deposited and redissolved at millions of sites on its surface.

Not all phosphate-rich foods are a source of this nutrient. Plants store phosphate in their seeds as inositol hexaphosphate (IHP) so that they can germinate and put down roots without relying on phosphate from the soil, although it is ultimately from this source that the plant gets its phosphate for growth and to provide its own seeds with IHP. Foods like nuts are rich in IHP phosphate, but this is not a form of phosphate that is digestible. Wholemeal flour has 340 milligrams of phosphate per 100 grams, but most of this is IHP. (White flour has much less, 130 milligrams per 100 grams, but this is mostly digestible.)

Since phosphate is an essential part of the diet, it is perhaps not surprising to find that phosphates are considered safe when used as food additives. The permitted phosphate-containing food additives used most frequently in foods are various sodium, calcium and potassium salts of phosphoric acid (H_3PO_4), and even the acid itself is added to cola drinks. Disodium phosphate is added to evaporated milk to emulsify the fat and it also gives the necessary calcium-to-phosphate balance that prevents this type of milk from becoming semi-solid when it is stored.

Phosphate food additives were first introduced in the nineteenth century as leavening agents, to give cakes, pastries and biscuits a lighter texture. They consisted of calcium dihydrogen phosphate and sodium bicarbonate, and are still the key ingredient in self-raising flour.

[111] To prevent the fat in cheese from separating, it has to be emulsified and the best way of doing this is to add around 2% of disodium phosphate, which is soluble in the melted cheese.

MEDICAL ELEMENT

As a poison, phosphorus quickly attacks the liver, and death will result within a week of taking in a sizeable amount. Breathing in phosphorus vapour over a long period of time led to a the industrial disease known as 'phossy jaw', which slowly ate away the victim's jaw bone. This condition afflicted those who worked making phosphorus matches in the nineteenth century, but cases of the disease continued to crop up in the twentieth century.

Despite the dangers, elemental phosphorus was a widely used pharmaceutical and books like *Free Phosphorus in Medicine*, published in 1874, extolled its supposed benefits. It was given only in doses of a twentieth of a grain (3 milligrams) but was particularly recommended for numerous conditions, such as nervous breakdown, depression, migraine, epilepsy, stroke, pneumonia, alcoholism, tuberculosis, cholera and cataracts. Phosphorus tonics and toothache treatments were sold as over-the-counter medicines well into the twentieth century, and phosphorus was still prescribed for certain bone conditions in the 1920s, although by 1930 it had been eliminated from medical pharmacopoeias. Phosphorus was useless as a medicament.

Much safer than prescribing phosphorus was to give a partially oxidized form known as hypophosphorous acid (H_3PO_2) or its salts, known as hypophosphites. These came into vogue after 1857 and were popular as a treatment for neuraesthenia, a condition characterized by a general lack of energy. They too were useless.

Today there is a new class of phosphorus-containing drugs, the bis-phosphonates, whose molecules consist of two phosphates linked through a carbon, i.e. $[O_3P\text{–}C\text{–}PO_3]^{4-}$, the carbon having two other groups attached to it (not shown). Bis-phosphonates are used to treat diseases that result in bone wastage of the type which occurs with Paget's disease and bone cancers. The most common bone disease is osteoporosis, and bis-phosphonates have been approved in many countries for treating this condition. These drugs work by binding strongly to the calcium in bone and this interferes with the action of the cells whose job it is to dissolve bone so that its components can be recycled to other sites in the body.

ELEMENT OF HISTORY

Phosphorus was first isolated by Hennig Brandt at Hamburg in 1669 when he evaporated urine and heated the residue until it was red hot, whereupon phosphorus vapour distilled from it; this he collected by condensing it in water. He tried to keep his discovery secret, thinking he might have discovered the fabled 'philosopher's stone' that could turn base metals into gold, but to no avail. Eventually he was reduced to selling it for a living and he supplied phosphorus to Daniel Kraft, who exhibited it around Europe. Brandt offered to sell the secret to Johann Kunckel of the University of Wittenberg, but Kunckel was unwilling to pay the price and went off and discovered how to make it himself. Brandt also earned money by showing Gottfried Leibnitz how to make it. By these various acts and his own reticence, he even forfeited the claim to be its discoverer, because these three men were often given credit for it.

The most successful phosphorus producer of the seventeenth and eighteenth centuries was Ambrose Godfrey, laboratory assistant of the great Robert Boyle, of London. Boyle was fascinated by phosphorus and he too discovered how to extract it from urine, but he carried out chemical investigations into it. Godfrey eventually went into business selling phosphorus and supplied most of Europe with it. Phosphorus was sufficiently expensive for its use to be mainly medicinal.

It was not until it was realized that bone was made of calcium phosphate, and that this too could be used to make phosphorus, that phosphorus became more widely available. In the early nineteenth century the phosphorus match was invented, and this spurred demand for the element. Most was manufactured by heating phosphoric acid (produced by dissolving bone in sulfuric acid) with charcoal. By the end of that century, phosphorus was being extracted from mineral phosphates by heating it with coke in an electric furnace.

ELEMENT OF WAR

Phosphorus was used in the wars of the twentieth century in tracer bullets, incendiary bombs and smoke grenades. Phosphorus shells, ostensibly fired to act as markers for aiming mortars, proved to be a weapon of terror because of the terrible wounds that burning phosphorus produced.

The scattering of phosphorus fire bombs over cities in World War II was to cause widespread destruction. The week of air raids on Hamburg in July 1943 is now notorious for the use of phosphorus bombs, when as many as 25 000 of them were dropped at a time, each weighing 14 kilograms (30 pounds). In total almost 2000 tonnes of burning phosphorus fell on that hapless city. Other cities suffered a similar fate.

Another group of warfare agents based on phosphorus are the nerve gases such as tabun, sarin and the VX agents which have been stockpiled in the past, and sometimes used by terrorists and dictators. These highly toxic organophosphates are capable of blocking the enzyme, acetylcholinesterase, which is a key part of the way the body transmits signals from one nerve ending to another. When the enzyme is blocked there is malfunctioning of major body organs, such as the heart, resulting in death.

ECONOMIC ELEMENT

There are many phosphate minerals, the most abundant being forms of apatite, $Ca_5(PO_4)_3X$, where X can be fluoride, hydroxy (OH) or chloride.[112] Fluoroapatite provides the most extensively mined deposits. The chief mining areas are Russia, USA, Morocco, Tunisia, Togo and Nauru. World production is 153 million tonnes per year and reserves amount to around 6 billion tonnes of easily accessible ore, although there are vast deposits of poorer quality ore.

Phosphate rock is processed in one of two ways: either it is dissolved in sulfuric acid, which converts the phosphate to phosphoric acid (H_3PO_4), or it is heated with coke and

[112] The gemstone turquoise is a copper–aluminium–iron phosphate mineral of composition $Cu(Al/Fe^{3+})_6(PO_4)_4(OH)_8$.

sand in an electric furnace, which releases phosphorus. Most phosphate rock ends up as phosphoric acid, destined to be used to make fertilizers; around 50 million tonnes of this acid are made each year. Some phosphoric acid produced directly from apatite can now be purified sufficiently to enable it to be used in foods, but previous food-grade phosphoric acid was made from phosphorus. This was converted to phosphorus pentoxide by burning, and then to the acid by dissolving the oxide in water. The pure form of the acid is used to make food additives, animal feed supplements, surface cleaners, detergents and dishwasher powders. Lower grade phosphoric acid is used industrially to give metals a protective layer and as a rust-remover and rust-preventer.

Early in the nineteenth century, it was found that treating bone or mineral phosphates with sulfuric acid produced a soluble form of calcium phosphate, called 'superphosphate', and that applying this to farmland boosted crop production. All fertilizers need to contain phosphate, in the form of either calcium hydrogen phosphate, which is soluble and therefore available to plants, or ammonium phosphate, which is also soluble and, in addition, provides the essential element nitrogen.

There was increased demand for phosphorus in the two World Wars of the twentieth century, but the biggest boost to production came after 1945 with the discovery of the benefits of adding sodium tripolyphosphate to washing powders. Eventually phosphorus production reached a million tonnes a year by the 1970s although it has since steadily declined from this high level. Sodium tripolyphosphate ($Na_5P_3O_{10}$), the phosphate in detergents and dishwasher powders, softens water by neutralizing calcium and keeping dirt in suspension once it has been washed off clothes and dishes.

Some phosphorus is converted to phosphorus trichloride (PCl_3) which then goes to making the phosphorous acid (H_3PO_3) derivatives, such as phosphite esters, which are needed for flame retardants, insecticides and weed-killers. A little is turned into phosphorus sulfides which go into oil additives (as corrosion inhibitors and to reduce engine wear) and strike-anywhere matches. Other compounds made from elemental phosphorus are the gas phosphine (PH_3), which goes to making flame-proofing agents and biocides; zinc phosphide (Zn_3P_2), which is used as a rat poison; and magnesium phosphide (Mg_3P_2), which is the basis of warning flares used at sea. In these flares, the magnesium phosphide is combined with calcium carbide. When this mixture gets wet, the phosphide forms the spontaneously flammable gas, diphosphine (P_2H_4), which ignites the acetylene gas (C_2H_2) given off by the calcium carbide.

Disodium phosphate (Na_2HPO_4) is used in making glass and ceramics, in leather tanning, as a water softener, and in dye manufacture. Calcium hydrogen phosphate is used in toothpaste and sodium monofluorophosphate (Na_2PO_3F) is added to this to provide fluoride.

Some synthetic fibres, such as polyester, can be made flame-retardant by incorporating a phosphorus-containing compound into the polymer chain itself.

High-grade phosphorus is used to make metal phosphides, such as those of gallium and indium which are used in light-emitting diodes. The phosphorus for this has to be 99.9999% pure, in other words have only one impurity atom per million atoms of phosphorus. Around 7 tonnes of such high-grade phosphorus is produced each year.

ENVIRONMENTAL ELEMENT

Phosphorus in the environment	
Earth's crust	1000 p.p.m.
	Phosphorus is the 11th most abundant element.
Soils	0.65 p.p.m.
Sea water	surface waters have 1.5 p.p.b.; deep ocean waters have 60 p.p.b.
Atmosphere	trace amounts

Phosphorus as phosphate is of crucial importance to the environment because this element is the limiting factor controlling the population of species in the sea and on land. In the oceans, the concentration of phosphate is very low, particularly at the surface. The reason lies partly with the insolubility of calcium, aluminium and iron phosphates, but in any case in the oceans phosphate is quickly used up and falls to the deep as organic debris. Nevertheless, there can be excess phosphate in rivers and lakes, resulting in excess growth of slimy green algae which can choke out other living things. For this reason there was once a campaign against the use of phosphates and laws were passed in many countries reducing the amount permitted in detergents and fertilizers – but see below.

The phosphorus cycle

This particular cycle is unlike the other natural cycles involving key elements, such as carbon and nitrogen, because phosphate cannot circulate via the atmosphere and its major environmental movement is from soil to rivers, to oceans, and then to bottom sediment, where it accumulates until it is moved by geological uplift to becoming dry land again, a process that can take hundreds of millions of years. During the course of this downward drift, the phosphate is cycled and recycled through the many plants, microbes and animals of the various ecosystems.

The phosphate on land amounts to around 200 million tonnes, of which around 3 million tonnes (1.5%) are in living things, and from where 20 000 tonnes a year of phosphate are washed to the seas. There it joins the 80 million tonnes already present, of which only 100 000 tonnes (0.12%) is in living things. Slowly the phosphate sinks to the bottom of the oceans to become part of the trillion tonnes of phosphate rock of the Earth's crust. While all these numbers seem large in absolute terms, on closer analysis their relative amounts show how sensitive is the planet to this element. The oceans of the world could support a vast population were all its phosphate to be accessible to living organisms, but most of this is at a depth to which sunlight cannot penetrate, so there is no energy input to enable marine plants to flourish and thereby sustain a food chain. In the seas, all the sunlight is at the top and all the phosphate is at the bottom. The result is that the oceans of the world are really marine 'deserts' in that they support relatively little life except where there are up-welling ocean currents which bring phosphate-rich water to the surface, enabling fish and other marine organisms to flourish.

CHEMICAL ELEMENT

Data file	
Chemical symbol	P
Atomic number	15
Atomic weight	30.973761
Melting point	44°C (white phosphorus)
Boiling point	280°C (white phosphorus)
Density	1.8, 2.2, and 2.7 kilograms per litre (1.8, 2.2 and 2.7 grams per cubic centimetre, respectively) for white, red and black phosphorus, respectively
Oxides	P_4O_6 and P_4O_{10}

Phosphorus is a non-metal and a member of group 15 of the periodic table of the elements. There are several forms of phosphorus, called white, red and black phosphorus, although their colours are more likely to be slightly different than those implied by these descriptions. White phosphorus is the kind manufactured industrially; it glows in the dark, is spontaneously flammable when exposed to the air, and is a deadly poison. Red phosphorus is made by heating white phosphorus for several days in a closed vessel; it does not glow, is stable in air and is not poisonous. Red phosphorus can vary in colour from orange to purple, due to slight variations in its chemical structure. The third form, black phosphorus, is made under high pressure, looks like graphite and, like graphite, has the ability to conduct electricity. Little is made because no use has yet been found for it.

Natural phosphorus consists entirely of the isotope phosphorus-31, which is not radioactive. Radioactive phosphorus-32 is produced synthetically for research purposes, and has a half-life of 14 days.

ELEMENT OF SURPRISE

Phosphorus, as phosphate, was for many years regarded as the worst polluter of lakes and inland seas, especially where these were fed by rivers whose waters had been enriched by industrial effluent, domestic sewage and run-off from over-fertilized land, all of which carried high levels of this nutrient. The result was eutrophication, which produced vast mats of blue-green algae (cyanobacteria) which multiplied to such an extent that they made life impossible for other aquatic creatures. Phosphate was assumed to be the cause as this is the nutrient that limits growth under normal conditions.

Rivers like the Rhine, and the Great Lakes of North America, suffered most. This led to campaigns against the use of phosphates in detergents, these being seen as the primary cause because the onset of the problem seemed to coincide with the introduction of these products, which contained up to a third of their weight as phosphate.

In fact, research in the 1990s showed that phosphate was not as environmentally

Pronounced pla-tin-uhm, the name is derived from the Spanish *platina*, meaning little silver.

French, *platine*; German, *Platin*; Italian, *platino*; Spanish, *platino*; Portuguese, *platina*.

HUMAN ELEMENT

Platinum in the human body	
Total amount in body	not known, but low

Platinum has no known biological role. The metal is non-toxic and platinum implants are generally tolerated by the body, although some platinum compounds are poisonous. In the muscles of mammals the level of platinum is around 0.2 p.p.b.

MEDICAL ELEMENT

About half of those who come into regular contact with platinum compounds eventually suffer an allergic reaction, known as platinosis, with symptoms akin to those of asthma and the common cold. Those afflicted are few in number because not many people handle such compounds. The metal itself does not have this effect.

In 1962, Barnett Rosenberg was investigating the effects of electromagnetic fields on cell division and noticed the formation of giant filaments growing from *Escherichia coli* (better known as *E. coli*) bacteria. Curiously, the cells were growing but not dividing. It took Rosenberg 3 years to find out why, but his research led to the development of a chemotherapy drug that was used in clinical trials in the early 1970s and approved for use in 1978. It has since saved the lives of thousands of cancer patients with tumours of the testicles, breast, head and neck; it is especially used in treating children. Its cure rate exceeds 90%.

Rosenberg had been using platinum because he assumed it would be completely inert. In fact a small amount of it had reacted with the chloride and ammonium ions in the culture medium to form the compound *cis*-diamminedichloroplatinum(II), a chemical which had first been made in the early nineteenth century and was known as Peyrone's chloride, $PtCl_2(NH_3)_2$. This drug now has a generic name, cisplatin, and a trade name, Platinol.

Cisplatin, and similar platinum-based anti-cancer drugs, work by latching on to DNA at any point where it has two guanine bases next to each other. This distorts the

DNA, thereby making replication impossible, nor can the drug molecules be dislodged by DNA repair enzymes. As might be expected, cisplatin affects not only cancer cells but the cells in other organs of the body, in particular the kidneys and bone marrow. Many of those treated experienced side-effects such as those associated with platinosis, plus vomiting, loss of sensation in the hands and feet, and inner ear damage. These side-effects can now be relieved by using less potent variants of the drug, and by other drugs that suppress vomiting. The drug variant known as carboplatin is less damaging to the kidneys and is the preferred treatment for ovarian cancer.

ELEMENT OF HISTORY

Probably the oldest worked specimen of platinum is that from an ancient Egyptian casket of the seventh century BC, which was unearthed at Thebes and was dedicated to Queen Shapenapit. However, it is the only known example from Egypt and there is no evidence that the Greeks or Romans (or the Chinese) knew of platinum. Across the Atlantic it was known and used: native people of South America, living on the Pacific coast of what is now the Columbia/Ecuador border, were able to work platinum, as burial goods dating back 2000 years show.

The Spaniards conquered this area in the sixteenth century but they were more interested in gold and it was not until the eighteenth century that they began to take an interest in the curious metal that could be polished like silver but, unlike silver, never tarnished. Some alluvial deposits in South America were rich in both metals, especially those of the Chocó region and around the river Pinto, in Columbia, and platinum was often called *platina del Pinto*. Nuggets from these locales were 85% platinum alloyed with around 8% iron, plus a small proportion of the other platinum-group metals such as palladium, rhodium and iridium.

It is impossible to know who discovered platinum. The question really is: who first recognized that it was a metal in its own right? In 1557 an Italian scholar, Julius Scaliger, wrote of a metal from Spanish Central America that could not be made to melt and this must have been platinum. A manuscript by José Sánchez de la Torre y Armas, dated 1726, described the separation of gold and platinum.

The man often credited with bringing platinum to people's attention is Antonio Ulloa (1717–95). In 1735 he set sail for Panama where he first encountered platinum; he described it in detail in his diary, later published. The French ship on which Don Antonio embarked to return to Spain was captured by the Royal Navy and he ended up in England. There he was entertained by members of the Royal Society, to which he was elected a fellow.

Platinum was also 'discovered' around this time by Charles Wood who came across it in Jamaica in 1741 and sent a sample of it to the Royal Society, but not until 1750. Others who noted the newly discovered metal were the Swede Henrik Theophil Scheffer in 1752, the Englishman William Lewis in 1754, the German Andreas Marggraf in 1757 and the Frenchmen Pierre Macquer and Antoine Baumé in 1758.

In the eighteenth century, the Spanish authorities tried to ban platinum because of the adulteration of gold with it, and for a time they closed the mines where it was to be found and decreed that existing stocks of the metal were to be disposed of into deep water. Only later did they realise that *platina* was a useful metal; in the years 1788–1805 they shipped 3 tonnes across the Atlantic to Spain. Platinum was given to several European scientists, who were keen to investigate it, in the hope that they might discover a way of casting it and finding uses for it.

At one time it looked as though platinum would be as abundant as silver and gold, especially when deposits were found in the Urals in 1822, and by 1828 the Russian government was even issuing rouble coins made of the metal. Almost 15 tonnes of platinum went into making around 1.5 million such coins before the project was abandoned as the price of the metal, which became rarer and rarer, eventually exceeded the value of the coins.

For a long time platinum could not be cast as a metal because its melting point is too high. In the eighteenth and nineteenth centuries it was worked by hammering hot platinum sponge (made by heating ammonium hexachloroplatinate). This produced objects that looked as if they had been cast from molten metal.

ECONOMIC ELEMENT

Platinum nuggets occur naturally, as the uncombined metal, as does an alloy of platinum and iridium known as platiniridium. Some platinum is extracted as a by-product of copper and nickel refining. Three-quarters of the world's platinum comes from South Africa, where it occurs as cooperite (platinum sulfide, PtS), while Russia is the second largest producer, followed by North America where sperrylite (platinum arsenide, $PtAs_2$) occurs in the nickel-bearing deposits of Ontario, but even the richest deposits contain only 5 grams of the metal per tonne of rock processed.

World production of platinum is around 155 tonnes a year (expressed in the trade as 5 million ounces) and reserves total more than 30 000 tonnes, three-quarters of which are in the 150-mile long Merensky Reef, in South Africa. This was discovered in 1924, and mining began in 1929. In the 1950s output increased and a lower layer, even richer in platinum was discovered.

The largest use (50%) of platinum is for jewellery; another 30% goes into catalytic convertors and 20% into industry. Platinum rings, earrings and necklaces are particularly popular in Japan and that society accounts for half the world demand, whereas once it was much more fashionable in Europe, during 'La Belle Epoque', the period before World War I. Platinum jewellery began to be hallmarked around 50 years ago as a guarantee of its quality.

It has been estimated that one in every five products which we purchase either contains some platinum or needs platinum for its manufacture. Often it is alloyed with a small amount of iridium which increases its durability and hardness. Such was the respect for platinum's stability and resistance to corrosion that it was chosen for the standard metre and the standard kilogram, both housed at the International Bureau of Weights and Measures in Paris.[113]

[113] A replica kilogram weight is kept in Russia, and was originally standardized against the Paris one. Strangely it was found to have decreased in weight by 17 micrograms (0.0017%) when it was checked after a century of resting in a sealed underground vault. The loss was probably due to traces of osmium which can be oxidized by the air to form volatile osmium tetroxide.

Of the platinum which goes into catalytic convertors, around a quarter is recovered, with the USA being particularly geared for such recycling. A typical catalyst contains less than 2 grams of platinum but this covers a surface area inside the convertor that is greater than the area of a football pitch.

Platinum is used in the chemical, electrical, glass and aircraft industries, each accounting for about 10 tonnes of the metal per year. In the chemical industry a fine platinum gauze catalyst is needed for the manufacture of nitric acid from ammonia. This has to be replaced regularly because the metal slowly becomes used up; about 0.3 grams of platinum is lost per tonne of nitric acid produced. Silicone production, and that of benzene and xylenes, also relies on platinum catalysts and the industry also uses platinum for the spinnerets consisting of fine holes through which molten glass and rayon are forced in order to convert them to fibres.

The electrical industries use platinum coatings for computer hard discs, thermocouples and fuel cells. The first of these takes the lion's share, with 80% of discs now using platinum. Modern PCs have a hard-disc drive with several discs, made either from aluminium or from high-quality glass coated with several layers, one of which is a cobalt–platinum alloy with magnetic properties in which information is stored in a series of circular tracks. Around 15 tonnes of platinum a year are consumed to satisfy demand for these discs.

Proton exchange membrane (PEM) fuel cells look likely to become important for electric vehicles this century, because they operate at low temperatures, start up from cold, and are compact. By 2010 there should be more than 100 000 such electric vehicles being produced each year, and every one will need about 10 grams of platinum. PEM fuel cells consist of platinum-coated carbon electrodes separated by a polymer membrane. At the anode, the fuel (hydrogen gas) releases its electrons and the stripped hydrogen nuclei (the protons) then migrate through the polymer to the cathode where they react with oxygen from the air to form water and pick up electrons. The net result is a flow of current.

The glass industry uses platinum for optical fibres and liquid crystal display glass, especially for laptop and hand-held computers. Platinum is used to coat turbine blades in jet engines and spark-plugs for cars, in balloon catheters and pacemakers – where a 90% platinum/10% osmium alloy is used – and in some dental alloys. Like gold, platinum can be hammered into sheets of less than a micrometre in thickness and such foil has been used to coat the cones of space missiles.

Some platinum, around 6 tonnes a year, is sold as bullion to investors and collectors in the form of small bars and coins. The American Eagle, the USA's first platinum coin and with a face value of $25, was launched in 1997 and is the most traded.

ENVIRONMENTAL ELEMENT

Platinum in the environment	
Earth's crust	approx. 1 p.p.b.
	Platinum is the 75th most abundant element; in other words, it is very rare.
Soils	varies from 0.5 to 75 p.p.b.
Sea water	approx. 20 p.p.t.
Atmosphere	nil

Platinum is present at such low levels in the environment that it poses no threat to plants or animals. Plants growing on the soil over platinum-bearing rocks can absorb up to 6 p.p.m., mainly in their roots.

CHEMICAL ELEMENT

Data file	
Chemical symbol	Pt
Atomic number	78
Atomic weight	195.078
Melting point	1772°C
Boiling point	approx. 3800°C
Density	21.5 kilograms per litre (21.5 grams per cubic centimetre)
Oxides	PtO_2 is the most stable one, but PtO and PtO_3 also exist

Platinum is a lustrous, silvery-white, malleable, ductile metal and a member of group 10 of the periodic table of the elements. It has the third highest density, behind osmium and iridium. Platinum is unaffected by air and water, but will dissolve in hot *aqua regia*, in hot concentrated phosphoric and sulfuric acids, and in molten alkali. It is as resistant as gold to corrosion and tarnishing. Indeed, platinum will not oxidize in air no matter how strongly it is heated.

There are six naturally occurring isotopes:[114] the most abundant are platinum-194, which accounts for 33%, platinum-195 (34%) and platinum-196 (25%). The others are platinum-198 (7%), platinum-192 (1%) and platinum 190 (0.01%). The latter is weakly radioactive, with a half-life of 700 billion years, while the other five are non-radioactive.

ELEMENT OF SURPRISE

Platinum has the capability, as a catalyst, to initiate some chemical reactions at room temperature. For example, if a platinum wire is held in methanol (methyl alcohol)

[114] There are 37 known isotopes with mass numbers ranging from 166 to 202.

vapour, it soon glows red hot as it catalyses the oxidation of this chemical to formaldehyde. Platinum will cause a mixture of oxygen and hydrogen to explode just as if it had been sparked. In the twentieth century a type of cigarette lighter was manufactured which did away with flints and instead had a rim of platinum around the fuel outlet which would induce the fuel to react instantly with the oxygen of the air.

Plutonium

Pronounced ploo-toh-nee-uhm, this element was named in line with the two elements before it in the periodic table: uranium had been called after the planet Uranus, and neptunium after Neptune, so this element was named after Pluto which had been discovered in 1930. There was debate about whether the name should be plutium or plutonium, but the latter was chosen because it sounded better. The element's chemical symbol should have been Pl, but the discoverers preferred Pu because it sounded rather rude!

French, *plutonium*; German, *Plutonium*; Italian, *plutonio*; Spanish, *plutonio*; Portuguese, *plutónio*.

HUMAN ELEMENT

Plutonium in the human body	
Total amount in body	Minute traces

Plutonium has no role to play in living things and is never encountered in bulk outside nuclear facilities or research laboratories. Nevertheless, because of nuclear power accidents and nuclear weapons testing, there has been enough environmental pollution by plutonium to ensure that we all have a few atoms of this element in our body. It is highly dangerous because of the α-particles it emits. Normally these pose no threat because they are unable to penetrate even the thinnest of material, so it is safe to handle a container of plutonium, and even to hold the bare metal in gloved hands.

If plutonium oxide is ingested, it is not likely to be retained because it is so insoluble – only about 0.04% is absorbed by the body – but the little that is absorbed tends to end up in the bone marrow, where it poses a serious threat.

MEDICAL ELEMENT

Plutonium-238 is used as the energy source in heart pacemakers.

ELEMENT OF HISTORY

Plutonium was first made on 14 December 1940 at Berkeley, California, by Glenn T. Seaborg, Arthur C. Wahl and Joseph W. Kennedy, who produced it by bombarding uranium-238 with deuteron ions (heavy hydrogen ions). This first produced an isotope of the element neptunium which had been discovered earlier that year. This new isotope, neptunium-238, had a half-life of 2 days and decayed with emission of β-particles, which suggested it was forming the next higher element, i.e. element 94. Within a couple of months, element 94 had been conclusively identified and its basic chemistry shown to be like that of uranium.

A paper was dispatched to the journal *Physical Review* announcing the discovery in March 1941, but it was immediately withdrawn when it was realized that another isotope of element 94, isotope-239, was capable of undergoing nuclear fission in a way that could fuel an explosive chain reaction and might be used to make an atomic bomb.[115]

To begin with, the amounts of plutonium produced were tiny, and invisible to the naked eye, but by August 1942 there was enough to see, and a few weeks later there was enough to weigh, although it amounted to only 3 micrograms. However, by 1945 the Americans had several kilograms, and enough plutonium to make three atomic bombs.

ELEMENT OF WAR

When a nucleus of plutonium-239 is hit by a neutron, it not only undergoes fission, releasing a lot of energy, but it also releases several more neutrons, each of which can strike a neighbouring atom causing it to undergo fission and release yet more energy and yet more neutrons, and so on. In a fraction of a second, the chain reaction runs out of control and a colossal explosion ensues, large enough to destroy a city. To produce such an explosion requires there to be a certain 'critical mass' of the metal.

In World War II, the US Government set up a 'Plutonium Project' as part of their atomic weapons research programme. As a result, a great deal of this element's chemistry was uncovered. That information was largely incidental to the project, however, and the research focused mainly on ways in which plutonium might best be separated from other elements. By the end of 1943, a pilot plant at Oak Ridge was in operation, making plutonium from uranium by the series of nuclear reactions:

$$^{238}\text{U} + \text{n} \rightarrow {}^{239}\text{U (half-life 24 min)} - \beta^- \rightarrow {}^{239}\text{Np (2.4 days)} - \beta^- \rightarrow {}^{239}\text{Pu (24000 years)}$$

To begin with, the production was about a gram a day but this soon increased. After 6 months the uranium fuel rods were removed and the plutonium extracted by dissolving them in nitric acid to form plutonium nitrate, a convenient way to handle and transport the element.

By the spring of 1945, several kilograms had been amassed, and the first atomic explosion, using plutonium, took place at Alamogordo, in the desert of New Mexico, on July 16. It detonated 6 kilograms of plutonium and was triggered by using small conventional explosives to force several pieces of plutonium together to give the critical mass necessary for a runaway chain reaction. At the same time, a sudden burst of neutrons was provided by an 'initiator' made of beryllium and polonium which, when mixed, produce a shower of neutrons.

[115] The paper was eventually published in 1946.

The second plutonium explosion was in the form of a bomb, code-named 'Fat Man', which was dropped by the USAAF on the Japanese city of Nagasaki at 11.02 a.m. on 9 August 1945. The explosive capacity was equivalent to several thousand tonnes[116] of TNT, killing about 70 000 citizens and wounding 100 000. (The first bomb used in Japan was a uranium bomb, code-named 'Little Boy', which had been dropped on Hiroshima 3 days earlier, on 6 August 1945.) A second plutonium bomb was to be dropped on Kumagaya 2 days later, but the bomb was not quite ready, so 6000 tons of conventional bombs were used instead to destroy that city. World War II ended the following day, and the city of Kokura, scheduled for nuclear obliteration on 17 August, was spared.[117]

The Imperial Japanese Army had its own atomic weapons programme, according to Tatsuasaburo Suzuki, one of the last remaining members of the research team who was still alive in the 1990s. The programme was thwarted by a shortage of materials, although the Army had deduced that such a bomb could be made using uranium-235. They even had a target for their first bomb: Saipan, the Japanese island captured by the Americans in 1944, and which was being used as a bomber base.

The even more destructive hydrogen bombs are themselves triggered by a plutonium bomb whose explosion generates temperatures high enough to cause hydrogen atoms, in the form of the heavier isotopes, deuterium or tritium, to fuse together, as hydrogen does in the Sun. Because of the fission-generated heat needed to initiate the fusion reactions, these are classified as thermonuclear weapons. As hydrogen atoms fuse, they emit vastly more energy than an atomic bomb and give explosions equivalent to millions of tonnes of TNT.

ECONOMIC ELEMENT

More plutonium has been produced than any other synthesized element, primarily because of its nuclear weapon capabilities. It is also being produced as a by-product of nuclear power plants. In theory it may one day be used to fuel such power plants. Annual world production of plutonium is probably in excess of 50 tonnes and there may be more than 1000 tonnes of the metal in storage, either as bombs or as metal rods.

A nuclear reactor relies on the fission of fuel rods enriched with uranium-235, which has to be partly separated from the more abundant uranium-238 isotope before it can be used. The nuclear fuel produces heat in abundance, but after about 3 years the fuel rods have to be reprocessed to remove the 3% of waste products that have accumulated. When this has been done, the uranium and the plutonium, which has also formed, can be used as fuel again.

In so-called 'breeder' reactors, which are used to make plutonium for bombs, uranium-238 is bombarded with neutrons to form plutonium-239. Many of the bombs

[116] At the time it was said to be equivalent to 22 000 tonnes, but this was probably a gross overestimate.

[117] Kokura had been the original target for the first plutonium bomb. Because it was obscured by clouds, the bomber went on to its back-up target, Nagasaki, where there were the Mitsubishi shipyards and torpedo works. Sadly the bomb fell a few miles to the north of the intended aiming point and exploded over the city's main Christian church, wiping out the largest Christian community in Japan.

stockpiled during the Cold War – and there were 65 000 of them, amounting to more than 200 tonnes of the metal – are now being dismantled, but the problem is what to do with the plutonium they contain. This has to be stored safely and prevented from falling into the wrong hands. It could be used as fuel for nuclear reactors, which would slowly use it up, but for a variety of reasons – political, economic and environmental – this is unlikely to happen for some time.

Plutonium-239 can also be used as a fuel in the new generation of fast-breeder nuclear reactors which burn a mixed oxide (MOX) fuel consisting of uranium and plutonium. MOX fuel contains around 60 kilograms of plutonium per tonne of fuel, and after its 4 year lifespan such a fuel element would have burned up three-quarters of its plutonium. One gram of plutonium used in a conventional nuclear reactor has the potential to release as much energy as a tonne of oil.

The plutonium in a spent nuclear fuel rod is also mainly plutonium-239 (around 58%), but there is also plutonium-240 (24%), plutonium-241 (11%), plutonium-242 (5%) and plutonium-238 (2%). This last isotope can be used as a power source, but in a different way. Although it would be possible to separate it from spent fuel rods, it is easier to obtain it by neutron bombardment of neptunium-237. It too is an α-emitter but without emitting γ-rays. Therefore, it is entirely safe so long as it is enclosed in a steel casing; in this guise, its heat is used to generate electric current in deep-sea diving suits and spacecraft.

Plutonium-238 was used on the *Apollo 14* lunar flight of 1971 to power seismic devices and other equipment left on the Moon, and it was also the power supply used by the two *Voyager* spacecraft launched in 1977. These were designed to take advantage of a rare celestial conjunction of planets that occurs only once every 175 years and which enabled scientists to study four of the outer planets, Jupiter, Saturn, Neptune and Uranus. These were visited over a period of 10 years, during which masses of data and pictures were transmitted back to Earth.

Plutonium-238 mixed with beryllium generates neutrons and is used for research purposes.

ENVIRONMENTAL ELEMENT

Plutonium in the environment	
Earth's crust	trace
	Plutonium is among the 10 least abundant elements.
Soils	virtually nil
Sea water	virtually nil
Atmosphere	virtually nil

Plutonium is present in minute quantities in uranium ores, formed when one of the neutrons from a uranium-235 atom undergoing fission is captured by the more abundant uranium-238, which then undergoes β-emission to give first neptunium (half-life 2.3 days) and then plutonium with a half-life of 24 000 years. Other isotopes, of higher atomic weight, are produced by neutron capture. Traces of primordial plutonium has been detected in bastnäsite ore and it has been part of the Earth's crust since the planet was formed. It is not this plutonium that poses a threat to the environment, but that which has been released from nuclear bombs tested above ground, a practice that was prohibited by international agreement among the major powers in 1963 by the Limited Test Ban Treaty. Not all nuclear-weapons testing nations signed the treaty, however. France continued to test in the atmosphere for many years after this, finally halting in the 1980s as a result of international pressure. Of the other nations who later developed nuclear weapons – namely China, India, Pakistan and Israel – China and Israel conducted tests in the atmosphere over the same period. Over the years there have been about 550 atmospheric nuclear explosions.

In the USA, the plutonium salvaged from unwanted atomic bombs is melted into glass logs as plutonium oxide, and these are to be buried in Nevada's Yucca mountains 100 miles North-East of Las Vegas. These 2-tonne logs are made from borosilicate glass containing cadmium and gadolinium which, like boron, are powerful neutron absorbers. The use of gadolinium zirconium oxide ($Gd_2Zr_2O_7$) has been proposed; this can safely immobilize plutonium for up to 30 million years. The plan is to encase the logs in stainless steel and store them deep underground, possibly 4 kilometres deep with their boreholes plugged with concrete. Some people fear that in the distant future slow corrosion and water seepage might leach enough plutonium and concentrate it in a rock cavity to produce a critical mass. This seems an unlikely event because plutonium oxide is the least soluble of oxides – a million litres of water will only dissolve one atom of it.

CHEMICAL ELEMENT

Data file	
Chemical symbol	Pu
Atomic number	94
Atomic weight	244.0642 (plutonium-244)
Melting point	640°C
Boiling point	3330°C
Density	19.8 kilograms per litre (19.8 grams per cubic centimetre); when the metal melts it becomes even denser, so that solid plutonium floats on it
Oxide	the main one is PuO_2 but there are others

Plutonium is a silvery, radioactive metal, a member of the actinide group of the periodic table of elements. The metal will ignite in air at 135°C and explodes in contact with the solvent carbon tetrachloride. A piece of plutonium feels warm to the touch because of the energy given off by α-decay, and a large lump of plutonium will produce enough heat to boil water. The metal itself, which is produced by reacting plutonium fluoride with either lithium or barium at 1200°C, is attacked by oxygen, steam and acids, but not by alkalis. Plutonium is one of three fissile elements, along with thorium and uranium.

The most abundant isotope is plutonium-239, with a half-life of 24 000 years. There are longer-lived isotopes such as plutonium-244 (half-life 80 million years) and plutonium-242 (376 000 years), but most isotopes, of which there are 19 known, have much shorter half-lives, such as the useful plutonium-238 (88 years).

ELEMENT OF SURPRISE

Plutonium oxide occupies 40% more volume than plutonium metal, and this is always a threat when storing the metal. At the Los Alamos National Laboratory in New Mexico, USA, a canister which held 2.5 kilograms of plutonium metal had not been properly welded air-tight in 1983, and the oxygen that seeped in reacted with the metal to form plutonium oxide. Over a period of 10 years this escaped as dust into the polythene bag in which the canister was contained. There the radioactivity caused the polythene to become brittle and give off some of its hydrogen atoms, which then diffused into the plutonium and formed plutonium hydride, which acted as a catalyst for the oxidation. When the polythene bag finally burst, it let in more air and further oxidation occurred even more rapidly. Had the process gone on much longer then the expanding plutonium oxide would have ruptured the outer cannister and contaminated the whole storage facility.

Pronounced pol-oh-niuhm, this element is named after Poland, the native country of Marie Curie who first isolated the element.

French, *polonium*; German, *Polonium*; Italian, *polonio*; Spanish, *polonio*; Portuguese, *polónio*.

HUMAN ELEMENT

Polonium in the human body
None

Polonium has no biological role and is never encountered outside nuclear facilities or research laboratories. It is highly dangerous because of its intense radioactivity (it is an α-emitter) and this overrides other toxicity considerations which might follow from its chemical similarity to tellurium, the noxious element immediately above it in the same group of the periodic table.

ELEMENT OF HISTORY

Uranium ores contain about 100 micrograms of polonium per tonne (0.00000001%), and it was from this source that, in 1898 in Paris, Marie Curie, working with her husband Pierre, obtained the first sample of polonium after months of painstaking work. They were motivated in wanting to find out what it was in the uranium mineral pitchblende that caused it to emit four times more radioactivity than could be explained by the uranium itself. They were able to purchase several tonnes of ore residues from a uranium mine at Joachimsthal, Bohemia, courtesy of the Austrian government to whom they had appealed for help because they could not afford to buy pitchblende.

The existence of this element had been forecast by the great Mendeleyev in 1891, who could see from his periodic table that there should be an element that followed bismuth and would resemble tellurium; he even predicted it would have an atomic weight of 212.

ECONOMIC ELEMENT

No one now extracts polonium from uranium ores. Instead, the isotope isolated by Marie Curie, pollonium-210, is made in gram quantities – probably about 100 grams per year – by bombarding bismuth with neutrons in a nuclear reactor. A capsule of polonium containing only 1 gram will reach a temperature of 500°C because of the intense α-radiation it emits, and for this reason polonium is used as a lightweight heat supply for space satellites. Such a capsule will generate 520 kilojoules of energy per hour.

Polonium was once used in textile mills and by the manufacturers of photographic plates. This is because the ionizing radiation it emits enables static electricity to dissipate into the air, which prevented static build-up in textile mills (where it could give operatives a nasty shock) and prevented dust from being attracted to photographic plates.

Polonium is used as a source of α-radiation for research and, alloyed with beryllium, it can even act as a portable source of neutrons, which normally only access to a nuclear reactor can provide.

ENVIRONMENTAL ELEMENT

Polonium in the environment	
Earth's crust	minute traces in some minerals
	Polonium is among the 10 least abundant elements.
Sea water	nil
Atmosphere	nil

Polonium poses no real threat to the environment because there is so little of it and its half-life is short.

CHEMICAL ELEMENT

Data file	
Chemical symbol	Po
Atomic number	84
Atomic weight	209
Melting point	254°C
Boiling point	962°C
Density	9.3 kilograms per litre (9.3 grams per cubic centimetre)
Oxides	PoO_2 and PoO

Polonium is a reactive, silver-grey radioactive semi-metal and a member of group 16 of the periodic table of the elements. It dissolves in dilute acids; the solution so formed is pink, the colour of the polonium(II) ion (Po^{2+}), but rapidly turns yellow as the α-radiation forms oxidizing species from water molecules and these convert it to polonium(IV) ions (Po^{4+}). Polonium is fairly volatile: about half of a sample of it will evaporate within 3 days unless it is kept in a sealed container.

The longest-lived isotope is polonium-209 with a half-life of 102 years. Polonium-208 has a half-life of around 3 years, and polonium-210, which is the one most commonly used, has a half-life of 138 days. The remaining isotopes, of which there are 30 with atomic weights ranging from 194 to 218, have half-lives of only days, hours, minutes or seconds. Polonium-209 and -208 are made by bombarding lead or bismuth with protons, deuterons or α-particles in a cyclotron, but they are costly to produce.

ELEMENT OF SURPRISES

1. Polonium is regarded as one of the deadliest substances known: the maximum safe body burden is only 7 picograms (7×10^{-12} grams), making polonium around a trillion times more toxic, weight for weight, than hydrogen cyanide.
2. We cannot escape having some polonium in our body because it is formed from radioactive radon gas. This gas may be chemically inert, but if breathing it in coincides with its decay to polonium, as can happen because of radon's short half-life, the polonium may lodge in the lungs and from there move into the bloodstream. Polonium targets no particular organ of the body but, because it is an α-emitter, wherever it ends up it has the potential to damage DNA and that can lead to cancer.
3. Marie Curie died of leukaemia in 1934, caused by her exposure to radium over many years. Her daughter, Irène Joliot-Curie, who also won a Nobel Prize, likewise died of leukaemia – in 1956 – but that was attributed to her exposure to polonium as a result of a capsule of the element exploding on her laboratory bench 15 years earlier.

Pronounced poh-tass-iuhm, the name is derived from the English word potash. The chemical symbol K comes from *kalium*, the Mediaeval Latin for potash, which may have derived from the Arabic word *qali*, meaning alkali.

French, *potassium*; German, *Kalium*; Italian, *potassio*; Spanish, *potasio*; Portuguese, *potássio*.

HUMAN ELEMENT

Potassium in the human body	
Blood	plasma, 400 p.p.m.; red blood cells, 4000 p.p.m. (0.4%)
Bone	2100 p.p.m. (0.21%)
Tissue	16 000 p.p.m. (1.6%)
Total amount in body	varies within range 110–140 grams depending on the weight of muscle in the body

Potassium is an essential element for almost all living things, except possibly a few bacteria. Because one of its commonly occurring isotopes is radioactive, it has been suggested that this element is the main cause of 'natural' genetic modification in plants and animals, and indeed around 2500 potassium atoms disintegrate in the human body every second. Red blood cells have most potassium, followed by muscles and brain tissue.

In the body, potassium has many functions, the more important of which are regulating intracellular fluids, solublizing proteins, operating nerve impulses and contracting muscles. Potassium, as the ion K^+, concentrates inside cells, and 95% of the body's potassium is so located, unlike sodium and calcium, which are more abundant outside cells.[118] The ratio of potassium between cell and plasma in the human body is 27 to 1. That which is in the plasma acts as an electrolyte, helping maintain plasma viscosity and osmotic pressure.

[118] The concentration of potassium ions inside cells is high relative to the concentration of sodium ions. This is thought to reflect the ratio of these elements in the seas of the pre-Cambrian period in which life evolved. Today there is much more sodium than potassium in sea water.

Cell membranes have channels through which sodium and potassium ions flow selectively and against a concentration gradient which would normally induce them to flow from locations of higher to those of lower concentration. There are more than 100 of these so-called sodium–potassium pumps per square micrometre of membrane surface,[119] and each can transfer 200 sodium and potassium ions per second in and out of the cell. Some channels only permit potassium to pass through. The movement of sodium and potassium ions across nerve cell membranes is responsible for the transmission of nerve impulses, because this lateral motion of charge passes like a wave along the direction of the nerve fibre, as if it were an electric current.

The toxin of the black mamba has been used to probe the layout of the potassium channels in the human brain because it specifically blocks these channels. By injecting volunteers with a radioactive form of the venom, and then mapping where it has collected, researchers showed that the highest concentration of these channels occurs in the hippocampus region, the part of the brain that is important in learning.

MEDICAL ELEMENT

There are some conditions which lead to potassium deficiency, such as starvation, kidney malfunction and the long-term taking of diuretics to stimulate urine loss. Diuretics often have extra potassium included in their formulation to counter this loss. Severe attacks of diarrhoea lead to a temporary lack of potassium and consequently a feeling of weakness. Chronic potassium deficiency leads to depression and confusion. However, it is rare for potassium deficiency to lead to ill health because this element is abundant in food, particularly in vegetables and fruits.

The human body needs a constant supply of potassium to make lean tissue and to keep the kidneys working. If there is a deficiency, then the person experiences muscular weakness, which can affect the heart muscle, causing irregular beating and even cardiac arrest. On the other hand, an excess of potassium in the body depresses the central nervous system, and large doses of several grams of potassium chloride will paralyse it, causing convulsions, diarrhoea, kidney failure and even heart attack. Injecting a concentrated solution of potassium chloride can be fatal, because if there is too much potassium outside nerve cells, that which is inside the cells cannot move as required and the electrical impulses these cells should be transmitting fade away. All body functions are affected, but none more dramatically than the heart, whose muscles stop beating.

Murders have been committed with potassium, and doctors and nurses have been known to end the lives of terminally ill patients by giving them injections of potassium chloride solution. This results in rapid death from a heart attack. Strangely, what is illegal in most parts of the world is performed legally in the USA, as a form of capital punishment. Condemned men who agree to donate their organs for transplants may be executed by what is curiously described as a 'non-toxic lethal injection' of potassium chloride. Unlike poison gas or the electric chair, this method of execution leaves all the body's organs undamaged.

[119] Equivalent to 100 trillion per square centimetre.

FOOD ELEMENT

It is not generally appreciated that the need for potassium salts in the diet is much greater than that for sodium salts, and indeed it has been suggested that increasing the former relative to the latter would be beneficial. Some studies have shown that taking extra potassium can reduce blood pressure by increasing sodium excretion, but other research has failed to confirm this. The usual range for potassium intake is between 1 and 6 grams per day; the average intake is 3 grams for men and 2.5 grams for women. The recommended daily intake is 3.5 grams for both men and women, much more than the 1.5 grams a day recommended for sodium. Vegetarians take in a lot more potassium than non-vegetarians because potassium is abundant in all plants.

Around 90% of dietary potassium is absorbed by the body's lower intestine. We must have a regular supply of dietary potassium because we have no mechanism for storing it in the body. The average adult loses about a gram a day in their urine, yet few people are affected by a deficiency of this metal. Some foods are particularly rich in potassium, such as raisins and almonds, which have 860 milligrams per 100 grams (0.86%), dates and currants (approx. 750), peanuts (680), rhubarb (430) and bananas (350). Other common foods with above-average amounts of potassium are potatoes, bacon, cantaloup melon, bran, mushrooms, chocolate and fruit juices. The only foods with virtually no potassium are sugar, vegetable oils, butter and margarine.

Potassium comprises more than 1% of the weight of some processed foods, such as All-bran (1.1%), butter beans (1.7%), dried apricots (1.9%), yeast extracts[120] (2.6%) and instant coffee (4.0%). Salt substitutes are 50% potassium chloride (KCl), mixed with common salt (sodium chloride). Although potassium chloride is often used as a nutrient or dietary supplement, there are rare cases of excess ingestion by humans proving fatal: a person who ate half an ounce of this salt (14 grams) died as a result, although the normal amount to cause a serious toxic response is more like 20 grams.

ELEMENT OF HISTORY

People knew of potassium salts long before potassium was known as an element. Native Americans used the ash of the saltwort to flavour food, and they preserved meats with wood ash, both of which have a high level of potassium carbonate (potash). Similar practices went on in Asia. Potassium salts in the form of saltpetre (potassium nitrate, KNO_3) and alum (potassium aluminium sulfate, $KAl(SO_4)_2$) were known and used. The seventeenth century diarist, John Aubrey, reported that spreading plant ash was a way of improving soil, a practice that had first been advocated by Edward Broughton of Kington in Herefordshire in the early 1600s.

When Lavoisier drew up the first list of chemical elements, he included among them various 'earths' which could not at the time be broken down into simpler components and so they qualified as elements according to his definition. We now know these earths as metal oxides. Humphry Davy turned his attention to these with a view to seeing whether they could be further decomposed. He attempted to do this using electricity;

[120] Known by various trade names such as Marmite and Vegemite.

at first he dissolved them in water but found that passing an electric current through the solutions gave off only oxygen and hydrogen. On 6 October 1807 he decided to do away with the water and simply took some fresh potash that had become moist by exposure to air and placed this on a platinum disc connected to the negative pole of a battery. He then put a platinum wire, which was connected to the positive pole of the battery, in contact with the potash and within minutes observed the formation of metallic globules on the platinum disc. These were globules of a hitherto unknown element, potassium, and the first time a metal had been isolated using electrolysis. To prove that these globules were not coming from the platinum, Davy repeated the experiment using other metals as the electrodes. The results were the same. The newly discovered element was immediately christened potassium by Davy, after the source from which it came.

Davy noted that when the metallic globules were thrown into water they skimmed about on the surface, burning with a lavender-coloured flame. This colour is typical of potassium and was the traditional flame test used to reveal the element, by putting a tiny sample of it in a Bunsen burner flame. (Burning logs in a hot fire also show the effect.)

The French chemists Louis-Josef Gay-Lussac and Louis-Jacques Thenard were also able in 1808 to prepare potassium by heating a mixture of potassium hydroxide and iron filings to a high temperature. This was the method subsequently used by Davy to obtain more of the metal.

ECONOMIC ELEMENT

Minerals mined for their potassium are the pinkish ore sylvite (KCl), carnallite (a mixed potassium magnesium chloride, $KCl.MgCl_2$) and alunite (potassium aluminium sulfate, $KAl_3(SO_4)_2(OH)_6$). The main mining area used to be Germany, which before World War I had a monopoly on potassium, mined as carnallite at Stassfurt from 1861.[121] Today most potassium minerals come from Canada, USA and Chile, while the Dead Sea brines of Israel/Jordan are also a source.

Orthoclase (potassium aluminium silicate, $KAlSi_3O_8$) is also mined extensively, not for its aluminium content but because of its use in making porcelain, ceramics and glass. Adding potassium to glass makes it stronger and more scratch-resistant.

The world production of potassium ores is about 50 million tonnes, and reserves are vast (in excess of 10 billion tonnes). The process of extraction from sylvite involves crushing the ore and then separating the potassium chloride from the other minerals present by using a concentrated solution of sodium chloride of exactly the right composition to ensure that only the impurities in the ore, and not the potassium chloride, dissolve.

Most potassium (95%) goes into fertilizers, and the rest goes mainly into making potassium hydroxide (KOH), by the electrolysis of potassium chloride solution, and then converting this to potassium carbonate (K_2CO_3). Potassium carbonate goes into

[121] Production ceased in 1972.

glass manufacture, especially the glass used to make televisions, while potassium hydroxide is used to make liquid soaps and detergents. A little potassium chloride goes into pharmaceuticals, medical drips and saline injections.

Other potassium salts are used in baking, photography and tanning leather, and to make iodized salt (for the prevention of goitre, see p. 197), but in all cases it is the negative anion, not the potassium, that is the key to their use, and that is also the case with potassium chlorate, produced to make the heads of matches and for fireworks, and with potassium nitrate, which is needed for gunpowder.

ENVIRONMENTAL ELEMENT

Potassium in the environment	
Earth's crust	21 000 p.p.m. (2.1%)
	Potassium is the eighth most abundant element.
Soils	approx. 14 p.p.m.
Sea water	380 p.p.m.
Atmosphere	traces

Most potassium occurs in the Earth's crust as silicate minerals, such as feldspars and clays. Potassium is leached from these by weathering, which explains why there is quite a lot of this element in the sea, although not as much as sodium because potassium tends to end up in ocean sediment. The concentration of sodium in the sea, which in terms of sodium chloride is 27 grams per litre, is more than 35 times the concentration of potassium chloride, which is only 0.75 grams per litre.

Potassium is a key plant element. Although it is soluble in water, little is lost from undisturbed soil because as it is released from dead plants or animal excrement, it quickly becomes strongly bound to clay particles, and is retained ready to be re-absorbed by the roots of another plant. Ploughing upsets this system of nutrient recycling and potassium is lost from the land, which is why it has to be replaced by fertilizers. Wood ash used to be the main source of fertilizer potassium, but today it is potassium chloride of mineral origin which accounts for almost all potassium in fertilizers.

A rather unusual environmental use for potassium chloride has been suggested, and that is to increase rainfall over regions prone to drought. Normally, clouds release only about a third of their moisture as rain, but this can be doubled if they are seeded with fine particles of potassium chloride from flares mounted on the wings of aircraft flying beneath the clouds. The potassium chloride drifts up into the cloud and down comes heavy rain.

CHEMICAL ELEMENT

Data file	
Chemical symbol	K
Atomic number	19
Atomic weight	39.0983
Melting point	64°C
Boiling point	774°C
Density	0.86 kilograms per litre (0.86 grams per cubic centimetre)
Oxides	K_2O, KO_2 and K_2O_2

Potassium is a soft, silvery-white metal and a member of group 1, also known as the alkali metal group, of the periodic table of elements. The metal is obtained from the reaction of sodium metal with potassium chloride. Potassium is silvery when first cut but it oxidizes rapidly in air and tarnishes within minutes, so it is generally stored under oil or grease. It is light enough to float on water with which it reacts instantly to release hydrogen, which burns with a lilac flame. It will even react with ice at –100°C. However, the metal will dissolve unchanged in liquid ammonia to form blue solutions that are used as strong reducing agents. Potassium has a low melting point of 64°C, but a potassium–sodium alloy, known as NaK and consisting of 78% potassium and 22% sodium, is liquid down to –10°C, and a three-component alloy of potassium, caesium and sodium will remain liquid down to –78°C.

The chemistry of potassium is almost entirely that of the potassium ion, K^+.

There are three naturally occurring isotopes: potassium-39 comprises 93% of the total, potassium-41 makes up almost all of the remaining 7%, and there is a tiny amount (0.012%) of potassium-40. The latter isotope is radioactive and is a pure β-emitter, with a half-life of 1.3 billion years. This decays by two routes, one of which (β^+-decay) produces argon-40, the gas that today accounts for about 1% of the atmosphere.[122] Potassium-40 can be used to date rocks by comparing the ratio of potassium to argon within them, the so-called potassium/argon dating method.

The energy released by the radioactive decay of potassium-40 contributes to the warming of the Earth and may play a part in the convection currents in the mantle.[123]

ELEMENT OF SURPRISE

World production of potassium metal itself is 200 tonnes per year. The metal is obtained by passing sodium vapour up a metal column at 870°C down which molten potassium chloride flows. They react and release potassium vapour which is condensed at the top of the column. Potassium metal is produced in order to manufacture potas-

[122] The other route converts potassium-40 to calcium-40.

[123] The mantle is the region that comes between the Earth's crust and the core. It extends some 3000 km below the crust.

sium superoxide (KO_2) which is formed when it burns in oxygen gas. This material is kept in mines, submarines and space vehicles in order to regenerate the oxygen in the air when this has become depleted. This oxide reacts with carbon dioxide to form potassium carbonate (K_2CO_3) and in so doing it releases oxygen gas.

Praseodymium

Pronounced prah-zee-oh-dim-iuhm, the name is derived from the Greek *prasios didymos*, meaning green twin, because it was one of two elements that made up the supposed element didymium – see below – and its compounds were a striking green colour.

French, *praséodyme*; German, *Praseodym*; Italian, *praseodimio*; Spanish, *praseodimio*; Portuguese, *praseodímio*.

Praseodymium is one of 15, chemically similar, elements referred to as the rare-earth elements, which extend from the element with atomic number 57 (lanthanum) to the one with atomic number 71 (lutetium). The term rare-earth elements is a misnomer because some are not rare at all; they are more correctly called the lanthanides, although strictly speaking this excludes lutetium. The minerals from which they are extracted, and the properties and uses they have in common, are discussed under lanthanides (p. 219).

HUMAN ELEMENT

Praseodymium in the human body	
Total amount in body	not known, but small

The amount of praseodymium in humans is tiny and the metal has no biological role, but it has been noted that praseodymium salts stimulate metabolism. It is difficult to distinguish the various amounts of the individual lanthanides in the human body, but they are present, and the levels are highest in bone, with smaller amounts being present in the liver and kidneys. There is no estimate of the amount of praseodymium in an average adult, and no one has monitored diet for praseodymium content, so it is difficult to judge how much the average person takes in, but it is probably only a milligram or so per year. Praseodymium is not taken up by plant roots to any great extent, and the amount in vegetables is on average only 1–2 p.p.b. (dry weight), so only a little gets into the human food chain.

Soluble praseodymium salts are mildly toxic by ingestion, but insoluble salts are non-toxic. They – and especially praseodymium metal powder and dust – are skin and eye irritants.

ELEMENT OF HISTORY

In 1841, Carl Mosander announced that cerium harboured two other elements: lanthanum and didymium. Whereas the first of these was a true element, the second was not, although it was accepted as such for more than 40 years. Chemists wondered whether didymium too might consist of more than one element, and their suspicions were confirmed when Bohuslav Brauner of Prague showed in 1882 that its atomic spectrum was not that of a pure metal. Karl Auer took up the challenge and in June 1885 he announced to the Vienna Academy of Sciences that he had succeeded in splitting didymium into its two components: neodymium and praseodymium in the form of their oxides. A pure sample of metallic praseodymium was first produced in 1931.

ECONOMIC ELEMENT

The most important ores in which praseodymium is to be found are monazite and bastnäsite; these are discussed in more detail together under lanthanides (see p. 222). The main mining areas are China, USA, Brazil, India, Sri Lanka and Australia. Reserves of praseodymium are estimated to be around 2 million tonnes. World production of praseodymium oxide is about 2500 tonnes per year, and the metal itself is produced by the reaction of praseodymium fluoride and calcium.

Praseodymium oxide is best known for its ability to give glass a 'pure' yellow colour. Such glass, which is still called didymium glass, is used for the goggles which protect the eyes of welders and glass-blowers because it filters out infrared (heat) radiation. Apart from didymium glass, praseodymium is also used in other ceramics and glazes, where it produces brilliant pastel colours, especially shades of green and yellow.

Praseodymium metal makes up the core material in carbon-arc electrodes for film studio lights, searchlights and flood lighting, and it is used in alloys for permanent magnets – see below.

ENVIRONMENTAL ELEMENT

Praseodymium in the environment	
Earth's crust	9.5 p.p.m.
	Praseodymium is the 39th most abundant element.
Soils	approx. 8 p.p.m. (dry weight) with range 1–15 p.p.m.
Sea water	1 p.p.t.
Atmosphere	virtually nil

Praseodymium is one of the more abundant of rare-earth elements and is four times more abundant than tin. It poses no environmental threat to plants or animals.

CHEMICAL ELEMENT

Data file	
Chemical symbol	Pr
Atomic number	59
Atomic weight	140.90765
Melting point	931°C
Boiling point	3510°C
Density	6.8 kilograms per litre (6.8 grams per cubic centimetre)
Oxide	Pr_2O_3

Praseodymium is a soft, malleable, silvery metal and a member of the lanthanide group of the periodic table of the elements. It reacts slowly with oxygen, forming a flaky green oxide layer, and while it is less readily oxidized than other rare-earth elements, it still needs to be stored under oil or coated with plastic. It reacts rapidly with water.

The only naturally occurring isotope, praseodymium-141, is not radioactive. Minute traces of other naturally occurring radioactive praseodymium isotopes are produced in uranium ores as fission fragments.

ELEMENT OF SURPRISE

Praseodymium alloyed with nickel, of composition $PrNi_5$, responds to being magnetized by getting colder. This unusual property has made it possible for scientists to approach to within one thousandth of a degree of absolute zero (–273.15°C).

Promethium

Pronounced proh-mee-thi-uhm, the name was suggested by Grace Mary Coryell, whose husband Charles was one of its discoverers – see below. She proposed that it be called prometheum after Prometheus, the Greek who stole fire from the Gods and gave it to humans, and this was accepted although later its spelling was changed to promethium in accordance with the spelling of most other metals.

French, *prométhium*; German, *Promethium*; Italian, *prometo*; Spanish, *prometio*; Portuguese, *promécio*.

Promethium is one of 15, chemically similar, elements referred to as the rare-earth elements, which extend from the element with atomic number 57 (lanthanum) to the one with atomic number 71 (lutetium). The term rare-earth elements is a misnomer because some are not rare at all; they are more correctly called the lanthanides, although strictly speaking this excludes lutetium. The minerals from which they are extracted, and the properties and uses they have in common, are discussed under lanthanides, on p. 219. Promethium is unlike the other lanthanides in that all its isotopes are radioactive with short half-lives so that it is absent from the ores that contain the other lanthanides.

HUMAN ELEMENT

Promethium in the human body
None

Promethium has no role to play in living things and is highly dangerous because of its intense radioactivity. Tests, in which samples of promethium were injected into animals, showed that it became localized on the surface of bones from which it was only slowly removed, behaving in a manner similar to other rare-earth elements.

ELEMENT OF HISTORY

Element 61 was first proposed in 1902 by John Branner, who speculated that there should be an element between neodymium and samarium. This was confirmed by Henry G. J. Moseley in 1914 when he proved that there must be an element with this atomic number. Attempts were made to discover element 61, and in the 1920s two groups of chemists, in the USA and Italy, announced that they had found it. Each group named it after the state and city where they worked, and so the elements illinium, named after Illinois, and florentium, named after Florence, appeared in the literature.

The former group, consisting of B. Smith Hopkins, J. Allen Harris and L. F. Yntema of the University of Illinois, thought they had found element 61 in 1926, in material extracted from neodymium (element 60) and samarium (element 62). They based their claim on the observation of inexplicable lines in the atomic spectrum, but other chemists found these could equally well be explained by minute traces of other elements, namely barium, bromine and platinum, and so the claim floundered.

Meanwhile, two Italians, Luigi Rolla and Rita Brunetti at the Royal University of Florence, claimed to have evidence of element 61 from samples of the rare-earth ore monazite obtained from Brazil. Rolla even claimed priority in its discovery, saying that he and Brunetti had found it in 1924 and locked their results away in a sealed vault in the Academy of Lincei. Again the evidence was based on spectroscopic data. Again it was shown to be incorrect.

We now know that neither claim could have been justified, for the simple reason that all the isotopes of promethium are radioactive with half-lives too short for them to have survived from the time the Earth was formed. There is no way that this element could have been successfully extracted from terrestrial sources. Tiny amounts of promethium do occur in uranium ores as a result of fission, so in theory it could be extracted from this source, but the task would have been technically impossible on account of the tiny amount that is present, calculated to be around a picogram (10^{-12} grams) per tonne of ore.

A more realistic claim to have obtained element 61 was made in 1938 by H. B. Law, M. L. Pool, J. D. Kurbatov and L. L. Quill at Ohio State University, who had bombarded praseodymium and neodymium variously with neutrons, deuterons and α-particles in a cyclotron and detected element 61 in the debris. They proposed the name cyclonium, but this was not accepted as a discovery because there was no *chemical* proof that the missing element had been made.

Such proof was at last forthcoming in 1945 from the work of J. A. Marinsky, L. E. Glendenin and Charles D. Coryell at Oak Ridge, Tennessee, USA, because they had the new technique of ion-exchange chromatography at their disposal and with it they were able to separate out samples of isotope 147 of the missing element. They wanted to call the element clintonium, after the Clinton Laboratories in which the work was done, until Coryell's wife suggested the name prometheum.

ECONOMIC ELEMENT

Promethium is obtained in milligram quantities from the fission products of nuclear reactors. There are no long-lived isotopes: promethium-145 (the longest lived isotope) has a half-life of 17.7 years, while the half-lives of promethium-146 and -147 are 5.4 years and 2.6 years, respectively. Most promethium is used only for research purposes. However, promethium-147 is commercially available and is employed as a low-energy β-emitter in luminous paint, where it glows with a pale blue or greenish glow. It is also used in so-called 'atomic batteries', which are specialized miniature batteries about the size of a drawing pin; these are used for guided missiles, watches, pacemakers and radios. They have a useful life of around 5 years.

Promethium-147's emission of low-energy β-particles is also used in meters that can measure the thickness of sheet steel – and even paper.

Promethium isotopes will generate X-rays when their β-rays impinge on heavy metals and it may one day be possible to produce portable X-ray units making use of this feature.

ENVIRONMENTAL ELEMENT

Promethium in the environment	
Earth's crust	minute traces
Sea water	nil
Atmosphere	nil

Promethium does occur in the Earth's crust in tiny amounts in certain uranium ores; it is constantly being produced as a result of nuclear fission, and continually disappearing as it undergoes radioactive decay. All the promethium which might once have existed on Earth when it formed, has long since radioactively decayed. Even if there had been as much promethium as cerium, which is the most abundant of the rare-earth elements, making up 68 p.p.m of the crust, we can calculate that it would all have vanished within 10 000 years.[124]

CHEMICAL ELEMENT

Data file	
Chemical symbol	Pm
Atomic number	61
Atomic weight	145
Melting point	1168°C
Boiling point	about 2700°C
Density	7.2 kilograms per litre (7.2 grams per cubic centimetre)
Oxide	Pm_2O_3

Promethium is a radioactive metal and a member of the lanthanide group of the periodic table of the elements. It is little studied because of its radioactivity and its rarity. A microgram sample of the metal was obtained in 1963 by F. Weigel of Munich who reacted promethium fluoride (PmF_3) with lithium metal at 800°C.

As stated above, all promethium isotopes are radioactive, and of the 35 that are

[124] Assuming it was made up of the longest-lived isotope, promethium-145, this would still mean that a million tonnes of promethium would become 500 000 tonnes after 17.7 years, 250 000 tonnes after 35.4 years, 125 000 tonnes after 53.1 years, and so on down to a single tonne after around 350 years. Even a million million tonnes would be reduced to a single tonne after only 700 years, and down to a mere gram in a little over 1000 years.

known, most have half-lives shorter than 10 minutes. The longest-lived isotope, promethium-145, undergoes a type of radioactive decay in which the nucleus captures one of the innermost electrons (this decay mode is known as electron capture) and emits γ-radiation, whereas the more common promethium-147 is a β-emitter with only weak γ-radiation.

Promethium salts generally have a pink or red colour, and all cause the surrounding air to glow with a pale blue-green light.

ELEMENT OF SURPRISE

When Grace Mary Coryell suggested that element 61 be named after Prometheus, for allegorical reasons, little did she realize that it would one day be detected in the heavens. Promethium may be missing from our solar system, but it has been detected in the spectrum of the star HR465 in Andromeda. Considering that no isotope longer-lived than promethium-145 can exist, this must mean that the star is manufacturing the element in vast quantities and on its surface. How this is happening is as yet unexplained.

Protactinium

Pronounced pro-tak-tin-iuhm, the name is derived from the Greek *protos*, meaning first, as a prefix to the name actinium, because it is a precursor to this element, forming it as it undergoes radioactive decay. The element was also called brevium and uranium-X_2 by those who first observed a short-lived isotope, but the name protoactinium was chosen for a longer-lived isotope and, by convention, took precedence, although it was shortened to protactinium in 1949 by the International Union of Pure and Applied Chemists.

French, *protactinium*; German, *Protactinium*; Italian, *protoattinio*; Spanish, *protactinio*; Portuguese, *protacnídio*.

HUMAN ELEMENT

Protactinium in the human body
None

Protactinium has no biological role and is never encountered outside nuclear facilities or research laboratories. It is highly dangerous because of its intense radioactivity, and this overrides other toxicity considerations, but conventional toxicity is thought to be low. Special precautions have to be taken when dealing with it, because it is an α-emitter.

ELEMENT OF HISTORY

In 1871, Mendeleyev suggested that there might be an element between thorium and uranium. However, unlike most of the other elements that he predicted, and which were discovered, this missing element was to remain undiscovered for another 40 years.

In 1900, William Crookes separated an intensely radioactive material from uranium, but could not identify it and simply named it uranium-X. In 1913, at Karlsruhe, Germany, Kasimir Fajans and Otto Göhring showed that this disintegrated by β-emission and named it brevium because it existed only fleetingly. We now know it is one of the members of the sequence of elements through which uranium-238 decays on its way to lead; brevium was the isotope protactinium-234.

A longer-lived isotope of the same element was physically separated from uranium ore in 1918 by Lise Meitner at the Kaiser-Wilhelm Institute for Chemistry, Berlin, where she worked with Otto Hahn, although he was in the army at the time. In 1917, Meitner had only 21 grams of pitchblende to work with and was unable to import more because of wartime restrictions. She ground up the sample and extracted it with hot concentrated nitric acid. This treatment dissolved all but 2 grams, which was in the form of an insoluble silicate. She dissolved this in hydrofluoric acid and was eventually to show that the precursor to actinium was present, in other words there was a radioactive element present that transmuted to actinium. (This isotope, protactinium-231, is part of the decay series of uranium-235 as it transmutes into lead.)

Meitner knew she had found a new element but needed more material if she was to obtain a sample of it. A colleague of hers contacted Friedrich Giesel, who was associated with a radium-producing firm in Braunschweig, and as a result Meitner obtained a further 100 grams of pitchblende residues (from which the uranium and radium had already been extracted). By early 1918, she had enough evidence for the 'new' element to suggest to Hahn that they submit a paper to the journal *Physikalishce Zeitschrift*, which they did in March that year. Its title was 'The mother substance of actinium, a new radioactive element of long half-life' and in it they proposed the name protoactinium.

In June of that same year, protoactinium was also reported by Kasimir Fajans, and also by Frederick Soddy, John Cranston and Andrew Fleck in Glasgow, Scotland.

In 1927, Aristid Grosse managed to extract 2 milligrams of protactinium as the oxide and in 1934 he reduced this to metallic protactinium by converting it to the iodide, PaI_5, decomposing this in a vacuum by means of a heated filament.

ECONOMIC ELEMENT

Protactinium-231 occurs naturally in uranium ores such as pitchblende, to the extent of 3 p.p.m. in the some ores. In 1961, the UK Atomic Energy Authority extracted 125 grams of 99.9% pure protactinium from 60 tonnes of spent uranium fuel elements; this is still the major world stock of this element. It was supplied to laboratories around the world, enabling the chemistry of the element to be studied. However, no commercial use has yet been found for it.

ENVIRONMENTAL ELEMENT

Protactinium in the environment	
Earth's crust	traces
	Protactinium is among the 10 least abundant elements.
Soils	traces
Sea water	0.00002 p.p.t., i.e. barely detectable
Atmosphere	nil

Protactinium-231 and thorium-230 can be used to date marine sediments. They derive from different isotopes of uranium but they have very similar properties and appear to be precipitated at the same rate. While each by itself would be a guide to the age of a sediment, the ratio of the two is a better measure of geological time because this ratio does not depend upon them having a uniform rate of sedimentation. Their respective half-lives are 32 500 and 80 000 years, which means they can be reliably used to date sediments as old as 175 000 years.

CHEMICAL ELEMENT

Data file	
Chemical symbol	Pa
Atomic number	91
Atomic weight	231.03588
Melting point	1840°C
Boiling point	about 4000°C
Density	15 kilograms per litre (est.) (15 grams per cubic centimetre)
Oxides	Pa_2O_5 and PaO_2

Protactinium is a silvery, radioactive metal that is part of the actinide group of the periodic table of the elements. It is attacked by oxygen, steam and acids, but not by alkalis. The element becomes superconducting below –272°C (1.4 Kelvin). It is of research interest only.

The radioactive decay series of uranium-235 and uranium-238 produce protactinium as a secondary decay product, so there are two naturally occurring isotopes found in uranium ores. These are protactinium-234 (with a half-life of 6 hours 42 minutes), which decays by β-emission and comes from the more abundant uranium-238 isotope, and protactinium-231 (with a half-life of 32 500 years), which decays by α-emission and comes from uranium-235.

Another 19 isotopes of protactinium are known, but none has a half-life longer than a month.

ELEMENT OF SURPRISE

When Lise Meitner and Otto Hahn first obtained protactinium, they investigated its chemistry and showed it to be like tantalum, the metal that came above an empty slot in group V of Mendeleyev's periodic table; indeed, when he had hinted that such an element should exist, he had named it eka-tantalum for this very reason.[125] In some respects the chemistry of protoactinium did resemble that of tantalum – for example, its more stable oxide has the same formula as that of tantalum oxide, Ta_2O_5 – but the chemical similarity was purely fortuitous and, in a modern periodic table, protactinium is put in its rightful place as a member of the actinide series of elements.

[125] The prefix 'eka' is the Sanskrit word for one, and was used by Mendeleyev to signify it came one place below a known element in his table, in this case below tantalum.

Radium

Pronounced ray-dee-uhm, the name is derived from the Latin *radius*, meaning ray.
French, *radium*; German, *Radium*; Italian, *radio*; Spanish, *radio*; Portuguese, *rádio*.

HUMAN ELEMENT

Radium in the human body	
Blood	0.007 p.p.t.
Bone	0.004 p.p.t.
Tissue	0.0002 p.p.t.
Total amount in body	30 picograms (30×10^{-12} grams)

Radium has no biological role and is rarely encountered outside nuclear facilities or research laboratories. Nevertheless it is part of the sequence of elements through which thorium and uranium decay and so it is part of the environment – and the human diet. We have an average daily intake of 2 picograms, although most passes through us unabsorbed.

Radium is highly dangerous because of its intense radioactivity as an α-emitter, and this greatly overrides other toxicity considerations, such as its similarity to barium. The maximum permissible body burden for radium-226 is not measured in weight but in the dose of radiation it delivers, and this is set at 7400 becquerels; 1 bequerel is one disintegration per second.[126] The natural amount of radium in the average person delivers around 1 bequerel.

One gram of radium will release 0.001 cubic centimetres of radon gas per day and this has to be taken into account when storing the metal. A gram of radium also releases 4000 kilojoules of energy per year.

MEDICAL ELEMENT

For many years, and especially in the first half of the twentieth century, radium was an essential part of the medical treatment for cancer. Surgeons implanted 'radium needles' (usually containing radium chloride or radium bromide) in tumours and then allowed

[126] The curie is another, older, unit of radioactivity, defined as that amount of radioactivity with the same disintegration rate as 1 gram of radium-226, which is 3.7×10^{10} disintegrations per second. A picocurie is a trillionth part of a curie, i.e. 1×10^{-12} Ci and is still sometimes used as a unit.

the intense radiation they emitted to destroy the cancerous cells. This form of treatment is now rarely used, partly because it put the lives of those who made the needles at risk. Workers at the London Radium Institute, where they were prepared, invariably had low white-cell counts and several of them eventually died of radium exposure.

The dangers of radium were apparent from the start. The first case of so-called 'radium dermatitis' was reported in 1900, only 2 years after the element's discovery. The French physicist Antoine Henri Becquerel carried a small ampoule of radium around in his waistcoat pocket for 6 hours and reported that his skin became ulcerated. Marie Curie experimented with a tiny sample that she kept in contact with her skin for 10 hours and noted how an ulcer appeared, although not for several days.

More serious illness came through the large-scale use of radium in the manufacture of luminous alarm clocks and watches, the dials of which were painted on by hand; those involved in their production were very much in danger. The work was done mainly by young girls who had a habit of licking the fine brushes they used in order to form a sharp point. The luminous paint contained only 70 micrograms of radium per gram but this was enough to cause severe problems when it got into the mouth and body, generally causing cancerous growths.

The dangers came to public notice in a well-publicized case of the 'Radium Girls' who sued their former employer, US Radium. The company tried devious methods to prevent the case coming to trial and it became a *cause célèbre* in the press; the plaintiffs were clearly dying, and all of them were quite young. In the end, the company settled out of court, awarding each girl $10 000, but they were all dead within a few years. Some of the girls from the US Radium factory were so contaminated with radium that their hair, faces, hands and arms glowed luminously in the dark. In the New Jersey factory, of the 800 who worked there in the years 1917–1924, 48 succumbed to radiation sickness of whom 18 died.

When luminous dials were needed for aircraft, compasses, gun sights and electrical instruments in World War II, the girls who painted them worked in a very different environment, protected by screens and good ventilation, with all surfaces in the workshops being scrubbed down daily.

ELEMENT OF HISTORY

Radium was discovered in 1898 by Pierre and Marie Curie at Paris. They managed to extract 1 milligram from 10 tonnes of the uranium ore pitchblende (more correctly known as uraninite), a considerable feat bearing in mind the chemically primitive methods of separation available to them. They knew that they had discovered a hitherto unknown element because its atomic spectrum revealed new lines in the red region (described as 'carmine' in colour). They named the element after the rays (*radii*) of faint blue light with which it glowed in the dark.[127] In the words of Marie Curie:

> One of our joys was to go into our workroom at night when we perceived the feebly luminous silhouettes of the bottles and capsules containing our products. It was really a lovely sight and always new to us. The glowing tubes looked like faint fairy lights.

[127] The glow is caused by its radioactivity exciting the surrounding air.

The metal itself was only isolated in 1911, by Marie Curie and André Debierne, by means of the electrolysis of $RaCl_2$. At Debierne's suggestion they used a mercury cathode, with which the radium that was liberated formed an amalgam. This was then heated to remove the mercury, leaving the radium behind.

ECONOMIC ELEMENT

Early in the twentieth century, radium was regularly extracted from uranium ores for use in luminous dials and in medical treatment. The amount of radium in uranium ores varies between 150 and 350 milligrams per tonne, depending on its source, with those from Zaire and Canada having the most. Extraction from ore is no longer carried out – production from spent nuclear fuels is now the source of the 100 grams or so per year that is required.

Luminous paint consisted of a mixture of radium bromide and zinc sulfide. The latter glowed by being activated by α-rays from the decay of the radium. The owner of a luminous watch or clock is in no danger because the α-rays cannot penetrate the glass face or the casing.

Radium and beryllium were once used as a portable source of neutrons. The beryllium-9 nucleus will capture an α-particle and then undergo spontaneous decomposition to carbon-12, releasing a neutron as it does. Today other α-emitters are preferred, such as polonium.

ENVIRONMENTAL ELEMENT

Radium in the environment	
Earth's crust	0.6 p.p.t.
	Radium is the 86th most abundant element, in other words it is very rare.
Soils	0.8 p.p.t. (although some have as much as 0.3 p.p.b.)
Sea water	less than 0.001 p.p.t. (but detectable)
Atmosphere	nil

Radium contributes to the background level of radiation that makes the Earth a naturally radioactive planet. It has been estimated that each square kilometre of the surface (to a depth of 40 centimetres) contains 1 gram of radium. Because radium is present in soils, some gets into vegetation: levels in plants have been measured at 0.03–1.6 p.p.t. The amount of radium in the environment has increased as a result of human activity, notably through the use of phosphate fertilizers because phosphate rock contains uranium and therefore radium. Other sources of radium contamination are cement production and coal burning.

In parts of North America there are places where well water has measurable amounts of radium to the extent that they exceed the Environmental Protection Agency's limit for safe drinking water, which is 5 picocuries per litre (approximately 0.2 becquerels per litre).

CHEMICAL ELEMENT

Data file	
Chemical symbol	Ra
Atomic number	88
Atomic weight	226
Melting point	700°C
Boiling point	1140°C
Density	approx. 5 kilograms per litre (5 grams per cubic centimetre)
Oxide	RaO

Radium is a silvery, lustrous, soft, radioactive metal and a member of group 2, also known as the alkaline-earth group, of the periodic table of elements. Although it is the heaviest member of that group it is the most volatile. It is bright when freshly prepared, but darkens on exposure to air. Radium reacts with oxygen and water.

There are three naturally occurring isotopes of radium: radium-223, which is part of the decay series of uranium-235 and which has a half-life of 11.5 days; radium-224, part of the thorium-232 decay series, with a half-life of 3.7 days; and radium-226, which is not only the longest-lived isotope – with a half-life of 1600 years – but comes from the decay of the most abundant uranium isotope, uranium-238. Altogether 25 radium isotopes are known, with atomic weights ranging from 213 to 230.

ELEMENT OF SURPRISE

Soon after its discovery, radium was universally regarded almost as if it were a new wonder drug, as radiation was seen as beneficial to the body. All kinds of quack cures were promised by those peddling radium treatments, and indeed the USA saw a 'radium craze' which began in 1903 and went on for almost 30 years. Typical of such radiation cures was the 'Cosmos Bag', which was to be applied to relieve arthritic joints. Most popular of all was 'Raithor', a weak solution of radium salts that was claimed to be a general preventative of disease, and even to cure stomach cancer and mental illness. Those who found Raithor too expensive could buy a 'Revigorator', a flask lined with radium in which you stored water overnight to be drunk every morning.

The most famous case of death from Raithor was that of Eben Beyers, a steel magnate of Pittsburgh, who drank a bottle a day for about 4 years, at the end of which he was suffering from severe radiation sickness and cancer of the jaw. This necessitated extensive surgery in 1931, but this failed to save his life. This widely publicized case marked the end of radium water cures and other over-the-counter treatments.

Pronounced ray-don, the name is derived from radium. This gaseous element was first called radium emanation, because it was emitted by radium. An alternative name was niton. A similar emanation given off by thorium was called thoron, and that from actinium was called actinon, although all three were isotopes of the same element. The confusion arose because they were all short lived, so little could be gleaned about their chemistry. In 1923, the situation was rationalized and the name radon became generally accepted, since it was clear that these were isotopes of a noble gas and should be named accordingly, i.e. that the name should end in '-on'.

French, *radon*; German, *Radon*; Italian, *radon (emanio)*; Spanish, *radón*; Portuguese, *rádon*.

HUMAN ELEMENT

Radon in the human body
None

Radon is highly unstable and highly dangerous because of its intense radioactivity. It would have no biological role even if this were not so, because it is an almost totally chemically inert noble gas. Nevertheless it may have an indirect biological role because it has been estimated that radon is the major contributor to natural background radiation. It is encountered in everyday life as part of the atmosphere and it diffuses out of rocks, where it is formed during the natural decay sequences of uranium and thorium.

Radon is a human health hazard for many underground miners, and not only those working in uranium mines, because it can seep from surrounding rock and build up appreciable concentrations in the air of mines. It had been noted in the Middle Ages that miners in some areas, such as Bohemia, died young of lung disease (cancer).

ELEMENT OF HISTORY

It is sometimes said that Friedrich Ernst Dorn at Halle, Germany, discovered radon in 1900 while investigating the pressure of gas which built up inside sealed ampoules of radium compounds, but this is to credit him with more than he himself claimed. Instead, credit for the discovery should go to Ernest Rutherford who studied the radioactive gas emanating from thorium, and who realised he was dealing with a new element.In 1902 at McGill University, Canada, Rutherford and his young colleague Frederick Soddy studied the gas, which they called radium emanation, and even showed that it was possible to condense it to a liquid, using liquid air.

In 1908, William Ramsay and Robert Whytlaw-Gray, at University College, London, collected enough of the gas to determine several of its properties, such as its density, noting that it was the heaviest gas known. They gave it the name niton, from the Latin *nitens* meaning 'shining', which is what it appeared to do in the dark. Eventually in 1923 the International Committee of Chemical Elements proposed the name radon by which it became known.

ECONOMIC ELEMENT

Radon collects over samples of radium-226 at the rate of around 0.001 cubic centimetres of radon per day per gram of radium. It is naturally present in some spring waters that contain dissolved radium, such as those at Hot Springs, Arkansas, which were once thought to be beneficial for this very reason.

Radon was sometimes used in hospitals to treat cancer and was produced as needed and delivered in sealed gold needles.

ENVIRONMENTAL ELEMENT

Radon in the environment	
Earth's crust	traces
	Radon is among the 10 least abundant elements.
Soils	traces
Sea water	10^{-8} p.p.t.
Atmosphere	10^{-9} p.p.t., i.e. one part in 10^{21} parts of air

There is a detectable amount of radon in the atmosphere. Natural radon is almost totally composed of the two isotopes radon-220 and radon-222. These are so abundantly produced that they, and their radioactive decay products, contribute to almost half of the ionization of the air near the surface of the Earth. Radon and its decay products contribute to atmospheric electricity. The danger of radon lies with the non-volatile radioactive isotopes to which it transmutes; these can adhere to dust particles and so enter the lungs. One of them, lead-210, has a half-life of 22 years and will tend to linger long in the body, like other lead atoms.

In the 1980s and 1990s, it became apparent that in certain localities the amount of radon escaping from the ground, or from buildings constructed of granite, was much higher than average and that this could accumulate indoors, putting the occupants at risk. Radon could build up in basement rooms or even in ground floor rooms with cracked concrete floors, unless there was adequate ventilation to ensure its dispersal. The result of breathing air contaminated with radon was that it exposed occupants to a higher risk of lung cancer.

Government agencies issued warnings of the dangers and booklets were distributed to those areas most at risk explaining how the problem could be prevented, such as by sealing cracks and gaps in floors and walls and around service pipes, and ensuring good ventilation in rooms. In severe cases of pollution the answer was to create a space below a concrete floor where the radon could collect and then to extract it with a fan. In the majority of homes the exposure to radon radiation amounted to around 20 becquerels per cubic metre, which is very low, but some homes have levels as high as the 400 becquerels per cubic metre that is considered the limit for industrial exposure. However, in some uranium mines levels of 10 million becquerels per cubic metre have been recorded.

One village in the Austrian Tyrol, Umhausen, has been built on a fall of granite rock that took place around 8700 years ago. In falling, the rocks fractured, making it easy for the radon produced from the uranium in them to escape. The average level of radon in homes built on this rock fall gives levels of around 2000 becquerels per cubic metre and in one it was more than 250 000 becquerels per cubic metre. The incidence of lung cancer in the village was said to be statistically higher than expected, but numbers involved were so small that this has been disputed.

The US Environmental Protection Agency released a comprehensive document produced by the US National Academy of Sciences in February 1998. It was Part VI of *The Biological Effects of Ionizing Radiation* (BEIR) series of reports and was entitled 'The health effects of exposure to indoor radon'. It confirmed that radon could be the cause of lung cancer, a conclusion borne out by epidemiological studies on humans and experimental studies on animals. An analysis of the medical records of 68 000 underground miners, and not only uranium miners, showed that 2700 had died of the disease. Radon is now thought to account for about 10% of cases of lung cancer, and this would explain why those who neither smoke, nor are exposed to cancer-causing fumes, are also prone to the disease.

Radon is not easily monitored. One analytical method employs a special polymer film that is exposed for up to a year and then analysed microscopically for the visible tracks left in the plastic by the intense α-particles emitted by the gas as it radioactively decays. A quicker method is to measure the radioactivity of the various isotopes to which radon-222 transmutes;[128] this is done by exposing charcoal to the air. This is left to absorb the radon for a set time, such as 48 hours, and then the γ-rays emitted by the 'daughter' elements into which it has transmuted can be detected. Home testing kits are available which are sent away for analysis after being exposed to the suspect environment.

[128] The sequence of elements is as follows: radon-222 (half-life 4 days) → polonium-228 (3 minutes) → lead-214 (27 minutes) → bismuth-214 (2 minutes) → lead-210 (22 years) → bismuth-210 (5 days) → polonium-210 (20 weeks) → lead-206 (stable).

CHEMICAL ELEMENT

Data file	
Chemical symbol	Rn
Atomic number	86
Atomic weight	222
Melting point	−71°C
Boiling point	−62°C
Density	9.7 grams per litre
Oxide	none

Radon is a colourless gas and one of group 18, also known as the noble gases, of the periodic table. From its position in the periodic table, one would expect it to behave like xenon and be chemically unreactive except towards fluorine, and indeed it forms radon bifluoride (RnF_2) and the cation RnF^+, although these have only a fleeting existence. (By analogy with xenon, radon should also form higher fluorides, an oxide and acids.) When radon is cooled below its freezing point, it phosphoresces brightly. It is little studied, partly because it is a noble gas and is therefore reluctant to form molecules, and partly because its intense radiation is likely to destroy any compound that it might form.

Radon is part of the radioactive decay series of naturally occurring uranium and thorium, and of actinium and plutonium. Of the natural isotopes, uranium-235 produces radon-219, which has a half-life of only 4 seconds, while thorium-232 produces radon-220, which is only slightly longer-lived, with a half-life of 56 seconds. It is the more abundant uranium-238 which produces the longest-lived radon isotope, radon-222, which has a half-life of 3.8 days. There are many more radon isotopes, 31 in all, but none has a half-life of more than a few hours.

ELEMENT OF SURPRISE

Not all epidemiological research supports there being a link between radon and lung cancer in the general population. While such a link is not disputed for underground miners, who are exposed to high levels of radon, it has been suggested that the very low levels to which the general public are exposed in homes and offices pose nothing like the same level of risk. Indeed, some studies in the USA, China and Scandinavia seem to show that there is an inverse relationship, with there being *less* lung cancer in regions with the higher levels of radon.

Some people still believe that breathing radon may be beneficial. In the little spa town of Le-Mont-Dore in the Auvergne, France, visitors can take radon-rich air in the form of 'nasal irrigation' with a tube inserted up one nostril so that they can breathe in a gas that is drawn from a nearby natural hot spring. While most of this gas is carbon dioxide, it also contains a level of radon well above average, and it is this which is supposed to activate the blood, combat allergies, improve digestion and stimulate the immune system.

Rhenium

Pronounced ree-ni-uhm, this element is named after *Rhenus*, the Latin name for the Rhine.

French, *rhénium*; German, *Rhenium*; Italian, *renio*; Spanish, *renio*; Portuguese, *rénio*.

HUMAN ELEMENT

Rhenium in the human body	
Total amount in body	not known, but very low

Rhenium has no biological role. Little is known about its toxicity, although this is believed to be minimal. Indeed, there are no reported cases of humans being affected by rhenium.

ELEMENT OF HISTORY

At the start of the twentieth century, the periodic table of the chemical elements had almost been completed, but the members of one particular group – group 7 – were, oddly, still missing, except for manganese, the element heading the group. The second member of the group, technetium, was to remain undiscovered for the simple reason that there are no technetium minerals occurring naturally on Earth because all its isotopes are radioactive and all would have long since disintegrated.[129] The third member of the group did exist, however, but defied all attempts to find it, even though its properties could be forecast from its position in the periodic table. For example, it would be a metal of high density and would have a range of oxides.

Despite several attempts to detect the missing element in manganese ores, it remained elusive, and indeed it was the last stable, non-radioactive naturally occurring element to be discovered. This was finally achieved in May 1925 by Walter Noddack and Ida Tacke,[130] working at the Physico-Technical Testing Laboratory in Berlin. There they concentrated the element from the ore gadolinite, in which it was an impurity, but only in trace amounts (10 p.p.m.).[131] That they had obtained the missing element was confirmed by Otto Carl Berg of the Siemens & Halske Company in Berlin,

[129] Tiny amounts of technetium are to be found among the fission fragments of uranium but these are not easy to detect.

[130] They were subsequently married.

[131] They also probably detected technetium in 1925 (see p. 423).

who examined the atomic spectrum of the concentrate they obtained and found several new lines that could only be due to a new element. The discovery was finally announced later that year by Tacke at a chemical meeting in Nuremberg.

Following the discovery, Noddack and Tacke continued researching rhenium and even obtained a gram of the metal from 660 kilograms of a Scandinavian ore, molybdenite (molybdenum sulfide). This was the most abundant source they could find. In the 1930s a molybdenum mine in northern Wisconsin was found to offer a better source of rhenium, although even the ore from this mine contained less than 2%.

ECONOMIC ELEMENT

Rhenium does not occur as the free uncombined metal, and no mineable mineral ore has been found. The ores gadolinite (see p. 154) and molybdenite (see molybdenum, p. 264) may contain a little rhenium and it is from the latter of these that rhenium is extracted via the flue dusts of molybdenum smelters. Although there was some production of rhenium in the years following its discovery, it was only in the 1950s that this became commercially worthwhile, when rhenium's use in alloys and in catalysts created a demand. World annual production is now around 5 tonnes and the estimated reserves of rhenium are 3500 tonnes, found mainly in ores from the USA, Russia and Chile.

Rhenium is a silvery metal, although usually produced as a grey powder. By pressure stamping this under vacuum and heating in the presence of hydrogen, it is possible to fabricate pure rhenium objects, although there is little demand for such items.

Rhenium is added to tungsten and molybdenum to form alloys that are used as filaments for ovens and lamps. It is also employed in thermocouples that can measure temperatures above 2000°C, and for electrical contacts which stand up well to electric arcs. It has occasionally been used for plating jewellery. Electroplating with rhenium was first achieved in 1934 and was shown to give a bright, hard deposit. However, the metal is susceptible to oxidation and its surface needs to be protected by a coating of iridium.

Rhenium is also used as a catalyst in the chemical industry, especially in processes involving the addition of hydrogen gas to other molecules, and it is particularly valued because, unlike other catalysts, it is not easily deactivated by traces of sulfur and phosphorus.

ENVIRONMENTAL ELEMENT

Rhenium in the environment	
Earth's crust	about 0.5 p.p.b.
	Rhenium is the 77th most abundant element, in other words it is very rare.
Soils	not known
Sea water	4 p.p.t.
Atmosphere	nil

There is so little rhenium in the environment that virtually nothing is known of how it would behave in soil, plants or animals. There are no instances of pollution by rhenium salts from mining or industry.

CHEMICAL ELEMENT

Data file	
Chemical symbol	Re
Atomic number	75
Atomic weight	186.207
Melting point	3180°C
Boiling point	5625°C
Density	21.0 kilograms per litre (21.0 grams per cubic centimetre)
Oxides	the most stable oxide is Re_2O_3, but ReO_2, Re_2O_5, ReO_3 and Re_2O_7 are also known

Rhenium is a silvery metal but rarely seen as such on account of its high melting point, which is the third highest after carbon and tungsten. It is a member of group 7 of the periodic table of the elements. Rhenium has the fourth highest density of all the elements, behind osmium, iridium and platinum. It resists corrosion but slowly tarnishes in moist air. It dissolves in concentrated nitric acid but, surprisingly, not in *aqua regia*. The metal's easy oxidation is demonstrated by its ready dissolution in bromine water.

There are two naturally occurring isotopes: rhenium-187, which comprises 63% and is weakly radioactive, and rhenium-185 (37%), which is not. Rhenium-187 is a β-emitter with a half-life of 45 billion years, which means that 96% of the element that was present when the Earth formed 4.57 billion years ago is still around. A gram of rhenium undergoes only 4.5 disintegrations a second.

ELEMENT OF SURPRISE

Until 1994, there was no evidence for any natural rhenium mineral and it was assumed that it only occurred as trace amounts in other ores, such as those of molybdenum. Then, in that year, four Russian mineralogists who were studying the Kudriavy volcano on Iturup, a Russian island off the North-East coast of Japan, came upon a soft, whitish-grey material that had a metallic lustre and was flaky, rather like graphite. Unlike graphite, though, it had a density of 7.5 grams per cubic centimetre, which is remarkably high for a mineral. It turned out to be almost pure rhenium sulfide.

Its formation could not be explained. Volcanic gases contain traces of rhenium, and the superheated steam that emerged from Kudriavy had a few parts per billion, but why this was depositing as rhenium sulfide at a rate of several grams per day could not be deduced. The volcano has since erupted and buried the mineral.

Rhodium

Pronounced roh-diuhm, the name is derived from the Greek *rhodon*, meaning rose-coloured.

French, *rhodium*; German, *Rhodium*; Italian, *rodio*; Spanish, *rodio*; Portuguese, *ródio*.

HUMAN ELEMENT

Rhodium in the human body	
Total amount in body	not known, but low

No biological role for rhodium has so far been discovered, nor is it ever likely to be, considering that this element is so rare in the environment. Most rhodium compounds are only slightly toxic by ingestion – rats had to be given doses of 200 milligrams per kilogram of body weight to be poisoned by it[132] – but mice, which were raised on drinking water containing 5 p.p.m. of dissolved rhodium, developed leukaemia. There are almost no reported cases of humans being affected by this element in any way.

ELEMENT OF HISTORY

Rhodium was discovered in 1803 by William Hyde Wollaston (1766–1828); the story is recounted in more detail under palladium (p. 305). Wollaston dissolved in acid a sample of platinum, which came from South America. From the solution he recovered first the platinum, and the palladium, by precipitating them. He was then left with a beautiful rose-red solid from which he obtained rose-red crystals of rhodium chloride and these he reduced to the metal itself by heating with hydrogen gas.

ECONOMIC ELEMENT

Rhodium occurs as rare deposits of the uncombined metal, for example in Montana, USA, and as rare minerals, such as the rhodium–lead sulfide, rhodplumsite ($Rh_3Pb_2S_2$). The metal, which is available commercially, comes as a by-product of the refining of certain copper and nickel ores which can contain up to 0.1% rhodium. Most rhodium

[132] Equivalent to 14 grams for an average 70 kilogram adult human.

comes from South Africa and Russia, and world production is around 16 tonnes per year. Estimated reserves are 3000 tonnes.

Most rhodium (85%) goes into catalytic converters for cars; this is increasing because rhodium is excellent at reducing emissions of nitric oxides. Reclaimed rhodium from exhausted converters now amounts to almost 2 tonnes per year.

The next biggest user, but small by comparison, accounting for only 6%, is the glass industry, which uses rhodium as a coating for optical fibres, optical mirrors and the reflectors of headlights. The reason for this use is the metal's excellent reflectivity as a thin surface film. A similar amount of rhodium is used as catalysts in the chemicals industry, e.g. in the making of nitric acid, acetic acid and oxo-alcohols, and in reactions which add hydrogen, when a rhodium compound discovered by the Nobel Prize-winner, Geoffrey Wilkinson (1921–96) is used, called Wilkinson's catalyst.

A little rhodium is used to coat electrical contacts, in spark plugs for aircraft and in laboratory crucibles.

ENVIRONMENTAL ELEMENT

Rhodium in the environment	
Earth's crust	approx. 0.2 p.p.b.
	Rhodium is the 79th most abundant element, in other words it is extremely rare.
Soils	not detected
Sea water	barely detectable
Atmosphere	nil

Rhodium is too rare for the amount of it in soils or natural waters to be assessed, and so its effect on the environment can be assumed to be nil. Tests on plants have shown that it is the least toxic member of the platinum group of metals.

CHEMICAL ELEMENT

Data file	
Chemical symbol	Rh
Atomic number	45
Atomic weight	102.90550
Melting point	1966°C
Boiling point	3725°C
Density	12.4 kilograms per litre (12.4 grams per cubic centimetre)
Oxides	the most stable is Rh_2O_3, but RhO and RhO_2 are also known

Rhodium is a rare, lustrous, silvery, hard metal and a member of the platinum group of metals and of group 9 of the periodic table of the elements. It is unaffected by air and water up to 600°C, and unaffected by acids including *aqua regia* below 100°C. It is attacked by molten alkalis.

The only naturally occurring isotope of rhodium, rhodium-103, is not radioactive.[133]

ELEMENT OF SURPRISE

When rhodium is heated up to its melting point, it absorbs oxygen from the atmosphere but does not become transformed into one of its oxides, because as it solidifies it releases the oxygen again.

[133] Tiny amounts of radioactive rhodium isotopes are produced as fission products in uranium minerals.

Pronounced roo-bid-iuhm, the name is derived from the Latin *rubidius*, meaning deepest red (ruby).

French, *rubidium*; German, *Rubidium*; Italian, *rubidio*; Spanish, *rubidio*; Portuguese, *rubídio*.

HUMAN ELEMENT

Rubidium in the human body	
Blood	2.5 p.p.m.
Bone	0.1–5 p.p.m.
Tissue	20–70 p.p.m.
Total amount in body	680 milligrams

Rubidium has no known biological role but has a slight stimulatory effect on metabolism, probably because rubidium is like potassium. The two elements are found together in minerals and soils, although potassium is much more abundant than rubidium. Plants will absorb rubidium quite readily. When stressed by a deficiency of potassium, some plants, such as sugar beet, will respond to the addition of rubidium.

Plants such as soya beans contain 220 p.p.m. rubidium, apples have 50 p.p.m., but sweetcorn has only 3 p.p.m. and onions even less, at 1 p.p.m. Tea and coffee also contain measurable amounts of rubidium. In these various ways rubidium enters the food chain and so contributes to a daily intake of between 1 and 5 milligrams, which is tiny compared with the recommended daily intake of potassium (3500 milligrams). Rubidium is absorbed easily from the gut and becomes distributed around the body, but there is no site where it prefers to accumulate, although only small amounts get into the bones and teeth. It is excreted in the urine.

Rubidium salts are not considered to be toxic.

MEDICAL ELEMENT

Rubidium is so like potassium that it is easily absorbed by the body. It is also slightly radioactive, on account of a long-lived radioactive isotope – see later – and in this way its progress can be monitored and used for medical research. For example, rubidium tends to be attracted to cancer cell membranes; surgeons have exploited this property to locate brain tumours at an early stage.

ELEMENT OF HISTORY

Rubidium was announced to the Berlin Academy at a meeting on 23 February 1861 by Robert Wilhelm Bunsen (1811–99) and Gustav Robert Kirchhoff (1824–87) of the University of Heidelberg, Germany. They had discovered it in lepidolite, a mineral that was first reported by a Jesuit priest, Nicolaus Poda, a century earlier. Lepidolite was known for its curious properties; for example, when it is thrown on to glowing coals it froths to begin with, and then becomes glass-like, indicating that its main component is silica.

Bunsen and Kirchhoff knew that lepidolite contained lithium, but suspected it may have harboured another alkali metal as well. They extracted the metal components into solution and then added platinum chloride, which precipitated the potassium that was present. It precipitated the rubidium as well but this was only revealed by washing the precipitate with boiling water many times. This slowly dissolved the potassium chloroplatinate and left a tiny amount of a residue whose atomic specturm showed two intense ruby-red lines that had never been seen before. These indicated a previously undiscovered element, which Bunsen and Kirchhoff named rubidium because of the colour of the lines.

A sample of pure rubidium was not prepared until 1928, when a chemist called Louis Hackspill produced it.

ECONOMIC ELEMENT

No minerals of rubidium are known, but rubidium is present in significant amounts in other minerals such as lepidolite, at about 1.5% (see lithium, p. 237), pollucite (see caesium, p. 81) and carnallite (see potassium, p. 336). It is also present in trace amounts in other minerals such as zinnwaldite (a potassium–lithium–iron–aluminium silicate) and leucite (a potassium–aluminium silicate). Some natural brines contain as much as 6 p.p.m. rubidium.

The amount of rubidium produced each year is small, and what demand there is can be met from a stock of a mixed carbonate by-product that is collected during the extraction of lithium from lepidolite. While this mixture is mainly potassium carbonate, it also contains around 23% rubidium carbonate and 3% caesium carbonate. When it is heated with metallic sodium it reacts to form sodium carbonate and liberates the other metals, which are then separated by distillation. This yields a product that is 99.5% pure rubidium.

The little rubidium that is produced is used for research purposes only. There is no incentive to seek commercial outlets for the metal, because what it can do can almost certainly can be done equally well by sodium, which is more than 5000 times cheaper, or even by potassium, which is 300 times cheaper. Rubidium can cost as much as US$20 000 per kilogram, making it more expensive than gold or platinum.

There have been occasional uses of rubidium as a 'getter' in vacuum tubes, i.e. to mop up minute traces of oxygen; and of rubidium carbonate to make special types of glass. It has also been demonstrated that rubidium could be used to turn heat energy into electricity via a so-called thermoelectric motor. In this, the rubidium is heated and loses an electron to form the positive ion Rb^+, which then passes through a magnetic field and in so doing generates an electric current in surrounding coils.

ENVIRONMENTAL ELEMENT

Rubidium in the environment	
Earth's crust	at least 90 p.p.m. and possibly more
	Rubidium is the 22nd most abundant element, but it may well be more abundant than thought – maybe as abundant as its neighbour, strontium (about 350 p.p.m.).
Soils	varies with a general range of 30–250 p.p.m., although some soils have very little and a few exceed 400 p.p.m.
Sea water	0.1 p.p.m.
Atmosphere	insignificant

The relative abundance of rubidium has been reassessed in recent years and it is now suspected of being more plentiful than previously calculated. It is very like potassium and there are no environments where it is seen as a threat.

Rocks that contain rubidium can be dated by the so-called rubidium–strontium dating method, which involves analysing the extent to which its radioactive isotope, rubidium-87, had transmuted into strontium-87.

CHEMICAL ELEMENT

Data file	
Chemical symbol	Rb
Atomic number	37
Atomic weight	85.4678
Melting point	39°C
Boiling point	688°C
Density	1.5 kilograms per litre (1.5 grams per cubic centimetre)
Oxides	Rb_2O is the most stable, but there is also a peroxide (Rb_2O_2) and a superoxide (RbO_2)

Rubidium is a soft, white, metal and a member of group 1, the so-called alkali metal group, of the periodic table of the elements. Rubidium is silvery when first cut but it soon ignites, so has to be stored under oil or grease. Unlike the other alkali metals, when rubidium is partially oxidized at low temperatures it forms two other oxides: Rb_6O and

Rb_9O_2. It reacts violently with water and even with ice at −100°C. Rubidium metal is obtained from rubidium chloride by the reaction of this with calcium or potassium metal.

There are two naturally occurring isotopes of rubidium: rubidium-85, which accounts for 72% and is not radioactive, and rubidium-87, which accounts for 28% and is radioactive, with a half-life of 50 billion years. It is a β-emitter but not a γ-emitter. While this suggests it is only weakly radioactive, it does contribute to the Earth's background radiation, and a sample of rubidium will emit enough radiation to blacken a photographic plate within 2 months.

ELEMENT OF SURPRISES

1. A rather curious property of rubidium–silver iodide ($RbAg_4I_5$) is that, unlike most salts, this solid is highly conducting. Indeed it has the capacity to conduct electricity like some electrolyte solutions.
2. In 1995, two thousand atoms of rubidium were cooled to a temperature of only a few billionths of a degree above absolute zero and this produced a unique state of matter in which all the atoms occupied the same quantum state.

Pronounced roo-thee-ni-uhm, the name is derived from the Latin name for Russia.

French, *ruthénium*; German, *Ruthenium*; Italian, *rutenio*; Spanish, *rutenio*; Portuguese, *ruténio*.

HUMAN ELEMENT

Ruthenium in the human body	
Total amount in body	not known, but low

Ruthenium has no known biological role. Most ruthenium compounds are non-poisonous and ingested ruthenium is retained in the bones for a long time. The volatile oxide, RuO_4, is highly toxic by inhalation, but is almost never encountered.

ELEMENT OF HISTORY

In 1807, the Polish chemist Jedrzej Andrei Sniadecki (1768–1838), who was based at the University of Vilno, began investigating some crude platinum ores from South America, hoping to detect yet another 'new' metal in them, following in the footsteps of Wollaston and Tennant who had isolated the elements rhodium, palladium, osmium and iridium from this source. In May 1808 his efforts were rewarded by the discovery of a 'new' metal which he called vestium, after the asteroid Vesta, which had been first observed the previous year.[134] Sniadecki published his finding in a Russian journal.

When leading French chemists tried to repeat his work they were unable to find any vestium in the platinum ore they had available and when Sniadecki learned of their work he dropped his claim to having discovered a 'new' element, believing that he had been mistaken in his findings. Then, in 1825, the great Berzelius, based at Stockholm, and G. W. Osann of the University of Dorpat (now Tartu) on the Baltic, investigated some platinum from the Ural mountains in Russia, and while Berzelius could only detect the metals already known, Osann reported finding *three* hitherto unknown metals which he named pluranium, polinium and ruthenium.

While the first two of these were never to be verified, the third was proved to be a

[134] The naming of elements after heavenly bodies was in vogue in the early 1800s, witness cerium after the asteroid Ceres and palladium after the asteroid Pallas.

'new' metal in 1840 by a former colleague of Osann's, Karl Karlovich Klaus (1796–1864) working at the University of Kazan. He showed that Osann's sample of ruthenium was very impure, and set about extracting and purifying his own sample and investigating its properties. The result is that today Klaus is often cited as the discoverer. He kept Osann's name for the element, ruthenium, which derives from the same root as Russia. As Klaus made each new compound of ruthenium, he sent a sample to Stockholm to Berzelius who was still sceptical of its existence, until the weight of evidence forced him to acknowledge that it was indeed a 'new' element.

ECONOMIC ELEMENT

Ruthenium metal has been found in the free state, and there are a few ruthenium minerals, such as laurite (ruthenium sulfide), which is the most common, ruarsite (ruthenium arsenic sulfide), and ruthenarsenite (ruthenium–nickel arsenide). All are rare and none acts as a commercial source of the metal. Ruthenium is obtained as a by-product of nickel refining in Ontario (Canada) and South Africa. World production is 12 tonnes per year and reserves are estimated to be around 5000 tonnes.

Ruthenium demand is rising. The metal finds use in the electronic industry (50% ends up this way) and the chemical industry (40%), with smaller amounts being used in alloying platinum and titanium. In electronics it used to be used mainly for electrical contacts but most now goes into chip resistors. In the chemical industry it is used in the anodes for chlorine production in electrochemical cells. These anodes, known as dimensionally stable anodes, or DSAs, are made from titanium coated with ruthenium oxide which gives them a long service life under corrosive conditions, as well as an improved release of chlorine gas. Ruthenium is also part of the catalysts for the production of ammonia from natural gas, and for the production of acetic acid from methanol.

Ruthenium is added to platinum for jewellery purposes, to make it harder, and to titanium deep-water pipes, to make them corrosion resistant. Both uses it serves extremely well.

ENVIRONMENTAL ELEMENT

Ruthenium in the environment	
Earth's crust	approx. 1 p.p.b.
	Ruthenium is the 74th most abundant element, in other words it is very rare.
Soils	0.5–30 p.p.b.
Sea water	less than 0.005 p.p.t. (barely detectable)
Atmosphere	nil

Ruthenium is one of the rarest of metals on Earth. Very few data are available on its impact on plants and estimates of its uptake have deduced levels of 5 p.p.b. or less, although algae appear to concentrate it – see 'Element of surprise'.

CHEMICAL ELEMENT

Data file	
Chemical symbol	Ru
Atomic number	44
Atomic weight	101.07
Melting point	2310°C
Boiling point	3900°C
Density	12.4 kilograms per litre (12.4 grams per cubic centimetre)
Oxides	most stable is Ru_2O_3, but RuO_2, RuO_3 and RuO_4* are also known.

* This oxide can react explosively and is very toxic.

Ruthenium is a lustrous, silvery metal of the so-called platinum group, and a member of group 8 of the periodic table of elements. It is unaffected by air, water and acids, but dissolves in molten alkalis.

There are seven naturally occurring isotopes of ruthenium, of which ruthenium-102 is the most abundant at 31.5%. The others are ruthenium-104 at 19%, ruthenium-101 at 17%, ruthenium-99 and ruthenium-100 both at 12.5%, ruthenium-96 at 5.5% and, least abundant, ruthenium-98 at 2%. None is radioactive.[135]

ELEMENT OF SURPRISE

The radioactive isotope ruthenium-106, which has a half-life of 372 days, is produced by nuclear reactors and nuclear explosions and can reach the human food chain, although the only food compromised by this has been the edible seaweed *Porphyra.* This grows in the Irish Sea and is harvested off the coast of South Wales to be eaten as laver bread. It became contaminated with radioactive ruthenium from discharges from the nuclear reprocessing plant at Sellafield in the UK.

[135] Minute traces of radioactive ruthenium isotopes are also formed as fission products in uranium ores.

Rutherfordium

This is element atomic number 104. Along with other elements above 100 it is dealt with under the transfermium elements on p. 457.

Samarium

Pronounced sam-ayr-iuhm, this element is named after samarskite, the mineral from which it was first extracted.

French, *samarium*; German, *Samarium*; Italian, *samario*; Spanish, *samario*; Portuguese, *samário*.

Samarium is one of 15, chemically similar, elements referred to as the rare-earth elements, which extend from the element with atomic number 57 (lanthanum) to the one with atomic number 71 (lutetium). The term rare-earth elements is a misnomer because some are not rare at all; they are more correctly called the lanthanides, although strictly speaking this excludes lutetium. The minerals from which they are extracted, and the properties and uses they have in common, are discussed under lanthanides, on p. 219.

HUMAN ELEMENT

Samarium in the human body	
Total amount in body	not known, but small

The amount of samarium in humans is tiny and the metal has no biological role, but it has been noted that samarium salts stimulate metabolism. It is difficult to separate out the various amounts of the different lanthanides in the human body, but they are present, and the levels are generally highest in bone, with small amounts being present in the liver and kidneys. The total amount of samarium in an average adult is thought to be around 50 micrograms, and its level in the blood is 8 micrograms per litre (8 p.p.b.).

No one has monitored diet for samarium content, so it is difficult to judge how much we take, but it is probably less than 1 milligram or so a year. Samarium is not taken up by plant roots to any great extent so does not get into the human food chain, although some vegetables can contain 100 p.p.b. (dry weight) and some plants as much as eight times this figure.

Soluble samarium salts are mildly toxic by ingestion, but insoluble salts are non-toxic. However, there are health hazards associated with these because exposure to samarium salts causes skin and eye irritation.

MEDICAL ELEMENT

The isotope samarium-153 is a β-emitter with a half-life of 2 days. It was approved for use as a radiopharmaceutical in the USA in 1997.

ELEMENT OF HISTORY

Samarium was discovered by Paul-Émile Lecoq de Boisbaudran (1838–1912) in Paris in 1879. Cerium, discovered in 1803, was suspected of harbouring other metals, and these were identified in 1839 by Carl Gustav Mosander as lanthanum and didymium. The latter in its turn was suspected of being a mixture of elements when its atomic spectrum was observed in 1879 and was noted to vary slightly according to the mineral from which it was obtained. In that year Boisbaudran extracted didymium from the mineral samarskite, which was named after the Russian engineer, Colonel V. E. Samarsky. He then made a solution of didymium nitrate, added ammonium hydroxide to it, and noticed that the precipitate which formed appeared to come down in two stages. He devised a way of separating the first precipitate and measured its spectrum. This proved to be the hydroxide of a previously unknown element which Boisbaudran named samarium.

However, impure samarium was also eventually to yield another rare-earth, europium, extracted from it by Eugène-Anatole Demarçay in 1901.

ECONOMIC ELEMENT

Samarium is found in minerals that include all the rare-earth elements; these are discussed under the lanthanides, on p. 222. Samarium-containing ores are found in China, USA, Brazil, India, Sri Lanka and Australia. The most important is monazite, which contains up to 3% by weight of samarium. Samarium is also present in bastnäsite and in the so-called ionic clays found in China. World-wide reserves of samarium are estimated to be around 2 million tonnes.

World production of samarium oxide is about 700 tonnes per year, and the metal is produced by heating the oxide with barium or lanthanum to temperatures high enough to drive off samarium as a vapour.

One outlet for samarium is in permanent magnets, where its alloy with cobalt, as either $SmCo_5$ or Sm_2Co_{17}, produces magnets that are not only ten thousand times more powerful than iron, but have the highest resistance to demagnetization of any known material. These magnets have allowed the miniaturization of devices like motors and headphones. Without samarium, personal stereos would not have been possible. In recent years, samarium–cobalt magnets have been overshadowed by NIB magnets (see under neodymium, p. 270) although they are still used in microwave applications because they retain their magnetism to temperatures above 700°C.

Samarium oxide finds specialized use in ceramics and for making glass that absorbs infrared rays. Calcium fluoride crystals doped with samarium are used in lasers and masers (the microwave equivalent of lasers); the latter are capable of cutting through steel and bouncing off the surface of the Moon.

In some countries the chemicals industry converts surplus ethanol to the gas ethene, which is the feedstock for many plastics, and this conversion is aided by samarium catalysts. Other uses for samarium are: in the core material in carbon-arc electrodes for film studio lights; in infrared sensitive phosphors; and as an excellent neutron absorber for the control rods which regulate the reactor core in nuclear power plants.

ENVIRONMENTAL ELEMENT

Samarium in the environment	
Earth's crust	8 p.p.m.
	Samarium is the 40th most abundant element
Soils	approx. 5 p.p.m. (dry weight), range 2–23 p.p.m.
Sea water	0.8 p.p.t.
Atmosphere	virtually nil

Samarium is the fifth most abundant of the rare-earth elements and is almost four times as common as tin. It poses no environmental threat to plants or animals.

CHEMICAL ELEMENT

Data file	
Chemical symbol	Sm
Atomic number	62
Atomic weight	150.36
Melting point	1077°C
Boiling point	1790°C
Density	7.52 kilograms per litre (7.52 grams per cubic centimetre)
Oxide	Sm_2O_3

Samarium is a silvery-white metal and one of the lanthanide group of the periodic table of the elements. It is relatively stable in dry air, but in moist air an oxide coating forms. When heated to 150°C, the metal ignites spontaneously. Like europium, samarium also has a lower, albeit less stable, oxidation state (II), and one of its compounds, the intense blue samarium iodide (SmI_2), is used by organic research chemists to make synthetic versions of natural products because of the way it can produce cyclic molecules with the right spatial arrangement of atoms. It is also good at dechlorinating pollutants such as polychlorinated biphenyls (PCBs) at temperatures as low as 60°C and so rendering them harmless.

Naturally occurring samarium has seven isotopes of which three are weakly radioactive and have very long half-lives: samarium-147, which accounts for 15%, has a half-life of 100 billion years; while samarium-148 (11%) and samarium-149 (14%) have half-lives of more than a million billion years. The most abundant isotopes are samarium-152, which accounts for 27%, and samarium-154, which accounts for 23%; the other non-radioactive isotopes are samarium-150, at 7%, and samarium-144, at 3%.[136]

ELEMENT OF SURPRISE

The oddest samarium compound is samarium(II) sulfide (SmS), which exists as black crystals with semiconductor properties. When these are scratched they immediately transform into golden crystals that conduct like a metal, and the same thing happens when they are put under high pressure.

[136] Altogether there are 24 known isotopes of samarium spanning the mass range 138–158.

Scandium

Pronounced skan-dium, the name derives from *Scandia*, the Latin name for Scandinavia.
French, *scandium*; German, *Scandium*; Italian, *scandio*; Spanish, *escandio*; Portuguese, *escândio*.

COSMIC ELEMENT

Scandium is more abundant in the heavens than it is down here on Earth. This observation was made as long ago as 1908 when Sir William Crookes and Gustav Eberhard examined the visible spectra of the Sun and certain other stars, and identified strong bands arising from this element.

HUMAN ELEMENT

Scandium in the human body	
Blood	approx. 8 p.p.b.
Bone	approx. 1 p.p.b.
Tissue	approx. 1 p.p.b.
Total amount in body	0.2 milligrams

Scandium has no known biological role. Only trace amounts reach the food chain, so the average person's daily intake is less than 0.1 microgram. Only about 3% of the plants that were analysed for scandium showed its presence, and even so amounts were tiny, with vegetables having only 5 p.p.b. although grass has 70 p.p.b. Tea leaves showed more than this, with an average of 140 p.p.b., which is perhaps understandable since this plant requires aluminium, and scandium is chemically very similar to aluminium.

Scandium is not toxic, although there have been suggestions that some of its compounds might be carcinogenic.

ELEMENT OF HISTORY

When Dimitri Mendeleyev devised his periodic table of the elements in 1869, he noticed that there was a large difference in atomic weights between calcium (40) and titanium (48) and he predicted that there should be another element of intermediate atomic weight (44) which he referred to as eka-boron, since he thought it would come below boron in the same group, group III of his table. Consequently he forecast that its oxide would have the formula X_2O_3.

Scandium was discovered 10 years later, in 1879, by Lars Fredrik Nilson (1840–99) who was the Professor of Analytical Chemistry at the University of Uppsala, Sweden. He extracted it from euxenite, a complex mineral containing as many as eight metals. He thought that this was only to be found in Scandinavia, so he called the newly discovered metal scandium.

Nilson had extracted erbium oxide from euxenite, and from this oxide he obtained both ytterbium oxide and another oxide of a lighter element that he could not immediately identify. Its atomic spectrum, however, showed lines not previously reported for any known metal, thus proving that he had stumbled across a 'new' one. He studied its chemistry, determined its atomic weight as 44, and showed that its oxide had the formula Sc_2O_3. His colleague at the university, Professor Per Theodor Cleve, pointed out that this was the missing eka-boron predicted by Mendeleyev.

Nilson never saw a sample of the metal itself. This was only made in 1937 by the electrolysis of molten scandium chloride dissolved in a melt of other metal chlorides at 800°C.

ECONOMIC ELEMENT

There is a greenish-black scandium ore called thortveitite ($Sc_2Si_2O_7$), which also contains yttrium. It is extremely rare and found mainly in Norway. Scandium is also present in small amounts in euxenite and gadolinite, but that which is used is extracted as a by-product of uranium mining and comes mainly from Norway and Malagasy (Madagascar). World production amounts to only about 50 kilograms per year. There is no estimate of how much is potentially available.

There are a few, rather specialized, uses for scandium such as in neutron filters for nuclear reactors. The artificially produced radioactive isotope, scandium-46, which has a half-life of 84 days, has been used in oil refineries to monitor the movement of various fractions as the oil is refined. In a similar manner it can detect leaks in underground pipes carrying liquids.

Another use of scandium is to induce germination of seeds. When it is applied as a dilute solution of scandium sulfate to corn, peas and wheat, it increases the number of seeds successfully germinating.

One US manufacturer of sports goods claimed to have made a baseball bat containing scandium saying that it had remarkable striking power, although this seems unlikely. (Apparently cricket bats incorporating the metal showed the same ability, but these were deemed unsporting and were forbidden.) Such uses indicate the potential of scandium to be a superior kind of aluminium, being equally lightweight, although much more costly. It might one day have industrial implications because it scores over aluminium in having a melting point that is 900°C higher; this could lead to its being used in aircraft and spacecraft.

ENVIRONMENTAL ELEMENT

Scandium in the environment	
Earth's crust	16 p.p.m.
	Scandium is the 35th most abundant element
Soils	approx. 7 p.p.m., but there is a wide range from 0.5 to 45 p.p.m.
Sea water	0.6 p.p.t. (barely detectable)
Atmosphere	nil

Because scandium precipitates from water – even from neutral water of pH 7 – as the insoluble hydroxide, $Sc(OH)_3$, there has been no geological process to concentrate this element to any extent and what there is tends to be widely distributed in many types of rock. It has been detected in more than 800 minerals so far. In the aquamarine variety of the gemstone beryl, the blue colour is thought to be due to traces of scandium, although normally the compounds of this element are colourless.

Ash residues from coal and crude oil contain more scandium than might have been expected – up to 0.1% in some cases – suggesting that environmental enrichment is likely to have occurred.

CHEMICAL ELEMENT

Data file	
Chemical symbol	Sc
Atomic number	21
Atomic weight	44.955910
Melting point	1541°C
Boiling point	2831°C
Density	3.0 kilograms per litre (3.0 grams per cubic centimetre)
Oxide	Sc_2O_3

Scandium is a soft, silvery-white metal and a member of group 3 of the periodic table of the elements. It tarnishes in air and burns easily, once it has been ignited. It reacts with water to form hydrogen gas and will dissolve in many acids. The pure metal is produced by heating scandium fluoride (ScF_3) with calcium metal.

Scandium forms a hydride, ScH_2, which is odd not only from the point of view of the known valency of this element (ScH_3 would be expected) but also because it is a good conductor of electricity for reasons that have yet to be deduced.

There is only one abundant naturally occurring isotope, scandium-45, and it is non-radioactive. Minute traces of radioactive scandium are present in the Earth's crust.

ELEMENT OF SURPRISE

When scandium iodide (ScI_3) is added to mercury vapour lights, it turns their intense but harsh light into something more akin to natural sunlight and consequently such lights are used in filming and for the floodlighting of outdoor sports arenas.

Seaborgium

This is element atomic number 106 and along with other elements above 100 it is dealt with under the transfermium elements on p. 457.

Selenium

Pronounced sel-een-iuhm, the name is derived from *selene*, the Greek name for the Moon.
French, *sélénium*; German, *Selen*; Italian, *selenio*; Spanish, *selenio*; Portuguese, *selénio*.

HUMAN ELEMENT

Selenium in the human body	
Blood	levels vary between 70 and 150 p.p.b.
Bone	levels vary between 1 and 9 p.p.m.
Tissue	approx. 0.1 p.p.m.
Total amount in body	approx. 14 milligrams

In 1975, selenium was proved to be an essential element for humans. Yogesh Awasthi, based at Galveston in Texas, discovered that it was part of the antioxidant enzyme, glutathione peroxidase, which eliminates peroxides before they can form dangerous free-radicals. Then, in 1991, Professor Dietrich Behne at the Hahn-Meitner Institute in Berlin found selenium in another enzyme, deiodinase, which promotes hormone production in the thyroid gland.

Every cell of the body contains more than a million atoms of selenium. The parts with the highest levels are hair, kidneys and testicles. The recommended maximum daily intake is 450 micrograms. Above this we risk selenium poisoning, the most obvious symptoms of which are extremely foul breath and body odour, sometimes experienced by those working with selenium in industry. The smell is caused by volatile methyl selenium molecules which the body produces in an effort to rid itself of the unwanted excess. It is possible to overdose on selenium: 5 milligrams, if taken as a single dose, would produce symptoms of toxicity.

FOOD ELEMENT

The dilemma about selenium, as with many other elements, is that if a person has either too little or too much of it in their diet their health will suffer. The daily intake of selenium varies between 6 and 200 micrograms, depending on the types of food that are eaten. The average Westerner takes in about 65 micrograms per day, which should be adequate to prevent selenium deficiency; this is adequate to meet the recommended 60 microgram intake for women although it is less than the recommended daily intake of 75 micrograms for men. On some days the body may lose more selenium than it takes in, but this does not pose a threat because the body can draw upon its store of selenium, and in particular that in the bones.

A deficiency of this element is said to be linked to all kinds of conditions: anaemia, high blood pressure, infertility, cancer, arthritis, premature ageing, muscular dystrophy and multiple sclerosis. The evidence is epidemiological and likely to be a secondary factor, rather than the primary cause, but in some cases the evidence is compelling, as in Keshan disease (see 'Medical element' below).

As a dietary supplement, selenium is usually taken in the form of sodium selenite (Na_2SeO_3), a white crystalline material which is soluble in water; the daily dose is 50 micrograms. An alternative way is to take brewer's yeast that has been grown on a selenium-enhanced culture medium.

Most people get their daily dose of selenium from breakfast cereals and bread, especially wholemeal bread, two slices of which will provide 30 micrograms although in the case of wheat – and meat – products the level depends upon the soil of the farm from which they came. Foods particularly rich in selenium are Brazil nuts and molasses (black treacle), both having more than 100 micrograms per 100 grams. Other foods with levels in excess of 30 micrograms per 100 grams are as follows: fish, such as tuna, cod and salmon; offal, such as liver and kidney; most nuts, such as peanuts; wheatgerm, bran and brewer's yeast.

A particularly rich source of selenium is the edible mushroom, *Albatrellus pes-caprae*, which is popular in Italy. This has 3700 micrograms of selenium per 100 grams fresh weight. A dish containing this amount would provide the eater with eight times the recommended daily *maximum*!

MEDICAL ELEMENT

According to some researchers, declining sperm counts and increased incidences of cancer can be explained by the decrease in selenium in the Western diet. For example, the average selenium intake fell by half in the last half century, due in part to the decline in popularity of selenium-rich foods such as kidney and liver. In a double-blind test carried out in 1993, men with low sperm counts were given either selenium or a placebo; the sperm counts of those given selenium went up by 100%, while counts for those given the placebo did not change.

In the UK, farmers have been giving their animals selenium supplements for years to keep them healthy, while in Finland there has been a programme to increase selenium levels in soil by adding the element to fertilizers. Large areas of China, such as the Keshan and Linxian regions, have selenium-depleted soils. Children of the Keshan region were particularly prone to congestive heart failure known as Keshan disease, and those living in the province of Linxian had a high incidence of stomach cancer. The incidence of both diseases was reduced when selenium supplements were issued to the population.

One theory as to why virus mutations often appear in China links this to low selenium levels in the population. This reduces immunity to disease, which in turn allows virus populations to grow much larger than usual, making it more likely that a virulent strain will emerge there.

ELEMENT OF HISTORY

Selenium was discovered by Jöns Jacob Berzelius at Stockholm in 1817. He had shares in a sulfuric acid works at Gripsholm about a hundred miles North-West of the Swedish capital at whose medical school he was Professor of Chemistry and Medicine. In the summer of that year he spent several weeks at the works as a consultant. Intrigued by a red-brown sediment which collected at the bottom of the chambers in which the acid was made, he took some it back to Stockholm for further study.

At first he thought it was the element tellurium, because it gave off a strong smell of radishes when heated, but Berzelius came to realize that, although it was chemically similar to tellurium, it was in fact a hitherto unknown element. He named it selenium (from the Greek *selene*, meaning Moon) to match the name tellurium (from the Latin *tellus*, meaning Earth). Berzelius noted that the newly discovered element was like sulfur in that it would vaporize on heating and re-deposit as beautiful crystals known as 'flowers' on a cool surface, but it differed in that, whereas 'flowers' of sulfur are bright yellow, those of selenium are red. This is the non-metallic form of the element. Selenium can have a metallic form also (see below).

Berzelius realized that selenium was also present in samples of tellurium and that it was the selenium that made tellurium smell of radishes. He also became affected by it personally – his housekeeper one day admonished him for his bad breath, saying he'd been eating too much garlic. On another occasion he was overcome by breathing in hydrogen selenide (H_2Se) gas, which is dangerously toxic.

Within a few years of its discovery, selenium turned up in many more minerals. In some instances the metallic form of the element was refined and cast into medallions carrying pictures of Berzelius.

ECONOMIC ELEMENT

Uncombined selenium is occasionally found and there are around 40 known selenium-containing minerals, some of which can have as much as 30% selenium – but all are rare, and generally they occur together with sulfides of metals such as copper, zinc and lead. The main producing countries are Canada, USA, Bolivia and Russia. The main consumers are the USA (30%), Europe (35%) and Japan (25%); the rest of the world accrounts for the remaining 10%.

Global industrial production of selenium is around 1500 tonnes a year, and about 150 tonnes of selenium are recycled from industrial waste and reclaimed from old photocopiers. Selenium is obtained as a by-product of the smelting of other metal ores, especially copper – in which copper selenide is an impurity. The most important source of selenium, accounting for 90% of production, is the slime that settles at the bottom of tanks when copper is refined electrolytically; this slime may contain up to 5% selenium.

The metallic form of selenium has the curious property of conducting electricity a thousand times better when light falls on its surface and so it is used in photoelectric cells, light meters, solar cells and photocopiers. These electronic uses account for about a third of all selenium production and require high-purity (99.99%) selenium.

The second largest use of selenium, accounting for about a quarter of production, is the glass industry, which adds sodium selenate (Na_2SeO_4) and sodium selenite (Na_2SeO_3) to glass to decolorize it; as cadmium selenide it can be added in larger amounts to produce a magnificent ruby colour. A lot is used to make bronze architectural glass for office building because this screens out the Sun's rays.

The third main use, taking about 15%, is sodium selenite for animal feeds and food supplements. Other uses of selenium are in metal alloys, such as the lead plates used in storage batteries and in rectifiers to convert a.c. to d.c. current. Selenium used to be used to vulcanize rubber as it produced a much more durable material that was needed for industrial belts. Some selenium compounds are added to anti-dandruff shampoos.

The red, non-metallic form of selenium is not used as such, but as cadmium selenide it makes a striking and permanent red pigment for paints, enamels and plastics. The toxicity of cadmium has curbed its use in recent years, but selenium is being reintroduced as a pigment in the guise of cerium selenide, which is safe.

ENVIRONMENTAL ELEMENT

Selenium in the environment	
Earth's crust	50 p.p.b.
	Selenium is the 67th most abundant element.
Soils	approx. 5 p.p.m.
Sea water	0.2 p.p.b.
Atmosphere	very low – of the order of 1 nanogram* per cubic metre of air

* One nanogram is a billionth (10^{-9}) of a gram.

Selenium is among the rarer elements on the surface of this planet, and is rarer than silver. Well fertilized agricultural soil generally has about 400 milligams per tonne since the element is naturally present in phosphate fertilizers and is often added as a trace nutrient.

Selenium is present in the atmosphere as its methyl derivatives; these are the counterparts to the sulfur compounds which are an important part of the sulfur cycle in Nature (see p. 415). There is a similar selenium cycle, whose volatile components – methyl selenide and dimethyl selenide – are released from soils, lakes and sewage by anaerobic bacteria. About 20 000 tonnes of selenium enter the atmosphere this way each year, of which two-thirds comes from human activity. This may even be beneficial if it is washed down by rain on to selenium-deficient land.

Some parts of the world have soils with high levels of selenium. This can be taken up by the roots of some plants to levels that make them toxic to grazing animals, which suffer a condition known as blind staggers. The famous traveller Marco Polo (1254–1324) observed that the animals of Turkestan behaved in a curious way as though they were drunk. The plant generally responsible for the staggers was the milk-vetch (*Astragalus*), which can concentrate selenium up to 1.4% of its dry weight. The Great Plains of the USA are also rich in selenium, and cowboys of the Wild West knew that their herds could be affected if they ate vetch, which they called 'loco weed' from the Spanish word *loco*, meaning insane. In 1934, Orville Beath proved that the staggers was caused by excess selenium in an animal's diet.

Selenium pollutes the environment around certain facilities, namely coal-burning power stations (since this fossil fuel can have as much as 10 p.p.m. selenium), metal smelters which use sulfide ores, and municipal incinerators burning paper, cardboard and tyres, which contain small amounts of selenium.

In 1983, the Kesterson Reservoir in the San Joaquin Valley, California, became so polluted with selenium from agricultural drainage that wildfowl chicks were being born deformed and some adults birds were even dying because of it. The 85-mile drainage canal that fed into the reservoir was cut off and reclamation undertaken. In 1989, the oil company Chevron stopped discharging selenium-polluted waste water into San Francisco Bay and pumped it instead to a 35-hectare wetland area that had been specially created to deal with the toxic waste. By the time the water had worked its way through this, most of the selenium had disappeared. However, when the company scientists sampled the sediment of the wetlands, they found only half the selenium that they expected. The rest had quite literally evaporated into thin air.

Renewable resource

Were the commercial sources of selenium ever to become exhausted, the element could be harvested from selenium-rich soils by growing crops of selenium-scavenging plants, so-called hyper-accumulators, such as milk-vetch. In this way 7 kilograms per hectare (3 kilograms per acre) could be extracted. The current world needs for selenium, of about 1500 tonnes per year, would require about 200 000 hectares to be farmed this way, but reserves of selenium in known ore deposits amount to over 100 000 tonnes, so it will be quite some time before the farming of selenium crops is introduced.

CHEMICAL ELEMENT

Data file	
Chemical symbol	Se
Atomic number	34
Atomic weight	78.96
Melting point	217°C
Boiling point	685°C
Density	4.8 kilograms per litre (4.8 grams per cubic centimetre)
Oxide	SeO_2

Selenium exists in two forms: a silvery metal or a red powder; the former is produced from the latter by heat. Selenium is a member of group 16 of the periodic table of the elements. The metal consists of spirals of linked selenium atoms while the red form has stacked rings each made up of eight selenium atoms. Selenium burns in air and is unaffected by water, but dissolves in concentrated nitric acid and alkalis.

There are six naturally occurring isotopes of selenium of which selenium-80 accounts for 50%. The others are selenium-79, which makes up 23.5% of the total; selenium-76 and selenium-82, which both account for 9%; selenium-77, which is 7.5%; and finally the rarest isotope, selenium-74, which amounts to 1%. None is radioactive. The radioactive isotope selenium-75, which has a half-life of about 17 weeks, is used for research and medical purposes.

ELEMENT OF SURPRISES

1. Although selenium is itself toxic in low doses, it can counter the effects of other metal toxins, especially cadmium, mercury, arsenic and thallium, and is known as an antagonist for these metals. Tuna fish, which accumulates higher than expected levels of mercury, is thought to be safe to eat because this fish protects itself by taking in an atom of selenium for every mercury atom it absorbs. In the 1970s, there was a food scare in the USA when analysis of canned tuna revealed higher-than-expected levels of mercury, which were assumed to be caused by environmental pollution. Overnight, canned tuna disappeared from supermarket shelves, and millions of cans were destroyed. Then, analysis of a sample of tuna from the nineteenth century that had been preserved in a museum display case showed the same level of mercury – and both showed a protective level of selenium. The protective role of selenium appears to be true for other marine mammals, such as seals, and may also be the way in which the men who work in mercury mines gain some protection from this dangerous metal.

   Other fish with high levels of mercury are shark, swordfish and king mackerel, and indeed pregnant women are advised to avoid eating such fish because the mercury thay contain could affect the foetus.

2. The idea of television was first proposed in the 1860s and in the 1880s selenium was tested to see if it could be used in transmitting motion pictures over wires. It was thought that because selenium increases its electrical conductivity when light falls on its surface, this might be a way to transmit a rapid sequence of impulses corresponding to light and dark down a wire, and to reconstruct them into moving pictures at the other end by stimulating patterns in selenium. The experiment was not very successful. [Walter Saxon]

Pronounced sil-i-kon, the name is derived from the Latin *silex* or *silicis*, meaning flint (but see text).

French, *silicium*; German, *Silicium*; Italian, *silicio*; Spanish, *silicio*; Portuguese, *silício*.

COSMIC ELEMENT

Silicon falls to Earth in the form of certain meteorites, known as aerolites, which are predominantly silicon dioxide, and in the form of glass-like tektites (which are also silicon dioxide) although these latter are of disputed origin.

HUMAN ELEMENT

Silicon in the human body	
Blood	4 p.p.m.
Bone	17 p.p.m.
Tissue	100–200 p.p.m.
Total amount in body	approx. 1 gram

Silicon as silicon dioxide (also more commonly known as silica, SiO_2) is essential to some species such as diatoms and sponges which use it to make their skeleton.[137] Nettle stings are tiny hypodermic needles of silica. Research in 1972 proved silicon to be necessary for bone growth, at least in chickens and rats, and it now seems likely that this element is an essential nutrient for humans as well. It concentrates in no particular organ of the body but is found mainly in connective tissue and skin. Silicon is non-toxic as the element and in all its natural forms, namely silica (sand) and silicates, which are the most abundant. However, some silicates with a fibrous structure, such as asbestos (a magnesium silicate), are carcinogenic.

The daily dietary intake of silicon varies quite markedly from as little as 20 milligrams to as much as 1200 milligrams, with cereals being the main source.

[137] Vast deposits of the mineral kieselguhr were laid down as the silica skeletons of diatoms millions of years ago.

MEDICAL ELEMENT

Louis Pasteur (1822–95) said that silicon would prove to be a treatment for many diseases and in the first quarter of the twentieth century there were numerous reports by French and German doctors of sodium silicate being used successfully to treat conditions such as high blood pressure and dermatitis.[138] By 1930, such treatments were seen to have been in vain and this medication fell out of favour. So things rested, until the discovery that silicon might have a role to play in human metabolism, and thence followed suggestions that it could have a role in conditions such as arthritis and Alzheimer's disease, but no new treatment based on these suggestions has yet emerged.

Meanwhile, silicon continues to be linked with a disease of its own: silicosis. Miners, stone-cutters, sand-blasters and metal-grinders develop this lung condition which is a recognized occupational disorder caused by the inhalation of minute particles of silica. Silicosis is one form of the condition known as pneumoconiosis, the symptoms of which are wheezing, coughing and shortness of breath. Asbestos miners, shipyard workers and those making insulation and brake-linings from asbestos contracted a more aggressive form of the disease which often developed into lung cancer, although certain forms of asbestos are more dangerous than others in causing it. Asbestos became the subject of extensive litigation in the USA and consequently is no longer used as a building material in Western countries, but it is still widely used in other parts of the world.

For a time it looked as if another kind of silicon material would fall foul of the law. The silicones, which are organosilicon–oxygen polymers, have been widely used as breast implants. In the USA, the majority of these were inserted purely for cosmetic reasons to enhance, rather than to replace lost tissue. They were introduced in the 1960s, and several million women – and a few former men – have had them inserted.

In the USA during the 1990s there were some well publicized court cases brought by women who claimed that silicone leaking from their implants had adversely affected their health, causing auto-immune disease and even cancer. The Food and Drugs Administration (FDA) had not investigated silicone implants because they were introduced before a 1976 US law requiring new medical devices to be approved, and in any case silicones had been thoroughly investigated in the 1940s and 1950s and found to have no adverse effect on living things. Nevertheless, successful law suits forced the leading silicone manufacturer, Dow Corning, to seek refuge in bankruptcy and offer a US$3 billion compensation package.

However, no scientific evidence has come to light showing that these compounds present any hazard to humans. The American Academy of Neurology in 1997, and the prestigious Karolinska Institute in Sweden in 1998, reported that women with silicone breast implants were no more likely to suffer adverse health effects than women without them. Outside America, silicones continue to be used successfully, and not only for breast implants – see 'Element of surprise'.

Another medical use of silicones is in indigestion tablets to relieve colic or gastric wind in humans, which can be especially painful after surgery; similarly, they are used

[138] Those who preferred a more natural, i.e. herbal, remedy were advised to drink tea made from the silicon-rich horsetail plant.

in veterinary medicine to relieve bloat in cows. The silicone acts as a defrothing agent, so allowing air or gas to escape.

One silicone, called Cisobitan, which has a ring of silicon and oxygen atoms, is being researched in Sweden as a possible drug for the control of prostate cancer.

ELEMENT OF HISTORY

Silicon, in the form of silica, was used by humans to make one of their first types of tool: sharp flints. The ancient civilizations used other forms of silica such as rock crystal and knew how to turn sand into glass, developing sophisticated technology for its production, colouring and shaping. Considering silicon's abundance, it is somewhat surprising that silicon aroused little curiosity among early chemists, even though it was clearly important to plants, and it had long been known that the ash from straw burning could be turned into glass.

In 1800, Humphry Davy said that silica was a compound, not an element as Lavoisier had assumed, but he was in no position to say what its composition really was, because attempts to reduce it to its components by electrolysis had failed. In 1811, Joseph Gay-Lussac and Louis-Jacques Thénard came near to isolating silicon by adding potassium metal to silicon tetrachloride. These chemicals reacted violently together and produced an impure form of silicon – which, strangely, Gay-Lussac and Thénard did not attempt to purify. For this reason the credit for discovering silicon usually goes to Jöns Jacob Berzelius of Stockholm, Sweden, who in 1824 obtained silicon by heating potassium fluorosilicate with potassium metal. Although his product was contaminated with potassium silicide, he removed this by stirring it with water, with which it reacts, and thereby obtained relatively pure silicon powder. This was a non-crystalline form known as amorphous silicon. Berzelius called the newly discovered element silicium, but in 1831 T. Thomson argued that it was a non-metallic element and should be named silicon to accord with the names of the similar non-metals, boron and carbon, and this became its name in English.

The other form of silicon, crystalline silicon, was first produced accidentally in 1854 by Henri Deville who electrolysed an impure melt of sodium aluminium chloride and produced aluminium silicide. When this was treated with water the aluminium dissolved, leaving shiny platelets of silicon.

ECONOMIC ELEMENT

Sand is to be found in abundance in all parts of the world and is used as such for many purposes, as well as being the source of the silicon produced commercially. A few silicate minerals are mined, e.g. talc (in Austria, Italy, India, South Africa and Australia) and mica (in Canada, USA, India and Brazil). Other mined silicates are feldspars, nepheline, olivine, vermiculite, perlite, kaolinite, etc. At the other extreme there are forms of silica so rare that they are desirable for this reason alone; they are better known by the names of the precious and semi-precious gems that they represent. The gemstone opal, which exhibits an attractive iridescence, is basically silica, as are rock crystal, agate and rhinestone. The more precious gemstone amethyst is essentially a purple variety of silica, coloured by traces of manganese.

Silicon is the basis of several industries. The construction industries rely on sand and cement, which is basically a type of calcium silicate. Glass manufacture is based on sand. Sand is also reduced to silicon as a raw material for the steel, chemical and electronics industries by heating it with coke (carbon) in an electric furnace with carbon electrodes, operating at 2000°C. Provided the silica is in excess, the product is silicon, in the amorphous form; most of this goes into making ferrosilicon for steel. Almost 4 million tonnes of this grade of silicon are made each year. Ferrosilicon is used to deoxidize steel and the alloys themselves are used in dynamo and transformer plates, springs and machine tools. The silicon that is needed for making ferrosilicon steels needs only to be around 96% pure, and these alloys contain between 8 and 13% silicon.

For other alloys, such as that with aluminium, the silicon needs to be at least 98% pure – and preferably 99%; this grade is known as metallurgical grade silicon and 500 000 tonnes of this are made each year. Adding silicon to aluminium improves its castability. Such alloys are used for engine blocks and cylinder heads.

Silicones

These consist of chains or rings of alternate silicon and oxygen atoms. Each silicon has two organic groups attached; these keep the chains stable and allow them to move smoothly over one another, which is why silicone oil is such a good lubricant. Depending on the lengths of the silicon–oxygen chains, it is possible to have a light mobile fluid with very short chains, or a viscous oil with long chains. Connecting these chains together crossways ('cross-linking' them) produces silicone rubber and resins which are noted for their water-repellency and resistance to oxidation and chemical attack.

The first silicones were made by the British chemist Frederic Kipping in the 1920s, but they were so difficult to produce it seemed unlikely they would ever find a use outside the laboratory. Then, in 1940, Eugene Rochow, a chemist working for the Corning Glass Works in New York State, found a cheap and easy way of making them by heating elemental silicon (metallurgical grade) with organochlorine compounds under pressure at 350°C, and treating the organochlorosilanes that formed with water. Today, half a million tonnes of silicones are made each year. Silicones remain stable when hot and are extremely versatile, finding use as high-temperature lubricants, insulating oils, waterproof sealants, rubber hose and plastic components for photocopiers and car engines.[139] Many silicone oils are used in cosmetics and hair conditioners because they not only act as a protective layer, but leave skin and hair feeling silky smooth.

Silicone rubber conducts heat rapidly so it is used for finger-touch contact switches. Silicone putty is ideal for sealing around roofs, windows, pipes and, especially, bathroom and kitchen fittings because it repels water. Silicones can be used for waterproofing fabrics such as canvas.[140]

[139] The imprints of silicone rubber boots are to be found in the dust on the surface of the Moon.

[140] However, when they were used in the 1960s to waterproof and protect ancient stone monuments against urban pollution, they proved to be a disaster. The silicone film slowly peeled away, sometimes removing the surface of the carvings it was meant to be protecting.

Silicon chips

The element itself, when ultrapure, is the semiconductor of transistors and other electronic devices. Pure silicon, known as polycrystalline silicon, is needed for microchips and can be produced in three ways:

- by reacting zinc and silicon tetrachloride;
- by reducing trichlorosilane ($SiHCl_3$) with hydrogen gas; or
- by the thermal decomposition of silane (SiH_4).

In all cases the starting materials must be as pure as can be made, to ensure that the product is not contaminated in any way. Such crystals are then suitable for use in microchip devices and solar panels, although they have to be 'doped' with other elements such as arsenic, gallium or boron, but only at a level of around 1 p.p.m. It is this type of silicon which revolutionized the worlds of commerce, communication and learning in the last quarter of the twentieth century, and indeed entered everyday life – and language – particularly in the form of personal computers. This is the silicon of Silicon Valley.

Carborundum

Silicon carbide (SiC), better known as carborundum, is almost as hard as diamond and is widely used as an abrasive in powders, pastes and sandpapers. It is produced by heating silicon dioxide and carbon in an electric furnace at 3000°C but without the silica being in excess. It was discovered accidentally in 1891 by E. G. Acheson. Pure silicon carbide is transparent and is used in lasers and gives light of wavelength 456 nanometres, while newer methods of synthesis produce single crystals that may be used for sophisticated equipment such as X-ray mirrors, solar cells and high-temperature transistors.

Quartz crystals

Quartz is pure silicon dioxide. While some of this occurs naturally as rock crystal and rhinestone, that which is used commercially is manufactured. Quartz crystals have a peculiar property of resonating (vibrating) at a very precise frequency when subjected to the pulsation of an electric current, and this is the technology behind the quartz of clocks and watches, which thereby keep exact time. Quartz crystals are also used in radio and television transmitters.

Sodium silicate

There are several forms of this, which are produced by heating soda (sodium carbonate) and sand to 1500°C. The product, a glass-like solid, is then dissolved in water under pressure, and sodium silicates can be crystallized from the resultant solution. They are used in detergents, adhesives and textile bleaching and to enhance oil recovery. By pumping a solution of sodium silicate down an exhausted oil well it is possible to force more oil out of a nearby production well. The alkalinity of the solution neutralizes acidic compounds in the oil-bearing strata and this releases more oil.

ENVIRONMENTAL ELEMENT

Silicon in the environment	
Earth's crust	28%
	Silicon is the second most abundant element, after oxygen.
Soils	the key component
Sea water	30 p.p.b at the surface, 2 p.p.m. in the deeper layers
Atmosphere	traces

The Earth's crust is composed primarily of silicate minerals, which explains why silicon is second only to oxygen in its abundance. The great variety of silicate rocks comes about as a result of the chemical versatility of the silicate ion, $[SiO_4]^{4-}$, which consists of a silicon atom attached to four oxygen atoms. Two of these ions can link together, sharing oxygens, to give disilicates, or several can join together in a chain to give metasilicates, which can take many forms. Two such chains can link oxygens to form amphiboles, or even share three oxygens to form sheets of layered silicates. Minerals of the latter type – such as mica, vermiculite or talc – often have a flaky appearance and their structure accounts, for example, for the smooth feel of talc and its former widespread use in skin care. (For all these silicates, there need to be counterbalancing positive metal ions.)

When all four oxygens of the silicate ion are shared, the result is silica itself, SiO_2, and there is no negative charge because each oxygen atom is bonded to two silicon atoms. Yet even silica can take several forms; for example, quartz has a spiral arrangement, whereas vitreous silica has no long-range order at all. When a tiny amount of aluminium is present, silicates form even more remarkable structures called zeolites, which are three-dimensional networks with large channels and cavities into which small molecules can enter and become trapped, such as the molecular sieves use for drying solvents and gases. Some zeolites act as catalysts and perform seemingly impossible reactions, such as converting waste gases into liquid fuels.

Silicates dissolve slightly in water to form silicic acid, $Si(OH)_4$, and in this way 200 million tonnes of silicates move around the environment in lakes and seas every year. In addition, 4 billion tonnes of silica particles are carried by rivers to the seas and oceans.

Solar panels promise great environmental benefits as a benign way of producing electricity. They consist of a layer of silicon doped with arsenic, a so-called *n*-semiconductor, next to a layer of silicon doped with boron, a *p*-semiconductor. When sunlight, even diffuse sunlight, falls upon such a wafer it provides the energy to cause electrons to flow from the *n*- to the *p*-layer, thereby creating an electric current. Such panels have been used for years to provide power for satellites in orbit, although their efficiency is relatively low, around 15%. Thin-film photovoltaic modules, which consist of amorphous silicon deposited on to a substrate glass of titanium dioxide, can convert sunlight to electricity at a commercially viable rate and such arrays are now generating power at several sites around the world.

CHEMICAL ELEMENT

Data file	
Chemical symbol	Si
Atomic number	14
Atomic weight	28.0855
Melting point	1410°C
Boiling point	2355°C
Density	2.3 kilograms per litre (2.3 grams per cubic centimetre)
Oxide	SiO_2

Ultrapure crystals of silicon have a blue-grey metallic sheen. The element is a member of group 14 of the periodic table of the elements. Bulk silicon is unreactive towards oxygen because a surface layer of silicon dioxide forms, which protects the bulk of the material. It is also unreactive towards water and acids, except hydrofluoric acid, but dissolves in hot alkali.

There are three naturally occurring isotopes of silicon: silicon-28, which accounts for 92%, silicon-29 (5%) and silicon-30 (3%). None is radioactive.

ELEMENT OF SURPRISES

Professor Henry Reid, at Canniesburn Hospital, Glasgow, Scotland, discovered the remarkable power of silicones to reduce scarring in badly burned patients. Applying a dressing coated with silicone gel to badly burned skin speeds up the body's ability to replace scar tissue with new skin, and the process of healing then takes weeks instead of years. The effect of the silicone gel is to increase the water content of the affected skin and so to aid its regeneration.

Pronounced sil-ver, the name is derived from *siolfur*, the Anglo-Saxon for silver. The chemical symbol, Ag, comes from *argentum*, the Latin for silver. Some elements take their name from the country where they originated (e.g. americium) or the homeland of their discoverer (e.g. polonium), and one (rhenium) has even been named after a river. Silver is unique in that the reverse has happened: it has given its name to a country (Argentina) and a river (the River Plate). The Spanish name for the latter was *Rio de la Plata*, river of silver, so called because it was a rich source of silver plundered from the native population in the sixteenth and seventeenth centuries. When the people of that region threw off Spanish rule in the nineteenth century, they changed the name of their country to Argentina, thereby severing the links with its colonial past but preserving the links with its mineral wealth.

French, *argent*; German, *Silber*; Italian, *argento*; Spanish, *plata*; Portuguese, *prata*.

HUMAN ELEMENT

Silver in the human body	
Blood	3 p.p.b.
Bone	varies between 10 and 40 p.p.b.
Tissue	varies between 10 and 250 p.p.b.
Total amount in body	approx. 2 milligrams

Silver has no biological role and indeed is especially toxic to lower organisms. Soluble silver salts irritate the skin and mucous membranes, and can cause death if ingested, although excess hydrochloric acid in the stomach protects the body by precipitating insoluble silver chloride. As much as 90% of ingested silver passes through the gut unabsorbed, but over a 50 year lifespan it has been estimated that a person could well accumulate up to 9 milligrams of silver. Most silver lodges in the liver and skin.

Foodstuffs contain measurable amounts of silver; for example, flour contains about 0.3 p.p.m. and bran about three times this level. Milk has between 25 and 50 p.p.b., beef, pork and mutton around 40 p.p.b. and fish can have as much as 10 p.p.m. Consequently, humans get an average daily intake of silver of between 20 and 80 micrograms, depending on the food they eat, but such tiny amounts pose no threat to health.

MEDICAL ELEMENT

Mediaeval pharmacists sold silver nitrate, which they called lunar caustic, as a remedy for rubbing on, and removing, warts; it actually worked, and was still available in the early twentieth century. Silver compounds became an important part of medical pharmacopoeias in the nineteenth century, taking many forms. In 1884, the German obstetrician, Dr F. Crede, had shown that it was possible to prevent the blindness that afflicted many babies shortly after birth, by using silver nitrate solution eye drops. We now know that this killed the microbial infection that caused the disease. Silver nitrate was also prescribed to treat severe burns, which it did by preventing infection. Silver acetate was given as anti-smoking tablets, effective because of the bitter taste they produced when a cigarette was smoked.

Treatment with silver medicines had its drawbacks. A high intake into the body caused the skin, hair and eyes to take on a grey tinge, a condition known as argyria, but now rarely met with. Apart from this side effect, argyria appeared not to be detrimental in the long term.

The silver ion, Ag^+, is deadly to both bacteria and viruses and silver preparations may return once again to hospitals because they kill antibiotic-resistant strains of infection. A polymer-protected form of silver, called Argleas, has been developed in the UK; it releases its silver at a controlled rate, thereby overcoming the neutralizing effect of the ever-present chloride ions which precipitate unprotected silver as ineffective silver chloride. This new form of silver has received approval from the US Food and Drugs Administration for use in antimicrobial dressings and with catheters.

Colloidal silver is still marketed as an over-the-counter treatment for acne, although one commercial preparation has a silver level of only 30 p.p.m.

ELEMENT OF HISTORY

Slag heaps near ancient mine workings in Turkey and Greece reveal that silver deposits were first mined as long ago as 3000 BC. Silver was known to all the ancient civilizations but, unlike gold, it is rarely found as the natural metal, so that even though it is much more abundant, it did not come into use until later. Indeed, when it first appeared in ancient Egypt, it was more valued than gold.

Silver was refined by cupellation, a process invented by the Chaldeans[141] around 2500 BC and described in the Bible (Ezekiel 22: 17–22). This consisted of heating the molten metal in a shallow, porous cup known as a cupel, over which blew a strong draught of air. This oxidized the other metals, such as lead, copper and iron, leaving only silver (and any gold that was present) as a globule of molten metal.

The rise of Athens, and its remarkable civilization, was made possible partly through its inhabitants' exploitation of local silver mines at Laurium, which they operated from 600 to 300 BC, producing about 30 tonnes a year of the metal. The ore was mainly galena (lead sulfide) but with a silver content of several per cent; this was easy to extract by cupellation. These mines continued to be operated right through the Roman Empire as

[141] They originally inhabited an area that is now southern Iraq.

well, although most Roman silver came from Spain. In the Middle Ages, German mines became important as the main supplier of silver to Europe.

Silver was also mined by the ancient civilizations of Central and South America, but trade in this metal really opened up after the Spanish conquest. Rich deposits were discovered in 1535 at Charcas, Peru, then at Potosí in Bolivia in 1545, and Zacatecas, Mexico in 1548. Production from these sources totalled more than 500 tonnes per year.

In the nineteenth century, silver was found in the USA, especially the Comstock Lode in Nevada, and by 1870 production worldwide amounted to 3000 tonnes a year. It was to continue to grow.

ECONOMIC ELEMENT

Metallic silver occurs naturally as crystals, but more generally as a compact mass; there are small deposits in Norway, Germany and Mexico. The chief silver ores are acanthite (silver sulfide, Ag_2S) and stephanite (silver antimony sulfide, Ag_5SbS_4); the former is more important and is mined in Mexico, Bolivia and Honduras, while stephanite deposits are worked in Canada. Silver is also found as the mineral horn silver (silver chloride, $AgCl$) and silver arsenide (Ag_3As). However, silver is mostly obtained as a by-product in the refining of other metals, such as copper, and techniques for its separation now include froth flotation and electrorefining.

World production of newly mined silver is around 17 000 tonnes per year, of which only about a quarter comes from silver mines. The rest is a by-product of refining other metals: 40% of new silver comes from lead–zinc mining, 22% from copper and 13% from gold. The chief producing countries are Mexico, Peru, USA and Australia, all of which produce more than 1500 tonnes per year. Reserves amount to around 1 million tonnes. In addition to new silver, about 10 000 tonnes of silver is reclaimed per year.

Silver coins were used for thousands of years but had the disadvantage of being easily worn away, despite the hardening effect of alloying the silver with a little copper, and no silver coins are used as circulating currencies today, although some countries continue to mint silver coins as commemorative pieces; the USA annually mints both 50 cent and 1$ denominations, for example.[142] Coins that look like silver are made of nickel–copper alloy.

The major outlets for silver are photography, the electrical and electronic industries and for domestic use as cutlery, jewellery and mirrors.[143] Tableware and trophies are made of the alloy sterling silver, which is 93% silver and 7% copper. Jewellery is more likely to be 80% silver, 20% copper. Electroplated nickel silver, known as EPNS, was once popular as cutlery; as its name implies, it was made by depositing a thin surface layer of silver by means of an electrolytic cell.

Silver salts continue to be a part of photography despite the rise in digital imaging, mainly because traditional methods have the advantages of low cost, ease of use, high-quality images and protection against illegal copying, which is especially important in

[142] The minting of silver coins often ceased because the metal cost more than the face value of the coin.

[143] Freshly deposited silver has the highest reflective ability of all metals.

movie films. The amount of silver going into prints, films and X-ray images totals 8000 tonnes annually and is still increasing.

Both colour and black and white images have relied on silver since the early days of photography. This is because silver bromide and silver iodide are sensitive to light; when light strikes a film coated with one of these compounds, some of the silver ions revert to the metal in the form of tiny nuclei. The film is then developed with a reducing agent which causes more silver to deposit on these nuclei. When the image (negative) has the desired intensity, the unaffected silver bromide or iodide is removed by dissolving in a fixing agent (thiosulfate), leaving the image behind.

Light-sensitive glass works on the same principle except in this case the electron that converts colourless silver ions (Ag^{+}) to metallic silver, and so darkens the glass, has to return whence it came once the sunlight fades. This is done by incorporating copper ions into the glass, and it is these ions which donate and accept the electrons as required.

Silver is widely employed in the electrical industry because of its excellent conductivity. It is the best conductor of heat and electricity of all the metals and hence is widely used for electrical and electronic devices, especially as it makes and breaks electric circuits cleanly and so is used in all electric keyboards. Even the primitive telegraph devices introduced in the 1830s had silver contacts.

Brazing, the soldering and bonding of one metal to another, is best done with silver–tin alloys, sometimes with a little copper or titanium added as well. These have generally replaced the silver–tin–lead alloys formerly used as solders. Such solders are not only used in domestic appliances, but are also a key part of the micro-world of printed circuits and the macro-world of power lines. Silver brazing is also used to stick industrial diamonds to grinding wheels, and it has the added advantage of dissipating more effectively the heat generated during their operation.

Other uses of silver are: in dentistry as silver–tin–mercury or silver–tin–copper–mercury amalgams; as the oxide in high-capacity zinc long-life batteries; and occasionally as the iodide for seeding clouds to promote rainfall.

ENVIRONMENTAL ELEMENT

Silver in the environment	
Earth's crust	70 p.p.b.
	Silver is the 66th most abundant element.
Soils	approx. 0.5 p.p.m.
Sea water	about 0.1 p.p.t. at the surface; about 2 p.p.t in the depths
Atmosphere	traces

Silver levels in soil are not usually high except in mineral-rich areas when they can sometimes be as much as 44 p.p.m. near old workings. Plants can absorb silver and measured levels come in the range 0.03–0.5 p.p.m. (dry weight). Plant ash generally contains some silver but less than 5 p.p.m. Fungi and green algae can have as much as

200 p.p.m. and horsetails, lichens, mosses and some trees will take in more than expected, although only in response to silver being in the soil at enhanced levels.

CHEMICAL ELEMENT

Data file	
Chemical symbol	Ag
Atomic number	47
Atomic weight	107.8682
Melting point	962°C
Boiling point	2212°C
Density	10.5 kilograms per litre (10.5 grams per cubic centimetre)
Oxide	Ag_2O

Silver is a soft metal with a characteristic silver sheen when polished; it is a member of group 11 of the periodic table of the elements. Silver, like gold, is malleable and ductile. It can be beaten so thin as to be almost transparent, and 1 gram of the metal can be drawn into a wire nearly 2 kilometres long. It is stable to water and oxygen but is slowly attacked by sulfur compounds in the air to form a black sulfide layer, which is why silver objects need regular cleaning.[144] Silver dissolves in sulfuric and nitric acids.

Silver occurs as the isotopes silver-107, which accounts for 52%, and silver-109, which accounts for 48%. Neither is radioactive.

ELEMENT OF SURPRISES

1. There are several anecdotes that indicate that silver has long been known to protect drinking water and may explain why silver coins were to be found at the bottom of many wells. It was reported by the Greek historian Herodotus (485–425 BC) that the Persian king, Cyrus the Great, travelled with his own supply of water. This was drawn from a special stream, boiled and then stored in silver vessels. It requires only a little silver to sterilize water: 10 p.p.b. will suffice. Silver is still used to sterilize water and even footwear. Swimming pools can be made safe by adding silver salts to the water rather than by chlorination. Socks for athletes are available in the USA that have silver fibres woven into them to prevent bacteria producing the sulfur molecules that cause smells. Not all microbes are killed by silver. In 1999, the bacterium, *Pseudomonas stutzeri*, was discovered in a Canadian silver mine. This protects itself by depositing the unwanted metal in tiny cavities in its cell walls.
2. Silver is being used to dispose of the stockpile of chemical weapons. An electrochemical cell has been devised by a British company, AEA Technology, which uses

[144] This is the reason why silverware should never come into contact with high-sulfur foods like egg, onion, garlic or mustard. A sulfide layer does not protect the rest of the silver, so staining can go deep. Some electroplated silverware is protected with a coating of rhodium.

Pronounced soh-di-uhm, the name is derived from the English word soda, although it has been said that its discoverer, Humphry Davy, may have named it after sodanum, a headache cure. The chemical symbol, Na, comes from the Latin *natrium*, meaning soda.

French, *sodium*; German, *Natrium*; Italian, *sodio*; Spanish, *sodio*; Portuguese, *sódio*.

COSMIC ELEMENT

The Sun and many other stars shine with visible light in which the yellow component dominates and this is given out by sodium atoms in a high-energy state.

HUMAN ELEMENT

Sodium in the human body	
Blood	plasma, 3500 p.p.m. (0.35%); red cells, 250 p.p.m.
Bone	10 000 p.p.m. (1%)
Tissue	varies between 2000 and 8000 p.p.m. (0.2–0.8%)
Total amount in body	100 grams

Sodium is essential for animals but less important for plants, and indeed these generally contain relatively little. Sodium in the body is mainly in the fluid outside cells, as the table above shows for blood, the situation being just the opposite to that which pertains with potassium (see p. 333). Blood needs a lot of sodium to regulate osmotic pressure and blood pressure, as well as to help solubilize proteins and organic acids. From the dawn of life, cells had to struggle to keep sodium out in the face of an osmotic pressure from a marine environment in which it was abundant.

Sodium's most important function is the movement of electrical impulses along nerve fibres, and this it does in conjunction with potassium, with both sodium and potassium ions being able to move through the cell membrane. Sodium ions move in and are then quickly expelled again by the 'sodium pump' while potassium ions move out and are then carried back. As much as 40% of a body's energy goes into operating the sodium pump.

It is necessary to take in a regular supply of sodium because it is continually lost from the bloodstream as it is filtered by the kidneys. Although a lot of sodium is recycled, much is lost in urine (which contains around 350 p.p.m.), faeces and sweat.

Sodium compounds are not hazardous insofar as their sodium content is concerned, but excess salt[145] (sodium chloride, NaCl) can provoke a toxic response, such as vomiting.

FOOD ELEMENT

The amount of sodium a person consumes each day varies from individual to individual and from culture to culture; some people get as little as 2 grams per day, some as much as 20 grams. Sodium may be essential but controversy surrounds the amount required – see 'Medical element' below.

The sodium that comes with a diet of fresh foods is equivalent to around 3 grams of salt a day and is all that a person requires. Foods with more than 100 milligrams of sodium per 100 grams are tuna, sardines, eggs, liver, butter, cheese and pickles. Vegetables generally have very little sodium, but celery and peas have quite a lot. The average intake of salt in a Western-style diet is about 9 grams a day, with 3 grams coming from processed foods, such as cereals and snacks, and 3 grams from table salt. A Japanese person can have double this amount because of the prevalence of seafood in the diet.

Sodium chloride triggers a specific reaction on the tongue and is one of the four basic taste sensations. This is rather puzzling because no other mineral component in the diet provokes such a response. Nor is it only humans who experience a craving for salt; animals will lick salt cakes if they are available. It has even been suggested that the relatively low level of sodium in plants may make this element a limiting factor to animal life on land.

Salt is called rock salt when it comes from mineral deposits, and sea salt when it is obtained by evaporating sea water. Both are refined by redissolving them in fresh water and evaporating the filtered solution until the salt recrystallizes as pure sodium chloride which is sold to food manufacturers and is available in supermarkets as cooking salt. So-called 'table salt' has a little magnesium carbonate added to keep it free-flowing.

MEDICAL ELEMENT

Sodium deficiency in animals and humans causes muscular spasms, popularly known as cramp, or charley horse in the USA. A sudden excess of sodium, in the form of salt solution, has traditionally been used as an emetic, but continued high levels of sodium in the diet lead to long term risks.

A person with heart trouble or kidney disease will be advised by their doctor to cut down on sodium. Heart and kidney malfunctions result in the body being less able to remove excess salt and so it compensates by retaining more water. This causes

[145] In chemistry, the term 'salt' covers a whole group of compounds, namely those consisting of positive and negative ions, but in ordinary speech salt refers specifically to one kind of salt, that of sodium chloride, Na^+Cl^-, and this is how it will be used in this section.

increased pressure on the arteries, resulting in high blood pressure with all the risks to health that this implies.

There are three grades of reduced salt diet, which are referred to as low-salt, no-salt, and salt-free. A low-salt diet means no table salt or salt used for cooking. Foods high in salt, such as potato snacks and certain cheeses, are also not allowed. Such a diet still provides about 6 g of salt a day, which comes from things like breakfast cereals, biscuits, cheese, and processed foods. A no-salt diet cuts out most of these and limits bread to three slices per day, so that salt intake is around 3 g a day. A salt-free diet excludes all salt except that which comes as a natural part of the food we eat, which may be less than 2 g per day; a patient on this diet eats mainly foods which have very little sodium anyway, such as boiled rice.

SALT AND HEALTH

The link between salt intake and the health of the general public has proved to be a controversial subject, with some believing excess salt causes illness, while others say it has no effect. In 1992, an analysis of several surveys, which together covered 47 000 people, was conducted by a group at St Bartholomew's Hospital, London; they found an epidemiological relationship between salt intake and high blood pressure. Graham MacGregor of St George's Hospital Medical School, London, came to the same conclusion, claiming that excess salt was the indirect cause of more than 30 000 deaths in the UK every year.

In 1996, the *Journal of the American Medical Association* published a report from the Mount Sinai Hospital, Ontario, which came to a very different conclusion. This group, led by Alexander Logan, found that salt posed no threat at all for people whose blood pressure was normal. Those for whom a restricted sodium intake was advisable were older people who already had high blood pressure. David McCarron of the Oregon Health Sciences University came to the same conclusion in 1998, while another survey, chronicling the diets of 11 000 US citizens from 1970 to 1990, even concluded that those who ate more salt tended to live longer. The longevity of the Japanese, which exceeds all other groups, also seems to bear this out.

Salt is a blessing in tropical countries, where it saves millions of lives. Diarrhoea, and the resulting dehydration it causes, are reported to kill more than 10 million children a year. The answer to this potentially fatal illness is to drink a solution of glucose and salt, and the United Nations children's organization, UNICEF, distributes millions of sachets for making up such solutions. They contain 20 grams of glucose, 2 grams of salt, 3 grams of sodium citrate and 1.5 grams of potassium chloride. For use, these ingredients are dissolved in a litre of boiled water. The same treatment can also help those suffering from cholera, which kills its victim by blocking the ion channels in the intestines that absorb water from the food in the gut. The result is severe dehydration and often death.

ELEMENT OF HISTORY

Sodium carbonate (Na_2CO_3; also known as soda or, industrially, as soda ash) is mentioned in the Bible as a cleaning agent (Jeremiah 2: 22) although there it is called nitre, a name that eventually came to mean another salt, potassium nitrate. In the ancient world, sodium carbonate was harvested in Egypt where it came from the Natron Valley in which flood water from the Nile gathered each year and which yielded a crop of soda crystals as it dried out. Later a surface deposit of trona, another form of sodium carbonate, was located in the Sahara, from where it was transported to Tripoli and exported.

The demand for sodium carbonate came partly from its use as a washing aid ('washing soda') and partly because it was needed to manufacture glass. In many areas it was produced from plant ash, especially that from certain types of fern which have high sodium levels. However, even these could not meet the demand created by the Industrial Revolution and, in 1783, Nicolas Leblanc devised a method for making soda from salt. Earlier in the eighteenth century, Henri-Louis Duhamel du Moncease (1700–1782) had shown that salt could be turned into soda by dissolving it in nitric acid, boiling to dryness and then heating with charcoal. Leblanc's process used cheaper sulfuric acid to form sodium sulfate which was then heated with limestone (calcium carbonate) and charcoal.

Even at the start of the nineteenth century, the elemental composition of salt and soda remained a mystery. The great chemist Antoine Lavoisier even thought salt might contain nitrogen. The answer came from the Royal Institution in London in October 1807 when Humphry Davy reduced caustic soda (sodium hydroxide, NaOH) to sodium. He had recently produced metallic potassium by the electrolysis of potassium hydroxide and he next turned his attention to caustic soda from which he obtained metallic globules of sodium. He reported that the electrolysis of caustic soda required a much higher current from his battery than that needed to generate potassium.

The following year, Joseph Gay-Lussac and Louis-Jacques Thénard made sodium by heating caustic soda and iron filings to red heat. Later that century, sodium was produced from sodium carbonate and carbon at 1100°C, and made commercially by the electrolysis of sodium hydroxide, the method used by Davy.

ECONOMIC ELEMENT

Sodium occurs in many silicate minerals but none is exploited solely for the sodium it contains. Those deposits which are extracted with this end in view are halite (rock salt) and trona, which is sodium carbonate bicarbonate,[146] of composition $Na_3(CO_3)(HCO_3)$. The main mining areas for halite are Germany, Poland, USA and UK; and those for trona are Kenya and USA. A newly discovered mineral deposit in Colorado is currently being developed; this is nahcolite, which is almost pure sodium hydrogen carbonate.

World production of salt is around 200 million tonnes per year; this huge amount is mainly extracted from salt deposits by pumping water down bore holes to dissolve it and pumping up brine. A growing amount of salt is now being produced as a renewable

[146] The correct name for bicarbonate is hydrogen carbonate; its formula is HCO_3^-.

resource from sea water, something that has been done on a small scale in many parts of the world for centuries. At Dampier, on the North-West coast of Australia, there are vast evaporation ponds which yield 5 million tonnes of salt from sea water each year.

Salt is used mainly by the chemical industry – 60% ends up there – where it is turned either into chlorine gas and sodium hydroxide, or into sodium carbonate. About 20% of salt goes to the food industry where it is used as a preservative and flavouring agent, and the remaining 20% finds various outlets, such as in de-icing roads in winter.

Sodium hydroxide is the main alkali of the chemical industry and it also finds domestic use in unblocking drains. Sodium carbonate production amounts to around 30 million tonnes per year; whereas formerly this was produced from salt and limestone ($CaCO_3$) in the Solvay process, developed by Ernest Solvay in Belgium in 1861, much now comes from the vast natural deposits found in the USA, which represent a reserve of almost a billion tonnes.

Solid sodium carbonate is needed to make glass; almost half ends up this way, while the rest is used as a feedstock for the chemical industry, to make detergents or in metallurgy. A solution of sodium carbonate is the main acid-neutralizing agent in water treatment, and it is added to fizzy drinks to boost the solubility of carbon dioxide. Sodium hydrogen carbonate also acts as a leavening agent in baking, as well as being used in the textile and leather industries and for making soaps and detergents. Fire extinguishers which produce a jet of water or foam do so by generating carbon dioxide gas from the action of sulfuric acid on sodium hydrogen carbonate.

Sodium metal is commercially important and is produced in so-called Downs cells wherein molten sodium chloride, mixed with calcium and barium chlorides, undergoes electrolysis. The other chlorides are there to lower the melting point of sodium chloride from 808°C to 600°C.[147] The molten salts are contained in a ceramic-lined vessel fitted with a graphite anode and a steel cathode, from which the molten sodium rises and can be skimmed off. As the molten metal cools it releases a few per cent of calcium which crystallizes out and is filtered off, leaving molten sodium of 99.9% purity. World production of sodium metal is about 80 000 tonnes per year.

Sodium metal is used by industry in the manufacture of various chemicals and for extracting metals, namely beryllium, thorium, titanium and zirconium which it does when it is heated with their halide salts. Other chemicals are produced using sodium metal. These include:

- lead tetraethyl and tetramethyl, which are still used as an antiknock agent in gasoline (albeit in declining amounts);
- sodium borohydride ($NaBH_4$), used for bleaching paper pulp;
- sodium azide (NaN_3), which goes into the air-bags in cars; and
- sodamide ($NaNH_2$), used in dye manufacture.

Sodium metal is also used in street lighting, where a tiny piece is sealed in neon-containing lamps. Such bulbs use less electricity and their light penetrates mist and fog better than other forms of street lighting. When the bulb warms up, the sodium vapour is excited and emits a characteristic yellow light of wavelength 589 nanometres.

[147] The composition of the molten salt electrolyte is 28% NaCl, 26% $CaCl_2$ and 46% $BaCl_2$.

ENVIRONMENTAL ELEMENT

Sodium in the environment	
Earth's crust	23 000 p.p.m. (2.3%)
	Sodium is the sixth most abundant element.
Soils	major component
Sea water	10 500 p.p.m. (1.05%)
Atmosphere	traces

Sodium has been leaching from the Earth's rocks and soils into its oceans for billions of years, and once there it remains for hundreds of millions of years because it is so soluble. Sodium draining to inland seas that eventually dried up accounts for the vast rock salt and trona deposits that are to be found around the world.

CHEMICAL ELEMENT

Data file	
Chemical symbol	Na
Atomic number	11
Atomic weight	22.989770
Melting point	98°C
Boiling point	883°C
Density	0.97 kilograms per litre (0.97 grams per cubic centimetre)
Oxides	Na_2O; also Na_2O_2 and NaO_2

Sodium is a soft, silvery-white, metal which is a member of group 1, also known as the alkali metal group, of the periodic table of the elements. It oxidizes rapidly when cut and has to be stored under paraffin. When it burns in air the main oxide it forms is sodium peroxide, Na_2O_2. The metal reacts vigorously with water to evolve hydrogen gas, which sometimes ignites. On the other hand, sodium dissolves in liquid ammonia to form a deep blue solution which gradually becomes metallic-looking as the amount of sodium increases; under such conditions it is believed that negative sodium ions, Na^-, are present. Normally the chemistry of sodium is that of the unreactive sodium ion, Na^+, whose salts are almost all very soluble in water.

The only naturally occurring isotope of sodium, sodium-23, is not radioactive. Other isotopes are made in nuclear reactors and are radioactive. One, sodium-24, with a half-life of 15 hours, is used in medical research.

ELEMENT OF SURPRISE

Liquid sodium metal is used in heat exchangers in some nuclear reactors. Potentially it is highly dangerous because sodium heated to above 125°C bursts into flames when exposed to the air, and the temperature quickly rises to 800°C whereupon the flames become almost impossible to extinguish. Yet the French Superhenic breeder reactor is cooled with almost 6000 tonnes of liquid sodium.[148] Researchers there have developed a special sodium fire extinguisher called Marcalina, which is composed of lithium and sodium carbonates and graphite. The liquid sodium coolant is contained in an inner loop round the reactor, but this tends to accumulate radioactive isotopes sodium-22 (half-life 2.6 years) and sodium-24 (half-life 15 hours). Consequently, a second loop of liquid sodium is used to transfer the heat from the inner loop to the exterior of the reactor.

[148] France generates a larger proportion (75%) of its electricity from nuclear energy than any other country.

Pronounced stron-tee-uhm, this element is named after the small town of Strontian in Scotland.

French, *strontium*; German, *Strontium*; Italian, *stronzio*; Spanish, *estroncio*; Portuguese, *estrôncio*.

HUMAN ELEMENT

Strontium in the human body	
Blood	30 p.p.b.
Bone	35–140 p.p.m.
Tissue	120–350 p.p.b.
Total amount in body	320 milligrams

Certain deep-sea creatures incorporate strontium into their shells as strontium sulfate, and stony corals require it, which is why it needs to be added to the water in aquariums. Although strontium has no function in the human body, it is present in sizeable amounts, resembling rubidium and bromine in this respect, and like them it is absorbed because of its similarity to another essential mineral, in this case calcium, which is why most ends up in the skeleton. It is not regarded as toxic.

FOOD ELEMENT

Foods containing strontium range from the very low, e.g. in corn (which has only 0.4 p.p.m.) and oranges (0.5 p.p.m.), to the high, e.g. in cabbage (which has 45 p.p.m.), onions (50 p.p.m.) and lettuce (74 p.p.m.) (all dry weights). Strontium-90 was found to concentrate in plants such as beans, but not in grain crops.

Normally it does not matter that we take in some strontium in our diet (about 2 milligrams a day) but it does matter when this is the radioactive isotope, strontium-90 – see below.

MEDICAL ELEMENT

Strontium-87m[149] is a radioactive isotope with a half-life of 2.8 hours, which is used for diagnostic purposes in medical research. It quickly disappears from the body, and within a day less than 0.3% of the dose remains.

The isotope strontium-89 is a β-emitter. It is employed as a radiopharmaceutical and was approved for use in the USA in 1995.

ELEMENT OF HISTORY

In 1787, a dealer in mineral specimens in Edinburgh, Scotland, was offered a newly-discovered specimen that had been found in a lead mine at Strontian, on the West coast. He called it 'aerated baryta' because he thought it was a barium mineral. It aroused the curiosity of a local doctor, Adair Crawford (1748–95), who was interested in baryta (barium oxide) as a possible medicament, but he realized that the two were not related and that the mineral was a new 'earth' which he named strontia (strontium oxide) after the name of the mineral, which was strontianite. He published a paper on his research in 1790 and this prompted Thomas Charles Hope of Edinburgh to begin a fuller investigation the following year. Hope proved conclusively that strontia contained a 'new' element and noted that it caused the flame of a candle to burn red, whereas baryta gave a green colour.[150]

In 1799, another strontium mineral was discovered in Sodbury, Gloucestershire, England, where the locals were using it as gravel for paths in ornamental gardens. This was strontium sulfate and the mineral was named celestite.

Strontium metal itself was isolated in 1808 by Humphry Davy in London by the same process by which he had recently isolated barium and calcium: i.e. the electrolysis of a mixture of the chloride and mercury oxide.

ECONOMIC ELEMENT

The chief ores are celestite (also known as celestine, $SrSO_4$) and strontianite (strontium carbonate, $SrCO_3$). The main mining areas are UK, Mexico, Turkey and Spain. World production of strontium ores is about 140 000 tonnes per year from an unassessed total of reserves.

Strontium metal is produced by heating strontium oxide and aluminium under vacuum conditions. These react to form strontium and aluminium oxide and the strontium is distilled off and collected.

Demand for the metal is limited; a little is used as a 'getter' in vacuum tubes to remove the last traces of air. Most strontium is used as the carbonate in special glass for

[149] The 'm' stands for metastable and signifies that this is an isomer of strontium-87. Just as the electrons in the orbits around an atom's nucleus can be in an excited state, and give off light as they return to their ground state, so the subatomic particles in the nucleus can be in an energetically excited state, referred to as an isomer.

[150] In fact this red colour had been noted by others soon after the mineral was discovered.

television screens and visual display units. It is best known for the brilliant reds that can be produced by strontium salts in fireworks and warning flares. In this respect strontium is unique and cannot be substituted for.

Although strontium-90 is a dangerously radioactive isotope, it is a useful by-product of nuclear reactors from whose spent fuel it is extracted. Its high-energy radiation can be used to generate an electric current, and for this reason it can be used in space vehicles, remote weather stations and navigation buoys. Some is used in gauges for measuring the thickness of paper, plastic film, fabrics and paints. The high electron flux it emits can be used to neutralize static electric charges of the kind that build up on machinery handling paper, plastics and fibres.

ENVIRONMENTAL ELEMENT

Strontium in the environment	
Earth's crust	370 p.p.m.
	Strontium is the 16th most abundant element.
Soils	approx. 200 p.p.m., range 18–3500 p.p.m. (0.35%)
Sea water	8 p.p.m.
Atmosphere	trace

Generally, strontium becomes immobilized in the environment by precipitation as strontium carbonate, or incorporated into invertebrate shell material. The highest levels of strontium are found in deserts and the soils of forests.

Strontium-90 caused a major worldwide pollution concern in the mid-twentieth century, being produced by above-ground nuclear explosions which contaminated the whole planet with it. The majority of these nuclear tests took place between 1945 and 1963. Strontium-90 is a β-emitter with a half-life of 29 years, and is a serious threat because it is one of the most powerful emitters of ionizing radiation and therefore capable of causing serious damage to dividing cells.[151] Its presence was detected in the milk teeth of infants in the 1950s, showing how prevalent it had become, having been washed out of the atmosphere on to grassland, to be eaten by cows, and so end up in milk and other dairy products. Those children have carried it as a burden for the rest of their lives.

[151] It is a pure β-emitter.

CHEMICAL ELEMENT

Data file	
Chemical symbol	Sr
Atomic number	38
Atomic weight	87.62
Melting point	769°C
Boiling point	1384°C
Density	2.5 kilograms per litre (2.5 grams per cubic centimetre)
Oxide	SrO

Strontium is a silvery, white, relatively soft metal, which is a member of group 2, also known as the alkaline earth group, of the periodic table of elements. The bulk metal is protected to a certain degree by an oxide film, but even so it has to be stored under paraffin. The metal will burn in air if ignited, and is attacked by water.

There are four naturally occurring isotopes of strontium: strontium-88, which comprises 82.5%; strontium-86, 10%; strontium-87, 7%; and strontium-84, only 0.5%. None is radioactive. Strontium-87 is the product of radioactive decay and derives from rubidium-87, so that the ratio of strontium-87 to the other isotopes varies slightly with location.[152]

ELEMENT OF SURPRISE

Crystals of strontium titanate ($SrTiO_3$) out-sparkle even diamond because they have a higher refractive index. However, unlike diamond, the surface of these crystals is easily scratched, so strontium titanate has not been widely used as a gemstone, although it has been used in optical devices.

[152] There are also traces of strontium-89 and strontium-90 as uranium fission products in uranium ores.

Pronounced sul-fer, the name may be derived from the Sanskrit *sulvere* or the Latin *sulfurium*, both ancient names for sulfur.

French, *soufre*; German, *Schwefel*; Italian, *solfo*; Spanish, *azufre*; Portuguese, *enxofre*.

Because sulfur has been known for so long, there is a word for it in almost every language. These names are usually very different from its chemical name, and include 'brimstone' in English, *rikki* in Finnish, *ken* in Hungarian, *iwo* in Japanese, *liu huang* in Chinese, *gundhuk* in Hindi, *whanariki* in Maori and *isibabule* in Zulu.

Chemists use the prefix 'thio' to denote the presence of sulfur in a molecule; for example, thioether is ether ($C_2H_5OC_2H_5$) with a sulfur instead of the expected oxygen ($C_2H_5SC_2H_5$). 'Thio' comes from the Greek *theo*, meaning 'of god', and this link is thought to arise because sulfur was a component of incense.

COSMIC ELEMENT

Sulfur is found in some meteorites and there appears to be a deposit of sulfur on the surface of the Moon, near the crater Aristarchus. The Earth's iron core is believed to contain about 5% sulfur. Io, one of the moons of Jupiter, has a surface covered with sulfur and has active volcanoes that spew it forth. It has been suggested that sulfur-using bacteria might live there, of the kind that survive in sulfur-rich environments on Earth.

Life on Earth may have been possible because of sulfur. Conditions in the early seas were such that simple chemical reactions could have generated the range of amino acids that are the building blocks of life. (Some microorganisms, which live near hydrothermal vents in the sea, still rely on reduced forms of sulfur, such as hydrogen sulfide (H_2S), as an energy source.) Experiments have shown that reactions between hydrogen sulfide, ammonia and simple carbon-containing molecules, such as formaldehyde, will yield a wide range of amino acids.

Sulfur tells us about the development of the Earth's atmosphere. Sedimentary rocks of the Precambrian period reveal the change that occurred around 2.4 billion years ago, before which there were only traces of oxygen. By 2.1 billion years ago the levels of oxygen were substantial, as shown by the sulfites and sulfates in the rocks laid down during this period. The ratio of sulfur isotopes reveals the geological processes and biological fractionations that were then in operation.

HUMAN ELEMENT

Sulfur in the human body	
Blood	1800 p.p.m. (0.18%)
Bone	varies between 500 and 2400 p.p.m. (0.05–0.24%)
Tissue	varies between 5000 and 12 000 p.p.m. (0.5–1.2%)
Total amount in body	140 grams

All living things need sulfur. It is especially important for humans because it is part of the amino acid methionine, which is an absolute dietary requirement for us (see below). The amino acid cysteine also contains sulfur. Cysteine is used to link peptide chains by forming sulfur-to-sulfur bonds between them; these bonds are abundant in the single-stranded protein keratin, which is the main constituent of hair and nails.

Elemental sulfur is not toxic; but many simple sulfur derivatives are, such as sulfur dioxide (SO_2) and hydrogen sulfide.

Dietary sulfur

The average person takes in around 900 milligrams of sulfur per day, mainly in the form of protein. Not all protein foods contain the essential amino acid, methionine, but all fish and meats contain it at levels up to 4%. Soya beans contain methionine, but many plant proteins – such as those of peanuts, lentils and kidney beans – are deficient in this amino acid.

Some sulfur in food and drink comes in the form of preservatives based on sulfur dioxide or its derivatives, the sulfites. Sulfur dioxide kills bacteria, is a good anti-oxidant and prevents food from browning. It is used on a large scale as a preservative, both as itself or as its sodium, potassium and calcium sulfite salts which form sulfur dioxide in solution. It is widely used with fruits and vegetables because it preserves their natural colour; for example, peeled potatoes stay white in its presence.

When grapes are crushed, the juice will spontaneously ferment due to the action of wild yeasts on the surface of the fruit. As little as 100 p.p.m. of sulfur dioxide is enough to prevent undesirable yeasts from multiplying while allowing desirable yeasts to flourish. These yeasts are often added as specially cultured yeasts and can survive high levels of sulfur dioxide.

More sulfite may be added to wines just before bottling, to prevent further fermentation, and this can result in levels of sulfur dioxide as high as 350 mg per bottle. Most of this sulfur dioxide eventually reacts with other components in the wine and disappears, but some young white wines have noticeable amounts of sulfur dioxide – up to 50 milligrams per bottle – which can cause problems (see 'Medical element').

Sulfur plays a part in other dietary components. Vitamin B_1, also known as thiamine, has a ring structure incorporating a sulfur atom as part of its molecule. Deficiency of this vitamin leads to the disease beriberi. The co-enzyme biotin also contains sulfur, as do natural flavour agents that give a characteristic aroma to cooked lamb, passion fruit, roasted coffee and peanuts. The flavour (and smell) of garlic and onions arises from the sulfur-containing chemicals di-2 propenyldisulfide (in garlic) and thiopropanal *S*-oxide (in onion).

Sulfur is also a component of the synthetic sweeteners, saccharin, cyclamate and acesulfam. In each case it is bonded to a nitrogen atom, a combination that is critical to activating the sweetness receptors of the tongue. These molecules pass through the body undigested.

MEDICAL ELEMENT

The use of sulfur in medicine goes back to prehistoric times and this element continued to be prescribed for nearly 3000 years. A popular remedy for digestive upsets was brimstone and treacle; this laxative worked by the sulfur being converted to alkali sulfides in the intestine, and these acting as an irritant. Sulfur was usually taken as tablets, the dose being 1–4 grams. Other sulfur preparations included ointments mixed with wool fat, for external application to treat acne or scabies, a parasitic infection. Sulfur was also applied as a solution in organic solvents, such as chloroform or ether.

Sulfur medicaments are no longer used, although sulfur atoms are present in some modern drugs, such as penicillin, a natural sulfur compound. Its forerunners, the sulfa drugs, started life as dyes but were found to have antibiotic properties when tested on rabbits and were then used on humans.

Sulfur dioxide gas is easily recognized by its choking smell, which irritates the nose and lungs. In those who are sensitive to it, it may induce release of histamines, which can trigger asthma attacks.

Halitosis (bad breath) is caused by three molecules, all derivatives of sulfur. They are hydrogen sulfide (H_2S), methyl mercaptan (CH_3SH) and dimethyl sulfide ($(CH_3)_2S$). The amounts of these gases which we can detect are very small: hydrogen sulfide can be detected at only 5 micrograms per litre of air[153] and dimethyl sulfide at even less, 1 microgram per litre, but methyl mercaptan is the world's smelliest molecule, requiring only 0.02 micrograms to register with the nose. These gases are produced in the mouth from bacteria acting on cysteine and methionine in food particles or even attacking the lining of the mouth itself.

Metalloproteins, i.e proteins which bind metals, rely on sulfur atoms to attach themselves to essential metals such as iron, copper, zinc and molybdenum. They also bind to harmful metals such as cadmium, which can then be safely locked away in the liver to await disposal.

ELEMENT OF WAR

Sulfur was probably a component of Greek fire, the liquid combustible agent which the Byzantine Empire used successfully for hundreds of years and which terrorized opponents. It was said to set the sea on fire and was usually projected at attacking ships. It was reputed not be extinguishable once it had started to burn. The exact composition

[153] Hydrogen sulfide is said to smell of rotten eggs and indeed these give off this gas, which is formed by bacteria decomposing the amino acid methionine in the yolks.

of Greek fire remained a closely guarded secret, and no record now exists of how it was made. It is thought to have been sulfur dissolved in a mixture of hydrocarbons, these being distilled from rock-oil, the black tarry material that oozed from the ground in the region between the Black and Caspian Seas.

Gunpowder, which was also made from sulfur, was discovered by the Chinese in around AD 950. Its recipe was a closely guarded secret but within 250 years was known in Europe. Gunpowder became widely used after Roger Bacon (1214–1292), an English Franciscan monk, revealed a recipe for it: charcoal, sulfur and potassium nitrate in the ratio of 5:5:7 parts by weight; these had to be finely ground together. Today the blend is charcoal (15%), sulfur (10%) and potassium nitrate (75%). Gunpowder only explodes if it is ignited in a confined space where the carbon dioxide and sulfur dioxide gases produced in the chemical reaction expand rapidly and can be used to project a bullet or cannon ball. Gunpowder is still the basis of firecrackers.

The mustard gas of World War I was a sulfur compound, *bis*-(2-chloroethyl)sulfide, which was a blistering agent as well as giving off a highly toxic vapour. Curiously, the antidote developed to protect troops against arsenical war gases is also a sulfur compound, known as British anti-lewisite (BAL). This is still used to treat people suffering from heavy metal poisoning.

ELEMENT OF HISTORY

Sulfur is mentioned 15 times in the Bible, and was best known for destroying the twin cities of Sodom and Gomorrah. It was also known to the ancient Greeks: in Homer's *Odyssey* the hero Odysseus says: 'Bring me sulfur, old nurse, that I may purify the house.' This is a reference to the practice of burning sulfur as a fumigant.

The Romans mined sulfur in Sicily and used it for making sulfur matches (which are easily lighted by touching them against a glowing ember). It was also used for bleaching cloth and making wine, which uses were achieved by burning it, to form sulfur dioxide, and allowing this to be absorbed by the wet cloth or the grape juice. (These uses continued up to the 20th century, and while bleaching of cloth is no longer done in this manner, wine is still protected by the use of sulfur dioxide.)

Sulfur, along with mercury, was believed to be a component of all metals, a theory put forward by the famous alchemist Jabir. It was therefore believed that, given the right conditions, it should be possible to transmute one metal into another. The result of this belief was centuries of futile attempts by alchemists to find the fabled 'philosopher's stone' that would turn base metals to gold.

In the eighteenth century, Antoine Lavoisier thought that sulfur was an element, but in 1808 Humphry Davy investigated a sample of sulfur and found it contained hydrogen, which appeared to put paid to the idea. However, his sample was impure and, when Joseph Gay-Lussac and Louis-Jacques Thénard proved it to be an element the following year, Davy obtained a better sample and convinced himself that it truly was an element.

ECONOMIC ELEMENT

Sulfur occurs naturally near volcanoes. This is why Sicily supplied much of the sulfur for the ancient world. Its sulfur mines were still employing more than 50 000 men in the nineteenth century. The earliest way of processing sulfur was to spread the mineral on sloping slabs and set fire to it at the top. As it burned it melted the sulfur below, which then ran off and was collected, but the process wasted about half the sulfur.

Native sulfur occurs naturally as massive deposits in Texas and Louisiana in the USA. The Louisiana deposit was discovered in 1865, during drilling for oil. This thick layer, at a depth of 150 metres, extended into Texas. Extraction proved difficult until Herman Frasch (1851–1914) studied the problem in the 1890s. He decided that the only way to get the sulfur out was to melt it and pump it to the surface as a liquid. Sulfur melts at 113°C, so Frasch decided to use superheated steam for the purpose. This was pumped down into the seam of sulfur and up came molten sulfur. This method of extraction became known as the Frasch process.

Many sulfide minerals are known: pyrite and marcasite are iron sulfides; stibnite is antimony sulfide; galena is lead sulfide; cinnabar is mercury sulfide; and sphalerite is zinc sulfide. Other, more important, sulfide ores are chalcopyrite, bornite, penlandite, millerite and molybdenite. A lot of sulfur occurs as sulfates, the most abundant ones being anhydrite and gypsum, which are calcium sulfates, and epsomite and kieserite, which are magnesium sulfates. Other sulfate minerals are barite (barium sulfate) and celestite (strontium sulfate).

The chief source of sulfur for industry is the hydrogen sulfide of natural gas (Canada being the main producer). It is imperative that the hydrogen sulfide be removed from natural gas before use. Removal is achieved by spraying the gas with a solvent, such as ethanolamine, in which hydrogen sulfide is soluble. This gas is then reclaimed and reacted with a measured supply of oxygen, which reacts to form water and leaves the sulfur behind. The sulfur can be used as a fuel in its own right or as chemical feedstock.

Sulfur comes also from mined sulfur and from pyrites (FeS_2), the roasting of which at 1200°C in the absence of air releases about half its sulfur. World production of sulfur exceeds 50 million tonnes a year, and reserves are of the order of 2.5 billion tonnes.

Sulfur is a key industrial chemical and the raw material for making sulfuric acid. Other applications for sulfur are in vulcanizing rubber, in making corrosion-resistant concrete which has greater strength and is frost resistant, in gunpowder and fireworks, for solvents (dimethyl sulfoxide and carbon disulfide) and in a host of other products of the chemical and pharmaceutical industries.

Dusting plants with powdered sulfur was introduced in the nineteenth century to control fungal diseases, especially powdery mildew. A solution of calcium polysulfide was also used as a pesticide spray.

Charles Goodyear (1800–60) accidentally discovered the benefits of reacting sulfur with natural rubber, thereby converting this latex polymer from a sticky, soft material of limited use, to something of wide application suitable for hundreds of uses, but especially tyres for vehicles.[154]

---

[154] The term 'vulcanization' was suggested because Goodyear made his discovery when he placed a sample of rubber mixed with sulfur on a hot stove, so his friend William Brockedon suggested he called the new material after Vulcan, the Roman god of fire.

Methyl mercaptan (also known as methane thiol) is the simplest member of a series of compounds in which there can be chains of up to 20 carbon atoms attached to sulfur. Mercaptans with three and four carbons are those we encounter when we smell a leak of gas because these are added purposely to give the gas a detectable odour. The mercaptan with 18 carbons is a wax used in silver polishes.

Methyl mercaptan is used in the production of pesticides, especially for weedkillers, and also to make methionine for fortifying animal feeds to increase the amount of this amino acid in their meat and milk.

Most sulfur ends up as sulfur dioxide gas, which is made in various ways: some is made by burning sulfur or hydrogen sulfide or by roasting metal sulfides in air, and some is even reclaimed from waste sulfuric acid and sulfates. Most sulfur dioxide goes into making sulfur trioxide, by heating it at 420°C with oxygen in the presence of a catalyst such as vanadium(V) oxide, and thence to sulfuric acid.

Sulfuric acid is produced on such a large scale as to make it the world's number one industrial chemical. It is needed for many other processes, the major ones being the production of phosphate fertilizers and the removal of rust from iron and steel (known as 'pickling') prior to their being painted. The chemical industry uses sulfuric acid in the production of explosives, fuels, rayon, Cellophane, paints, enamels, paper, dyestuffs, antifreeze and detergents. The original surfactants that were introduced into detergents in the 1940s were derivatives of sulfuric acid (H_2SO_4) in which one of the hydrogens was replaced by a hydrocarbon group and the other hydrogen was neutralized as the sodium salt. These surfactants made much better cleaning agents than traditional soaps because they did not form soap scum in hard water.

ENVIRONMENTAL ELEMENT

Sulfur in the environment	
Earth's crust	260 p.p.m.
	Sulfur is the 17th most abundant element.
Soils	major component
Sea water	870 p.p.m.
Atmosphere	various gases, totalling about 1 p.p.b.

There are many sulfur-containing compounds in the air. The most abundant one is carbon oxygen sulfide (COS) which comes mainly from swamps and resists oxidation. Its concentration is 0.5 p.p.b. Sulfur dioxide, on the other hand, is emitted in vast quantities but is rapidly washed out of the atmosphere and its concentration is 0.2 p.p.b. Other gases are hydrogen sulfide, carbon disulfide, which also come from swamps, and dimethyl sulfide, which comes from the oceans. These are easily oxidized and washed out of the air, so their levels are only about 0.01 p.p.b. each.

Sulfur as sulfur dioxide is emitted in large volumes from volcanoes, but this is now matched by that emitted by human activity, mainly from the burning of coal and heating oil, and is partly responsible for acid rain. Up to 5% of the weight of a fossil fuel may be sulfur; coal contains the most. Over 300 million tonnes of sulfur dioxide are released to the Earth's atmosphere each year.

Sulfur dioxide from power stations can be removed by 'scrubbing' the flue gases with sprays of water and neutralizing the acid with lime, but this is an expensive answer to the problem. As an alternative, the coal can be converted to a mixture of hydrogen gas and carbon monoxide gas (known as synthesis gas), leaving the sulfur behind, and the gas burned separately to release the energy. Alternatively, powdered coal can be burned with powdered limestone ($CaCO_3$), which traps the sulfur as calcium sulfate.

Sulfur cycles

Sulfur is like nitrogen in that it passes endlessly between soil, the atmosphere and the sea. It is absorbed and released by all living things. Most organisms take in sulfur as sulfate and reduce this to sulfide so that it can be incorporated into the amino acids methionine and cysteine needed for protein.

Marine algae take in sulfate and convert it to dimethylsulfonium propionate which they use to maintain their osmotic balance with sea water. When the algae die, or are eaten, this compound is converted to dimethyl sulfide, which is released from the oceans in vast quantities (transferring as much as 50 million tonnes of sulfur a year to the atmosphere). In the atmosphere it is converted to dimethyl sulfonic acid by reacting with hydroxy radicals (HO·) and then to sulfur dioxide and eventually to sulfuric acid, which may act as a planetary regulator by promoting cloud formation and so have a cooling effect on the Earth's climate.

The amounts of sulfur in the main compartments of the environment are:

- on land: 3 billion tonnes as sulfate and 600 million tonnes in living things;
- in the seas: 1.3 million billion tonnes as dissolved sulfate and 24 million tonnes in living things;
- in the atmosphere: around 3.5 billion tonnes of gaseous sulfur compounds; these are continually being added to and removed.

The bacterium *Pyrobaculum islandicum* lives in hot springs and gets its energy from the reaction of sulfur and hydrogen. This sulfur-loving bacterium laid down the vast deposits of uncombined sulfur during the Permian and Jurassic periods between 280 and 130 billion years ago. At the bottom of shallow seas conditions favoured these anaerobic bacteria which convert metal sulfates to metal sulfides and then to elemental sulfur.

CHEMICAL ELEMENT

Data file	
Chemical symbol	S
Atomic number	16
Atomic weight	32.066
Melting point	113°C
Boiling point	444°C
Density	2.1 kilograms per litre (2.1 grams per cubic centimetre)
Oxides	SO_2 and SO_3

Sulfur is a non-metal and comes below oxygen in group 16 of the periodic table of the elements. There are several forms of elemental sulfur, of which the yellow orthorhombic form is the most common. This consists of rings of eight sulfur atoms (S_8). These rings can pack together in different ways. Other ring sizes – S_4, S_6, S_7, S_9, S_{10}, S_{11}, S_{12}, S_{18} and S_{20} – are also known. When sulfur is heated it melts and will boil to deposit fine yellow orthorhombic crystals, which are the most recognizable form of the element, and are called 'flowers of sulfur'.[155] Sulfur is stable to air and water, but burns if heated. It is attacked by oxidizing acids.

The chemistry of sulfur is complex because of the many oxidation states in which it can exist, extending from –II to VI, i.e. from the reduced state of hydrogen sulfide through to the fully oxidized state of sulfate (SO_4^{2-}). Sulfur dioxide is an intermediate oxidation state and acts as a reducing agent as it becomes oxidized to sulfate.

Sulfur is notable for its ability to bond to itself to form long chains and rings of atoms. When the common form is melted and then heated to around 175°C, the rings open and join together to form long chains, making the liquid very viscous; when this liquid is cooled it forms a rubbery solid called plastic sulfur.

Naturally occurring isotopes of sulfur are sulfur-32, which makes up 95% of the total, sulfur-33 (0.8%), sulfur-34 (4.2%) and sulfur-36 (0.02%). None is radioactive.

The isotopic composition of sulfur varies, depending on its source, and this is why its atomic weight is given less precisely than for other elements. The ratio of the two most abundant isotopes sulfur-32/sulfur-34 is 22.22 for sulfur that arrives from outer space and is found in material of meteorite origin. This is taken as the standard ratio for the element as it existed when the Earth was formed. Since then, so-called isotopic fractionation has occurred, so that the ratio varies; it is smaller for sulfate in the seas and larger for sulfur deposits on land. It is smaller in the oceans because the loss of sulfate there favours loss of the lighter sulfur-32, leaving the seawater enriched with sulfur-34. Meanwhile sedimentary deposits on land have slightly more sulfur-32 because the bacteria that produced these showed a slight preference for reducing the lighter sulfur isotope.

[155] When sulfur boils it forms a green vapour of S_8 molecules, but if the temperature of this is raised to 720°C and the pressure reduced, the gas becomes violet-coloured due to its dissociating into S_2 molecules.

ELEMENT OF SURPRISES

1. If something smells awful then it probably contains sulfur. For example, skunk odour arises from three sulfur compounds, 2-butenethiol, isopentanethiol and 2-butenyl methyl disulfide, all of which smell truly awful.
2. The titan arum lily (*Amorphophallus titanium*) only flowers once every 4 years. It produces an attractive bloom but a repulsive smell, at least as far as humans are concerned. However, a species of bee finds it irresistible and the plant relies on this to carry out its pollination. The smell, which comes from dimethyl disulfide and dimethyl trisulfide, has been described as the aroma of rotting fish. In Indonesia the lily is known as the 'corpse plant'.
3. A Venezuelan plant related to the mimosa produces carbon disulfide which it releases from its roots to protect them against nematodes (parasitic worms) and fungi.

Tantalum

Pronounced tan-ta-lum, the name is derived from the name of the legendary Greek figure, Tantalus, King of Sisyphus in Lydia. Tantalus stole the food of the gods, thereby becoming immortal, and then invited them to a feast at which he served up the flesh of his son, Pelops. The gods were not amused and as a punishment they decreed that Tantalus was to be tormented forever by thirst and hunger. In the Underworld he was to stand up to his neck in water, which receded whenever he tried to drink, and to be surrounded by the branches of vines heavy with ripe grapes, which swayed out of reach as he stretched to pick them. The discoverer of this element evidently felt similarly tormented by its elusive nature and named it accordingly.

French, *tantale*; German, *Tantal*; Italian, *tantalio*; Spanish, *tántalo*; Portuguese, *tântalo*.

HUMAN ELEMENT

Tantalum in the human body	
Blood	not known, but low
Bone	approx. 30 p.p.b.
Tissue	not known, but low
Total amount in body	about 0.2 milligrams

Tantalum has no biological role to play and appears to be non-toxic. When rats were given doses of tantalum oxide (Ta_2O_5) it passed through them completely without effect and without any being absorbed. Radioactive tantalum-182 (half-life 114 days) showed that when the element was injected into the body it concentrated in the bones and kidneys. There have only been a few cases of industrial injury caused by tantalum or its compounds and these mainly involve skin rashes.

MEDICAL ELEMENT

Tantalum metal is well tolerated by the body and consequently has been used in many ways during surgery: as plates in the repair of skull fractures; as bolts to fasten broken bones; and as wire for sutures for torn ligaments. A few people who have had tantalum implants have suffered skin rashes, but these cleared up when the implants were removed.

Tantalum is used for equipment in the pharmaceutical industries because it does not contaminate the materials with which it comes in contact, and it allows the same apparatus to be used for many applications.

Some tantalum compounds are used in diagnostic medicine because they are more visible on X-rays than other agents, especially when examining the lungs.

ELEMENT OF HISTORY

Tantalum was reported as a 'new' metal in 1802 by Anders Gustav Ekeberg (1767–1813), who was Professor of Chemistry at Uppsala University, Sweden. However, when Wollaston, the eminent British chemist, analysed minerals from which it had been extracted, he declared it was identical with the niobium, discovered the year before. These two elements often occur together and are chemically very similar, which makes them difficult to separate even today, and especially so by the methods available two centuries ago.

More than 40 years were to pass before Heinrich Rose (1795–1864) separated tantalum and niobium and proved conclusively that they were different; that was in 1846. This was a remarkable achievement, and yet *pure* tantalum was not produced until Werner von Bolton of Charlottenburg finally accomplished it in 1903.

No one doubts that Ekeberg had isolated tantalum and so was the discoverer of a hitherto unknown element, thanks to his exceptional skills as a chemist, yet even his samples must have contained quite a lot of niobium.

ECONOMIC ELEMENT

Occasionally, tantalum is found as the uncombined metal. The chief tantalum ores are tantalite,[156] which also contains iron, manganese and niobium, and samarskite, which contains seven metals. Another mineral which contains both tantalum and niobium is pyrochlore, which occurs as large deposits in Canada and Brazil.

The main mining areas are Thailand, which is the single largest producer, Australia, Democratic Republic of Congo, Brazil, Portugal and Canada, but most is obtained as a by-product of tin extraction – some tin slag can contain as much as 10% of tantalum. The demand for tantalum is around 2300 tonnes a year. There is no expectation that there will ever be a shortage of this element and no assessment of total reserves of extractable metal have been reliably calculated.

The difficulty in refining tantalum is separating it from niobium; this is done by converting both metals to their potassium fluoro-derivatives which can then be separated by solvent extraction or ion exchange. Tantalum is then reclaimed by the electrolysis of the potassium fluorotantalate or by reacting this salt with sodium metal.

Tantalum finds use in four areas: high-temperature applications; electrical devices; surgical implants; and for handling corrosive chemicals. It would be used more but for its high cost. It is rarely used as an alloying agent because it tends to make metals

[156] If the amount of niobium exceeds that of tantalum, the ore is generally known as columbite.

brittle, although with some metals it has dramatic strengthening effects and it is used in turbine blades, rocket nozzles, heat shields, and nose caps for supersonic aeroplanes.

The metal is easily worked cold by rolling, forging or drawing into wire. It makes wire that is very strong. For a time, tantalum was used for light-bulb filaments on account of its ductility and high melting point, which is exceeded only by that of tungsten and rhenium. The metal became available commercially in 1922 and was soon being used to make all kinds of things, from corrosion-resistant laboratory apparatus to jewellery. Accurate weights for analytical chemistry were made from tantalum, which was seen as a suitable replacement for platinum.

Tantalum resists corrosion and is almost impervious to chemical attack; for this reason it has been employed in the chemical industry, e.g. for heat exchangers in boilers where strong acids are vaporized. It is ideal for making parts for vacuum furnaces. It is possible to coat other metals with tantalum and even a layer as thin as 100 micrometres will provide excellent corrosion resistance. The oxide layer which forms on the surface of the metal consists of tightly packed atoms, not only protecting the bulk of the metal, but also acting as an insulating layer, which is why a major use of tantalum is in capacitors.

Other ways in which tantalum has been employed are as the electrodes in neon lights, in a.c./d.c. rectifiers, for the glass of special lenses, and for the spinneret dies through which cellulose solutions are forced in the process of making rayon fibres.

ENVIRONMENTAL ELEMENT

Tantalum in the environment	
Earth's crust	2 p.p.m.
	Tantalum is the 51st most abundant element.
Soils	approx. 1 p.p.m.
Sea water	2 p.p.t.
Atmosphere	nil

Because tantalum oxide is so insoluble, there is almost no tantalum to be found in natural waters. Few attempts have been made to measure its level in soils; what work has been done suggests a range of from 0.1 to 3 p.p.m., with most soils being at the lower end of this range. Only tiny amounts of tantalum are taken up by plants: the amount in vegetation rarely exceeds 5 p.p.b., and in food plants the most that has been measured is only 0.5 p.p.b. (fresh weight).

CHEMICAL ELEMENT

Data file	
Chemical symbol	Ta
Atomic number	73
Atomic weight	180.9479
Melting point	2996°C
Boiling point	5425°C
Density	16.7 kilograms per litre (16.7 grams per cubic centimetre)
Oxides	TaO_2 and Ta_2O_5

Tantalum is a shiny, silvery metal which is soft when pure. It is one of the transition metals and a member of group 5 of the periodic table of the elements. Tantalum is virtually resistant to corrosion due to an oxide film on its surface; it is unaffected by *aqua regia*, but is attacked by hydrofluoric acid and molten alkalis.

Tantalum is predominantly the isotope tantalum-181, which accounts for 99.99%, the remainder being tantalum-180 which is incredibly weakly radioactive, with a half-life of more than 10 trillion years (10^{16} years).

ELEMENT OF SURPRISE

A form of tantalum carbide, TaC, has been made at the Los Alamos National Laboratories, in the USA, which is even harder than diamond. TaC has a melting point of 3738°C and is used in special cutting tools.

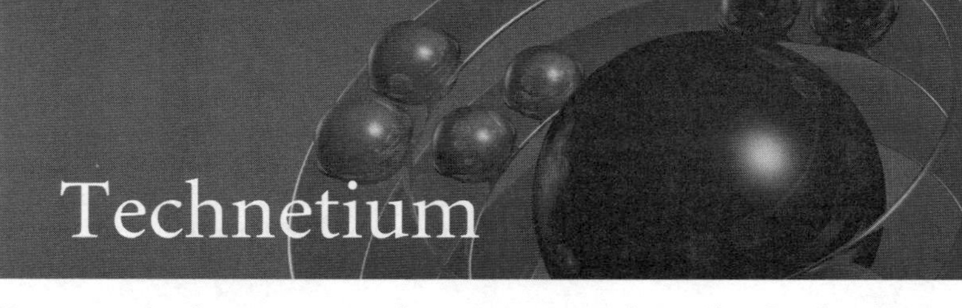

Technetium

Pronounced tek-nee-see-um, the name comes from the Greek *tekhnetos*, meaning artificial. The element might well have been called masurium – see 'Element of history'.

French, *technétium*; German, *Technetium*; Italian, *tecneto*; Spanish, *tecnecio*; Portuguese, *tecnécio*.

COSMIC ELEMENT

The longest-lived isotope of the element has a half-life of only 4 million years, which means than any technetium around when the Earth formed 4.57 billion years ago would be long gone. So it was all the more surprising when, in 1952, Paul Merril, an astronomer at the Mount Wilson and Palomar Observatories in California, examined the light given off by stars known as red giants, and found that the spectra contained lines showing that the stars were rich in technetium. Because all isotopes of technetium are relatively short lived compared with the age of stars, and especially compared with the age of red giants which are stars at the end of their lives, this could only mean that this element was being formed within the star itself. This proved that the stars are the furnaces in which the chemical elements are being manufactured.

Stars like the Sun are converting hydrogen and helium into carbon, oxygen, nitrogen, neon, magnesium, silicon and iron, but nothing heavier than iron (atomic number 26). Red giants make most of the other elements, while it falls to supernova to create the heaviest atomic nuclei of all, such as gold, iodine and uranium.[157]

HUMAN ELEMENT

Technetium in the human body	
Total amount in body	nil

Since all the technetium present when the Earth formed had disappeared by the time life emerged on the planet, there can never have been a biological role for it to play.

[157] Because the human body needs elements heavier than iron, such as copper, zinc and selenium, we know these must have been created by a red giant at some time in the past, and because we also need iodine, then part of us must once have been scattered into the cosmos by a supernova. Truly, we are made of star dust.

Only those individuals who have had medical treatment with technetium have any of this element in their body.

MEDICAL ELEMENT

The isotope technetium-99m is used in medical diagnosis. It is produced by the radioactive isotope molybdenum-99, which can be extracted from nuclear fission products, or produced specifically by bombarding molybdenum-98 with neutrons to form molybdenum-99. This decays with a half-life of 67 hours to form technetium-99m, and the two elements are easily separated. The technetium-99m has a half-life of 6 hours and emits only γ-rays as it reverts to technetium-99.

Technetium-99m is used in a medical diagnostic technique known as immunoscintigraphy, in which it is incorporated into a monoclonal antibody that has sites capable of binding to cancer cells. By mapping the γ-ray emission it is possible to diagnose the disease within a few hours. The method is especially useful for diagnosing difficult-to-detect cancers in the intestine. The product was developed by the German chemical company, Hoechst, and is sold under the trade name Scintimun.

When technetium-99m, together with a tin compound, is injected into the body, it binds to red blood cells and can be used to map disorders associated with the circulatory system; when it is accompanied by a diphosphate ion it binds strongly to heart muscle and this can be used to assess the damage done by a heart attack.

The by-product of technetium-99m treatment is technetium-99, which is quickly eliminated from the human body, but in any case its long half-life of 212 000 years means that it adds only a little to the body's burden of naturally occurring radioactive isotopes.

ELEMENT OF HISTORY

Technetium occupies the place below manganese in the periodic table, and it long tantalized chemists because it could not be found despite reports to the contrary. Some claimed to have discovered it and gave it names such as davyum, lucium and nipponium, but none of the claims was substantiated. A report of its discovery by a team of German chemists in 1925 appeared to have more credibility than most – they named it masurium – but it has always been assumed that they were wrong, although it now appears they might have been correct.

'Masurium' was detected by bombarding a sample of the mineral columbite with a beam of electrons and from the faint X-rays that were emitted the presence of element number 43 was deduced. The work was done by Ida Tacke, Walter Noddack and Otto Berg.[158] Their method of analysis was repeated in 1999 by David Curtis of the Los Alamos National Laboratory in the USA, using a sample of uranium from a Canadian ore deposit. A few of the uranium atoms that undergo spontaneous nuclear fission will produce technetium and Curtis was able to detect and estimate that there was about 1 nanogram (1×10^{-9} grams) of technetium per kilogram of uranium. Moreover, he was able to show that the method used by the German workers would have been sensitive

[158] Tacke and Noddack later married.

enough to detect these small amounts also. As columbite has been known to contain as much as 10% uranium, there must also be some technetium atoms in this mineral. In theory at least, Tacke, Noddack and Berg might have been the first chemists to find technetium.

Technetium really only came to light in 1937 when Emilio Segrè and Carlo Perrier at the University of Palermo in Sicily separated it from a sample of molybdenum that had been bombarded with deuterons for several months in the cyclotron at the University of California, Berkeley.[159]

ECONOMIC ELEMENT

Technetium is rarely encountered outside nuclear facilities or research laboratories, although some is used in medical diagnosis. World production of technetium runs to many tonnes because this element is extracted from spent nuclear fuel rods. However, only tiny amounts of this have any commercial use. One isotope, technetium-95m, with a half-life of 61 days, is used as a source of γ-rays.

ENVIRONMENTAL ELEMENT

Technetium in the environment	
Earth's crust	only in minute traces naturally, but some is stored in nuclear waste depositories
Soil	virtually nil
Sea water	nil
Atmosphere	nil

Because technetium is produced in tonne quantities in nuclear reactors, it is adding to the planetary burden of unwanted radioactive waste. A little technetium escapes to the environment via its use in medical diagnosis.

CHEMICAL ELEMENT

Data file	
Chemical symbol	Tc
Atomic number	43
Atomic weight	98
Melting point	2172°C
Boiling point	4877°C
Density	estimated to be.11.5 kilograms per litre (11.5 grams per cubic centimetre)
Oxides	TcO_2 and Tc_2O_7

[159] Segrè was dismissed from his academic post in 1938 because he was opposed to Mussolini's fascist regime, so he emigrated to the USA.

Technetium is a radioactive metal that is unusual in being a relatively light element (atomic number 43), yet having no stable isotopes. The longest-lived isotope is technetium-98, with a half-life of 4.2 million years. Technetium is a member of group 7 of the periodic table of the elements. It occurs naturally in uranium ores, where it is produced by spontaneous fission, but only in trace amounts.

As the bulk metal it is silvery, but it is more commonly obtained as a grey powder. Technetium resists oxidation but slowly tarnishes in moist air, and burns in oxygen when in powder form. It dissolves in nitric and sulfuric acids, but not in hydrochloric acid.

ELEMENT OF SURPRISE

The best way to protect steel that is in contact with aerated water is to dissolve a little potassium pertechnetate(VII) ($KTcO_4$) in the solution: 55 p.p.m. is sufficient to prevent any corrosion, even when the system is under pressure and the temperature is 250°C. Remarkable as this protection is, potassium pertechnetate can only be used in sealed systems because all its isotopes are radioactive.

Tellurium

Pronounced tel-oor-iuhm, the name is derived from the Latin *tellus*, meaning Earth. French, *tellure*; German, *Tellur*; Italian, *tellurio*; Spanish, *teluro*; Portuguese, *telúrio*.

HUMAN ELEMENT

Tellurium in the human body	
Blood	6 p.p.b.
Bone	not known
Tissue	approx. 15 p.p.b.
Total amount in body	about 0.7 mg (700 micrograms)

There is no biological role for tellurium. Elemental tellurium has low toxicity but unpleasant side effects, producing extremely bad breath and body odour. Those who worked with it often developed what was called 'tellurium breath', even when there was as little as 10 micrograms of the element per cubic metre of air. Those who come into contact with tellurium are advised to take extra vitamin C, which reduces considerably the smell of tellurium on the breath.

The daily intake for the average person is judged to be around 600 micrograms, most of which is absorbed and excreted in the urine, with most of the remainder passing through the gut unabsorbed, apart from around 10 micrograms which is exhaled on the breath. The body does this by converting some tellurium to volatile dimethyl telluride, $(CH_3)_2Te$, which it can expel via the lungs or the sweat glands.

In 1884, tests were carried out on volunteers who were given 0.5 micrograms of tellurium oxide, TeO_2, by mouth. In an hour it was detectable on their breath and the smell did not disappear for 30 hours. Some men, given doses of 15 mg, were still found to have 'tellurium breath' 8 months later.

As little as 2 grams of sodium tellurite (Na_2TeO_3) can be fatal. This was discovered accidentally in 1946 when three soldiers were given it from a mislabelled bottle as part of their medical treatment. Two of them died within 6 hours. Acute tellurium poisoning results in vomiting, inflammation of the gut, internal bleeding and respiratory failure. Chronic poisoning produces 'garlic breath', tiredness and indigestion.

ELEMENT OF HISTORY

Tellurium was discovered in 1783 by Franz Joseph Müller von Reichenstein (1740–1825) at Sibiu, Romania. Müller, as he was known, had amassed a collection of rare minerals during his trips around Europe and was particularly intrigued by one which came from a mine near Zalatna. It had a metallic lustre. Müller thought it was a sample of uncombined antimony, but preliminary tests showed that it was not antimony, and he announced that it was bismuth sulfide. However, further investigation showed that neither bismuth nor sulfur was present, and Müller finally concluded that the mineral was a compound of gold with an unknown element.[160]

For 3 years Müller researched the new mineral, often calling it *metallum problematum* or *aurum paradoxum* because of its puzzling nature. Finally he was convinced that it was a hitherto undiscovered element, and he sent a sample of the mineral to other chemists asking them to confirm his findings, which they did. He published his discovery in an obscure journal in Vienna.

In 1796, Müller sent a sample of the mineral to the eminent German chemist Martin Klaproth (1743–1817) in Berlin, who also investigated it, confirmed it was a 'new' element, obtained a pure sample and decided to call it tellurium. Rather strangely, this was not the first sample of tellurium to pass through his hands. In 1789, he had been sent some by a Hungarian scientist, Paul Kitaibel (1757–1817), who also claimed to be its discoverer. There is no doubt that Kitaibel had independently discovered it, and naturally he was somewhat upset when he read Klaproth's account of tellurium in which Müller was mentioned but he was not. A pained correspondence followed.

There is no doubt that Müller should get the credit for discovering tellurium, although he seemed reluctant to have it widely known, hence his decision to publish news of it in a little-read journal. Because he had already once wrongly identified the mineral from which it came, this is perhaps understandable.

ECONOMIC ELEMENT

Samples of uncombined tellurium are sometimes found, but they are extremely rare. There are some tellurium minerals, such as calaverite ($AuTe_2$), sylvanite ($AgAuTe_4$) and tellurite (TeO_2), but none is mined as a source of the element. Commercial tellurium is obtained from the anode slime of copper refining, which can contain as much as 8%. World production is around 220 tonnes per year, coming mainly from the USA, Canada, Peru and Japan. The reserves of this element have not been assessed, but since demand is small it is unlikely there would ever be a shortage.

Tellurium is a semi-metal usually obtained as a grey powder by the electrolytic reduction of sodium tellurite solution. It is used in alloys with copper and stainless steel to improve their machinability, and is added to lead (0.05%) to make it harder and more acid-resistant, especially that which is used in batteries. At various times, tellurium has been used to vulcanize rubber, to tint glass and ceramics, in electronic devices, and as an industrial catalyst in oil refining.

[160] The mineral was gold telluride, $AuTe_2$, and is one of the few naturally occurring compounds of gold.

When added to rubber, tellurium speeds up the curing process and makes the product less susceptible to ageing and less likely to be affected by oil, which softens normal rubber. Tellurium is also used in photoreceptors and other microelectronic devices, in the form of bismuth telluride.

ENVIRONMENTAL ELEMENT

Tellurium in the environment	
Earth's crust	5 p.p.b.
	Tellurium is the 72nd most abundant element.
Soils	ranges from 0.05 to over 30 p.p.m., although generally at the lower end
Sea water	0.15 p.p.t.
Atmosphere	traces

Tellurium is present in coals at up to 2 p.p.m. and this is probably the major source of release of this element to the environment. Micro-organisms can absorb tellurium and then emit it in a volatile form by methylating it to dimethyl telluride, which is detectable in air, especially near copper refineries.

Plants can take in tellurium from soil and have been found with levels as high as 6 p.p.m. although few food plants have more than 0.5 p.p.m. and most have less than 0.05 p.p.m. (fresh weight), the level being lowest in apples. Onion and garlic, however, can have up to 300 p.p.m. (dry weight).

CHEMICAL ELEMENT

Data file	
Chemical symbol	Te
Atomic number	52
Atomic weight	127.60
Melting point	450°C
Boiling point	990°C
Density	6.2 kilograms per litre (6.2 grams per cubic centimetre)
Oxides	TeO, TeO_2 and TeO_3

Bulk tellurium is a semi-metal, silvery-white and metallic-looking, but is usually available as a dark grey powder. It is a member of group 16 of the periodic table of the elements, where it lies on the boundary of metals and non-metals, and has properties of both. As a metal it is brittle and easily ground to powder. Tellurium burns in air or oxygen, is unaffected by water or hydrochloric acid, but dissolves in nitric acid.

Tellurium has eight naturally occurring isotopes, a number exceeded only by xenon, which has 9, and tin, which has 10. Tellurium-130 is the most abundant, accounting for 34% of the total. This isotope is also marginally radioactive, with a half-life of 2 billion trillion years (2×10^{21} years).[161] The next most abundant isotope is tellurium-128 (32%) which is also radioactive, with a half-life of 7.7×10^{24} years. The others are tellurium-126 (19%), tellurium-125 (7%), tellurium-124 (4.5%) and tellurium-122 (2.5%), none of which is radioactive. The least abundant isotopes are tellurium-120 (0.1%) and tellurium-123 (0.9%); the latter is also radioactive with a half-life of 13 trillion years. There are traces of other, radioactive, isotopes produced as fission fragments.

ELEMENT OF SURPRISE

Dimitri Mendeleyev found tellurium rather vexing when he was drawing up his first periodic table of the elements in 1869. The list of atomic weights showed tellurium to be heavier than the element that followed it, iodine. Tellurium's atomic weight was 128 whereas iodine's was 127. Mendeleyev concluded that one of the atomic weights must be wrong because tellurium clearly preceded iodine in his periodic table. Despite intensive work to reassess their relative atomic weights, they remained as 128 and 127, respectively.

The explanation of the anomaly had to wait almost 50 years until it was realized that elements could have numerous isotopes. The dominant isotopes of tellurium are 130 and 128, which together account for two-thirds of the mass and give an overall atomic weight of 127.6, whereas for iodine there is only one dominant isotope, iodine-127; its atomic weight is 126.9.

[161] A 1 gram sample of tellurium sees only one atom undergo radioactive decay every 2½ years.

Pronounced ter-bi-uhm, this element was named after Ytterby, Sweden – see 'The village of the four elements' under yttrium (p. 496).

French, *terbium*; German, *Terbium*; Italian, *terbio*; Spanish, *terbio*; Portuguese, *térbio*.

Terbium is one of 15, chemically similar, elements referred to as the rare-earth elements, which extend from the element with atomic number 57 (lanthanum) to the one with atomic number 71 (lutetium). The term rare-earth elements is a misnomer because some are not rare at all; they are more correctly called the lanthanides, although strictly speaking this excludes lutetium. The minerals from which they are extracted, and the properties and uses they have in common, are discussed under lanthanides, on p. 222.

HUMAN ELEMENT

Terbium in the human body	
Total amount in body	not known, but very small

Terbium has no biological role. It is present in humans but it is difficult to know how much there is, because little work has been done to separate out the various amounts of the different lanthanides in the body. The levels are highest in bone, with smaller amounts being present in the liver and kidneys. No one has monitored diet for terbium content, so it is difficult to judge how much we take in, but it is probably only a few micrograms or so per year. Very little terbium is taken up by plant roots so it does not get into the human food chain, and those vegetables that have been analysed have less than 1 p.p.b. (dry weight).

Terbium may be mildly toxic by ingestion, but its insoluble salts are non-toxic. Terbium powder and terbium compounds are very irritating if they come into contact with the skin and eyes.

ELEMENT OF HISTORY

Terbium was first isolated in 1843 by Carl Gustav Mosander (1797–1858) at Stockholm, Sweden. He suspected that the earth yttria (yttrium oxide), which had been discovered in 1794, might harbour other elements, just as ceria (cerium oxide) did – he had studied the latter a few years earlier and extracted lanthanum from it. So it proved to be, and Mosander was able to show that although yttria was mainly yttrium oxide, its colour was due to two other oxides: terbium oxide (yellow) and erbium oxide (rose pink). There is more of this story in the section on yttrium – see p. 496.

ECONOMIC ELEMENT

The most important ores are monazite and bastnäsite, but neither contains more than a fraction of a per cent of terbium – monazite contains just 0.05% and bastnäsite even less, 0.02% – although enough is present for it to be extractable. Terbium is also found in the minerals cerite, gadolinite and xenotime, and one ore, euxenite, contains as much as 1%.

The main mining areas are China, USA, Brazil, India, Sri Lanka and Australia and reserves of terbium are estimated to be around 300 000 tonnes. World production of terbium is around 10 tonnes per year. The metal itself is produced by heating terbium fluoride and calcium metal in a tantalum crucible under vacuum conditions.

Terbium has few commercial uses, partly because it is such a rare element and therefore costly, being four times more expensive than platinum. An unusual property of terbium alloys is that they will lengthen or shorten when exposed to a magnetic field; one such is a mixture of terbium, dysprosium and iron, of composition $Tb_{0.3}Dy_{0.5}Fe_2$. These so-called magnetostrictive alloys can store a lot of strain energy and research is currently going on into how this might be used to work tiny motors, pumps and injection systems.

Terbium is used in solid-state devices, lasers (sodium–terbium borate emits green laser light at 546 nanometres) and low-energy light bulbs (see europium, p. 140) and mercury lamps.

There was once a scheme to use a ceramic containing terbium for false teeth, because it mimics the gleam of real teeth, but nothing came of it.

ENVIRONMENTAL ELEMENT

Terbium in the environment	
Earth's crust	1 p.p.m.
	Terbium is the 57th most abundant element.
Soils	approx. 0.7 p.p.b., range 0.5–1.8 p.p.b.
Sea water	0.2 p.p.t.
Atmosphere	virtually nil

Terbium is one of the rarer rare-earth elements although it is still twice as common in the Earth's crust as silver. It poses no environmental threat to plants or animals.

CHEMICAL ELEMENT

Data file	
Chemical symbol	Tb
Atomic number	65
Atomic weight	158.92534
Melting point	1356°C
Boiling point	3120°C
Density	8.23 kilograms per litre (8.23 grams per cubic centimetre)
Oxide	Tb_4O_7, which is a mixture of Tb_2O_3 and TbO_2

Terbium is a silvery metal, soft enough to be cut with a knife, and a member of the lanthanide group of the periodic table of the elements. It is reasonably stable in air, but is slowly oxidized and reacts with cold water.

The only naturally occurring isotope is terbium-159, which is not radioactive.

ELEMENT OF SURPRISE

Terbium has led to increased safety in X-ray diagnosis by reducing the time for which a patient is exposed to these dangerous rays. This has been achieved by using the more responsive terbium-doped phosphors for the imaging screens. These allow the same quality image to be produced in a quarter of the time previously required.

Pronounced thal-iuhm, the name is derived from the Greek *thallos*, meaning a green twig.

French, *thallium*; German, *Thallium*; Italian, *tallio*; Spanish, *talio*; Portuguese, *tálio*.

HUMAN ELEMENT

Thallium in the human body	
Blood	0.5 p.p.b.
Bone	2 p.p.b.
Tissue	approx. 4–70 p.p.b., with the majority in muscle tissue
Total amount in body	approx. 0.5 milligrams

There is no biological role for thallium. Nevertheless, the average person takes in about 2 micrograms of thallium a day, as part of the diet, and this accumulates in the body over time, with the highest concentrations in the kidneys and liver. This element is a toxic heavy metal, which can even be absorbed through the skin; handling its compounds directly can lead to loss of the fingernails.

MEDICAL ELEMENT

Thallium was once part of the medical pharmacopoeia, being prescribed by doctors as a depilatory (hair remover) and given as a pre-treatment for ringworm of the scalp. Thallium did not kill the ringworm,[162] but it caused a patient's hair to drop out, so that the condition could be more easily treated. Its ability to do this was discovered by a Dr R. Sabourand, the chief dermatologist at the St Louis Hospital in Paris, in 1898 when he was searching for something that might cure the night sweats experienced by patients with tuberculosis. Thallium had no effect on those symptoms, but it caused the patients' hair to fall out and so it became the standard treatment for hair removal for almost 50 years.[163]

[162] Ringworm is a fungal disease that is highly contagious and often transmitted by contact with animals.

[163] In the 1930s, thallium acetate was openly sold as 'Koremlou creme' for removing unwanted facial hair.

Thallium is an insidious poison because it mimics the essential element potassium, the element it resembles in size and ionic charge. Consequently, it affects potassium-activated enzymes in the brain, muscles and skin, producing the characteristic symptoms of thallium poisoning, namely lethargy, numbness, tingling of the hands and feet, blackouts, slurred speech, general debility and hair loss. Symptoms take many days to appear, peaking after about 2 weeks from the day of ingestion. Hair loss begins after 10 days and occurs all over the body. A fatal dose for an adult would be around a gram and yet doses of half a gram (500 mg) of thallium salt were prescribed in cases of ringworm.

The body tries to rid itself of thallium, and does so by excreting it into the intestines, but there it is soon reabsorbed in mistake for more potassium. The antidote is Prussian blue, the dye of blue ink, which breaks this cycle of excretion and reabsorption by exchanging its potassium for thallium, to which it clings more strongly, carrying it from the body. A German pharmacologist, Horst Heydlauf of Karlsruhe, discovered this cure in the 1970s at a time when thallium poisoning was thought to be incurable.

ELEMENT OF MURDER

Agatha Christie is often blamed for bringing thallium sulfate to the attention of would-be-poisoners. Her detective story, *The Pale Horse*, which was published in 1961, described the symptoms of thallium poisoning. It showed how these could easily be attributed to other causes – in this case to black magic curses. In the 1980s and 1990s the Iraqi regime used thallium sulfate to dispose of its opponents, who appeared to die of natural causes.

Agatha Christie's fictional story may also have suggested thallium as a suitable poison to Graham Young who, in 1971, put thallium sulfate into his workmates' coffee at a factory at Bovingdon in Hertfordshire, England. Several workers were taken ill and two died of a mysterious illness. It was only when Young himself suggested that the cause might be thallium, that the strange illness was correctly diagnosed.

Young was arrested, tried and found guilty of murder in 1972 and sentenced to life imprisonment. He had previously been on trial for poisoning his family and ordered to be detained in a secure prison for the criminally insane, but he had been released after appearing to respond successfully to psychiatric treatment. He committed suicide in 1990. See also 'Element of surprise'.

Not all thallium deaths are malicious: some are accidental, such as the incident in Guyana in 1987 which affected hundreds of people, of whom 44 died, when they drank the milk from cows that had eaten molasses which had been poisoned with thallium sulfate in order to kill sugar-cane rats.

A radioactive synthetic isotope, thallium-201, which has a half-life of 73 hours, is used in the diagnosis of heart disease. This will displace some of the potassium in the heart muscle, but only if there is an adequate supply of blood to carry it there, and then the penetrating γ-rays that it emits as it decays can be monitored from outside the body.

The patient is given an injection of the isotope and then scanned by a scintillation counter before and after physical exercise. The uptake of thallium-201 by the heart, and its distribution within the heart, will reveal the extent of the damage.

ELEMENT OF HISTORY...

...and element of controversy. With thallium, the question arose: what counts as 'discovery'? Was this when William Crookes (1832–1919), of the Royal College of Science in London, first observed a green line in the spectrum of some impure sulfuric acid, and realized that this meant a hitherto unknown element was present? That was in 1861. He named it thallium because the colour reminded him of a fresh green shoot. He began to investigate it, albeit very slowly, but produced a few simple salts of it.

Meanwhile, in 1862, Claude-Auguste Lamy (1820–78), a physicist of Lille, France, began to research thallium more thoroughly, even to the extent of extracting the metal itself and casting a small ingot of it. The French Academy consequently credited him with the discovery. He sent the ingot to the London International Exhibition of that year, where it was acclaimed as a 'new' metal and he was awarded a medal.

Crookes was furious when he learned what had happened and used *Chemical News*, the weekly newspaper he founded and edited, to campaign to have the award withdrawn and given to him, on the grounds that he was the true discoverer.[164] Accusations and counter-accusations, tinged with xenophobia, were made until eventually the exhibition committee felt obliged to award Crookes a medal also.

In the year following its discovery, thallium was found to be quite widespread; it was detected in spring waters, wine, tobacco and sugar beet. A thallium mineral was found in Sweden in 1866, and named crookesite in honour of Crookes. In 1867, thallium oxide was found to be a component of the flue dust of pyrite-roasting kilns. In all cases its characteristic flash of bright green, when a sample was put into a flame, enabled it to be easily detected.

ECONOMIC ELEMENT

Thallium minerals are rare, but a few are known, such as crookesite (a copper–thallium selenide, Cu_7TlSe_4) and lorandite (a thallium–arsenic sulfide, $TlAsS_2$). Thallium is mainly obtained as the by-product of zinc and lead smelting, but some is extracted from pyrites when the sulfur from this ore is to be used to make sulfuric acid. World production of thallium compounds is around 30 tonnes per year, although only a tiny fraction of this is converted, electrolytically, to the metal itself. There has been no assessment of how great the reserves are, although this would be an academic point since demand is never likely to be large.

Thallium is used for making low-melting point special glass for highly refractive lenses; some is turned into thallium sulfide, selenide and arsenide for photoelectric cells, and some into thallium bromide–iodide crystals for infrared detectors because these are

[164] He had announced his discovery of thallium in the issue of *Chemical News* published on 30 March 1861.

transparent to this kind of radiation. Thallium salts are used as reagents in chemical research; for example, thallium(III) nitrate is a selective oxidizing agent.

Thallium metal is not used as an alloying agent although 8% of thallium in mercury enabled thermometers to record temperatures down to –58°C, which is 20°C lower than with mercury alone.

Thallium sulfate is still sold in developing countries where it is still permitted as a pesticide, although banned in Western countries. It is particularly effective as a sugar syrup against rats, cockroaches and ants.

ENVIRONMENTAL ELEMENT

Thallium in the environment	
Earth's crust	0.6 p.p.m.
	Thallium is the 59th most abundant element.
Soils	0.02–2.8 p.p.m. but mostly in the range 0.1–0.3 p.p.m.
Sea water	approx. 10 p.p.t.
Atmosphere	virtually nil

Thallium is not a rare element; it is 10 times more abundant than silver. The element is widely dispersed, mainly in potassium minerals such as sylvite and the caesium mineral pollucite. There has been no significant contamination of the environment by thallium from industry, unlike that caused by its neighbours in the periodic table, mercury and lead. It has been estimated that around 600 tonnes a year are emitted by smelters and metal processing industries, and a similar amount from coal-fired power stations.

Trace amounts of thallium find their way into crops such as grapes, sugar beet and tobacco, although in this last case too much tends to damage the plant. Most roots have no difficulty in absorbing thallium from the soil in which they grow, and the more thallium that is present the more they absorb. Sometimes quite high levels of thallium in vegetation have been measured; pine trees can have 100 p.p.m. while some flowers have reached 17 000 p.p.m. (1.7%). Normally, vegetable plants and meats have between 0.02 and 0.12 p.p.m.

In 1980, vegetation growing around a cement works in Germany was found to contain raised levels of thallium, this being emitted from the cement kilns in which pyrite, with unsuspected high levels of thallium, was being used to make a new type of cement. Cabbages and grapes were found to have the highest levels of thallium, with up to 45 p.p.m. (fresh weight) and 25 p.p.m., respectively. Normally these plants would have less than 0.1 p.p.m. Even the eggs of local hens had more than 1 p.p.m. of thallium, although most of this was concentrated in their shells.

CHEMICAL ELEMENT

Data file	
Chemical symbol	Tl
Atomic number	81
Atomic weight	204.3833
Melting point	304°C
Boiling point	1457°C
Density	11.9 kilograms per litre (11.9 grams per cubic centimetre)
Oxides	Tl_2O and Tl_2O_3

Thallium is a soft, silvery-white metal, and a member of group 13 of the periodic table of the elements. It tarnishes readily in moist air and reacts with steam to form the hydroxide (TlOH). The metal is attacked by acids, rapidly so by nitric acid.

Thallium can exist in two oxidation states, (I) and (III). In the lower state as the ion Tl^+ it resembles the potassium ion, K^+. The higher oxide, Tl_2O_3, is formed by heating the lower oxide in the presence of oxygen, and when treated with concentrated nitric acid it forms the thallium(III) nitrate used in research.

Thallium occurs naturally as only two isotopes, thallium-203, which accounts for 30%, and thallium-205, which accounts for 70%. Neither is radioactive.[165]

ELEMENT OF SURPRISE

Murderers who poison people may think they are in the clear once their victim has been cremated. However, in the case of Graham Young (see box above) it provided him with no such protection. His case became a milestone in forensic detection because the ashes of Bob Egle, one of his victims, were reclaimed and analysed by a technique known as atomic absorption spectrometry; this revealed a level of 5 p.p.m. of thallium. This was well above any natural level of this element in cremated remains, and proof that he had been poisoned with it.

[165] There are also minute amounts of radioactive thallium isotopes produced in a decay sequence of uranium and other heavy radioactive elements.

Thorium

Pronounced thor-iuhm, this element was named after Thor, the Scandinavian god of war.

French, *thorium*; German, *Thorium*; Italian, *torio*; Spanish, *torio*; Portuguese, *tório*.

COSMIC ELEMENT

Much of the internal heat of the Earth is thought to come from the radioactive decay of thorium.

HUMAN ELEMENT

Thorium in the human body	
Blood	0.2 p.p.b.
Bone	2–12 p.p.b.
Tissue	not known
Total amount in body	40 micrograms

There is no biological role for thorium, nor would we expect there to be, because this element is radioactive in all its forms. We take some in with our diet: the average amount is estimated to be around 3 micrograms a day. Plant roots can absorb thorium; it is present only at p.p.b. levels in most plants, although sometimes levels of over 1 p.p.m. have been recorded (based on dry weight). The levels in vegetables are in the range 5–20 p.p.b. (again dry weight).

Research has shown that 99.98% of ingested thorium passes through the body unabsorbed, and of the 0.02% which is absorbed, three-quarters deposits in the skeleton. Thorium compounds are thought to be moderately toxic chemically and those who worked with its compounds sometimes suffered from a form of dermatitis, while prolonged exposure was likely to increase the risk of cancer.

ELEMENT OF HISTORY

In 1815, Jöns Jakob Berzelius, who was Professor of Chemistry at the Karolinska Institute, Stockholm, Sweden, thought he had extracted a 'new' element from a recently discovered mineral, and he named it thorium. The mineral later turned out to be the phosphate of a metal that was already known: yttrium. Nevertheless, Berzelius so liked the new name that he used it again when, in 1829, he really did discover a 'new' element.

This he had extracted from a curious rock specimen sent to him by Hans Esmark, a Norwegian pastor and amateur mineralogist, who had found it near Brevig. Esmark's father, the Professor of Mineralogy at the University of Christiania, Oslo, realized that the specimen had not previously been recorded, so he suggested his son should send it to Berzelius. The mineral was thorium silicate, and is now known as thorite. Berzelius even produced a sample of metallic thorium by heating thorium fluoride with potassium, and confirmed it as a 'new' metal.

The radioactivity of thorium was first demonstrated by Gerhard Schmidt in 1898, and confirmed by Marie Curie later that same year. It was the second element, after uranium in 1896, to have its radioactivity discovered.

ECONOMIC ELEMENT

Thorium occurs naturally as the minerals thorite, $ThSiO_4$, uranothorite $(Th,U)SiO_4$, and thorianite, ThO_2. It is a major component of the lanthanide ore monazite, which can contain up to 12% thorium oxide, and this is currently the source of the world's supply. Thorium is also present in significant amounts in the minerals zircon (see zirconium, p. 507), titanite (see titanium, p. 453), gadolinite (see yttrium, p. 496) and betafite (see uranium, p. 479). World production of thorium is in excess of 30 000 tonnes per year. Known reserves exceed 3 million tonnes.

Until the inherent dangers associated with its radioactivity were realized, thorium and its compounds found some important retail uses, the best known of which was in gas mantles. These were made in their hundreds of millions and consisted of thorium oxide (plus 1% cerium oxide) which, when heated, emitted a bright white light. It was Karl Auer's[166] invention of these which enabled town gas to compete successfully with electricity for domestic and street lighting well into the twentieth century. Another use of thorium compounds was in toothpaste, which was first marketed in the early 1900s by the Berlin Auer company and sold in Germany for many years.

While these uses of thorium have now been discontinued, the metal and its compounds are still employed in other ways, although not ones which expose the general public to its radiation.

Thorium oxide is used in the manufacture of refractory materials for the metallurgical industries. It is also used as a catalyst in the process that converts ammonia into nitric acid, and in oil refining. Thermionic emitting devices, and photoelectric cells that measure the wavelength of ultraviolet light, both contain thorium.

Thorium metal can be extracted by heating thorium oxide with calcium, or by reacting thorium tetrafluoride with magnesium, the process being carried out under an atmosphere of argon. Thorium reacts rapidly with the oxygen of the air and becomes coated with a protective layer of thorium oxide; this is why thorium crucibles are used in high-temperature research. Thorium is also added to the tungsten that goes into the

[166] For more about Auer see neodymium (p. 268) and lutetium (p. 241).

filaments of incandescent light bulbs. As a metal, it is soft and ductile, but its alloys can be very strong; for example, adding thorium to magnesium enables this metal to be fabricated into components that can withstand high temperatures.

Like uranium, thorium could be a source of nuclear fuel (thorium-based reactors have been operated experimentally) and there is probably more energy locked up in the world's thorium than in all known fossil fuels deposits.[167] When thorium-232 is bombarded with neutrons, it transforms first to thorium-233, which has a half-life of only 22 minutes, decaying to uranium-233. This then undergoes fission in the same way as uranium-235, liberating lots of energy plus enough neutrons to start a controlled chain reaction to keep the process going.

Thorium needs to be 99.999% pure it if it to be used as nuclear fuel, but the problems of producing such high-grade material have been solved. Thorium can be 'burnt' in a nuclear reactor either as the metal itself or as the oxide, nitrate, fluoride or chloride. One of the advantages of a thorium-powered nuclear reactor would be that it would not generate plutonium. Although such power plants have been built, they were beset with economic and mechanical problems and have not proved commercially viable.

ENVIRONMENTAL ELEMENT

Thorium in the environment	
Earth's crust	12 p.p.m.
	Thorium is the 37th most abundant element.
Soils	range 0.8–11 p.p.m.
Sea water	approx. 10 p.p.t.
Atmosphere	virtually nil

Thorium is surprisingly abundant in the Earth's crust, being almost as abundant as lead. It is three times more abundant than uranium, five times more abundant than tin and 200 times more abundant than silver. Granite contains up to 80 p.p.m. of thorium. Because thorium oxide is highly insoluble, very little of this element circulates through the environment, although the thorium ion, Th^{4+}, is soluble, especially in the presence of organic acids in soils, and the level of thorium around coal- and oil-burning power plants can be relatively high due to its release from these fossil fuels. Near one such Italian power plant soils were found to have 40 p.p.m. of thorium

Thorium is an α-emitter, albeit only weakly so, which explains why it has commercial applications even today. It decays eventually to lead-208, passing through a series of short-lived isotopes, although one of them is radon, and for this reason thorium materials need proper ventilation. However, this isotope of radon is radon-220, with a half-life of only 55 seconds, which means that it does not have time to escape from subsoil and into houses so it is not an environmental hazard.

[167] There are three fissile elements, thorium, uranium and plutonium, of which thorium has the lowest atomic number.

CHEMICAL ELEMENT

Data file	
Chemical symbol	Th
Atomic number	90
Atomic weight	232.0381
Melting point	1750°C
Boiling point	about 4800°C
Density	11.7 kilograms per litre (11.7 grams per cubic centimetre)
Oxides	ThO and ThO_2

Thorium is a radioactive, silvery metal, and the most abundant of the actinide series of elements of the periodic table. As a bulk metal it is protected by an oxide coating, but it is attacked by steam, and will slowly dissolve in acids. The powdered metal has been known to ignite spontaneously. Thorium oxide, ThO_2, has the highest melting point of any oxide, 3300°C.

Thorium-232 is the main isotope that occurs naturally; because it has a half-life of 14 billion years, the amount on Earth is still about 85% of that which was present when the planet formed. Although it is an α-emitter, this is of such weak intensity that it poses no real threat compared with other sources of radiation.

Of the 24 other known isotopes of thorium, the longest lived is thorium-230, with a half-life of 75 000 years; along with other thorium isotopes it occurs naturally in trace amounts in actinide ores where it is generated as part of the radioactive decay sequences of other elements. Uranium-235 decays via thorium-231 (half-life 26 hours) and thorium-227 (half-life 19 days), while uranium-238 decays via thorium-234 (24 days) and thorium-230. Thorium-232 itself also decays via another of its isotopes, thorium-228 (half-life 2 years).

ELEMENT OF SURPRISE

In the early years of X-ray diagnosis, patients were injected with a colloidal suspension of thorium oxide to highlight the blood stream and organs of the body. The 10 000 individuals who underwent this treatment were later found to have a higher than expected incidence of leukaemia, bone cancers and abnormal chromosomes due to radiation, while some suffered liver damage due to the toxic effects of thorium.

Pronounced thyoo-li-uhm, the name comes from Thule, the ancient name for Scandinavia.

French, *thulium*; German, *Thulium*; Italian, *tulio*; Spanish, *tulio*; Portuguese, *túlio*.

Thulium is one of 15, chemically similar, elements referred to as the rare-earth elements, which extend from the element with atomic number 57 (lanthanum) to the one with atomic number 71 (lutetium). The term rare-earth elements is a misnomer because some are not rare at all; they are more correctly called the lanthanides, although strictly speaking this excludes lutetium. The minerals from which they are extracted, and the properties and uses they have in common, are discussed under lanthanides, on p. 219.

HUMAN ELEMENT

Thulium in the human body	
Total amount in body	not known, but small

Thulium has no biological role, although it has been noted that thulium salts stimulate metabolism. Soluble thulium salts are regarded as slightly toxic if taken in large amounts, but the insoluble salts are completely non-toxic. It is difficult to separate out the various amounts of individual lanthanides in the human body, but research on animals showed levels are highest in bone, liver and kidneys, and this is no doubt how thulium itself will distribute itself in the body, but in any case it is reasonable to assume that actual amounts are very low indeed. The thulium we take in with our food is probably only a few micrograms a year. Thulium is not taken up by plant roots to any extent and so does not get into the human food chain. Vegetables typically have only 1 p.p.b. (dry weight) of thulium.

ELEMENT OF HISTORY

Thulium was first isolated as its oxide by Per Teodor Cleve (1840–1905) at the University of Uppsala in 1879. The discovery of one group of the lanthanide elements began with yttrium which, unknown at the time, was contaminated with traces of other rare-earth elements. In 1843, erbium and terbium were extracted from it; then, in 1874, Cleve looked more closely at erbium and realized that it must contain yet other elements because he observed that its atomic weight varied slightly depending on the source from which it came. He extracted holmium from erbium in 1878 and thulium a year later, in 1879.

In 1911, the American chemist Theodore William Richards performed 15 000 recrystallizations of thulium bromate in order to obtain a pure sample of the element and so determine exactly its atomic weight. Richards won the 1914 Nobel Prize for Chemistry.

ECONOMIC ELEMENT

The most important thulium-containing ores are monazite, which contains 0.002%, and bastnäsite, which contains even less, 0.0008%. The chief ores from which thulium is obtained come mainly from China, USA, Brazil, India, Sri Lanka and Australia. Reserves of thulium are estimated to be about 100 000 tonnes.

World production of thulium is around 50 tonnes per year as thulium oxide. The metal itself was first obtained by heating thulium oxide with lanthanum, although it is now produced by heating thulium fluoride with calcium. Because it is so expensive, and has little to offer that other rare-earth elements cannot provide, thulium finds little application outside chemical research. For example, although thulium-doped zinc sulfide fluoresces with a blue colour, this does not justify its use over other phosphors. However, thulium-doped calcium sulfate has been used in personal radiation dosimeters because it can register, by its fluorescence, especially low exposure levels.

Thulium which has been irradiated in a nuclear reactor can be used as a portable X-ray source for medical use.

ENVIRONMENTAL ELEMENT

Thulium in the environment	
Earth's crust	0.5 p.p.m.
	Thulium is the 61st most abundant element.
Soils	approx. 0.5 p.p.m., range 0.4–0.8 p.p.m.
Sea water	0.25 p.p.t.
Atmosphere	virtually nil

Thulium is the second rarest lanthanide element, after promethium, although it is still seven times as abundant in the Earth's crust as silver. It poses no environmental threat to plants or animals.

CHEMICAL ELEMENT

Data file	
Chemical symbol	Tm
Atomic number	69
Atomic weight	168.93421
Melting point	1545°C
Boiling point	1947°C
Density	9.32 kilograms per litre
Oxide	Tm_2O_3

Thulium is a bright silvery metal and a member of the lanthanide group of the periodic table of the elements. It is soft enough to be cut with a knife. It slowly tarnishes in air, but is more resistant to oxidation than most rare-earth elements. It reacts with water. There is only one naturally occurring isotope, thulium-169, which is not radioactive.

ELEMENT OF SURPRISE

What is surprising about thulium is that, unlike the other rare-earth elements, there is nothing at all surprising about it. While all these elements are very similar to one another, there was generally some feature of note about each one, but there is nothing unique that draws one's attention to thulium. This is probably due to its rarity and cost, which may have deterred people from investigating it or seeking uses for it.

Tin

Pronounced tin, the name comes from the Anglo-Saxon *tin*. The chemical symbol Sn comes from the Latin *stannum*, which may be derived from the Sanscrit word *stan*, which means hard. This is also where the word stannery, the name for a tin mine, comes from. Some tin compounds are also called stannates.
French, *etain*; German, *Zinn*; Italian, *stagno*; Spanish, *estaño*; Portuguese, *estanho*.

HUMAN ELEMENT

Tin in the human body	
Blood	approx. 0.4 p.p.m.
Bone	1.4 p.p.m.
Tissue	0.3–2.4 p.p.m.
Total amount in body	30 milligrams

There is no evidence of humans lacking, or indeed needing, this metal, although at one time it was thought to be part of the stomach hormone, gastrin. It may be essential to some creatures; rats fed on a tin-free diet failed to grow properly, but recovered when a tin supplement was given, suggesting it has some essential role.

Plants easily absorb tin, but it is not essential or even beneficial for them to do so, and most that is absorbed is retained in their roots. Plants growing in uncontaminated soil can have up to 30 p.p.m. of tin; for example tin in wheat grains is 7 p.p.m, and that in corn is 3 p.p.m. Plants growing on contaminated soil can have very high levels of tin: sugar beets grown adjacent to a chemical plant had 1000 p.p.m. (dry weight) while vegetation near a tin smelter reached levels of 2000 p.p.m.

Tin compounds can be poisonous by ingestion, but to what extent the tin they contain is responsible for this is debatable, and inorganic tin compounds are generally regarded as non-toxic. However, organotin compounds are toxic, especially when they have three organic groups attached. Such organotin compounds are able to penetrate biological membranes and once inside a cell they upset various metabolic processes with fatal results. Trimethyl-tin, and especially triethyl-tin, are particularly toxic to humans, but larger organic groups are much less toxic, which is why some, such as tributyl-tin (TBT), are used to protect the hulls of ocean-going ships against marine encrustation. Triphenyl-tin is also not toxic to humans and is used in agricultural crop sprays. Ethyl-tin derivatives are no longer manufactured.

FOOD ELEMENT

Relatively little of the tin in food is absorbed by the body – only about 3% in the case of tin(II) compounds and less than 1% in the case of tin(IV) compounds. That which is absorbed is mainly excreted in the urine, but some tends to collect in the skeleton and liver.

Tin has always been part of the human diet, but the amount in the diet greatly increased when canned food was introduced in the nineteenth century. Attempts have been made to reduce the leaching of tin from the can into the food it contains. The average person takes in around 0.3 milligrams of tin a day, of which 0.2 milligrams comes naturally from food. Some people take in much as 1 milligram per day if they eat a lot of canned foods, in which tin levels of more than 700 p.p.m. have been recorded (this was a sample of tomato purée that had been stored for some time). The lacquering of the inside of cans has considerably reduced the amount of tin that leaches into the food they contain. The USA has imposed limits of 300 p.p.m. for tin in canned food; the UK limit is 200 p.p.m.

The first containers for preserving food were made of glass and produced by the Frenchman Nicholas Appert in 1810. The jars were filled, made airtight, and then sterilized by heating. John Hall and Bryan Donkin in Deptford, England, copied the idea but used tinplate cans instead. Their canned meats were particularly popular with officers of the Royal Navy, and were an essential part of the food stores taken on long voyages of exploration, such as those undertaken to find the North-West Passage around the North of Canada and Alaska. Such voyages could end in disaster; this was attributable to the canned food, not because of the tin, but because of the lead solder used to seal the cans – see lead (p. 233).

MEDICAL ELEMENT

In France in the 1960s and early 1970s, diethyl-tin di-iodide was prescribed as the drug Stalinon by doctors for the treatment of staphylococcal skin infections. This resulted in widespread toxicity and several deaths, which were thought to be due to contamination by triethyl-tin iodide.

ELEMENT OF HISTORY

Tin by itself was of little use but a tin ring and pilgrim bottle have been found in an Egyptian tomb of the eighteenth dynasty (1580–1350 BC) despite the fact that there was no tin ore to be found in Egypt itself. Pure tin has also been found at Machu Picchu, the mountain citadel of the Incas.

When copper was alloyed with a small amount of tin – around 5% was the usual amount – it produced bronze, which not only melted at a lower temperature, so making it easier to work with, but also produced a metal that was much harder, and ideal for tools and weapons. This discovery is commemorated in the term 'Bronze Age', which was a recognized stage in the development of civilization. How bronze was discovered we do not know, but the peoples of Egypt, Mesopotamia and the Indus valley were using bronze around 3000 BC.

It was not known where these early bronze smiths got their tin until, in 1989, an ancient tin mine was discovered in the Taurus mountains of Turkey. Later, tin was traded around the Mediterranean by the Phoenicians who obtained it from Spain, Brittany and Britain, especially from the Scilly Isles and Cornwall. Julius Caesar mentions British tin in his book *Commentaries on the Gallic War*. Tin from Cornwall was appreciated for its quality and was first mined in the early Bronze Age around 2000 BC. Mining continued there until very recently, when it became uneconomic.

Throughout history, bronze was used not only for weapons, but also to create great works of art. Among the most famous was the third century BC Colossus of Rhodes, a 30 metre high statue of the Sun God (destroyed in an earthquake in 224 BC). Other monumental bronzes were the Great Buddha of Kamakura, Japan, made in AD 1252 and the East Door of the Baptistry of St John, Florence Italy, made by Lorenzo Ghiberti in the fifteenth century. The sixteenth century Benin bronzes of Nigeria are another fine example of the art of the bronze smiths.

Tinplate was first mentioned by Theophrastus in 320 BC, and there was a large industry in the Middle Ages in Bohemia and Saxony. In the seventeenth century, Saxony was the centre for tin-plating, after tin mines were established in the area. Tin-working employed upwards of 80 000 men and the goods they made were exported all over Europe. By the mid-eighteenth century, tin was being used to make plates, pans, tankards, teapots and coffee pots as well as for plating lots of other domestic utensils such as cauldrons and basins. About this time the tin smelters of Cornwall were adding bismuth to make tin harder and give it a brilliant lustre.

Tin can be hammered into thin sheets, known as tin foil; this was used for hundreds of years until aluminium foil replaced it.

ECONOMIC ELEMENT

There are a few tin-containing minerals, but only one ore is of commercial significance and that is cassiterite. The main mining areas are to be found in the 'tin belt' which stretches from China through Thailand, Burma and Malaysia to the islands of Indonesia. Malaysia produces around 40% of the world's tin. Other important tin-mining areas are Bolivia and Brazil, which has the largest tin mine of all. Global production is in excess of 140 000 tonnes per year and workable reserves amount to more than 4 million tonnes. Tin concentrates are also produced in large quantities, around 130 000 tonnes per year.

Tin is produced by heating cassiterite with coke in a furnace. The metal itself is used in alloys for solders, bronze, pewter and dental amalgams, although most ends up as tinplate. Tin compounds are used as polymer additives and anti-fouling paints.

Tinplate is made by hot-dipping sheet steel into a solution of tin(II) chloride or tin(II) sulfate, or by electrolytic deposition; this accounts for about a third of the tin that is produced. A tin can is coated with a layer of tin only about 15 micrometres thick. More than 100 billion cans of food are produced every year and about 30% of used cans are collected for recycling.

Tin alloys are employed in many ways: as solder (33% tin, 67% lead); pewter (85% tin, 7% copper, 6% bismuth and 2% antimony); bell metal (76% copper, 24% tin); Babbitt metal (e.g. 89% tin, 7% antimony, 4% copper) which is used for heavy bearings; and dental amalgams (a typical modern one being 60% silver, 27% tin and 13% copper; see under mercury p. 259).[168]

The niobium–tin alloy that corresponds to the formula Nb_3Sn is used for superconducting magnets. As little as 0.1% tin added to cast iron improves its strength and wear resistance. Tin is also added to many other castings such as engine cylinder blocks, crankshafts and brake drums.

A tank of molten tin is used to produce float glass by the Pilkington process. In this, a perfectly smooth and even quality of window glass is obtained by allowing the molten glass to float on the liquid metal. Tin is also deposited electrically on to glass as a thin film, which increases its resistance to breaking and allows thinner-walled containers to be manufactured. Applied as a thicker film it makes glass conducting, and such glass is used for panel lighting.

Tin(IV) oxide is used for ceramics and has been for millenia; the ancient Babylonians produced wall tiles in the ninth century BC covered with a tin-based glaze. Some tin pigments give beautiful colours; for example, tin–vanadium is yellow and cobalt stannate is sky blue (known as Cerulean Blue). Tin oxide is used in gas sensors because as it absorbs a gas its electrical conductivity increases and this can be monitored.

Tin(II) chloride ($SnCl_2$) is used as a mordant in dyeing calico and natural silk. The tin becomes attached to the fabric and then the dye to the tin. In the case of silk dyes the mordant and dye can more than double the weight of the fabric.

Tin(II) fluoride (SnF_2) can be used in fluoride toothpaste because tin(II) is not toxic.

The use to which organotin compounds are put depends upon the number of organic groups attached to the metal. Tin with four organic groups attached is used as catalysts in the production of polyolefins such as polyethylene. Tin with three groups attached, of which TBT is the best known, are used as wood preservatives and anti-fouling paints, and to prevent unwanted growths of moulds on stone structures. They also are effective against fungi and mites, and while they might pose a risk if such fungicides and pesticides were to contaminate food supplies, there is little risk because organotin compounds are rapidly degraded during food processing and cooking. Tin with one or two organic groups is added to PVC to make it less heat-sensitive. Organotin compounds used to be produced on a large scale – more than 40 000 tonnes per year in the 1970s and 1980s – but this has fallen as use of tributyl-tin anti-fouling paint has been restricted.

Zinc stannates are preferred as fire-retardants because they also suppress smoke emission.

[168] Tin can be used to adulterate silver, and in AD 1124, in the reign of King Henry I of England, the men who produced silver coins at the Royal Mint were accused of doing just this and were sentenced to have their right arms amputated. However, forensic investigations in the twentieth century failed to find any of the silver coins of that period with more than a tiny amount of tin, suggesting that the workers may have been wrongly convicted.

ENVIRONMENTAL ELEMENT

Tin in the environment	
Earth's crust	2 p.p.m.
	Tin is the 49th most abundant element.
Soils	approx. 1 p.p.m.
Sea water	4 p.p.t.
Atmosphere	virtually nil

Tin(IV) oxide is insoluble and the ore strongly resists weathering, so the amount of tin in soils, and particularly in natural waters, is low. The concentration in soils is generally within the range 1–4 p.p.m. but some soils have less than 0.1 p.p.m. while peats can have as much as 300 p.p.m. Tin became an environmental problem when TBT was widely used in marine paints which were introduced in the 1960s, proving to be immensely popular since it prevented the drag caused by marine growths on hulls. It reduced the time spent in dry-dock, to the extent that some ships needed repainting only once every 5 years. The TBT in the paint was released at a controlled rate and kept the boat or ship free of growths. Previously copper-containing paints had been used but these were much less effective.

Although TBT saved energy and other resources, to an estimated US$7 billion per year, by the 1980s it was clear that species such as oysters, marine snails and dog-whelks, which flourish in coastal waters, were undergoing strange sexual changes or becoming infertile. The cause was TBT: as little as 1 nanogram per litre of sea water (1 p.p.b.) was capable of causing the observed changes. Countries began banning organotin anti-fouling paints from 1987 onwards for use on boats of less than 25 metres in length, and in harbours and dockyards. In 1988 the US Congress also outlawed TBT.

CHEMICAL ELEMENT

Data file	
Chemical symbol	Sn
Atomic number	50
Atomic weight	118.710
Melting point	232°C
Boiling point	2270°C
Density	white tin, 7.3 kilograms per litre (7.3 grams per cubic centimetre); grey tin, 5.8 kilograms per litre (5.8 grams per cubic centimetre)
Oxides	SnO and SnO_2

Tin is a soft, pliable, silvery-white metal that is a member of group 14 of the periodic table of the elements. It is unreactive to oxygen because it is protected by an oxide film on the surface. It dissolves in acids and bases. Pure tin undergoes a transition when cooled below 13°C from the metal form (white tin, known as β-tin) to a powdery form, known as grey tin or α-tin (see 'Element of surprise').

There are 10 naturally occurring stable isotopes of tin, more than for any other element; of these, tin-120 accounts for 32.6%, tin-118 for 24% and tin-116 for 14.5%. The less abundant isotopes are tin-119 (8.6%), tin-117 (7.7%), tin-124 (5.8%), tin-122 (4.6%), tin-112 (1%), tin-114 (0.6%), and tin-115 (0.4%).[169] Other elements have more naturally occurring isotopes, e.g. xenon and tellurium, but these include radioactive isotopes among their number.

ELEMENT OF SURPRISE

Tin can decay to dust. Below 13°C the metal slowly changes to a powder, an effect known as 'tin plague', and this has been used to explain some famous disasters. However, tin converts from the normal metal to the grey powder only very slowly, even at –10°C, although the conversion becomes much quicker at –33°C, and such low temperatures are not unknown. (The conversion can be prevented by the addition of small amounts of antimony or bismuth to the tin.)

Tin plague really came to public notice during the hard Russian winter of 1850 when the temperature dropped to record lows for weeks on end. During this time tin buttons were found to crumble but the most noticeable effect was on organ pipes in churches which were generally made from tin and which became covered with scaly patches that crumbled when touched. It was even suggested that Napoleon's Grand Army froze to death on its retreat from Moscow in 1812 because the buttons on their uniforms were made from tin which disintegrated. This was only a myth, because at the time the buttons on a soldier's uniform were made of bone. Only officers had metal buttons, and those were made of brass.

The deaths of the members of the Scott expedition to the South Pole in 1912 have also been attributed to tin. The explorers left stores, including cans of paraffin, at various points as they trekked to the South Pole to be used on the way back. When these were needed, however, the cans were found to be empty, the fuel having drained away through tiny holes in the tin-soldered joints. The result was that Scott and his comrades eventually died of exposure. The cans of food in the stores were unaffected and were still found to be edible when some were opened more than 50 years later, but even if the tin of such cans had flaked off, the food inside them would have been preserved by the intense cold.

[169] There are also traces of radioactive tin isotopes in uranium minerals, produced as fission products.

Pronounced tit-ayn-iuhm, this element was named after the Titans, the sons of the Earth goddess of Greek mythology. The word titanate refers to the grouping TiO_3^{2-}.

French, *titane*; German, *Titan*; Italian, *titanio*; Spanish, *titanio*; Portuguese, *titânio*.

COSMIC ELEMENT

Titanium oxide bands are prominent in the spectra of M-type stars. These are the coolest type of star, with a surface temperature of less than 3200°C, and are red in colour.

HUMAN ELEMENT

Titanium in the human body	
Blood	approx. 50 p.p.b.
Bone	approx. 40 p.p.m.
Tissue	1–2 p.p.m.
Total amount in body	approx. 700 milligrams

There is no known biological role for titanium. There is a detectable amount of titanium in the human body and it has been estimated that we take in about 0.8 milligrams a day, but most passes through us without being absorbed. It is not a poisonous metal and for this reason white titanium dioxide has replaced white lead in paint (see below). The human body can tolerate titanium in large doses, and, if anything, titanium has a stimulatory effect.

Most plants contain some titanium, generally about 1 p.p.m. (dry weight), although a few, such as horsetail and nettle, have up to 80 p.p.m. Food plants contain 2 p.p.m. or less. Titanium can have a beneficial effect on plants, stimulating the production of carbohydrates, so there may be some biological role for this element, but the mechanism by which it makes plants grow bigger has yet to be identified.

MEDICAL ELEMENT

In the 1950s, surgeons noted that titanium metal was ideal for pinning together broken bones. It resists corrosion, bonds well to bone and is not rejected by the body. Hip and knee replacements, pace-makers, bone-plates and screws, and cranial plates for skull fractures, can be made of titanium and remain in place for up to 20 years.[170] Titanium implants have been used for attaching false teeth, via metal plugs inserted into the jaw bones using a technique developed by Professor P. I. Branemark of Gothenburg, Sweden. There are people with such implants who have had them for more than 30 years.

Before inserting a titanium device into the body, the device has to be prepared by exposing it to a high-temperature plasma arc which strips off the surface atoms, thereby exposing a fresh layer of the metal which is instantly oxidized. It is to this oxide film that tissue bonds strongly.

ELEMENT OF HISTORY

The first titanium mineral was discovered in 1791 at a remote village in Cornwall, south-western England, by the Reverend William Gregor (1761–1817) who was the vicar of Creed. He was intrigued by a black sand that he had come across at the side of a stream in the nearby parish of Menachan. The sand was odd in that it was attracted to a magnet. Gregor analysed it as best he could and deduced it was made up of two metal oxides – iron oxide, which explained the magnetic properties, plus one that was not so affected, but which he was unable to identify. He realized he had stumbled upon a previously unknown metal and reported it to the Royal Geological Society of Cornwall, and he also announced his discovery in the 1791 edition of the German science journal, *Crell's Annalen*, in which he suggested calling the newly discovered mineral menachanite after the name of the parish where it had been found.[171] Had he been able to isolate the metal, and prove it was a 'new' element, he is said to have thought of calling it menaccin.

Four years later, in 1795, the eminent German scientist Martin Heinrich Klaproth, of Berlin, rediscovered the same element and it was he who gave it the name titanium, derived from the Titans of Greek mythology; he said that he chose this name because in itself it had no meaning so 'could give no rise to any erroneous idea'. Klaproth produced the metal oxide from a specimen of a red ore known as *Schörl*,[172] which he had been sent from Hungary. When he was told of Gregor's earlier discovery, he investigated a sample of menachanite and confirmed it too yielded the same metal oxide.

Neither Gregor not Klaproth lived to see the metal itself. Although Klaproth tried hard to reduce the oxide, he did not succeed. Indeed, titanium is impossible to extract from its oxide ore by heating with carbon: while this does react to remove the oxygen, it reacts further to produce titanium carbide, which is also very intractable. Others in the nineteenth century, who tried to isolate titanium and thought they had succeeded, were invariably disappointed when they discovered that they had obtained only titanium nitride. Those who produced samples of the metal in which there was only a few per cent of impurities found it to be brittle and unworkable.

[170] Prince Charles had his broken elbow repaired with a titanium support.

[171] The mineral is now known as ilmenite.

[172] This is now known to be a form of tourmaline.

It was not until 1910 that M. A. Hunter, working for General Electric in the USA, made pure titanium by heating titanium tetrachloride and sodium metal under high pressure in a sealed vessel. This yielded 99.8% pure titanium, which now revealed itself to be a rather remarkable metal: easily worked, incredibly strong, retaining this property at high temperatures, and very resistant to corrosion.

ECONOMIC ELEMENT

Important titanium minerals are rutile, brookite, anatase (which are all forms of titanium dioxide, TiO_2), ilmenite (iron titanate, $FeTiO_3$) and titanite (calcium titanate silicate, $CaTiSiO_5$). Perovskite (calcium titanate, $CaTiO_3$) may be rich in niobium, cerium and other lanthanide elements and may be a source of these metals rather than titanium. The stars in star sapphires and star rubies are due to titanium dioxide impurities.

The chief mined ore, ilmenite, occurs as vast deposits of sand in Western Australia, Norway, Canada (where it is mixed with haematite) and the Ukraine (where it is mixed with magnetite). Large deposits of rutile in North America and South Africa also contribute significantly to the world's supply of titanium. World production of the metal is around 90 000 tonnes per year, and that of titanium dioxide is 4.3 million tonnes per year. Reserves of titanium amount to more than 600 million tonnes.

Titanium is used either as the metal or as its oxide. Each has become highly important in human affairs in the past 50 years.

Titanium metal

Titanium metal can cope with all kinds of extreme conditions due to the impervious layer of titanium dioxide which immediately forms on the surface of the metal. This layer is only 1–2 nanometres thick to begin with, although it continues to grow slowly, reaching 25 nanometres after about 4 years. This protective layer can resist almost all forms of chemical attack – with the few exceptions noted in the 'Chemical element' section below – and even when it is damaged it instantly repairs itself.[173]

The metal has to be obtained via titanium tetrachloride ($TiCl_4$), which is reduced to the metal by heating it with magnesium metal at 1300°C. The process was first shown to be viable by Justin Kroll in Luxembourg in 1932, although he used calcium metal as the reducing agent. In the modern process the chemical reaction between the titanium tetrachloride and magnesium generates enough heat to keep the process going and the slag of molten magnesium chloride is tapped from the bottom of the reactor and is recycled electrolytically to magnesium metal, and chlorine gas, which goes to making more titanium tetrachloride. Titanium tetrachloride is a crystal-clear, volatile liquid which boils at 136°C and is consequently easy to purify.[174]

Titanium metal is sold in four certified grades which vary according to their oxygen

[173] If the oxide film on titanium is artificially enhanced, by anodic oxidation, it produces an iridescent surface and is then suitable for jewellery, particularly for earrings.

[174] It has been used to create smoke screens because when it reacts with water it produces dense clouds of titanium dioxide particles, and hydrochloric acid vapour around which water droplets condense.

content: grade 1 has 0.18%; grade 2 has 0.25%; grade 3 has 0.35%; and grade 4, the strongest, has 0.40%.

As much as two-thirds of titanium metal ends up in aircraft engines and frames. The remainder goes into in chemical plants, heat exchangers, replacement hip joints and bone pins.

Power plant condensers worldwide contain millions of metres of titanium piping and it is claimed that none has ever failed due to corrosion. Titanium is a strong as steel but 45% lighter and it is used for lightweight alloys for the aerospace industries. It has the added bonus of being immune to metal fatigue. The fan blades of an aircraft engine are likely to be made of an alloy consisting of 90 parts titanium to six parts of aluminium and four parts of vanadium.[175]

Titanium has its drawbacks, such as low thermal and electrical conductivity. It lacks the elasticity of steels, becomes brittle when exposed to hydrogen gas, and tends to stick when in contact with other metals. Despite all this, its uses are increasing because of its inherent benefits, i.e. excellent strength-to-weight ratio and high temperature properties, which make it the preferred metal for gas turbine engines and their housing.

Although it is difficult to produce, titanium metal justifies its cost. The thin oxide layer on the surface of titanium enables it to resist the corroding action of sea water and so it is used in off-shore oil rigs, in the hulls of some submarines, and for propeller shafts, rigging and other parts of ships exposed to sea spray and water. It is also used on board ships for fire pumps, heat exchangers and piping. Titanium metal is ideal for heat transfer applications in which the coolant is sea water or even polluted waters. Desalination plants rely particularly on titanium components. Titanium pipe is used in oil exploration and its light weight and flexibility make it the preferred material for deep-sea exploration

Architects are beginning to use titanium cladding for buildings, an example of which is the Guggenheim Museum. This occupies a waterfront site in Bilbao, Spain, and is sheathed in 33 000 square metres of pure titanium sheet which is guaranteed to resist corrosion for more than 100 years. The roof of the new central station of Hong Kong's railway consists of 6500 square metres of titanium sheeting.

Engineers use titanium in chemical plants where corrosive acids, chlorine and sulfur compounds are handled. Titanium electrodes, often coated with a precious metal, are used in chlorine production, electroplating and other similar applications.

Titanium dioxide

Titanium dioxide is the modern equivalent of white lead, but much safer because titanium compounds are not poisonous. More than half the titanium dioxide manufactured goes into paints, a quarter into plastics and the rest into paper, fibres, ceramics, enamels, food colouring, printing inks and laminates.

The titanium oxide industry started in the 1930s when paint manufacturers were seeking a replacement for white lead and found titanium dioxide to have excellent

[175] The engines of a Boeing 747 jet aircraft contain more than four and a half tonnes of titanium.

covering properties. It is non-toxic, does not discolour and has a very high refractive index,[176] which explains the brilliant whiteness titanium dioxide imparts to domestic appliances such as fridges, washing machines and tumble driers. Titanium dioxide ointment is used as a sunscreen because it prevents ultraviolet rays from reaching the skin.

Titanium dioxide is traditionally made by dissolving titanium ore in sulfuric acid, precipitating the wet oxide and heating it at 1000°C. The more modern process uses chlorine gas to convert the ore to titanium tetrachloride which is then oxidized with oxygen at 1000°C. This regenerates the chlorine gas which is recycled. In an improved chloride process, the titanium tetrachloride is oxidized in a plasma arc at 2000°C. The titanium dioxide is milled to form crystals of the right size; sometimes these are coated with aluminium oxide to make them mix with liquids more easily.

Apart from titanium metal and titanium dioxide, there are other commercial products, albeit of relatively small demand. Titanium carbide, TiC, is made by the action of carbon black on titanium dioxide at 2000°C. It is the most important metallic hard material after tungsten carbide, and in fact is the hardest of all the metal carbides, but it is rarely used by itself since it has a relatively high oxygen content which makes it brittle. It is used mixed with the carbides of tungsten, tantalum and niobium.

Another material is the alloy nitinol, which can 'remember' a previous shape and return to it (see nickel, p. 282).

ENVIRONMENTAL ELEMENT

Titanium in the environment	
Earth's crust	4400 p.p.m.
	Titanium is the ninth most abundant element.
Soils	approx. 0.33% (3300 p.p.m.), range 0.02–2.40%
Sea water	approx. 1 p.p.b.
Atmosphere	traces in dust

Titanium is widespread in rocks and soils and the majority of minerals – especially silicate ones – contain it. Titanium dioxide and titanates are the most stable of all soil components: they are so resistant to weathering that very little titanium is leached into rivers.

In itself, titanium poses no threat to the environment. At times, however, the manufacture of titanium dioxide has been responsible for large-scale pollution of rivers and coasts by acidic effluent, although this is no longer permitted.

[176] Refractive index measures a substance's ability to scatter light. Titanium dioxide's refractive index (2.7) is even greater than that of diamond (2.4).

CHEMICAL ELEMENT

Data file	
Chemical symbol	Ti
Atomic number	22
Atomic weight	47.867
Melting point	1660°C
Boiling point	3287°C
Density	4.5 kilograms per litre (4.5 grams per cubic centimetre)
Oxides	TiO_2 is the most stable, but there are also TiO and Ti_2O_3

Titanium is a hard, lustrous, silvery metal which heads group 4 of the periodic table of the elements, and it is one of the transition metals. The powdered metal will burn if ignited and will also burn in nitrogen, forming titanium nitride. Titanium is unaffected by many acids (except hydrofluoric acid, phosphoric acid and concentrated sulfuric acid) and alkalis.

There are five naturally occurring isotopes of titanium, of which titanium-48 accounts for 74%, followed by titanium-46 with 8%, titanium-47 with 7.5%, titanium-49 with 5.5% and titanium-50 with 5%. None is radioactive.

ELEMENT OF SURPRISES

1. Titanium dioxide is used in paints because of its superb covering power and for the same reason it is added to lipstick. A typical lipstick contains 10% of titanium dioxide, along with oil, wax, emollient, preservative and the all-important dye, which should preferably be one that binds chemically to the lips and so is indelible. Commonly used dyes are the deep red eosins; lighter shades are produced by increasing the amount of titanium dioxide.
2. To date, the shortest laser pulses have been achieved using a titanium-doped sapphire laser, which produced pulses of only 4 femtoseconds (4 million billionths of a second, i.e. 4×10^{-15} seconds). [Walter Saxon]

The transfermium elements

Elements of atomic number 101 and above are called the transfermium elements. All have short half-lives, sometimes as short as fractions of a second, with the result that few of their properties have been recorded. Indeed, for some of them all that is known is that they can be made, and some of them have yet to be named. The naming of these elements has caused controversy in the past, mainly because those who claimed to have been the first to have made them have sometimes not always had those claims verified by the International Union of Pure and Applied Chemistry (IUPAC) whose job it is to approve the names and symbols of the chemical elements. The ones that have been named are as follows: element atomic number 101 is mendelevium; 102, nobelium; 103, lawrencium; 104, rutherfordium; 105, dubnium; 106, seaborgium; 107, bohrium; 108, hassium; 109, meitnerium, and 110, darmstadtium.

None of these elements has a human role, a biological role, an economic role or an environmental role. Even the chemical data are limited to atomic number, isotopes and their half-lives, and sometimes a little information about one or two compounds formed by them.

As Table 4 shows, elements beyond fermium have very unstable nuclei, and they becomes progressively harder to make and detect. The half-lives of the longest-lived isotopes become briefer until we reach element 111, when they begin to increase. Elements 112 to 118 are referred to as the 'island of stability' although their stability is relative and only to be compared with those elements immediately preceding them. The island of stability had been predicted by physicists many years ago, and long before there was any possibility of exploring it. At the centre of the island there should be a peak corresponding to element 114, and especially its isotope 298. This would have 114 protons and 184 neutrons, and nuclear theory has it that this would be particularly stable, maybe for thousands of years, because it corresponds to energy levels within the nucleus being complete and hence more stable. Element 114 would come below lead in group 14 of the periodic table.

Elements adjacent to element 114 would also have enhanced stability of the nucleus, so that element 112 (below mercury), element 113 (below thallium), and element 115 (below bismuth) might all have isotopes that were stable enough for them to be collected, and from which chemical compounds might even be made. This island of stability is expected to extend to element 118, which would be a noble gas and so complete the bottom row of the periodic table.

Table 4 The half-lives of the longest-lived isotopes of the transfermium elements*

Atomic number	Symbol	Isotope	Half-life
101	Md	mendelevium-258	52 days
102	No	nobelium-259	58 minutes
103	Lr	lawrencium-262	216 minutes
104	Rf	rutherfordium-261	65 seconds
105	Db	dubnium-262	34 seconds
106	Sg	seaborgium-266	21 seconds
107	Bh	bohrium-267	17 seconds
108	Hs	hassium-277	16.5 minutes
109	Mt	meitnerium-268	0.07 seconds
110	Ds	darmstadtium-269	270 microseconds
111		272	1.5 milliseconds
112		285	0.28 milliseconds
113			
114		289	30.4 seconds
116		292	50 microseconds

*The longest-lived isotope of fermium is fermium-257 with a half-life of 101 days.

ELEMENT 101: MENDELEVIUM

Pronounced men-del-eev-iuhm, the element is named after Dimitri Mendeleyev, who produced one of the first periodic tables – see p. 519.

French, *mendelévium*; German, *Mendelevium*; Italian, *mendelevio*; Spanish, *mendelevio*; Portuguese, *mendelévio*.

Chemical data: chemical symbol, Md;[177] atomic number, 101; atomic weights of known isotopes range from 245 to 261; the most stable of these are mendelevium-258 (half-life of 52 days) and mendelevium-260 (32 days).

The first atoms of mendelevium were made in 1955 by Albert Ghiorso, Bernard G. Harvey, Gregory R. Choppin, Stanley G. Thompson and Glenn T. Seaborg. Only 17 atoms of mendelevium were produced during an all-night experiment on 18 February of that year. Using the 60-inch cyclotron at Berkeley, California, they bombarded einsteinium-253 with α-particles (helium ions) and mendelevium-256 was detected with a half-life of around 78 minutes. In later experiments, several thousand atoms were made, and now millions of atoms can be produced, enough to demonstrate that the chemistry of this metal would be based on oxidation states mendelevium(II), Md^{2+}, and mendelevium(III), Md^{3+}. It has also been possible to estimate this element's melting point as 827°C.

[177] Mv was used at first but changed to Md in 1963 by IUPAC.

ELEMENT 102: NOBELIUM

Pronounced no-beel-iuhm, and named after Alfred Nobel, Swedish chemist, industrialist and founder of the Nobel Prizes.

French, *nobélium*; German, *Nobelium*; Italian, *nobelio*; Spanish, *nobelio*; Portuguese, *nobélio*.

Chemical data: chemical symbol, No; atomic number, 102; atomic weights of 15 known isotopes range from 249 to 262;[178] the longest-lived isotope is nobelium-259 with a half-life of 58 minutes.

In 1956, the Joint Institute for Nuclear Research in Dubna, USSR, reported element 102, naming it joliotium (symbol Jo) after Irene Joliot-Curie. They produced it by bombarding plutonium-239 and -244 with oxygen-16. Their claim was essentially ignored, but in 1997 IUPAC awarded the discovery of element 102 to the JINR (see Rutherfordium for the background to the decision).

Element 102 was announced, with the proposed name of nobelium, in 1957 by a team of international scientists at the Nobel Institute of Physics in Stockholm, who bombarded curium-244 with carbon-13 ions in a cyclotron and found what they thought was an α-emitting isotope of the new element, with a half-life of 10 minutes. Unfortunately, Russian and American scientists were unable to confirm their findings – and indeed no nobelium isotope with such a half-life has been found. However, their original choice of name was never contested, probably because it was immediately approved by IUPAC in 1957.

Element 102 was made the following year at Berkeley, California, in the Heavy Ion Linear Accelerator, in which atoms of curium-244 and curium-246 were bombarded with carbon-12 ions, and element 102, isotope 254, with a half-life of 55 seconds, was identified. The experiment was performed by Albert Ghiorso, Torbjorn Sikkeland, John R. Walton and Glenn T. Seaborg. Later they proved that it was possible to make several thousand atoms of nobelium within 10 minutes by bombarding californium-249 with carbon-12 ions. Also in 1958, a group of scientists at the Joint Institute for Nuclear Research (JINR) in Dubna, USSR (now Russia), produced atoms of nobelium by bombarding plutonium-239 with oxygen-16.

Research has shown, somewhat unexpectedly, that the chemistry of nobelium in solution would be based on a stable nobelium(II) state, No^{2+}, and a less stable nobelium(III) state, No^{3+}, rather than the other way round.

In 1998, physicists at the Argonne National Laboratory, Illinois, were able to make an atom of nobelium-254, which decayed after 55 seconds, by bombarding lead with calcium ions, and then trap it in a silicon detector and study its γ spectrum. From this they were able to deduce that the nucleus of the atom was a squashed sphere with a high spin.

ELEMENT 103: LAWRENCIUM

Pronounced law-rence-iuhm, the element was named after Ernest O. Lawrence, the inventor of the cyclotron, the research instrument with which several new elements have been first produced.

[178] Nobelium-254 has two isotopic states.

French, *lawrencium*; German, *Lawrencium*; Italian, *lawrentio*; Spanish, *lawrencio*; Portuguese, *laurêncio.*

Chemical data: chemical symbol, Lr; atomic number, 103; the atomic weights of 11 known isotopes range from 252 to 262; the longest lived isotope, lawrencium-262, has a half-life of 216 minutes, undergoing decay by spontaneous fission or electron capture.

Element 103 was first made, an atom at a time, by Albert Ghiorso and co-workers in 1961 at the Lawrence Radiation Laboratories at the University of California at Berkeley, by bombarding a 3 microgram target of californium with boron ions. They claimed to have identified isotope-257 with a half-life of 8 seconds, proposing the name lawrencium and symbol Lw. However, this symbol was not accepted and was changed to Lr in 1963 by IUPAC, when the name and symbol were approved.

The researchers at the JINR in Dubna, USSR (now Russia) said they were unable to register such an isotope; it later transpired that it was probably isotope 259, which has a half-life of 6 seconds. In 1965, the Russian team made lawrencium-256 with a half-life of 35 seconds, and this was confirmed by the Berkeley squad.

The workers at Berkeley were able to extract lawrencium atoms from an aqueous solution into an organic solvent and deduce that, as expected, its preferred oxidation state was lawrencium(III), i.e. Lr^{3+}; there were indications that the lower oxidation state, Lr^{2+}, could also exist, but was much less stable.

ELEMENT 104: RUTHERFORDIUM

Pronounced ruth-er-ford-iuhm, the element is named after the New Zealand chemist Ernest Rutherford, who was one of the first to explain the structure of atoms and who won the Nobel Prize for Chemistry in 1908.

French, *rutherfordium*; German, *Rutherfordium*; Italian, *rutherfordio*; Spanish, *rutherfordio*; Portuguese, *ruterfórdio.*

Chemical data: chemical symbol, Rf; atomic number, 104; the atomic weights of 14 known isotopes range from 253 to 266; the longest lived isotope, rutherfordium-261, has a half-life of 65 seconds, undergoing decay by one of three processes: α-emission, spontaneous fission or electron capture.

The first claim for element 104 came in 1964 from a group of scientists headed by Georgy Flerov at the JINR, Dubna, USSR (now Russia). They reported isotope 260 with a half-life of 0.3 seconds, which they had obtained by bombarding plutonium-244 with neon-22 ions. They named the element kurchatovium (symbol Ku) after Igor Kurchatov, the head of the Russian nuclear research programme. The claim was disputed in 1969 by a group of scientists led by Albert Ghiorso at the Lawrence Radiation Laboratory of the University of California at Berkeley (later named the Lawrence Berkeley National Laboratory (LBNL)).

They used the Heavy Ion Linear Accelerator to make isotope-257 and isotope-259 which they named rutherfordium (symbol Rf) in honour of Ernest Rutherford. By bombarding californium-249 with carbon-12 and carbon-13 nuclei, they made several thousand atoms of element 104, isotopes 257 and 259, with half-lives of 4.7 seconds and 3.1 seconds, respectively (decaying to nobelium-253 and nobelium-255, respectively). They also made isotope 258 in the same way, and this has a half-life of 12 milliseconds.

Element 104 had been given two names, but which should be chosen? It had always been accepted that the first person to discover an element had the right to propose a name for it. A protracted period of controversy over priority of discovery followed, one that came to include later-discovered elements 105, 106 and 107. This wrangling was not completely resolved until 1997. By the 1980s it had become clear that since elements 104–109 had been discovered without being officially named, something had to be done to resolve the controversies and get names approved for these elements. Two unofficial names were already in use for elements 104 and 105.

IUPAC, in 1979, proposed a temporary compromise for names, and set up a committee to look into the disputes over the rival claims. It was decided that elements should be named solely on their atomic number, according a scheme in which 0 would be called 'nil', 1 would be called 'un', 2 'bi', 3 'tri', 4 'quad', 5 'pent', 6 'hex', 7 'sept', 8 'oct' and 9 'enn'. Since all the elements in this part of the periodic table are metals, their names were all given the ending '-ium'. Accordingly, the IUPAC name for element 104 was thus unnilquadium (un-nil-quad-ium). Similarly, elements 105 and 106, whose claim to discovery were also contested, became unnilpentium and unnilhexium.

In 1985, IUPAC, in conjunction with the International Union of Pure and Applied Physics (IUPAP), set up the Transfermium Working Group to look into the rival claims to elements 104–107, and to review all elements with atomic numbers higher than fermium. In 1992, the working group issued its report and recommendations. Its conclusions were somewhat vague, and they said that both the Russian and American researchers had some justification for making their claims. Meanwhile, in 1994, the Lawrence Berkeley National Laboratory brought matters to a head by naming element 106 seaborgium, which IUPAC rejected on the grounds that it was wrong to name an element after a living scientist.

Later that year, IUPAC's Commission on Nomenclature for Inorganic Chemistry suggested a new set of names for these elements: element 104 was to be called dubnium, element 105 was to be joliotium[179] (which had been originally proposed by the Dubna group), element 106 to be rutherfordium (which had previously been the name for element 104), element 107 to be bohrium and element 108 to be hahnium (which previously had been the name for element 105).

These names did not go down at all well with the American Chemical Society which defied IUPAC by approving the name seaborgium for element 106 at its annual meeting in 1995, and ordering all its publications to use only this name. The confrontation led to 2 years of negotiation between the two groups. In the end the names were agreed as follows: element 104 would be called rutherfordium, element 105 dubnium and element 106 seaborgium. These names were approved by IUPAC in 1997 along with names for elements 107–109.

[179] After the French physicist Frédéric Joliot (1900–58) who, with his wife Irène Curie, produce the first synthetic isotopes in 1933. The Russians had proposed this name in 1956 for element 102, naming it after Irène Joliot.

ELEMENT 105: DUBNIUM

Pronounced dub-nee-um, the element is named after Dubna, the Russian town where the JINR is situated.

French, *dubnium*; German, *Dubnium*; Italian, *dubnio*; Spanish, *dubnio*; Portuguese, *dúbnio*.

Chemical data: chemical symbol, Db; atomic number, 105; the atomic weights of nine known isotopes range from 255 to 263; the longest-lived isotope, dubnium 262, has a half-life of 34 seconds, undergoing decay by one of three processes: α-emission, spontaneous fission or electron capture.

Two isotopes, 260 and 261, were reported in 1967 by a group of scientists led by Georgy Flerov at the JINR at Dubna. They produced the isotopes of element 105 by bombarding americium-243 with neon-22 ions. The element was not given a name, as they waited to do more research to confirm their finding. In 1970 they proposed calling it nielsbohrium (symbol Ns) in tribute to Niels Bohr, the Danish atomic physicist and winner of the 1922 Nobel Prize for Physics.

The claim was disputed in 1970, when a team headed by Albert Ghiorso at the Lawrence Radiation Laboratory of the University of California at Berkeley, reported isotope 260 of this element. This had been made by bombarding a target of californium-249 with nitrogen-15 using the Heavy Ion Linear Accelerator. They measured its half-life at 1.5 seconds and named it hahnium (symbol Ha) after Otto Hahn, the German chemist who first reported uranium fission. Since then several atoms of element 105 have been made in the same way.

Element 105 was also caught up in the controversy over names, and for several years was called both nielsbohrium and hahnium in the literature, with hahnium becoming quite well-known. The issue was only settled in 1997 by naming it dubnium, in acknowledgement of the work of the Russian group, while their name for the element, nielsbohrium, was given to element 107, albeit shortened to bohrium.

A first attempt at assessing dubnium's chemical behaviour was made on isotope 262, produced by bombarding berkelium-249 with oxygen-18. It belongs to group 5 of the periodic table, and should especially resemble tantalum in having a preferred oxidation state of V, but these first experiments suggested this was not the case.

ELEMENT 106: SEABORGIUM

Pronounced see-borg-iuhm, the element is named after Glenn T. Seaborg who had been instrumental in producing several transuranium elements.

French, *seaborgium*; German, *Seaborgium*; Italian, *seaborgio*; Spanish, *seaborgio*; Portuguese, *seabórgio*.

Chemical data: chemical symbol, Sg; atomic number, 106; the atomic weights of 10 known isotopes range from 258 to 266 plus 269; the longest-lived isotope is seaborgium-266, with a half-life of 21 seconds.

The first atom of seaborgium was reported to have been synthesized at the JINR in Dubna in June 1974, by a team headed by Geogry Flerov and Yuri Oganessian. A target of lead-207 and lead-208 was fused with chronium-54 in the cyclotron there, to yield isotope-259. This was the first use of the so-called cold fusion method as a way of discovering an element. The result, however, could not be confirmed.

At the same time, a team headed by Albert Ghiorso, at the Lawrence Berkeley National Laboratory in California, reported isotope 263 with a half-life of 0.8 seconds, which they made in the Heavy Ion Linear Accelerator by bombarding californium-249 with oxygen-18 ions. This experiment was repeated in 1993 using the 88-inch-diameter cyclotron, and the result confirmed the earlier findings.

Neither the Russians nor the Americans proposed a name for element 106 – the years of controversy over elements 104 and 105 had cautioned them to wait until their respective discovery claims had been adjudicated. In 1992 the Transfermium Working Group, set up by the IUPAC and IUPAP (see above), issued its report which suggested that although the Russian group had discovered the element first, the American work was the more convincing and that the discovery should be shared. The Lawrence Berkeley National Laboratory people refused to accept this, wanting sole recognition, and this they were accorded by IUPAC in 1993 when they submitted their confirming results.

This cleared the way for Ghiorso to choose a name for the element and in 1994 he decided to call it seaborgium, even though IUPAC rejected it on the grounds that it was inappropriate to name an element after a living scientist. What followed, and how the issue was resolved, is detailed under rutherfordium above.

Several atoms of seaborgium have since been made by bombarding californium-249 with oxygen-18 ions using an 88-inch-diameter cyclotron, which produces one seaborgium atom per hour. Seaborgium-269 was also detected as part of the α-decay chain of elements formed by element 118; seaborgium-265 was also detected in the decay of element 112.

The little research that has been carried out on seaborgium's chemistry suggests that it prefers oxidation state VI and forms an oxy-anion, SgO_4^{2-}, and a compound SgO_2Cl_2, which is entirely in line with its position in group 6 of the periodic table, below tungsten. The work was done using only seven atoms of the element to study its chemistry in the gas phase and in solution.

ELEMENT 107: BOHRIUM

Pronounced bor-iuhm, the element is named after Niels Bohr, the Danish atomic physicist and winner of the 1922 Nobel Prize for Physics. He was the first person to merge atomic and quantum theory to give the correct explanation of atomic structure, which he did in 1913.

French, *bohrium*; German, *Bohrium*; Italian, *bohrio*; Spanish, *bohrio*; Portuguese, *bóhrio*.

Chemical data: chemical symbol, Bh; atomic number, 107; the atomic weights of eight known isotopes range from 260 to 267; the longest-lived of these isotopes is bohrium-267, with a half-life of 17 seconds.

The credit for the discovery of bohrium was accorded to the Laboratory for Heavy Ion Research – the Gesellschaft für Schwerionenforschung (GSI) – in Darmstadt, Germany. There was, however, a controversy over who first discovered element 107. In

1976, a team headed by Yuri Oganessian, at the JINR in Dubna, USSR (now Russia), claimed its discovery after bombarding a target of bismuth-209 with chromium-54 ions, but their evidence was weak. In 1981, the GSI also bombarded bismuth-209 with chromium-54 and produced element 107. Only in 1983 was the JINR experiment convincingly successful, when they synthesized bohrium-260 in their 157-inch cyclotron.

In 1985, the IUPAC, in conjunction with the IUPAP, set up the Transfermium Working Group to study the controversies over elements 104–107 and in 1992 they issued a report of assessments and recommendations. In a strangely worded conclusion, they said the GSI should be awarded the discovery for bohrium because they had the more credible submission, but the JINR had probably discovered it first. As a result they recommended that the two laboratories share in the discovery by conferring with each other on the choosing of a name. The two groups agreed to do this, and they submitted the name nielsbohrium in honour of Niels Bohr, and then agreed to shorten it to bohrium in line with other elements named after individuals which were based on the surname alone. In 1997 the name was finally approved by IUPAC.

In the year 2000, the chemical characterization of bohrium was carried out on atoms generated in the Philips cyclotron at the Paul Scherrer Institute, at Berne in Switzerland. Although only six atoms of bohrium were made, three of these were swept into an atmosphere of oxygen and hydrogen chloride, there to react to form the oxychloride, BhO_3Cl, whose 'volatility' was assessed to be 180°C. In this respect the research confirmed that bohrium really was a member of group 7 of the periodic table, coming below rhenium, whose oxychloride, ReO_3Cl, has a boiling point of 131°C.

ELEMENT 108: HASSIUM

Pronounced hass-iuhm, the element is named after the German state of Hesse where it was first made at the GSI in Darmstadt, Germany.

French, *hassium*; German, *Hassium*; Italian, *hassio*; Spanish, *hassio*; Portuguese, *hássio*.

Chemical data: chemical symbol, Hs; atomic number, 108; the atomic weights of eight known isotopes range from 263 to 269 and 273 to 277; the longest-lived isotopes are hassium-277, with a half-life of 16.5 minutes and hassium-285 with a half-life of 15.4 minutes.

Hassium was first synthesized in 1984 by Peter Armbruster, Gottfried Münzenberg and co-workers at the GSI in Darmstadt, in the German state of Hesse by bombarding lead-208 with iron-58 ions. Two isotopes were produced: hassium-264, with a half-life of 0.080 milliseconds, and hassium-265, with a half-life of 1.8 milliseconds, both of which decayed by α-emission. Armbruster and colleagues decided to call the new element hassium, whose name they derived from Hassia, the Latin name for Hesse.[180]

The Russian group at the JNIR in Dubna also produced element 108 in the same year, using the same method, in which a target of lead-208 was bombarded with atoms of iron-58 to give an atom of hassium.

[180] The IUPAC committee thought the German state of Hesse did not merit having an element named in its honour and suggested hahnium instead, after the late German radiochemist, Otto Hahn, but the discoverers rejected this suggestion and the name hassium was ratified by IUPAC in 1997.

A later joint venture by Russian and American scientists at Dubna, and led by Yuri Lazarev, produced hassium-267 by the reaction of uranium-238 and sulfur-34. The new isotope had a half-life of 19 milliseconds and also decayed by α-emission.

The GSI in 2001 studied the chemical properties of hassium, producing hassium tetroxide (HsO_4); it was similar to osmium tetroxide (OsO_4), indicating its chemical relationship to osmium, which comes above it in the periodic table. The study was done on seven atoms of hassium.

Due to the controversies surrounding the discovery of elements 104–107, no name was proposed for this element until 1992, when an IUPAC evaluation committee issued its report. The name hassium was approved in 1997; see under Rutherfordium for details.

ELEMENT 109: MEITNERIUM

Pronounced mite-neer-iuhm, the element is named after the Austrian physicist Lise Meitner, who discovered the element protactinium and was the first to suggest that radioactive atoms could undergo nuclear fission. As a member of the team that discovered fission in 1938, she was the first to recognize that this nuclear event had occurred.

French, *meitnerium*; German, *Meitnerium*; Italian, *meitnerio*; Spanish, *meitnerio*; Portuguese, *meitnério*.

Chemical data: chemical symbol, Mt; atomic number, 109; the atomic weights of six known isotopes range from 265 to 270; the longest-lived isotope is meitnerium-268, with a half-life of 0.07 seconds.

Meitnerium was first made in 1982 at the Laboratory for Heavy-Ion Research at the GSI in Darmstadt, Germany, by a group headed by Peter Armbruster and Gottfried Münzenberg. They bombarded a target of bismuth-209 with accelerated iron-58 ions. After a week, a single atom of element-109, isotope 266, was detected. This decayed after 5 milliseconds by α-emission to give bohrium and then dubnium, and then, via electron capture, to rutherfordium-258.

Other isotopes of meitnerium have been observed as part of the decay sequence of even heavier elements – see below.

None of meitnerium's chemistry has been researched, but it should resemble other elements of group 9, i.e. be like iridium.

Due to the controversies surrounding the discovery of elements 104–107, no name was proposed for this element until 1992, when an IUPAC evaluation committee issued its report. The name meitnerium was approved in 1997; see under Rutherfordium for details.

ELEMENT 110: DARMSTADTIUM

Pronounced darm-stat-iuhm, the element is named after Darmstadt in Germany where it was first made at the GSI.

Chemical data: chemical symbol, Ds; atomic number 110; the atomic weights of the four reported isotopes are 269, 271, 273 and 281; the longest-lived isotope is darmstadtium-269 with a half-life of 270 microseconds.

The first claim to having made element 110 came, in 1987, from an international team headed by Yuri Oganessian and based at the Joint Institute for Nuclear Research (JINR), in Dubna, USSR (now Russia).

They said they had made it by two routes: bombarding a uranium target with argon-40 ions, and by bombarding thorium with calcium-44 ions. The new element, isotope-276, had a half-life of 9 ms. The claim was disputed. In 1991, Albert Ghiorso and co-workers at the Lawrence Berkeley National Laboratory claimed to have made element 110, isotope 267, but again the proof was less than convincing.

Then in 1994 Peter Armbruster, heading a team of German, Russian, Finnish, and Slovakian scientists at the GSI in Darmstadt, Germany, bombarded a lead target with nickel ions of energy 311 million electron volts at a rate of 3 trillion ions per second. They produced three atoms of element 110, isotope 269, which had a half-life of 270 microseconds, decaying by a chain of α-emissions to hassium, seaborgium and rutherfordium. Atoms of darmstadtium-269 were also produced using the same process at Berkeley in 2003.

A second isotope, darmstadtium-271, was also produced at the GSI in 1994 by bombarding lead-208 with nickel-64 ions.

In 1995, a team of Russian and American physicists, headed by Yuri Lazarev, working at the JINR in Dubna, made isotope 273 by bombarding plutonium with sulfur-34 ions. Its half-life may be around 300 microseconds and it decayed via a similar cascade of α-emissions as darmstadtium-269.

Isotope 273 was also detected as part of the α-decay chain of elements from element 112. In 1998, the JINR detected isotope 281 as part of the α-decay chain from its discovery of element 114.

From its position in the periodic table, below platinum, this element should have the physical properties of a platinum metal and, were it long-enough lived, it should be possible to make a wide variety of compounds, including several oxides: DsO, DsO_2 and DsO_3.

UNNAMED (AS YET) ELEMENTS

Those yet to be named have atomic numbers between 111 and 116. The only data about them is when, by whom and how they were made, their half-lives and the mode of their radioactive decay. (In what follows the symbol M is used for the unnamed element in each case.) They are provisionally named according to the IUPAC rules, explained under rutherfordium above, although these names are not particularly attractive. Eventually they will be superseded by names proposed by the researchers and approved by IUPAC.

Element 111: Unununium (Uuu)

Three atoms of element 111 were made in 1994 by an international team led by Peter Armbruster at the GSI in Darmstadt, Germany, by bombarding a bismuth target with high-energy nickel-64 ions. Isotope 272 was formed with a half-life of 1.5 milliseconds. It decayed through a chain of α-emissions to meitnerium-268, bohrium-264, dubnium-260 and lawrencium-256.

From its position in the periodic table, in group 11 below gold, this element should have the physical properties of a noble metal and, were it long-enough lived, it should be possible to make compounds of it although, like gold, it might be reluctant to form them.

Element 112: Ununbiium (Uub)

In 1996, scientists at the GSI in Darmstadt, Germany, reported element 112. The group, headed by Peter Armbruster and Sigurd Hofmann, detected an atom of this in an experiment in which they bombarded for 2 weeks a lead target with high-energy zinc ions travelling at 30 000 kilometres per second. It was isotope 277 and had a half-life of 0.28 milliseconds, decaying by a chain of α-emissions to element 110, hassium, seaborgium, rutherfordium, nobelium and fermium.

In 1998, isotope 285 was produced in the α-decay chain resulting from the discovery of element 114 by the JINR. Isotope 284 was discovered by the JINR 2 years later as an α-decay product from their synthesizing isotope 292 of element 116.

From its position in the periodic table, in group 12 below mercury, this element should have the physical properties of a heavy metal and, were it long-enough lived, it should be possible for it to have two kinds of chemistry, corresponding to oxidation states M(I) and M(II), with the latter more stable.

Element 113: Ununtriium (Uut)

This has yet to be reported. From its position in the periodic table, in group 13 below thallium, this element should have the physical properties of a main group metal and, were it long-enough lived, it should be possible for it to have two kinds of chemistry, corresponding to oxidation states M(I) and M(III) of roughly equal stability.

In 1998 the GSI tried to synthesize element-113 by bombarding bismuth-209 with zinc-70 nuclei in an unsuccessful 42-day run. [Walter Saxon].

Element 114: Ununquadium (Uuq)

It has long been predicted that an atom with 114 protons and 184 neutrons, i.e. element 114, isotope 298, would be particularly stable. In early 1999, a team headed by Yuri Oganessian and Valdimir Utyonkov at the JINR at Dubna, Russia and Kenton Moody at the Lawrence Livermore National Laboratory (LLNL), Livermore California, USA announced that they had made atoms of element 114 the previous year in December 1998, albeit not the 298 isotope, and that these were many times more stable than the elements preceding it.

They reported isotope 289, made by bombarding a plutonium-244 target with a high-energy beam of ions of the rare, naturally occurring, isotope calcium-48. The work was done in collaboration with the LLNL, who supplied the plutonium target material. A single atom of element 114 was produced, and it decayed after 30.4 seconds. The researchers calculated that 5 billion billion (5×10^{18}) atoms of calcium had been fired at the target during the course of the 40 day experiment and that only one atom of element 114 had been formed!

In 1999, an international team based at Dubna made two atoms of isotope 287 by bombarding plutonium-242 with calcium-48. This new isotope had a half-life of 5 seconds and decayed by α-emission to give element 112, which then underwent spontaneous fission.

In the year 2000, isotope 288 was discovered by the JINR, as part of an α-decay product of their successful synthesizing of a new isotope of element 116, isotope 292.

From its position in the periodic table, in group 14 below lead, element 114 should have the physical properties of a heavy metal and it should be possible for it to have two kinds of chemistry, corresponding to oxidation states M(II) and M(IV), with the former more stable.

Element 115: Ununpentium (Uup)

This has yet to be reported. From its position in the periodic table, in group 15 below bismuth, this element should have the physical properties of a heavy metal and it should be possible for it to have two kinds of chemistry, corresponding to oxidation states M(III) and M(V), with the former being more stable.

Element 116: Ununhexium (Uuh)

Element 116 was discovered in 2000 at the JINR in Dubna, Russia, by a team headed by Yuri Oganessian, Vladimir Utyonkov, and Kenton Moody. A curium-248 target was bombarded with calcium-48 nuclei to get isotope-292. This was done in collaboration with the LLNL, who supplied the target material. At the time, the researchers simply thought that they were synthesizing a new isotope of an already discovered element.

In 1999, the LBNL had announced the discovery of elements 118 and 116, the latter being produced by the alpha decay of the former. However, the LBNL withdrew its claim in 2001 after five other experiments, at various laboratories around the world, had attempted to make 118 but had failed to confirm its existence. Re-analysis of the original LBNL data uncovered a fraud perpetrated by one of its head researchers – see also element 118. From its position in the periodic table, in group 16 below polonium, this element should have the physical properties of metal and it should be possible to make various compounds of it, such as oxides MO_2 and MO_3, and fluorides, such as MF_4 and MF_6

Element 117: Ununseptium (Uus)

This has yet to be reported. From its position in the periodic table, in group 17 below astatine, this element should have the physical properties of a halogen or a metalloid and, were it long-enough lived, the element should exist as a volatile M_2 molecule, form salts with many of the other elements, as does iodine, and several other compounds, including acids, oxy acids, and fluorides such as MF_5 and MF_7.

When it has been made and christened, this element is likely to be given a name ending in '-ine' to fit with the other members of group 17 of the periodic table, the halogen elements.

Element 118: Ununoctium (Uuo)

In 1999, the LBNL bombarded lead-208 with krypton-86 nuclei, claiming to have made element-118 isotope-293 (and element 116 as its alpha-decay product). LBNL re-ran the experiment in 2000 and 2001 but could find no evidence of element-118. Nor was it found when others searched for it at the GSI in Germany, the Institute of Physical and Chemicals Research (RIKEN) in Japan, and the Great National Heavy

Ion Accelerator (GANIL) in France. In 2001, LBNL retracted its claim and investigated, uncovering a fraud by head researcher Victor Ninov who had faked data to make it appear that three atoms of element-118 had been synthesized. [Walter Saxon]

From its position in the periodic table, in group 18 below radon, this element should have the physical properties of a noble gas and, were it long-enough lived, it should be possible to make a few compounds of it, such as fluorides. It might also be expected to take a name ending in '-on', in accord with the noble gas elements above it in the periodic table.

Element 119: Ununennium (Uue)

This has yet to be reported. From its position in the periodic table, in group 1 below francium, this element should have the physical properties of an alkali metal and, were it long-enough lived, the element should display the singly-charged ion M^+ as its most favoured chemical state.

Robert Smolanczuk, the nuclear theorist who suggested a possible method for making atoms of elements 116 and 118, has also indicated that it should be possible to synthesize element 119, isotope 294 using a lead-208 target bombarded by rubidium-87 nuclei. If this were to prove a success, the α-decay chain from element 119 would produce elements 117, 115 and 113, thereby allowing three additional elements also to be discovered at the same time.

Tungsten *or* Wolfram

Pronounced tung-sten, the name is derived from the Swedish *tung sten*, meaning heavy stone. The chemical symbol, W, is derived from wolfram, the old name of the tungsten mineral wolframite. The International Union of Pure and Applied Chemistry (IUPAC) allows wolfram as an alternative name for the element, and indeed it is the preferred one. The element got the name wolfram ('wolf dirt') from German tin miners who found that the presence of certain stones greatly interfered with the smelting of tin and produced a lot more slag than usual; they explained this in terms of the tin being 'devoured,' like a wolf devours sheep. Tungstate refers to the ion WO_4^{2-}.

French, *tungstène*; German, *Wolfram*; Italian, *wolframio* (*tungsteno*); Spanish, *wolframio*; Portuguese, *tungsténio*.

HUMAN ELEMENT

Tungsten in the human body	
Blood	1 p.p.b.
Bone	0.2 p.p.b.
Tissue	not known
Total amount in body	20 micrograms

Tungsten is only mildly toxic, although tungsten dust is a skin and eye irritant. The estimated daily intake of tungsten is around 12 micrograms per day but only a small amount of this is absorbed, as shown by tests using a radioactive tungsten tracer. That which is absorbed tends to deposit in the bones and spleen.

Plants will pick up tungsten from the soil in which they grow and some trees in the Rockies have been found to have up to 100 p.p.m. Food plants generally have very little and some none at all; and those in which it was detected rarely exceeded 0.1 p.p.m. (dry weight). It was observed that watering with sodium tungstate solution increases the growth and yields of grapes and alfalfa. Tests using radioactive tungsten-185 showed that barley removed a large part of that which was applied to a sample of soil and appeared to absorb it in the form of the tungstate. There has even been a suggestion that some plants may have a biological need for tungsten, in that it might substitute for molybdenum in nitrogen-fixing enzymes, but the evidence is slim. In addition, enzymes that occur in anaerobic bacteria, such as *Clostridium,* seem to work quite well using tungsten. However, in 1973 tungsten was positively identified in the enzymes of hyperthermophilic archaea, which are organisms that thrive around hydrothermal vents under the sea. Bacteria which live in these environments also need tungsten enzymes to generate ATP.

ELEMENT OF HISTORY

Minerals containing tungsten came to the attention of investigators in the mid-eighteenth century. Some, like the German chemist Johann Gottlob Lehmann in 1761, analysed wolframite without recognizing that it contained two unknown metals, tungsten and manganese. In 1779, the Swedish chemist, Peter Woulfe, examined the mineral called *tung sten*, which was to be found in Sweden and in Germany, and concluded that it probably contained a hitherto unknown metal, but got no further.

Two years later, his fellow countryman, Carl Wilhelm Scheele, investigated the mineral and succeeded in isolating an acidic white oxide which he concluded was not that of molybdenum, which he had suspected the mineral might contain. He rightly deduced that it was the oxide of a 'new' metal. However, Scheele left off investigating it before he could show conclusively that it was an unknown element.

The credit for discovering tungsten went to two brothers, Juan José Elhuyar (1754–96) and Fausto Elhuyar (1755–1833), who came from northern Spain. They were interested in mineralogy and studied at some of the main European centres in Sweden and Germany. It was at the Seminary at Vergara, in Spain, that they carried out their researches into *tung sten* and wolframite in 1783, showing that both yielded the same acidic metal oxide, some of which they were even able to reduce to the metal itself by heating it with carbon. Fausto wanted to call the newly discovered element wolfram, while Juan preferred tungsten, the name by which the ore was better known, and this became this preferred name for the metal in England and France. In Germany, Spain and Italy it was wolfram.[181]

Across the world in China, porcelain manufacturers of the seventeenth century produced a unique peach colour with the aid of a tungsten pigment, although without realizing its nature.[182]

ELEMENT OF WAR

Just as adding tin to copper produced bronze, which made stronger swords and axes, so adding tungsten to iron produced a superior type of steel, much used in weaponry in the wars of the twentieth century. It all began in 1864 when the Englishman, Robert Mushet, added about 5% of tungsten to steel and produced a metal that was not only

[181] The older Elhuyar brother, Juan, became Director of Mines in the Spanish colony of Colombia, where he died in 1796, while his younger brother, Fausto, was appointed to the same position in Mexico. He set up the School of Mines in Mexico City, and as a member of the government he oversaw the development of mining in that country before returning to Spain in the early 1820s, where he became Director General of Mines for the whole of Spain.

[182] Tungsten is still used in pigments as bright yellow tungsten oxide or bright white barium and zinc tungstates.

harder and stronger, but could withstand red heat without deforming. Mushet's self-hardening steel, as it was called, first found use in machine tools, enabling metal cutters to work much faster and to last longer.

Soon, tungsten steels were being used to make guns and cannons, and in World War I the German armament makers perfected some that could fire up to 15 000 rounds – in contrast, Russian and French weapons became unserviceable after firing about half this number of shells or bullets. As the benefits of tungsten steel were appreciated, the demand rose rapidly, so that whereas only a few hundred tonnes of tungsten were produced in the late nineteenth century, by 1918 demand had soared to 35 000 tonnes, creating a chronic shortage of the metal in war-torn Europe.

The far-sighted German gun makers had stockpiled the metal early in the twentieth century against such an eventuality, but this was soon used up. The shortage became so acute that even the slag heaps of old tin mines and smelters were reworked to extract the tungsten that had so bedevilled the Mediaeval miners of Germany. In World War II, the mines in neutral Portugal, at Panasqueira and Borralha, became major producers.

Today, tungsten heavy metal, also known as heavimet or densalloy, is produced for armaments; it is made from tungsten sintered with either iron–copper or iron–nickel and has a density of around 18 grams per cubic centimetre. This is used for high-velocity penetrating shells as anti-tank weapons.

ECONOMIC ELEMENT

There are several recognized tungsten minerals, such as ferberite, $FeWO_4$, scheelite, $CaWO_4$, and wolframite, $(Fe,Mn)WO_4$. Only the last two are important and the main mining area is China, which today accounts for more than two-thirds of the world's supply. Other places with active tungsten mines are Russia, Austria, Bolivia, Peru and Portugal. Many other countries have had tungsten mines but most of these became uneconomical to work. World production is around 40 000 tonnes per annum and reserves are estimated to be around 5 million tonnes, most of which is in China (3.5 million tonnes), Canada (600 000 tonnes) and the USA (400 000 tonnes), with the Canadian deposits as yet virtually untouched. Tungsten is also recycled and this meets 30% of demand.

Tungsten is generally obtained as a dull grey powder, which is difficult to melt. As the pure metal, it is easily worked, can even be cut with a hacksaw and is very ductile (a gram of the metal can be drawn into a wire 400 metres long), but tungsten generally contains a small amount of carbon and oxygen, which makes it much harder.

Pure metallic tungsten is used in light bulbs, electric contacts and arc-welding electrodes, and for heating elements in high-temperature furnaces. Tungsten does not easily volatilize because it has the highest boiling point of any material, 5700°C.[183]

In 1903, W. D. Coolidge succeeded in preparing tungsten wire by doping tungsten oxide before reducing it to the metal and then pressing the tungsten powder he obtained into thin rods, which he was able to draw out into fine wire; this became the ideal material for filaments in incandescent light bulbs. Not only is it high melting, it

[183] It would only just be molten on the surface of the Sun.

has the lowest expansion factor of all metals and the same as that of glass, so it can be sealed into the glass. Moreover, its vapour pressure is the lowest of all metals, which means that the filament does not evaporate and redeposit on the cooler parts of the light bulb. Although incandescent light bulbs are becoming less popular for domestic and indoor lighting, on account of the energy they waste by producing heat as well as light, they are still used for speciality lighting in film projectors, video camera lights, airport runway markers, photocopiers and stage lighting.

Tungsten is used in alloys, such as steel, to which it imparts great strength. So-called high-speed steels contain around 7% tungsten and other metals, and these are used for various tools such as taps, milling cutters, gear cutters, saw blades, punches and dies. However, these grades of steel are declining as cemented carbide tools take their place.

Cemented carbide, also known as hardmetal or cermet, is the most important use for tungsten. Its main component is tungsten carbide (WC), which is made by mixing tungsten powder with pure carbon powder and heating to 2200°C in a furnace. Various grades of cemented carbides are made, with between 75% and 96% of tungsten carbide.

Cemented carbide accounts for 40% of the world production of tungsten. It was invented in 1923 by K. Schroter, who combined cobalt and WC. Cemented carbide has the strength to cut cast iron and it makes excellent cutting tools for the machining of steel; indeed it revolutionized productivity in many industries when it was introduced in the 1930s. For example, drills with cemented carbide tips increased the lifetime of rock drills by a factor of 10 and these were particularly useful in drilling for oil. At the other end of the scale are the ultra-high-speed tungsten-tipped dental drills, and the drills used to bore very fine holes in printed circuit boards, which are tipped with a particularly fine grade of cemented carbide.

Tungsten carbide is reclaimed by immersing worn-out tools in molten zinc at 900°C, which dissolves the cobalt and releases the carbide.

X-ray tubes for medical use have a tungsten emitter coil and the screens used to view X-rays rely on calcium or magnesium tungstate phosphors to convert the X-rays into blue visible light. Tungsten is also used in microchip technology and liquid crystal displays where ultra-high purity molybdenum–tungsten targets provide display panels with improved definition.

A few tungsten compounds have found limited use. Tungsten oxide catalysts are used for oil refining, and for the removal of sulfur and nitrogen compounds from oil-based products, helping to increase the yields of petrol (gasoline) and other light hydrocarbons in crude oil processing, thereby making the products more environmentally acceptable. Tungsten disulfide is a high-temperature lubricant, able to act at temperatures higher than the more usual molybdenum disulfide. It has also been used on razor blades.

ENVIRONMENTAL ELEMENT

Tungsten in the environment	
Earth's crust	1 p.p.m.
	Tungsten is the 58th most abundant element.
Soils	1–2.5 p.p.m.
Sea water	90 p.p.t.
Atmosphere	very low: of the order of 1 nanogram per cubic metre of air

Very little tungsten has been detected in the few soils that have been analysed for it, although around an ore-processing plant in Russia levels as high as 2000 p.p.m. (0.2%) were found. Even so, this element poses no real threat to the environment. Indeed, in one respect it has been entirely beneficial: fishing weights made from plastic-covered tungsten have now replaced the poisonous lead sinkers previously used (see p. 232).

CHEMICAL ELEMENT

Data file	
Chemical symbol	W
Atomic number	74
Atomic weight	183.84
Melting point	3407°C
Boiling point	5700°C
Density	19.3 kilograms per litre (19.3 grams per cubic centimetre)
Oxides	WO_2, W_2O_5 and WO_3

Tungsten is a lustrous and silvery white metal that is a member of group 6 of the periodic table of the elements, a group it shares with the lighter metals chromium and molybdenum. The bulk metal resists attack by oxygen, acids and alkalis.

There are five naturally occurring isotopes of tungsten, of which tungsten-184 accounts for 31%, tungsten-186 for 29%, tungsten-182 for 26%, tungsten-183 for 14% and tungsten-180 for a mere 0.1%. None is radioactive.

ELEMENT OF SURPRISE

There are many ways in which tungsten enters everyday life, in ways quite unexpected. It is present, but unseen, in television sets and microwave ovens. More visible is the tungsten carbide which forms the tips of ball-point pens, while cemented tungsten carbide is to be seen as darts and golf clubs.

Ununbiium

This is element atomic number 112 and along with other elements above 100 it is dealt with under the transfermium elements on p. 457.

Ununhexium

This is element atomic number 116 and along with other elements above 100 it is dealt with under the transfermium elements on p. 457.

Ununoctium

This is element atomic number 118 and along with other elements above 100 it is dealt with under the transfermium elements on p. 457.

Ununquadium

This is element atomic number 114 and along with other elements above 100 it is dealt with under the transfermium elements on p. 457.

Unununium

This is element atomic number 111 and along with other elements above 100 it is dealt with under the transfermium elements on p. 457.

Pronounced yoo-rayn-iuhm, this element was named after the planet Uranus, which was itself named after the Uranus of Greek mythology, who was the god of the sky. Uranium often occurs in nature as the positively charged species UO_2^{2+}, which is known as the uranyl ion.

French, *uranium*; German, *Uran*; Italian, *uranio*; Spanish, *uranio*; Portuguese, *urânio*.

COSMIC ELEMENT

A significant part of the internal heat of the Earth comes from the radioactive decay of uranium atoms.

HUMAN ELEMENT

Uranium in the human body	
Blood	0.5 p.p.b.
Bone	varies between 0.2 and 70 p.p.b.
Tissue	varies between 1 and 3 p.p.b.
Total amount in body	approx. 0.1 milligrams (range 0.01–0.4 milligrams)

There is no biological role for uranium, nor ever likely to have been because it is radioactive. However, this does not prevent certain micro-organisms from absorbing it to the extent of concentrating it to levels 300 times higher than in the surrounding environment, although why this happens is not understood. The bacterium *Citrobacter* readily absorbs uranyl ions, provided it is supplied with a source of organic phosphate such as glycerol phosphate. The bacteria form crystals of uranyl phosphate until they become encrusted with them, and within a day 1 gram of *Citrobacter* will grow 9 grams of these crystals around itself. There are even suggestions that these microbes could be used to decontaminate water because of this ability.

Plants absorb a little uranium from the soil in which they grow, and the leaves of some plants have higher levels of uranium than seems necessary. The ash from burnt wood can have up to 4 p.p.m of this element. The range for most plants is 5–60 p.p.b. (dry weight) but food plants tend to have less; for example, corn and potatoes have 0.8 p.p.b. (dry weight).

The amount of uranium taken in with the diet is around 1–2 micrograms daily. The amount of uranium that is absorbed from food is around 5% if it is in the form of the soluble uranyl ion but about 0.5% if it is present as insoluble compounds, such as the oxide. The uranium that enters the bloodstream generally becomes lodged in the skeleton because of the affinity of uranium for phosphates, and once in the bone it will reside there for many years. Tests on dogs showed that sodium citrate solution boosted the excretion of uranium via the urine.

The radiation hazard of uranium generally overrides other toxicity considerations, but uranium compounds are *chemically* poisonous, and may cause irreversible kidney damage because uranium salts lodge in its tubules, eventually causing renal failure with the production of proteins and glucose in the urine. Soluble compounds, which can be absorbed through the skin, pass quickly through the body causing little damage, whereas insoluble ones, and particularly those which lodge as dust in the lungs, pose a more serious threat.

ELEMENT OF HISTORY

In the late Middle Ages, the mineral pitchblende (uranium oxide, U_3O_8) often turned up in the silver mines of Joachimsthal (now part of the Czech Republic, but then part of the kingdom of Bohemia) and was thought to contain zinc or iron, but it was of little use. The name was derived from *pech*, meaning pitch (or ill luck), and *blende*, meaning deceiver. In 1789, Martin Heinrich Klaproth, a pharmacist who had his own experimental laboratory in Berlin, Germany, investigated pitchblende and found it dissolved in nitric acid and precipitated a yellow compound when the solution was neutralized with sodium hydroxide.[184] He was convinced it was the oxide of a hitherto unknown element and when he heated the precipitate with charcoal and obtained a black powder, he assumed he had produced the metal itself. He based its name on that of the recently discovered planet Uranus and gave it the name 'uran', which later became uranium.

Later tests showed that Klaproth's 'metal' was in fact one of the oxides of uranium, and it fell to Eugène Peligot, who was Professor of Analytical Chemistry at the Central School of Arts and Manufactures, in Paris, to isolate the first sample of uranium metal, which he did in 1841 by heating uranium tetrachloride with potassium.

For most of the nineteenth century, uranium was not regarded as particularly dangerous and commercial uses were found for it. For example, in 1855 a factory was set up in Austria to manufacture uranium pigments for colouring pottery and glass. Glass to which uranium oxide had been added had a fluorescent yellow-green colour.

The discovery that uranium was radioactive came only in 1896 when Henri Becquerel in Paris found that a sample of uranium left in a drawer on top of an unexposed photographic plate caused this to become 'fogged' as if it had been partly exposed to light. From this he deduced that uranium was emitting invisible rays; and thus began a new chapter in science – and human history.

[184] The precipitate was probably sodium diuranate ($Na_2U_2O_7$).

ELEMENT OF WAR

The first atomic bomb used in warfare was a uranium bomb, code named 'Little Boy', which was dropped on the Japanese city of Hiroshima at 8:16 a.m. on 6 August 1945. This bomb contained enough of the uranium-235 isotope to start a runaway chain reaction, which in a fraction of a second caused a large number of the uranium atoms to undergo fission, thereby releasing a fireball of energy.[185] It had the destructive power of 12 500 tonnes of TNT; it destroyed almost 50 000 buildings and killed about 75 000 people, who either died immediately, or died later that month as a result of their burns.

To make this bomb it was necessary to enrich the isotope uranium-235, which accounts for less than 1% of uranium atoms, because this is a radioactive isotope that can sustain a chain reaction. The impact of a neutron on the nucleus of the atom causes it to split apart (undergo fission), and as it does it releases not only energy but several more neutrons. These in their turn can trigger other nuclei to disintegrate, releasing yet more energy and more neutrons, and so the chain reaction continues; in the case of an atomic bomb it runs out of control.

Uranium-235 was separated from its more abundant sister isotope, uranium-238, by converting the metal to uranium hexafluoride (UF_6), a volatile solid which sublimes at 57°C. This compound is made by reacting uranium oxide with hydrogen fluoride to form uranium tetrafluoride (UF_4), which is then reacted with fluorine gas to form UF_6. Because the two isotopes differ slightly in molecular weight, they diffuse at different rates through a silver–zinc membrane, with molecules of uranium-235 diffusing slightly more rapidly.

After repeated diffusions, a sufficient concentration of uranium-235 was obtained to provide the so-called 'critical mass' which ensures that the majority of released neutrons are captured by other atoms, rather than escaping from the mass of metal. The critical mass for a runaway chain reaction is only a few kilograms, and a gigantic explosion will ensue. In the 'Little Boy' bomb, explosive charges forced together two sub-critical pieces of uranium to create such a critical mass.[186]

There are other ways of separating uranium-235, such as using a high-speed gas centrifuge when the lighter UF_6 concentrates near the centre of the centrifuge. This is a much cheaper method and most uranium is now enriched this way. Another method exposes this vapour to laser radiation of precise energy to sever the uranium–fluorine bond of the lighter isotope but not of the heavier one, the result being that uranium-235 metal deposits out from the mixture.

Depleted uranium from nuclear power plants, and from which a lot of the uranium-235 has been removed, is one of the densest of metals. For this reason, it will deliver a much heavier blow to a target than other metals, and so it is used to make armour-piercing shells. Recently there has been concern that the dust from these weapons could cause cancers in those exposed to it.

[185] Most of the uranium atoms in an atomic bomb are there to provide the critical mass, and do not experience fission.

[186] A uranium bomb can also be used as the trigger for a hydrogen bomb – see p. 186.

ECONOMIC ELEMENT

There are many uranium minerals, the chief ones of which are the oxide ores, uraninite (UO_2) and pitchblende. Other minerals are autunite (calcium uranyl phosphate) and carnotite (potassium uranyl vanadate). Some varieties of samarskite (a complex oxide) may contain up to 23% uranium (see tantalum, p. 419); torbernite (calcium uranyl phosphate) and betafite (calcium, uranium, titanium, etc., oxide) are minor ores.

The chief uranium ore deposits occur in Australia, USA, Canada, Gabon, Congo, South Africa, Russia and China. Uranium is extracted in several ways: by open pit mining, underground mining or leaching low grade ores, or as a by-product of processing other ores. For example, mineral phosphates contain 100–200 p.p.m. of uranium oxide and this can be selectively extracted with organic esters dissolved in kerosene (paraffin). The esters form compounds with uranium that are soluble in this hydrocarbon solvent.

World production is 40 000 tonnes per annum, and reserves are thought to exceed 3 million tonnes as uranium ore, plus about the same amount in phosphate ores. Monazite sands also contain enough uranium to make them a potential source of the element. Although the concentration of uranium in sea water is low, the total amount there is around 5 billion tonnes, and it could be extracted using ion exchangers. Such a method was shown to be feasible in Japan in the 1980s.

Uranium metal is obtained by first converting it into uranium chloride and then reacting that with sodium metal, or by converting it to the oxide and reacting that with calcium or aluminium.

Most uranium is used to fuel nuclear reactors to generate electricity. This requires the uranium to be enriched with the uranium-235 isotope, and the chain reaction to be controlled so that the energy is released in a more manageable way. In theory a kilogram of uranium-235 can release around 20 trillion joules of energy (20×10^{12} joules), equivalent to more than 1500 tonnes of coal. The controlled release of energy can be done by increasing the proportion of uranium-235 to around 3%, or by using a moderator to slow down the neutrons that are released so more of them are captured by other uranium atoms. Heavy water is an ideal moderator.

Nuclear powered naval vessels and submarines were built by the US, USSR and Royal navies during the Cold War, and they are still in service. However, some, with their nuclear reactors, are now sitting lifeless on the bottom of the world's oceans as a result of accidents.

Nuclear reactors use rods of uranium or uranium oxide that are enriched with uranium-235 and these are clad in zirconium metal. As the fuel is consumed, the dissociation products accumulate and these may absorb the neutrons necessary to sustain the reactor, so after about 5 years a rod has to be replaced. The spent fuel rod is sent for reprocessing, but only after being left for a year to allow the intense radiation from some of the products to decay away.

Depleted uranium is also used as ship's ballast, as counterweights for aircraft and (as the uranyl ion) to add colour to ceramic glaze on tiles. Depleted uranium is preferred over other dense metals because it is easily cast and machined.

ENVIRONMENTAL ELEMENT

Uranium in the environment	
Earth's crust	2 p.p.m.
	Uranium is the 48th most abundant element.
Soils	range 0.7–11 p.p.m.*
Sea water	3 p.p.b.
Atmosphere	virtually nil

* The addition of phosphate fertilizers has increased the level of uranium on most farmed land to around 15 p.p.m.

Although uranium is radioactive, it is not particularly rare and is more abundant than tin, and 10 times more abundant than silver and mercury together. It is widely spread throughout the environment and so it is impossible to avoid uranium. Soils near fossil fuel power stations and phosphate fertilizer plants can be enriched in uranium due to emissions, since both types of raw material tend to have relatively high levels of uranium as impurities.

While uranium itself is not particularly dangerous, some of its decay products do pose a threat, especially radon, which can build up in confined spaces such as basements. Even the building material gypsum, used to make plaster, was found to be contaminated with uranium if it had been made from the waste products of phosphate mining.

When a uranium-235 atom splits in two, it can produce all kinds of other elements, but most tend to be in the weight ranges of 90–105 (krypton to ruthenium) and 130–145 (antimony to europium). Among the former is strontium-90, which was particularly worrying when above-ground nuclear weapons testing was carried out in the 1950s and early 1960s,[187] because this element is easily absorbed by humans and yet is dangerously radioactive. Another radioactive element released by uranium fission is iodine-131, which is equally worrying. This was widely scattered over the North of England in 1957 by an accident at the Windscale (now called Sellafield) nuclear power plant, and over Europe by the Chernobyl nuclear plant explosion in the Ukraine in 1986. Again, because iodine is an element essential to human health, this dangerously radioactive isotope is a problem when absorbed by the body.

The natural nuclear reactors

Although uranium oxides are very insoluble, when they are oxidized to uranyl ions (UO_2^{2+}) they become very soluble, and very mobile in the environment. In one locality, Oklo in Gabon, this allowed uranium to concentrate to such a level that it formed several natural nuclear reactors, probably as many as 20. This was thought to have occurred when oxygen levels first built up in the Earth's atmosphere, thereby allowing uranium oxides to be converted to uranyl ions. The Oklo event occurred at a time when the natural level of uranium-235 was probably around 3% and high enough to sustain

[187] Some countries, such as France and Israel, continued above-ground tests in the 1970s and 1980s.

a chain reaction, especially with water present. The reason we know this happened is that there is much less uranium-235 in the uranium ore mined in Gabon than in other uranium ores, showing that at some time in the past it has undergone fission, and there are still some of these fission-produced isotopes around.

These natural nuclear reactors started up around 2 billion years ago and ran for around a billion years until the concentration of uranium was too low to sustain them any longer. Water rich in organic matter that percolated through the rocks probably helped to moderate the reactors, and this eventually became tar-like as it decomposed. Traces of this bitumen can still be found. The radioactive waste products from the Oklo reactors were not entirely safely locked away during the millennia that followed; they appear to have been disturbed by a volcanic eruption around 720 million years ago.

CHEMICAL ELEMENT

Data file	
Chemical symbol	U
Atomic number	92
Atomic weight	238.0289
Melting point	1132°C
Boiling point	3754°C
Density	19.0 kilograms per litre (19.0 grams per cubic centimetre)
Oxides	UO_2, U_3O_8 and UO_3

Uranium is a silvery, ductile, malleable, radioactive metal and a member of the actinide series of the periodic table of the elements. It tarnishes in air and is attacked by steam and acids, but not by alkalis. It is one of the three fissile elements, the others being thorium and plutonium.

Uranium consists mainly of three radioactive isotopes: uranium-238, which accounts for 99.3% and has a half-life of 4.5 billion years, uranium-235 (0.7%, 700 million years) and uranium-234 (0.005%, 245 000 years).[188] Uranium-234 is produced as part of the decay sequence of uranium-235. Both uranium-235 and uranium-238 undergo radioactive decay via a sequence of other elements, all radioactive, until they arrive at lead-206 and lead-207, respectively, which are stable atoms. The amount of uranium and lead in various rocks has been used to date them, and indeed to date the Earth itself to 4.57 billion years old.

ELEMENT OF SURPRISES

1. In 1912, R. T. Gunther of Oxford University excavated a first century AD Roman villa on Cape Posilipo on the Bay of Naples, and came across a mosaic in which there were curiously coloured glass pieces. When these were analysed it was discovered

[188] Minute traces of other isotopes, uranium-232, -233, -236, -237, -239, and -240 also occur naturally.

that their tint was due to the presence of 1% uranium oxide, and that this mineral must have been deliberately added to colour the glass.

2. In 1998, researchers from the Natural History Museum, London, discovered that the lichen *Trapelia involuta* was happily growing on spoil heaps from an abandoned uranium mine in Cornwall, UK. X-ray investigations showed that the lichen was absorbing, and storing, uranium in the walls of its outer fruit cells, although for what purpose is not known. Yet the lichen appeared to be unaffected by it.

Vanadium

Pronounced van-ay-di-uhm, the element is named after Vanadis, the Scandinavian goddess of beauty. Other names were suggested by its original discoverer – see below. Oxygen-containing negative ions, such as VO_4^{3-}, are known as vanadates, while the positive ion, VO^{2+}, is known as vanadyl.

French, *vanadium*; German, *Vanadium*; Italian, *vanadio*; Spanish, *vanadio*; Portuguese, *vanádio*.

HUMAN ELEMENT

Vanadium in the human body	
Blood	0.02–0.9 p.p.b.
Bone	3.5 p.p.b.
Tissue	approx. 20 p.p.m.
Total amount in body	0.1 milligram

Vanadium is essential to some species, including humans. Microbes that convert atmospheric nitrogen to ammonia need vanadium. Feeding tests on chickens and rats showed that vanadium has a growth-promoting effect, and the same may be true for humans, although we are unlikely to be short of this metal – see below.

Vanadium compounds are not regarded as a serious hazard, and indeed this element appears to stimulate metabolism. However, workers exposed to vanadium pentoxide (V_2O_5) dust were found to suffer severe eye, nose and throat irritation and such workers must now be protected.

FOOD ELEMENT

The average daily intake of vanadium is 40 micrograms, which is more than enough to meet the body's requirements; indeed there are no cases of anyone ever suffering a deficiency. The absolute requirement has been estimated to be as little as 2 micrograms per day.

Only about 0.5% of dietary vanadium is absorbed by the body, and most of this is rapidly excreted. The element does not accumulate in any organ, but there is much more vanadium in the human body than we appear to need. Its role is thought to be as a regulator of one of the enzymes that govern the way sodium operates in the body, but it may have other roles as well.

Vanadium first came to the attention of nutritionists in 1977 when some commercial preparations of adenosine triphosphate (ATP) were found to disturb the sodium–potassium balance that works the nervous system. The reason turned out to be traces of vanadium which was acting as an enzyme inhibitor of ATPases, the enzymes that bring about the reaction whereby ATP releases its energy, and this it appears to do by substituting vanadate for phosphate. This discovery sparked off an interest in this element, and yet it is still not completely clear why vanadium is essential for humans.

Typical vanadium levels in foods are between 0.01 and 0.1 p.p.m., e.g. wheat has 0.08 p.p.m. and potatoes 0.02 p.p.m. (fresh weight), although rice has relatively high levels (up to 0.8 p.p.m.) and lettuce can even have 1 p.p.m. if it is grown on contaminated soil. Plants that produce linoleic acid, such as sunflowers, appear to absorb a lot of vanadium, up to 50 p.p.m. (dry weight). Seafood and liver are the sources of dietary protein that contain most vanadium.

ELEMENT OF HISTORY

Vanadium was discovered twice. The first time was in 1801 by Andrés Manuel del Rio (1764–1849) who was Professor of Mineralogy in Mexico City. He found vanadium in a specimen of *plomo pardo de Zimapan* (brown lead ore of Zimapan), a mineral now known as vanadite, which came from a mine in Hidalgo, Mexico. He was surprised at the varied colours of its salts and so he called it panchromium (meaning 'all colours'), but later changed its name to erythronium (meaning 'red') when he found that all its salts turned red on heating or treating with acid.

Del Rio gave some of his material to a visitor, the well-known traveller Alexander von Humbolt, who sent it to the French Institute in Paris, with a brief note saying how like chromium the new element was, on account of its multicoloured salts. (A longer letter describing its chemistry in more detail never arrived, the ship carrying it being wrecked in a storm.) The result was that a French chemist, H.V. Collet-Descotils, was asked to examine the new specimen, and he came to the conclusion that it really was a chromium mineral and published his report. Del Rio became convinced that all he had done was to rediscover chromium.

The second time vanadium was discovered was 30 years later, in 1831, by Nils Gabriel Sefström (1787–1845), Professor of Chemistry at the Karolinska Institutet at Stockholm, Sweden. He separated it from a sample of cast iron made from ore that had been mined at Småland. The iron makers were puzzled by the fact that sometimes their cast iron was brittle, while at other times it was strong. Part of the sample dissolved more readily in acid than the bulk of the metal and Sefström concentrated on this material. Although he had only a tiny amount, he was able to show that it was a previously unknown element, and in so doing he beat a rival chemist, Friedrich Wöhler, to the discovery.

Wöhler had been working on a sample of the mineral from Zimapan, but had been badly affected by hydrogen fluoride gas and had been ill for several months, during which time Sefström announced his results. The young Wöhler, however, was also aware of del Rio's work of 30 years previously, and his work confirmed the original discovery. Nevertheless, Selfström's name for the element, vanadium, was the one which became accepted.

Early attempts to extract vanadium from its compounds produced an indifferent material. The first reasonably pure sample was produced by Henry Roscoe in Manchester, England, in 1869, and he showed that previous samples of the metal must have been vanadium nitride (VN). Even his sample was not very pure, and had around 4% of impurities. A purity of 99.99% was only achieved in the 1920s by J. W. Marden and M. N. Rich of the Westinghouse Lamp Company in the USA.

ELEMENT OF WAR

Vanadium made steel stronger and lighter and made it possible for the first war plane to be equipped with a cannon, rather than just a machine gun. This was produced for the French airforce in World War I, and caused consternation when it first appeared. The type of warfare of the twentieth century required extra protection for combatants and vehicles. Nickel steel was no use, being easily penetrated by bullets and shrapnel. Vanadium steel, on the other hand, was found to offer better protection against bullets and shrapnel. As a result, it became the standard material from which soldier's helmets and other forms of protective armour were made.

ECONOMIC ELEMENT

Various vanadium ores are known, such as vanadinite (lead chloride vanadate), patronite (vanadium sulfide) and carnotite (potassium uranyl vanadate), but none is mined as such for the metal, which is generally obtained as a by-product of other ores. Vanadium is also abundant in Venezuelan oil, to the extent that the ash from it is a commercial source of vanadium pentoxide (V_2O_5).[189] The largest resources of vanadium are to be found in South Africa and Russia.

World production of vanadium ore is around 45 000 tonnes a year. Production of the metal itself comes to about 7000 tonnes per year. This is extracted from the red-brown oxide by heating with calcium in a pressure vessel. It cannot be extracted by heating with carbon, which reacts with the metal. Very pure vanadium is obtained by reacting vanadium trichloride (VCl_3) with magnesium, and this gives a metal that is much more ductile. Since most vanadium is added to steel, it is not necessary to use such high-grade metal; vanadium is usually added as ferrovanadium, which is produced from vanadium pentoxide by heating with ferrosilicon.

Vanadium is used mainly as alloys, especially for vanadium steels, which contain 0.1–3%, depending on the intended use. Such steels are rustproof, shock-resistant and vibration-resistant, and are used for springs, tools and jet engines, as well as armour plating. Vanadium alloys are also used in nuclear reactors because vanadium has low neutron-absorbing abilities and it does not deform by 'creeping' under high temperatures.

[189] The ash from Venezuelan oil is 45% vanadium pentoxide. It has been suggested that plants living at the time when these oil deposits were laid down absorbed much more vanadium than today's plants.

The oxide, V_2O_5, is used in ceramics to produce a golden colour, and is added to glass to produce a green or blue tint. Industry uses the oxide as a catalyst for the oxidation of sulfur dioxide (SO_2) to sulfur trioxide (SO_3), which is the precursor to sulfuric acid. Vanadium compounds are also used as catalysts in polymer production.

ENVIRONMENTAL ELEMENT

Vanadium in the environment	
Earth's crust	160 p.p.m.
	Vanadium is the 19th most abundant element.
Soils	approx. 100 p.p.m., ranging from 10 to 500 p.p.m.
Sea water	1.5 p.p.b.
Atmosphere	0.02 nanograms per cubic metre over the oceans, but around 1 microgram per cubic metre over populated regions, due to the burning of heating oil

Weathering is an important way in which vanadium is redistributed around the environment because vanadates are generally very soluble; the annual movement of vanadium this way is thought to exceed 2 million tonnes. This metal is not regarded as a threat.

Vanadium is abundant in most soils, in variable amounts, and is taken up by plants at levels that reflect its availability. Some species, such as toadstools, take up a lot of vanadium. The fly agaric toadstool, *Amanita muscaria*, has 180 p.p.m. (dry weight).

Marine worms absorb vanadium, and one species, *Ascidia nigra*, has blood cells with 1.5% vanadium. It has the ability to concentrate vanadium to levels 10 000 times higher than in the surrounding sea water. There was once a scheme in Japan to harvest such ascidians for their vanadium, from the vast tracts which live on the coastal shelf of that country. Vanadium makes up 10% of the blood cell pigment of the sea cucumber, *Holothuroidea*.

CHEMICAL ELEMENT

Data file	
Chemical symbol	V
Atomic number	23
Atomic weight	50.9415
Melting point	1887°C
Boiling point	3377°C
Density	6.1 kilograms per litre (6.1 grams per cubic centimetre)
Oxides	V_2O_5 is the most stable but there are also VO, V_2O_3 and VO_2

Vanadium is a shiny, silvery metal, which comes at the head of group 5 of the periodic table of the elements, and is a transition metal. It resists corrosion due to a protective film of oxide on the surface. The metal is attacked by concentrated acids, but not by alkalis, not even when these are molten.

There are two naturally occurring isotopes of vanadium: non-radioactive vanadium-51, which accounts for 99.8%, and weakly radioactive vanadium-50 (0.2%), which has a half-life in excess of 100 quadrillion years (10^{17} years), which is 20 million times the age of the Earth.[190]

ELEMENT OF SURPRISES

1. The fame of Damascus steel was due to the presence of vanadium. This steel was fashioned into the much prized weapons of daggers, scimitars and swords that were unsurpassed in their day. Recent research by John Vanhoeven, a materials scientist and engineer, has shown that minute traces of vanadium were responsible for giving Damascus steel its great strength and durability, with the especial advantage of taking and holding a lasting sharp edge. Amazingly, as little as 0.02% of vanadium was responsible for the steel's high quality, yet its presence was not appreciated by the Damascus steelmakers. The vanadium just happened to be an impurity in the source of iron ore they were using, and when the supply of this ran out, with it went their reputation of making the world's finest steel. The steelmakers could not understand why their tried-and-tested methods no longer worked, and by the 1800s Damascus steel was no more. [Walter Saxon]
2. 'But for vanadium there would be no automobiles!' So said Henry Ford who based his popular model-T motor car on lightweight vanadium steel. He came across this alloy in 1905 when he attended a racing car event at which there was a crash that completely wrecked a French vehicle. When Ford picked up a fragment, which was part of a valve spindle, he was surprised how hard, yet light, it was. He sent it for analysis and was told that it was steel which contained a few per cent of vanadium. Although such steel was not available in the USA, he decided nevertheless to make it an essential part of his new car and in 1908 the first new model-T was produced. Mass production began in 1913 and, by 1927, when the model was discontinued, around 15 million had been made, all with components such as gears, axles, shafts, springs and suspension made from vanadium steel.

[190] In a 1 gram sample of vanadium, only two atoms a week undergo radioactive decay.

Pronounced zee-non, the name is derived from the Greek *xenos*, meaning stranger. French, *xénon*; German, *Xenon*; Italian, *xeno*; Spanish, *xenón*; Portuguese, *xénon*.

COSMIC ELEMENT

Higher than expected levels of the xenon isotope, xenon-129, are to be found in some stony meteorites; these are thought to have come from the radioactive decay of iodine-129, which has a half-life of 17 million years.

HUMAN ELEMENT

Xenon in the human body	
Blood	minute trace
Total amount in body	very small

Xenon can have no biological role, and the gas is harmless, although it could asphyxiate if it were to exclude oxygen from the lungs. In the early 1950s, xenon was tested as a possible anaesthetic gas for use during surgery and was found to be superior to nitrous oxide, in that it produced fewer side effects. However, its expense rules out its use.

ELEMENT OF HISTORY

Xenon was discovered on 12 July 1898 by William Ramsay and Morris William Travers at University College, London. They had already isolated neon, argon and krypton from liquid air. As a result of their research and the interest it aroused, they were given a new liquid-air machine by the industrialist Ludwig Mond. They used it to extract more of the rare gas krypton from liquid air and, by repeated fractional distillation of this, they isolated an even heavier gas: xenon. When they examined it in a vacuum tube it gave a beautiful blue glow and they realized at once that here was yet another 'noble' gas. Because it had proved so difficult to find, they named it xenon.

In 1920 the French chemists C. Moureu and A. Lepape found that all the noble gases were present in the natural gas of Alsace-Lorraine and in many spring waters found in France, including the most famous, Vichy water.

ECONOMIC ELEMENT

The only commercial source of xenon is from industrial liquid-air plants. World production is less than 1 tonne per year,[191] although reserves of xenon gas in the atmosphere amount to 2 billion tonnes.

Xenon has relatively little commercial use, and what use it has generally comes from its visible spectrum. When an electric discharge is passed through the gas it produces a gentle blue light, which extends into the 'safe' ultraviolet region as well, and so can be used for sunbeds and the biocidal lamps seen in food preparation areas. Xenon 'blue' headlights and fog lights are used on some vehicles and are said to be less tiring on the eyes. They also illuminate road signs and markings better than conventional lights.

Other types of lamps use xenon's ability to deliver an intense burst of light when pulsed with a very high voltage, and so it is part of the built-in flash of modern cameras, as well as being essential for the strobe lights used in high-speed photography, to record such events as a bullet emerging from the barrel of a gun or striking a target. Xenon flash lamps are also used to activate ruby lasers.

Some xenon is employed as a supercritical fluid[192] in research laboratories where its chemical unreactivity makes it an ideal medium for studying molecules under extreme conditions.

ENVIRONMENTAL ELEMENT

Xenon in the environment	
Earth's crust	2 p.p.t.
	Xenon is the 85th most abundant element.
Soils	nil
Sea water	100 p.p.t.
Atmosphere	90 p.p.b.

The xenon of the air, and that which is dissolved in the sea, poses no threat to the environment. However, the radioactive xenon-133, with a half-life of 5 days, is a potential danger because it continually leaks from nuclear power and reprocessing plants, albeit in tiny amounts. It has been released in larger quantities during nuclear accidents, such as that at Chernobyl, Ukraine, in 1986.

[191] Amounting to around 175 000 litres, or 175 cubic metres, of the gas.

[192] See also water, p. 191.

CHEMICAL ELEMENT

Data file	
Chemical symbol	Xe
Atomic number	54
Atomic weight	131.29
Melting point	−112°C
Boiling point	−107°C
Density	5.9 grams per litre
Oxides	XeO_3 (explosive) and XeO_4 (unstable)

Xenon is a colourless, odourless gas, and one of the noble gas elements of group 18 of the periodic table. It was regarded as completely inert until, in 1962, Neil Bartlett produced a compound of it with platinum and fluorine. Other fluorine compounds soon followed and today there are more than 100 of them, including oxides, acids and salts, some of which are commercially available.

Even compounds with xenon–hydrogen, xenon–sulfur and xenon–gold bonds have been made, although these are stable only at low temperatures. The most exotic compound of xenon has four xenon atoms attached to a central gold atom, and this ion, $[AuXe_4]^{2+}$, will form stable crystals below −40°C,[193] a relatively high temperature for something so intrinsically unstable.

There are nine naturally occurring isotopes of xenon, of which the most abundant are xenon-132 (27%), xenon-129 (26.5%) and xenon-131 (21%). The others are xenon-134 (10.5%), xenon-136 (9%), xenon-130 (4%), xenon-128 (2%), and xenon-124 and xenon-126 (each with 0.1%). None is radioactive.[194], although xenon-136 appears to be so with a half-life of 9.3×10^{19} years.

ELEMENT OF SURPRISE

Although it is almost totally inactive chemically, xenon is used for space flights because it makes the best fuel for ion engines. In such an engine, a beam of ions is accelerated by an electric or electromagnetic field and then expelled from the vehicle at 30 kilometres per second (about 100 000 kilometres per hour), thereby giving the vehicle a powerful thrust in the opposite direction. To be suitable for an ion drive, atoms must be easily ionized (i.e. lose an electron, thereby acquiring a positive charge) and have as high a mass a possible. Mercury, caesium and xenon are the most likely contenders for ion engines; of these three, xenon is the safest to handle. Being a gas, it poses problems of storage on Earth, but not in space, where conditions are cold enough to freeze it solid. Xenon is preferable to caesium, which is corrosive, and to mercury, which poses

[193] The counterbalancing negative ions are two antimony fluoride ions, $[Sb_2F_{11}]^-$.

[194] There are minute traces of naturally occurring radioactive isotopes, with short half-lives, that are produced as spontaneous fission products of uranium atoms in uranium ores.

a threat to those exposed to its vapour. An ion engine is 10 times more efficient than a conventional propulsion unit. So-called xenon ion propulsion systems (XIPS[195]) are now used by at least 10 orbiting satellites to keep them in the correct orbit, and each has enough xenon fuel on board to last more than 12 years. The space probe *Deep Space I* is powered by a xenon ion engine.

[195] Pronounced 'zips'.

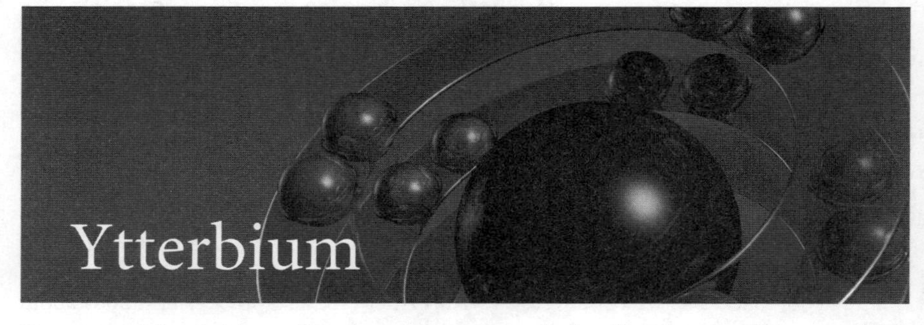

Ytterbium

Pronounced i-terb-iuhm, this element was named after Ytterby, Sweden – see 'The village of the four elements' under yttrium (p. 496).

French, *ytterbium*; German, *Ytterbium*; Italian, *itterbio*; Spanish, *yterbio*; Portuguese, *itérbio*.

Ytterbium is one of 15, chemically similar, elements referred to as the rare-earth elements, which extend from the element with atomic number 57 (lanthanum) to the one with atomic number 71 (lutetium). The term rare-earth elements is a misnomer because some are not rare at all; they are more correctly called the lanthanides, although strictly speaking this excludes lutetium. The minerals from which they are extracted, and the properties and uses they have in common, are discussed under lanthanides, on p. 222.

HUMAN ELEMENT

Ytterbium in the human body	
Total amount in body	not known, but very small

Ytterbium has no biological role, but it has been noted that ytterbium salts stimulate metabolism. Its soluble salts are mildly toxic by ingestion, but its insoluble salts are non-toxic. Ytterbium is a skin and eye irritant, and is also a suspected teratogen.[196]

It is difficult to separate out the various amounts of the different lanthanides in the human body, but they are present, and the levels are highest in bone, with smaller amounts being present in the liver and kidneys. There is no estimate of the amount of ytterbium in an average adult, and no one has monitored diet for ytterbium content, so it is difficult to judge how much we take in, but it is probably only a milligram or so per year. Ytterbium is not taken up by plant roots to any extent, so does not get into the human food chain, and the amount found in vegetables can be as low as 0.08 p.p.b. (dry weight). Some organisms, however, such as lichens, can absorb ytterbium and have 900 p.p.b.

[196] This is a technical term which literally means to generate a monster, by deforming a fetus.

ELEMENT OF HISTORY

Ytterbium was isolated in 1878 by the Swiss chemist Jean Charles Galissard de Marignac (1817–1894) at the University of Geneva, Switzerland. The story begins with yttrium which, unknown at the time it was first reported, was contaminated with traces of other rare-earth elements. In 1843 erbium and terbium were extracted from it, and then in 1878, Galissard de Marignac separated ytterbium from erbium. He had heated erbium nitrate until it decomposed and then extracted the residue with water and obtained two oxides: a red one which was erbium oxide, and a white one which he knew must be a hitherto unknown element, and this he named ytterbium oxide.

Ytterbium oxide was also discovered by Karl Auer (see also pp. 241 and 268) about this time and he named the 'new' element aldebaranium. (The ytterbium oxide that he and Galissard de Marignac had isolated was itself to yield another rare-earth oxide, in 1907, when George Urbain extracted lutetium oxide from it.)

Ytterbium metal was first made in 1937 by heating ytterbium chloride and potassium together, but it was not pure enough for its properties to be reliably measured. This had to wait until 1953 when a pure sample of the element was made.

ECONOMIC ELEMENT

Ytterbium is extracted commercially from monazite, which contains 0.1%, whereas the other common rare-earth ore, bastnäsite, has only 0.0006% ytterbium. Other ores are euxenite, a mixed oxide ore found in Greenland and Brazil, and xenotime, which is mainly yttrium phosphate. Ytterbium was also found in gagarinite, an yttrium mineral discovered in Kazakhstan in 1961, and which was named after the first astronaut, Yuri Gagarin. The main mining areas are China, USA, Brazil, India, Sri Lanka and Australia; and reserves of ytterbium are estimated to be around a million tonnes.

World production of ytterbium is around 50 tonnes per year, reflecting the fact that it finds little commercial application, although it has been shown to have some benefits, such as strengthening stainless steel, and some alloys have even been suggested as useful in dentistry. Like other rare-earth elements, it can be used to dope phosphors, or for ceramic capacitors and other electronic devices, and it can even act as an industrial catalyst.

Ytterbium has a single absorption band at 985 nanometres in the infrared and this has been used to convert radiant energy into electrical energy in devices that couple it to silicon photocells.

ENVIRONMENTAL ELEMENT

Ytterbium in the environment	
Earth's crust	3 p.p.m.
	Ytterbium is the 43rd most abundant element.
Soils	approx. 2 p.p.m. (dry weight), range 0.08–6 p.p.m.
Sea water	1.5 p.p.t.
Atmosphere	virtually nil

Ytterbium is one of the rarer rare-earth elements although it is still twice as common in the Earth's crust as tin. It poses no environmental threat to plants or animals.

Ytterbium salts are being introduced into the chemicals industry as catalysts in place of ones that are regarded as toxic and polluting.

CHEMICAL ELEMENT

Data file	
Chemical symbol	Yb
Atomic number	70
Atomic weight	173.04
Melting point	824°C
Boiling point	1193°C
Density	6.97 kilograms per litre (6.97 grams per cubic centimetre)
Oxide	Yb_2O_3

Ytterbium is a soft, silvery-white metal and one of the lanthanide group of the periodic table of the elements. It is slowly oxidized by air, but the oxide forms a protective layer on the surface. The metal reacts slowly with water and dissolves readily in dilute acids.

There are seven naturally occurring isotopes of ytterbium, of which ytterbium-174 is the most abundant, accounting for 32%; ytterbium-172 is second at 22%. The others are ytterbium-173 (16%), ytterbium-171 (14.5%), ytterbium-176 (12.5%), ytterbium-170 (3%) and the least abundant, ytterbium-168, of which there is only 0.1%. None of the isotopes is radioactive.

ELEMENT OF SURPRISE

The electrical conductivity of ytterbium varies rather oddly as pressure on the metal increases. At 16 000 atmospheres it becomes a semiconductor and thereafter its electrical resistance increases as the pressure mounts until it reaches 40 000 atmospheres when the trend suddenly reverses and ytterbium starts to become more conducting. This feature of ytterbium was put to use in stress gauges designed to monitor the intense shock waves of nuclear explosions.

Pronounced it-ree-uhm, it is named after Ytterby, Sweden – see 'The village of the four elements' below.

French, *yttrium*; German, *Yttrium*; Italian, *ittrio*; Spanish, *ytrio*; Portuguese, *itrio*.

Yttrium is often classified as one of the rare-earth elements because it is found in all rare-earth minerals and is very like them chemically – see p. 219. Nevertheless yttrium is not strictly speaking one of this group, and it is certainly not rare, being twice as abundant as lead.

HUMAN ELEMENT

Yttrium in the human body	
Total amount in body	approx. 0.5 milligrams

No biological role has been identified for yttrium although it occurs in most species, including humans, where it concentrates in the liver and bone. Indeed there is evidence that it is to be found in every living organism, sometimes in relatively high amounts, and it is associated with nucleic acids. Human breast milk contains 4 p.p.m.

Soluble yttrium salts, such as the nitrate, are regarded as mildly toxic, while insoluble ones, such as the sulfate or oxide, are regarded as non-toxic. However, the element is suspected of being carcinogenic for some animals.

Edible plants can have quite a range of yttrium levels, from 20 to 100 p.p.m. (fresh weight) with the highest values being recorded for cabbage. The seeds of woody plants have the highest amounts of all: 700 p.p.m. Although coal contains yttrium (7–14 p.p.m.) this is not thought to indicate its selective absorption by the organic substances from which it was derived.

MEDICAL ELEMENT

The radioactive isotope, yttrium-90, has been used in several types of medical treatment, generally for cancers. It attaches itself to monoclonal antibodies and these attach themselves to cancer cells. In this way the cells get a deadly dose of the intense β-radiation that the isotope emits when it decays. Yttrium-90 therapy has been employed in the treatment of lymphoma, T-cell leukaemia and ovarian, colorectal, pancreatic and bone cancers. Needles of yttrium-90 have been also used, as a precision alternative to the scalpel, by surgeons seeking to sever pain-transmitting nerves in the spinal cord.

Yttrium-90 is produced when strontium-90 decays. Strontium-90 is a fission product of nuclear reactors and has a half-life of 29 years, decaying to yttrium-90 which has a half-life of 64 hours. For this to be used in medical treatment it has to be extracted from the contaminating strontium which would be retained by the body (see p. 407). In the 1990s, chemical separation techniques were developed that could achieve complete separation of the yttrium-90 from the strontium-90.

ELEMENT OF HISTORY

Yttria (yttrium oxide) was first isolated and recognized as that of a 'new' element in 1794 by the chemist Johan Gadolin at the University of Åbo, Finland. The story of how

THE VILLAGE OF THE FOUR ELEMENTS

In 1787, a 30 year old army lieutenant, Karl Axel Arrhenius (1757–1824), came across an odd piece of black rock, which looked like a lump of coal, in an old quarry near the village of Ytterby, located near Vauxholm in Sweden. (The name Ytterby simply means 'outer village' and signifies merely its location, suggesting it has no noteworthy feature otherwise; Ytterby is now part of the Stockholm Archipelago.)

Arrhenius knew the rock was too heavy to be coal and he thought he had found a hitherto unknown ore of tungsten, a metal which had been discovered in Spain only 4 years previously. He gave the specimen over to the chemist, Johan Gadolin (1760–1852) of the University of Åbo, at Turku, Finland, who investigated the new mineral and in 1794 announced that it contained a new 'earth' which made up 38% of its weight. He called the new earth yttria, from the name of the village, and though he spelt it wrongly, the name stuck.

Yttria was in fact yttrium oxide, and was called an earth because it could not be separated from the oxygen it contained by the conventional method of heating it strongly with charcoal. The metal itself was first isolated by Friedrich Wöhler (1800–82) in 1828 by heating yttrium chloride with potassium metal.

The mineral from which yttrium oxide came was named gadolinite in honour of Gadolin. Gadolinite is essentially a silicate mineral of beryllium, iron and yttrium; the fact that it contains beryllium was missed by Gadolin and was discovered by Nicolas-Louis Vauquelin in Paris 4 years later. In any case we now realize that Gadolin's new earth must have been contaminated with many other elements, but how many were present in his sample of yttrium oxide we shall never know because his collection was lost when the university of Åbo was destroyed by fire in 1827.

In 1843, Carl Gustav Mosander investigated yttrium oxide more thoroughly and found that it was made up of three oxides: yttrium oxide, which was white; terbium oxide, which was yellow; and erbium oxide, which was rose-coloured. The names of these two newly discovered elements, erbium and terbium, were again derived from the name Ytterby. The fourth element whose name was to be based on the name of the village was ytterbium.

this and the other elements it contained came to be discovered is given in the box 'The village of the four elements'.

ECONOMIC ELEMENT

The yellow-brown ore xenotime can contain as much as 50% yttrium phosphate (YPO_4) and is mined in Malaysia. Yttrium is found in the rare-earth mineral monazite, of which it makes up 2.5%. The other abundant rare-earth mineral, bastnäsite, contains much less. Yttrium is also part of fergusonite, a glassy black ore found in Madagascar, and samarskite. The output of yttrium is about 600 tonnes per year, measured as yttrium oxide, and world reserves are estimated to be around 9 million tonnes. Yttrium metal is produced by heating yttrium fluoride with calcium metal, but only a few tonnes are made each year.

Yttrium is used in alloys; it improves magnesium castings, and gives a finer grain to chromium, molybdenum and zirconium metals. When added to cast iron it makes the metal more workable. Although metals are generally very good at conducting heat, there is an alloy of yttrium with chromium and aluminium which is heat resistant.

Yttrium oxysulfide, doped with europium, is used as the standard red component in colour televisions. Yttrium oxide in glass makes it heat- and shock-resistant, and is used for camera lenses. Yttrium oxide is suitable for making superconductors, which are metal oxides that conduct electricity without any loss of energy; one of the best known of these is the mixed yttrium–barium–copper oxide, $YBa_2Cu_3O_{7-9}$, commonly referred to as '1-2-3'. This was one of the first superconducting materials to work at liquid nitrogen temperatures (–183°C).

Yttrium iron garnet (YIG) can be made into large crystals for optics, or deposited as a thin layer, and as such is used in magnetic recording. It is also a transducer of acoustic energy, in other words turning sound into electrical energy, and is a component of microwave filters for radar. The aluminium version, known as YAG, is used in lasers and these can operate at high power output and be used for cutting and drilling metals. YAG is very hard and makes a sparkling diamond-like gemstone, sometimes used to make large-scale replicas.

ENVIRONMENTAL ELEMENT

Yttrium in the environment	
Earth's crust	30 p.p.m.
	Yttrium is the 28th most abundant element.
Soils	approx. 23 p.p.m. (dry weight)*, range 10–150 p.p.m.
Sea water	9 p.p.t.
Atmosphere	virtually nil

* Cultivation lowers this to around 15 p.p.m.

Yttrium is one of the rarer rare-earth elements although it is still 400 times as common in the Earth's crust as silver. It poses no environmental threat to plants or animals.

CHEMICAL ELEMENT

Data file	
Chemical symbol	Y
Atomic number	39
Atomic weight	88.90585
Melting point	1522°C
Boiling point	3338°C
Density	4.47 kilograms per litre (4.47 grams per cubic centimetre)
Oxide	Y_2O_3

Yttrium is a soft, silvery-white metal that is a member of group 3 of the periodic table of the elements. It is stable in air because it is protected by the formation of a stable oxide film on its surface. It burns if ignited, and is attacked by water which it decomposes to release hydrogen gas. The only naturally occurring isotope, yttrium-89, is not radioactive.[197]

ELEMENT OF SURPRISE

Samples of material from the Moon, amounting in all to 300 kilograms of dust, rock and drilled cores, which were brought back from the six *Apollo* Moon landings between July 1969 and December 1972, were found to have relatively high yttrium contents.

[197] There are also traces of radioactive yttrium isotopes as fission products in uranium minerals.

Pronounced zink, the name is derived from the German *zink*. The word may have originated in the Persian word *sing*, meaning stone.

French, *zinc*; German, *Zink*; Italian, *zinco*; Spanish, *cinc*; Portuguese, *zinco*.

HUMAN ELEMENT

Zinc in the human body	
Blood	7 p.p.m.
Bone	75–170 p.p.m.
Tissue	50 p.p.m. (muscle)
Total amount in body	2.3 grams

Zinc is generally thought to be non-toxic, and is an essential element for animals and humans. It also has an essential role for plants and is believed to protect them against drought conditions and diseases. In humans, zinc concentrations are greatest in the eye, prostrate, muscle, kidney and liver. Semen is particularly rich in zinc, and there is evidence that a lack of zinc is responsible for the low sperm counts in men.

Surplus zinc can be stored in the bones and spleen, but it is not easily released from these to make up for a dietary lack. Zinc is lost from the body at about 1% of the total per day, excreted via the intestines (90%), urine (5%) and sweat (5%). Zinc differs from metals like iron and copper in that it exists in its compounds in one oxidation state, zinc(II), as Zn^{2+}. As such it is non-toxic and it can attach itself to many other kinds of molecule.

The amount of zinc in plants varies according to the level in the soil. In soil with adequate zinc levels, the food plants that contain the highest levels of zinc are wheat (20–60 p.p.m.), sweetcorn (around 20 p.p.m.) and lettuce (12 p.p.m.); fruits have the lowest zinc contents, with apples and oranges having only 1 p.p.m. or less (all fresh weights). Some crops so deplete the zinc in soil that a subsequent crop may suffer; this has been observed with maize when it follows sugar beet.

Zinc is important in two key groups of proteins: enzymes and transcription factors.[198] There are more than 200 hundred kinds of zinc-containing enzymes, and transcription

[198] Transcription is the process whereby RNA is synthesized from a DNA template.

factors appear to be equally numerous, some having protrusions referred to as zinc 'fingers'.

Zinc-containing enzymes regulate growth, development, longevity and fertility. They are also important in digestion, nucleic acid synthesis and the immune system. One of the most important ones is carbonic anhydrase, without which the body could not rid itself of carbonate which it converts to carbon dioxide. Zinc is especially needed where there is rapid turnover, as in the immune system, bone marrow and the lining of the gut. Zinc is also known to activate those parts of the brain that govern taste and smell.

Zinc metal is a human skin irritant; breathing the fumes of hot zinc leads to 'fume fever', with sore throat, cough and sweating, progressing to more severe symptoms if the exposure is heavy.

FOOD ELEMENT

The modern Western diet is more likely to be deficient in zinc than in iron. Red meats are the best source of zinc in the diet. The average adult takes in anything from 5 to 40 milligrams per day, depending on their liking for meat. The minimum intake of zinc, to avoid deficiency symptoms, is 2–3 milligrams per day and, assuming a 30% absorption rate, the dietary intake for men should be around 7.5 milligrams per day and that for women 5.5 milligrams per day. The average man takes in around 10 milligrams per day and the average woman about 8 milligrams per day. Research shows that during pregnancy this amount is also sufficient, because a higher zinc absorption rate comes into operation, and this continues during lactation. Then, a nursing mother produces around 850 cubic centimetres of milk a day containing the 2 milligrams of zinc that her baby needs. (Cow's milk that is used to make formula feed for babies has to be fortified with zinc.)

Beef, lamb and liver have some of the highest zinc levels, as do herrings and most cheeses. Oysters have more than 7 milligrams of zinc per 100 grams, liver 6 milligrams, beef 4 milligrams, wheat 4 milligrams, cheese 3 milligrams, shrimp 2 milligrams, eggs 1 milligram and milk 0.4 milligrams per 100 grams. Fruit generally has relatively little zinc, the average being around 0.15 milligrams per 100 grams (fresh weight). Strict vegetarians can get their zinc from sunflower or pumpkin seeds, brewer's yeast, maple syrup, and bran.

Zinc salts can be purchased over-the-counter in many healthfood shops and pharmacies, and are recommended by some dietary advisers for conditions which do not respond to conventional treatments, such as anorexia nervosa, premenstrual tension, post-natal depression, acne and the common cold. However, evidence in support of zinc remedies being successful is not particularly convincing.

Zinc in food is absorbed through the gut, but can react with phytic acid in food to form insoluble zinc phytate, which cannot be absorbed by the body. It is estimated that almost half the world's population is at risk of zinc deficiency, especially those whose diet is mainly based on cereals and beans, because these contain a lot of phytic acid, or those whose diet consists mainly of edible tubers, such as cassava or potatoes, because these have low levels of zinc. In some countries where the population relies on such diets, foods are fortified with zinc; this occurs in Mexico, Indonesia and Peru.

MEDICAL ELEMENT

Over the years zinc compounds played a part in the treatment of illness: zinc sulfate was used as an emetic, zinc chloride as an antiseptic, zinc oxide as an astringent (to stop bleeding) and zinc stearate as an ointment. So-called calamine lotion, which is a mixture of zinc and iron oxides, has been sold by pharmacies since the seventeeth century and used to treat skin conditions, as have zinc ointments. These are made by mixing zinc oxide with plant oils, wool oil (lanolin), or even mineral oil (also called liquid paraffin) and are applied to eczema, nappy rash, scalds and sunburn.

Zinc oxide is still used as a sunblocker[199] against damaging ultraviolet rays from the Sun, the so-called UVB rays, which can cause skin cancer. It is applied to the face, and especially the nose and lips, of sportsmen like cricketers, and those who work in the open air, who may have to spend many hours at a time in the blazing sun.

The importance of zinc for certain enzymes was realized early in the last century, but it was not until 1968 that the first cases of human zinc deficiency came to light, in Iran. This was identified by Ananda Prasad who had carried out zinc deficiency audits on animals, and was then confronted with the same symptoms, of stunted growth and retarded sexual development, in humans. Prasad was to become the leading world authority on human zinc deficiency, finding many cases in Egypt where the soil is lacking in this nutrient, and showing that supplementing the diet with zinc sulfate could effect a cure.

Lack of zinc was also shown to be a result of the genetic disorder acrodermatitis enteropathica, which was previously considered to be fatal for all babies born with it, but which can now be treated with doses of zinc.

Too much zinc can also be damaging, causing vomiting, stomach cramps, diarrhoea and fever. Mass poisonings by zinc have occurred following the storage of fruit juices in galvanized containers.

Zinc also plays a role in the way the body metabolises alcohol. This is broken down in the liver by an enzyme called alcohol dehydrogenase, which has a zinc atom at its centre. Excessive consumption of alcohol damages the liver, and it has long been known that those with cirrhosis have lower than normal levels of zinc in that vital organ. Again, if the damage is not severe, zinc supplements can restore the liver's function.

ELEMENT OF HISTORY

According to Western tradition, zinc was discovered in Europe by the German chemist Andreas Marggraf in 1746. It would be more correct to say that he was the first to *identify* it as an element, because zinc had been known and used long before that date. Strabo (66 BC – AD 24) of Asia Minor mentioned an ore from Cyprus that, when refined, would distil 'mock silver' and that when this metal was alloyed with copper it produced brass. Alternatively, brass was made by heating powdered calamine mineral, which is zinc silicate, with charcoal and copper, a process that continued for almost 2000 years. The Roman writer Pliny the Elder (AD 23–79), who wrote an encyclopaedia,

[199] Now widely known by the Australian name of 'zincy'.

Historia Naturalis, mentioned an ointment that was used for healing wounds and sore eyes and which was probably zinc oxide. A statuette made of zinc has been found in the Roman province of Dacia (modern Romania) and may possibly pre-date the Roman occupation.

Marco Polo described the manufacture of zinc oxide in Persia in the thirteenth century, but it was almost certainly in India that zinc was first recognized as a metal in its own right. Care had to be taken not to lose the zinc, which is quite volatile, during the refining process and a process was developed in which organic materials, such as wool, were heated with zinc ores in a closed crucible. Waste from a zinc smelter at Zawar, in Rajasthan, testifies to the large scale on which the metal was refined, with one estimate being that more than a million tonnes of metallic zinc and zinc oxide were produced there from the twelfth to the sixteenth centuries.

The knowledge of zinc refining spread eventually to China where it was smelted on a large scale by the sixteenth century, and imported from there into Europe. An East India Company ship that sank off the coast of Sweden in 1745 was carrying just such a cargo of zinc; analysis of reclaimed ingots showed them to be almost the pure metal.

A Flemish metallurgist, P. M. de Respour, reported the extraction of metallic zinc from zinc oxide in 1668, and the early chemists began to take an interest in zinc around the turn of the eighteenth century when Geoffroy the Elder (1672–1731) described it, and spoke of it as if it were something new. He described how it would condense as curious yellow crystals – which were clearly zinc oxide – on bars of iron placed above the ore that was being smelted. Zinc could also turn up a by-product of the lead industry, sometimes being collected as the metal from crevices in the chimneys of furnaces, and referred to as *zinc* or *counterfeit.*[200]

Andreas Marggraf (1709–82) investigated calamine from Poland, England, Germany and Hungary in 1746 and obtained zinc by heating the ores with carbon in closed retorts. He realized that he was dealing with a previously unknown metal element. He was unaware that 3 years earlier the Swedish chemist Anton von Swab (1703–1768) had also distilled zinc from calamine and even extracted some from zinc blende (zinc sulfide, ZnS). Meanwhile, in Bristol, England, a zinc smelter was already in operation by the end of 1743, producing the metal at the rate of about 200 tonnes per year.

ECONOMIC ELEMENT

The dominant ore is zinc blende, also known as sphalerite. Other important zinc ores are wurtzite (which is another form of zinc sulfide), smithsonite (zinc carbonate, $ZnCO_3$) and hemimorphite, which is also known as calamine and is a zinc silicate (Zn_2SiO_4). Hydrozincite, which is basic zinc carbonate, is also mined for the metal when it is found in economic amounts.

The main zinc mining areas are Canada, Russia, Australia, USA and Peru. World production exceeds 7 million tonnes a year and commercially exploitable reserves exceed 100 million tonnes. Zinc is made by roasting the sulfide ore to form the oxide and then reducing this by heating it in a furnace with coke, under an inert atmosphere to prevent the zinc from oxidizing as it distils off. An alternative method of winning the

[200] This had even been remarked on in the sixteenth century by Georgius Agricola and Paracelsus.

metal is to convert the oxide to zinc sulfate and then extract the zinc electrolytically. More than 30% of the world's need for zinc is met by recycling.

Zinc is used either as the metal or the oxide. More than 50% of metallic zinc goes into galvanizing steel, while other large uses are in die-casting of components, such as carburettors, and for making brass. Galvanizing is done by dipping a steel object into molten zinc, or depositing a layer of zinc electrolytically. The zinc coating prevents oxygen and moisture from reaching the steel and even when the coating is scratched through it is the zinc which corrodes, and the steel does not rust. For a similar reason, a disc of zinc bolted to an iron rudder will protect it – at least until all the zinc has corroded away, and zinc is used as a sacrificial anode for underground pipelines made of steel. Galvanized steel is used for car bodies, fencing, motorway guard rails and lighting posts, suspension bridges, heat exchangers and roofing tacks.[201]

The metal is also used in alloys and batteries. The first dry batteries had a zinc anode, a carbon cathode and an alkaline electrolyte of ammonium chloride paste, and delivered 1.5 volts. Such batteries are still popular on account of their cheapness and, although not rechargeable, they are regarded by some as more environmentally friendly because they contain no toxic metal.

The main zinc alloy is brass, which consists of 33% zinc and 67% copper, but there are also the nickel silvers (typically with 20% zinc, 60% copper and 20% nickel) having names like China silver, German silver, nickel brass, white copper, etc., which are often used for tableware. The alloy Prestal (78% zinc, 22% aluminium) is almost as strong as steel but as easy to mould as plastic.

As the oxide, zinc's biggest uses are in the rubber industry and in pigments. As a pigment it is used in plastics, cosmetics, photocopier paper, wallpaper, printing inks, etc., while in rubber production its role is to act as a catalyst during manufacture and as a heat disperser in the final product. In rubber and plastics it also prevents damage to the polymer from ultraviolet rays. The use of zinc oxide in paints has mainly been supplanted by titanium dioxide. Zinc oxide will conduct electricity when light shines on it, and use has been made of this in some photocopiers.

Zinc sulfide, doped with other metals, is used as a phosphor in X-ray screens, televisions and fluorescent lighting. When the phosphor is bombarded with electrons, its atoms vibrate and as they return to their original state they emit visible light.

ENVIRONMENTAL ELEMENT

Zinc in the environment	
Earth's crust	75 p.p.m.
	Zinc is the 24th most abundant element
Soils	approx. 64 p.p.m., range 5–770 p.p.m.
Sea water	30 p.p.b.
Atmosphere	ranges from 0.1 to 4 micrograms per cubic metre; much of this comes from the wearing away of motor tyres

[201] Small items like tacks are galvanized by tumbling them in zinc dust at 400°C.

For most soils the amount of zinc which arrives via the atmosphere generally exceeds that which is lost by leaching. Zinc is soluble as the zinc ion (Zn^{2+}) but this is easily absorbed by minerals, clays and organic matter, and strongly held. In the past, rivers passing through industrial and mining districts could have zinc in excess of 20 p.p.m., but in most areas sewage treatment now removes most of it and levels are much less, e.g. the Rhine has only 50 p.p.b.

Some soils are heavily contaminated with zinc, and these are to be found in areas where zinc has been mined or refined, or where sewage sludge from industrial areas has been used as fertilizer. This can have several grams of zinc per kilogram of dry matter. At levels in excess of 500 p.p.m., zinc in soil interferes with the uptake of other essential metals such as iron and manganese, and also of boron. Yet there are soils with levels of zinc in excess of 2000 p.p.m. and a few with levels of 80 000 p.p.m. (8%) and 180 000 p.p.m. (18%) have been recorded, the former in the USA, the latter in Belgium. However, for these it is the cadmium, which invariably accompanies the zinc, that poses more of a threat.

Fruit- and nut-bearing trees in the western states of the USA were afflicted by disease in the nineteenth century, and the reason was thought to be that they were growing on soil deficient in copper. Curiously, some trees responded to the application of a copper solution while others did not. It eventually transpired that those which did benefit were in fact responding to the addition of zinc, which was coming from some of the galvanized buckets in which the copper solution was prepared.

CHEMICAL ELEMENT

Data file	
Chemical symbol	Zn
Atomic number	30
Atomic weight	65.39
Melting point	420°C
Boiling point	907°C
Density	7.1 kilograms per litre (7.1 grams per cubic centimetre)
Oxide	ZnO

Zinc is a bluish-white metal that heads group 10 of the periodic table of the elements. It tarnishes in air and reacts with acids and alkalis.

There are five naturally occurring isotopes, of which zinc-64 accounts for 49%, zinc-66 for 28%, zinc-68 for 19% and zinc-67 for 4%. The least abundant isotope is zinc-70 at only 0.6%. None is radioactive.

ELEMENT OF SURPRISES

(1) Rolled zinc sheeting has been used in roofing, especially in Paris where it was made mandatory in the 1860s when the city underwent a transformation, which resulted in

Zirconium

Pronounced zer-koh-ni-um, it is named from the Arabic *zargun*, meaning gold coloured.

French, *zirconium*; German, *Zirkonium*; Italian, *zirconio*; Spanish, *circonio*; Portuguese, *zircónio*.

COSMIC ELEMENT

Zirconium appears to be relatively abundant in certain types of stars, known as S-type stars, in which heavier elements are formed by neutron capture followed by β-decay, and it is also present in the spectrum of the Sun. Meteorites contain zirconium and samples of rock brought back from the Moon in the *Apollo* space missions were found to have a surprisingly high zirconium content.

HUMAN ELEMENT

Zirconium in the human body	
Blood	10 p.p.b.
Bone	less than 0.1 p.p.m.
Tissue	approx. 0.1 p.p.m.
Total amount in body	1 milligram

Zirconium has no known biological role, and its salts are generally regarded as being of low toxicity. The estimated daily dietary intake is around 50 micrograms (although for one individual it was measured at more than 100 micrograms). Most passes through the gut without being absorbed, and that which is absorbed tends to accumulate slightly more in the skeleton than in tissue. Zirconium cannot be detected in urine.

While aquatic plants have a rapid uptake of soluble zirconium, land plants have little tendency to absorb it, and indeed 70% of plants that have been tested showed no zirconium to be present at all. However, some plants do absorb a little from the zirconium that is present in all soils: the leaves of deciduous trees can have as much as 500 p.p.m. (ash weight), but food plants have less than 2 p.p.m. (dry weight) and often as little as 5 p.p.b.

MEDICAL ELEMENT

In the USA, lotions containing zirconium carbonate used to be sold as treatments for reactions to poison ivy. The zirconium combined with the irritant urushiol from the plant and rendered it inactive, but these lotions were discontinued in the 1960s after a few people appeared to react badly to them. Yet a group of 22 workers, who had been exposed for up to 5 years to the fumes of molten zirconium, were monitored and found not to suffer any adverse effects, even after 40 years.

ELEMENT OF HISTORY

Gems that contain zirconium were known in biblical times and called hyacinth, jacinth, jargon and zircon. Colourless varieties were thought to be an inferior kind of diamond, but that was shown to be false when a German chemist, Martin Heinrich Klaproth (1743–1817),[202] analysed a zircon in 1789 and discovered zirconium. He ran a pharmacy in Berlin, and investigated a specimen of jargon, a semi-precious stone from Ceylon (Sri Lanka),[203] and he began by heating it with alkali. From the product of that reaction he extracted a new oxide, which he called zirconia, and which he realized must be that of a 'new' element, zirconium. In the year he discovered zirconium he also discovered uranium, and this coincidental link between the two was to be strangely echoed 150 years later when the fates of these two elements became truly linked in the nuclear power industry – see below.

Klaproth was unable to isolate the pure metal itself, and sadly he did not live to see this achievement. Humphry Davy, in London in 1808, tried to separate zirconium from its oxide by electrolysis, which was the way in which he had successfully isolated potassium and sodium metals, but he failed. It was not until 1824 that the element was isolated, when the Swedish chemist Jöns Jacob Berzelius heated potassium hexafluorozirconate (K_2ZrF_6) with potassium metal in a sealed iron tube, and he obtained it as a black powder. Even so, little use was found for zirconium; it had no commercial application as a metal, and its chemical compounds were without any noteworthy features.

As the bulk metal, impure zirconium is hard and brittle whereas the pure metal is easily worked and can be drawn into wire. Metal of this quality was first produced only in 1925 by the Dutch chemists Anton van Arkel and J. H. de Boer, by the decomposition of zirconium tetraiodide (ZrI_4). In the 1940s, William Justin Kroll of Luxembourg developed a commercial process for making pure metal; in this zirconium tetrachloride ($ZrCl_4$) is reduced with magnesium.

ECONOMIC ELEMENT

The chief ores are zircon ($ZrSiO_4$), which is mined in Australia, USA and Sri Lanka, and baddeleyite (zirconium oxide, ZrO_2) which is mined in Brazil. World production is in excess of 900 000 tonnes per year of zircon, and 7000 tonnes of the metal are pro-

[202] Klaproth eventually became the first Professor of Chemistry at the University of Berlin when he was 60 years old.

[203] This is a form of zircon (zirconium silicate).

duced. The estimated reserves exceed a billion tonnes. Australia (East and West coasts), South Africa (Richards Bay), India (Kerala State), Sri Lanka and the USA (Trail Ridge, Florida) have vast deposits of zircon and zirconia sands. Australia and South Africa account for about 80% of zircon mining. Often the zircon is a by-product of the mining in these regions, which is primarily directed at recovering titanium from mineral-rich sands.

Zirconium, as zircon, zirconia and as the metal, ends up in ceramics, refractory materials such as furnace linings, glass, chemicals and alloys.

Zircon sand is traditionally a refractory material, making heat-resistant linings for furnaces, giant ladles for molten metal, and foundry moulds, because it resists the penetration of molten metal and gives a uniform finish. While these long-established markets have declined, others have developed to take their place; for example, ceramic glazes account for almost half of zircon consumption. It is also used to make the glass for televisions, although this may soon be less in demand as flat screens are introduced, and used for the feedstock for zirconium chemicals, an area which is expanding. Zircon mixed with vanadium or praseodymium makes blue and yellow pigments for glazing pottery, and zircon glazes are particularly used on sanitary ware.

Larger crystals of zircon are sold as a semiprecious gems; they come in a variety of colours, stones with a golden hue being most admired. When zircon is cut and polished, it shines with an unusual brilliance because it has a high refractive index.

Zirconia, the common name for zirconium oxide, only melts above 2500°C, and is used to make heat-resistant crucibles,[204] ceramics and abrasives. World production of pure zirconia is almost 25 000 tonnes per year. Zirconia also goes into various chemicals that end up as cosmetics, antiperspirants, food packaging,[205] and even fake gems – see also below. Zirconium hydroxychloride is the preferred antiperspirant component in roll-on deodorants. The paper and packaging industries are finding that zirconium compounds make good surface coatings because they have excellent water resistance and strength. Equally important is their low toxicity; zirconium carbonate has been approved for treating wrapping that comes in contact with food.

Zirconium metal had some hidden assets which suddenly brought it to prominence in the 1940s: it was found to be the ideal metal for inside nuclear reactors and nuclear submarines. It does not corrode at high temperatures, nor absorb neutrons to form radioactive isotopes. Even today the nuclear industry buys almost all of the metal that is produced and some nuclear reactors have more than 100 000 metres of zirconium tubing. Zirconium metal, as Zircaloy alloy (consisting of 98% zirconium with 1.5% tin plus 0.1% of iron, chromium and nickel), is used to make the cladding for uranium oxide fuel elements. As mined, zirconium contains 1–3% of hafnium, which is chemically very similar, and although it is difficult to separate the two elements this has to be done for the zirconium that is used by the nuclear industry because hafnium absorbs neutrons very strongly.

Zirconium metal is produced by heating zircon with carbon and chlorine to convert

[204] A red-hot zirconia crucible can be plunged into cold water without cracking.

[205] At one time it looked as if zirconium compounds would replace aluminium compounds in many applications because aluminium was believed, wrongly, to be the cause of Alzheimer's disease.

it to zirconium chloride, then separating this and heating it with magnesium. The metal is protected by an oxide layer which forms instantly on its surface by reaction with the oxygen of the air or with water vapour. This oxide film is impervious to both strongly acidic and alkaline media and mechanical attack, even up to 400°C, so that it finds use for handling corrosive chemicals in industry.[206] It is also used for surgical implants and prosthetic devices because it is compatible with body fluids. Because the metal resists high temperatures, it has been used for space-vehicles like the space shuttle which heat up on re-entry to the Earth's atmosphere.

Zirconium metal is used in alloys – it makes steel stronger and improves its machinability. In 1998, the world's largest petrochemical processing unit, 3 metres in diameter, 50 metres tall and costing US$6 million, was installed at a plant near Houston, Texas. It was made from zirconium alloy and, it is claimed, will last for more than 30 years, far longer than any stainless steel tower would last.

Early bulbs for flash photography used magnesium or aluminium foil. These were ignited electrically, but they were prone to exploding. In the USA, the Speed Midget bulb was developed which used zirconium metal, and was much more reliable. As photographic film become more sensitive, the size of the bulbs became smaller, and from 1960 to the 1980s the zirconium flash bulbs were the only ones in use, requiring tonnes of the metal. With the advent of electrical discharge bulbs, built into cameras, such flash bulbs are now rarely used.

Zirconium metal is used as a getter in vacuum tubes to remove traces of gases because it combines readily with oxygen, hydrogen and even nitrogen.

ENVIRONMENTAL ELEMENT

Zirconium in the environment	
Earth's crust	190 p.p.m.
	Zirconium is the 18th most abundant element.
Soils	range 35–550 p.p.m. (200–270 p.p.m. in garden soils)
Sea water	9 p.p.t.
Atmosphere	traces

Zirconium is not a particularly rare element but because its most common mineral, zircon, is highly resistant to weathering it is only slightly mobile in the environment. Zirconium is more than twice as abundant as copper and zinc, and more than 10 times more abundant than lead. Zirconium is regarded as completely non-toxic and environmentally benign, so its use will continue to grow. For example, it is being added to paints to replace the small amounts of lead compounds that are still included to improve drying.

[206] Some reagents, such as hydrofluoric acid, chlorine gas and solutions of iron(III) chloride and copper(II) chloride, will attack it.

CHEMICAL ELEMENT

Data file	
Chemical symbol	Zr
Atomic number	40
Atomic weight	91.224
Melting point	1852°C
Boiling point	4377°C
Density	6.5 kilograms per litre (6.5 grams per cubic centimetre)
Oxide	ZrO_2

Zirconium is a hard, lustrous, silvery metal, known as a transition metal and it is a member of group 4 of the periodic table of elements. It is very resistant towards corrosion due to an oxide layer on the surface. Zirconium does not dissolve in acids (except hydrofluoric acid) or alkalis. Powdered zirconium is black and it will burn in air if ignited, and zirconium dust is regarded as a very dangerous fire hazard.

There are five naturally occurring isotopes of zirconium, of which zirconium-90 accounts for 52%, zirconium-92 and -94 for 17% each, zirconium-91 for 11%, and zirconium-96 for 3%. This last isotope may be radioactive but this has been difficult to assess because its half-life is reputed to be more than 300 quadrillion years (300×10^{15} years) or around 60 million times longer than the age of the Earth.[207]

ELEMENT OF SURPRISE

The most dramatic use of zirconium, as zirconia, is in ultra-strong ceramics. The research which led to this was the need to develop a tank engine that was not made of metal, and so would need neither lubricating oil nor a cooling system. From this research came a new generation of tough, heat-resistant ceramics that are stronger and sharper than toughened steel and which make excellent high-speed cutting tools for industry. They are also sold as knives, scissors and golf irons. Bioceramics for joint replacement parts are also made of zirconia, and these are formed in precision presses and then fired at 1700°C.

[207] There are also a number of radioactive isotopes which occur naturally, in minute amounts in uranium minerals, produced by the spontaneous fission of atoms of uranium.

The Periodic Table

The Russian chemist Dimitri Mendeleyev drew up the first truly successful periodic table of the chemical elements one winter's morning in 1869. Today he is rightly seen as its *discoverer* because he based it on scientific principles, rather than its being merely a table of the elements. Chemists had been groping towards a classification of the elements before this date, as they continued to find more and more of them, and some came close to discovering the periodic table itself.

Since the time of Mendeleyev, other chemists have redesigned his periodic table, sometimes in response to new elements being discovered or synthesized, sometimes in response to advances in our knowledge about the nature of atoms. Of course it was inevitable that a periodic table would be devised sooner or later as the science of chemistry reached maturity, and once the nature of the atom had been revealed then it would have been an imperative. The surprise was that Mendeleyev's work preceded these advances by more than 30 years.

Three subatomic particles – protons, electrons and neutrons – determine the chemistry of an atom and consequently the layout of the periodic table. They are important in that order:

- *Protons:* the number of positively charged protons in the nucleus of an atom is the so-called atomic number, and it determines the element itself: hydrogen (atomic number 1) has one proton, helium two, lithium three, beryllium four, boron five, and so on up to element 118 and possibly beyond.
- *Electrons:* there are the same number of negatively charged electrons in the orbits that surround the nucleus as there are protons in the nucleus, but it is the electrons in the outermost orbits that are responsible for the chemical behaviour of an element. Elements that have the same arrangement of electrons in the outer orbit should resemble one another chemically. This explains the periodic repetition of chemical properties and behaviour that is observed in the groups of the periodic table. (Electrons weigh very little and contribute less than 0.05% to the mass of an atom.)
- *Neutrons:* the neutrally charged neutrons in the nucleus are also important – they have almost the same mass as protons, so contribute to the overall atomic weight. They are the key to explaining anomalous atomic weights and isotopes, and were discovered in 1932 by James Chadwick (1891–1974) who was awarded a Nobel Prize in 1935 for his work.

The remarkable achievement of Mendeleyev was that he produced his periodic table 27 years before Joseph Thomson (1856–1940) discovered the first subatomic particle, the electron, in 1897, and 42 years before Ernest Rutherford (1871–1937) published his discovery of the positively charged atomic nucleus in 1911.[208]

[208] Discovered in experiments in 1909.

When Mendeleyev heard of the discovery of electrons in 1897, he rejected the idea that they came from atoms. He believed that atoms were indivisible. Indeed he predicted that the electron would disappear from science in the way that phlogiston had (see p. 300). This may strike us as mildly eccentric now, but in those days some eminent scientists were still not convinced of the existence of atoms. One such sceptic was the respected electrochemist Wilhelm Ostwald (1853–1932). Today we can photograph them.

ATOMS AND ELEMENTS

The idea of atoms can be traced back to the ancient Greek philosophers who deduced that the world must be composed of them; they did this by applying arguments of logic, one of which went as follows. Consider what happens when you slice horizontally through a cone. If matter is continuous, then both faces of the slice must be exactly the same size, but since they are part of a cone then they must be different. The lower face must be larger, but that can only be so if we have sliced between two planes of atoms. The chief 'atomists' were Leucippus and his pupil, Democritus, who lived in the fifth century BC, and their ideas were developed by Epicurus (341–270 BC). We know of their theories because of the book *De rerum natura* [*Natural things*], which was written around 55 BC by the Roman poet Titus Lucretius. The book is still in print, thanks to a single copy that survived the Dark Ages.

Another philosophical debate concerned *elements*. Again we owe the idea to the Greek philosophers who speculated that although the world appeared extremely complex, all that we were seeing was just different forms of some primordial element. What might it be? For example, Heraclitus (died 460 BC) thought this element was water, while others suggested air, fire or earth. Empedocles (who also lived in the fifth century BC) proposed there were four elements, and this view was developed by the greatest Greek philosopher of them all, Aristotle (384–322 BC).[209]

This seemed a reasonable theory, and one that was supported by commonsense observations such as what happens when a stick burns in a fire. It can be seen to break down into the four elements: fire, water, air and earth. For 2000 years, Aristotle's ideas were accepted in Europe, almost without question, until the dawn of modern science in the seventeenth century.

Chemistry emerged from alchemy around AD 1700 and as it developed so too came the discoveries of previously unknown metals and gases, and the slow realization that the Greek concept of four elements was fundamentally wrong. There were elements, but these had to be defined in terms of their chemistry. Indeed, had the ancient philosophers but known it, they already were familiar with 10 of them:

- **carbon**, which as charcoal came with fire;
- **sulfur**, deposits of which were discovered near volcanoes;
- **copper**, the first metal to be worked, around 5000 BC and probably much earlier;[210]

[209] In the East, Chinese philosophers came to the same conclusion, but with the addition of a fifth element: wood.

[210] There is evidence that copper was known to Neanderthal Man.

- **gold** and **silver** objects, which were first produced about 5000 BC;
- **iron** smelting led to the Iron Age, which began around 2500 BC;
- **tin** as such was smelted around 2100 BC, but had been used earlier for bronze;
- **antimony** objects date from 1600 BC, although this metal was rarely used;
- **mercury** was first obtained from its sulfide ore, cinnabar, about 1500 BC;
- **lead** appeared around 1000 BC and was the most important metal domestically.

And there things rested for two millennia. In the West, the Roman Empire concentrated more on technological developments rather than scientific ones, although alchemy flourished for a while until the Emperor Diocletian suppressed it at the end of the second century. He feared that if the alchemists were to find a way of turning lead into gold, it would undermine the Empire's currency. However, the Dark Ages were about to descend in Europe and progress came to an end for 500 years. This traditionally began with the sack of Rome in 410 AD, and the mass migration of peoples from the North, moving South as the Earth entered one of its cooling periods.[211] Science and technology in the West declined, and its study became the province of the Arabs and only returned as they advanced westwards, bringing with them their store of ancient wisdom and their new knowledge of the world.

Alchemy still had its adherents and some elements were discovered by them during this period. Arsenic appears to have been isolated by the German alchemist, Magnus, in the middle of the thirteenth century; bismuth appeared towards the end of the fifteenth century; and phosphorus was made by Hennig Brandt of Hamburg in 1669. This last element had amazing properties: it burst into flames spontaneously when exposed to air and it glowed in the dark. The discovery of phosphorus is seen as a turning point between alchemy and chemistry or, rather, its investigation turned the English alchemist Robert Boyle into the first modern chemist. Meanwhile the development of mining had uncovered zinc and the exploration of the New World brought platinum to people's attention.

The eighteenth century saw the emergence of chemistry in Europe and the discovery of several new metals – cobalt (1735), nickel (1751), magnesium (1755), manganese (1774), chromium (1780), molybdenum (1781), tellurium (1783), tungsten (1783), zirconium (1789) and uranium (1789) – and a few gases, namely hydrogen (1766), nitrogen (1772), oxygen (1772) and chlorine (1774).

It was left to the great French chemist, Antoine Laurent de Lavoisier (1743–94), to bring order to the fledgling science, which he did with his remarkable book *Traité Elémentaire de Chemie* [*Elements of Chemistry*] in 1789. In this he defined a chemical element as something that could not be further broken down (decomposed) and he listed 33 substances that came within the terms of his definition. He classified them into four categories: gases, non-metals, metals and earths:

[211] The Dark Ages lasted until around AD 1000, when the Earth warmed up again and the Middle Ages began.

Category	Members
Gases	[light], [heat], oxygen, nitrogen, hydrogen
Non-metals	sulfur, phosphorus, carbon, chloride, fluoride, borate
Metals	antimony, arsenic, bismuth, cobalt, copper, gold, iron, lead, manganese, mercury, molybdenum, nickel, platinum, silver, tin, tungsten, zinc
Earths	lime, magnesia, baryta, alumina, silica

Lavoisier had given the idea of chemical elements scientific respectability. While his list contains some components that were never destined to be considered elements, such as 'light' and 'heat', it is still remarkable because he included things he called 'radicals', which he knew were non-metal elements but which could not be separated from other components, i.e. chloride, fluoride and borate. He also had a group called 'earths' which likewise were oxygen compounds but from which it had proved impossible to remove the oxygen using the methods then available. We now know that lime is calcium oxide, magnesia is magnesium oxide, barytes is barium oxide, alumina is aluminium oxide and silica is silicon dioxide.[212]

Another of Lavoisier's far-reaching contributions to chemistry was to advocate that chemicals should be named after the elements of which they were composed. This demystified the language of chemistry of its alchemical names, and the result was that chemicals like 'butter of antimony' became renamed as antimony chloride, and 'lunar caustic' as silver nitrate.

A milestone in the development of chemistry came with the discovery of oxygen, the gas given off when calx of mercury (mercuric oxide) was heated. The discovery is attributed to Joseph Priestley (1733–1804) in 1774, but oxygen had already been isolated by Carl Wilhelm Scheele (1742–86) 2 years earlier, although his work was not reported until 1777. Oxygen delivered the greatest blow to Aristotle's theory of four elements, when, in 1781, Priestley demonstrated that water was formed from hydrogen and oxygen.

Another intellectual revolution was initiated by John Dalton (1766–1844) who was a science teacher living in Manchester, England. He pondered on the atomic nature of matter and realized that the Law of Fixed Proportions, which recognized that elements combined with one another in fixed ratios of weights, could only be explained if the elements were composed of atoms. In 1803 Dalton presented a paper to the Manchester Literary and Philosophical Society entitled 'On the absorption of gases by water and other liquids'. In this he attempted to explain the different solubilities of gases by comparing their relative weights. This led him to propose a scale of atomic weights; when he published his talk in the newly launched *Proceedings* of the Society in 1805, he explored the idea further and appended a list of 20 elements with their atomic weights, and suggested atomic symbols – see Figure 1.[213] At a stroke, Dalton's theory not only

[212] This convention of naming oxides simply by replacing the 'um' or 'ium' ending of a metal's name with an 'a' continues to this day in names like zirconia.

[213] Dalton's full theory of atoms was published in 1808 as *A New System of Chemical Philosophy*. Dalton also devised a set of symbols to represent atoms, but this was not taken up. Jöns Jakob Berzelius proposed that elements should be represented by an alphabetical system derived by abbreviating the name – see p. 3. This proposal became generally accepted and is used today.

Fig. 1 Dalton's table of the elements.

proposed the existence of atoms, but also suggested that atoms of each element had individual weights and that these could be calculated relative to one another.

By 1810, chemistry was an established science, producing remarkable discoveries almost every year, not least of which was a constant supply of newly discovered elements. In the years between the publication of Lavoisier's *Traité* and Dalton's talk, another 11 had been discovered: titanium (1791), yttrium (1794), beryllium (1797), vanadium (1801), niobium (1801), tantalum (1802), rhodium (1803), palladium (1803), osmium (1803), iridium (1803) and cerium (1803). This pace was to continue, especially when Humphry Davy showed that some of Lavoisier's 'earths' could be decomposed by electrolysis; he then isolated the metals potassium (1807), sodium (1807), calcium (1808) and barium (1808). In the same year, 1808, Crawford discovered strontium.

Boron was also produced in 1808, and then came iodine in 1811. The year 1817 was particularly fruitful, with thorium, lithium, selenium and cadmium being announced.

ORDER OUT OF CHAOS

By now the chemical similarity of groups of elements began to be commented on. In 1829 Johann Wolfgang Döbereiner (1780–1849) announced his Law of Triads, in which he noted that many elements came in groups of three ('triads'), in which the weight of the middle element was the average of the lighter and heavier members.

Lithium–sodium–potassium were one such triad and others were chlorine–bromine–iodine and sulfur–selenium–tellurium. By the time Leopold Gmelin published the first edition of his compilation of essential chemical data, *Handbuch der Chemie*, in 1843, he noted there were 10 triads, three tetrads (groups of four) and even a pentad consisting of nitrogen, phosphorus, arsenic, antimony and bismuth. Today we recognize this pentad as group 15 of the modern periodic table.

Occasionally there were curious insights into the underlying nature of matter – and one in particular was well in advance of its time. This was the hypothesis put forward by William Prout (1785–1850) in 1815. He proposed that all the elements were multiples of the atomic weight of hydrogen, which is the lightest. If this is taken as 1, then it explained why all the other elements had weights that were whole numbers, or nearly so.

Prout's hypothesis was, tantalizingly, almost correct in assuming that hydrogen was the element of elements. In a way he was right. Most hydrogen atoms consist of a single proton, and it is the number of protons that determines the nature of all the elements. The nucleus of an atom is where 99.98% of its weight resides. Because this is composed of protons and neutrons, which have almost exactly the same weight, and because most elements have one dominant isotope, then it is not surprising that their weights will tend to whole numbers. Most elements have atomic weights that fall within 0.1 above or below a whole number, with very few having fractional numbers. Nevertheless it was these few exceptions that 'proved' the rule, i.e. tested it, and led to its rejection. It appeared to be merely a coincidence that many elements had almost whole-number atomic weights, and there were several quite important exceptions, notably magnesium (24.3), chlorine (35.5), nickel (58.7) copper (63.5), zinc (65.4) and tin (118.7).

The problem of atomic weights was to dog chemistry for several decades. Clearly atoms were too small to be weighed, or even counted, and so only comparative weights could be meaningful. But against what standard were they to be compared? The best one appeared to be oxygen, which forms compounds with almost all of the other elements. Oxygen's atomic weight was taken to be 8, which followed from a comparison with hydrogen, and the 1:8 weight ratio of these elements in water. At the time the composition of water was assumed to be HO, not H_2O.

In fact, the true chemical formula of water could have been deduced as early as 1800, when William Nicholson and Anthony Carlisle electrolysed water and showed it decomposed into two volumes of hydrogen and one of oxygen. The significance of this was not realized, even though in 1811 Amadeo Avogadro (1776–1856) suggested that equal volumes of gases contained equal numbers of particles (molecules). Eventually, it was realized that the formula for water was H_2O, and the weight of oxygen was correspondingly adjusted to 16.

From 1830 to 1860 only three elements were discovered: lanthanum (1839), erbium (1842) and terbium (1843). These metals were later to yield other metals because they were part of the group known as the rare earths, all of which have very similar properties, making them difficult to separate from one another. In 1853, John Hall Gladstone arranged all the known elements in order of their relative atomic weights and noted relationships between them.

Other chemists were also groping towards a logical arrangement of the elements around this time. In 1850, the German Chemist, Max von Pettenkofer, had extended Dobereiner's and Gmelin's ideas, noting that there was a connection between the atomic weight of an element and its chemical behaviour. However, he was hamstrung by the lack of reliable atomic weights to verify his theory. In 1857, the French chemist Jean Baptiste André Dumas observed horizontal relationships between metals. He too was thwarted by the lack of reliable atomic weight data, although his work was often cited as germane by those who devised early forms of the periodic table. Gustav Detlef Hinrichs, a Danish-born scientist who emigrated to the USA in 1861 and took a position at the University of Iowa, devised a kind of periodic table in 1867, in which he arranged the elements in a spiral form.[214]

The next era of element discovery came with the atomic spectroscope, which revealed that each element had a characteristic fingerprint pattern of lines in its visible spectrum. The technique was to spray a tiny sample of a solution of the element into the colourless flame of a Bunsen burner and observe the colours that ensued by viewing them through a prism so that they became separated across the spectrum, when it could be seen that they consisted of many bands and lines of colour. Because each element always gave the same pattern of lines, no matter what its source, and because each gave a different pattern of lines from any other element, it was realized that here was a technique for uncovering hitherto unknown elements. Merely subjecting a mineral to atomic spectroscopy, as it was called, immediately showed whether a previously unknown element was present. As a result rubidium, caesium, thallium and indium were announced in the years 1860–63.

By then the total of known elements exceeded 60 and chemists were beginning to ask whether there was a limit to the number. To answer this question required a theory that could explain the known and predict the unknown. The 'triads', 'tetrads' and 'pentad' were governed by regular increases in relative weight. Clearly chemists might use this universal property to rank all the elements, if only they could talk in terms of *atomic* weights rather than the more general comparative weights (known as equivalent weights). Equivalent weights depended upon other factors; for example, if a metal had two kinds of oxide it had two equivalent weights.

ATOMIC WEIGHTS AND THE FIRST PERIODIC TABLES

In 1858, the Italian chemist Stanislao Cannizzaro (1826–1910) published his 'Outline of a course of chemical philosophy', in which he showed how it was possible to assign an *atomic* weight to an element if Avogadro's Law was accepted. This said that equal volumes of different gases, at the same temperature and pressure, had the same number of molecules. Cannizzaro argued that this law led directly to the atomic weights of gaseous elements, and thence to other elements. A given volume of oxygen gas is 16 times heavier than the same volume of hydrogen. If the atomic weight of hydrogen is taken as 1 then it followed that the atomic weight of oxygen was 16. Cannizzaro presented a paper on his ideas at the First International Chemical Congress, which was

[214] The information in this paragraph was kindly provided by Walter Saxon.

held at Karlsruhe in 1860, and they were quickly accepted. Conference members eagerly sought copies of his table of atomic weights. One fell into the hands of a young Russian student, Dimitri Mendeleyev, who was then engaged in postgraduate research in Germany, and he took it back to St. Petersburg.

It was this conference that set the stage for the discover of the periodic table. For the first time the elements could be arranged in an ascending scale from the lightest to the heaviest. The first attempt to arrange them in a regular pattern was made in 1862 by a French geologist, Alexandre Emile Beguyer de Chancourtois (1820–86). He wrote the list on a piece of tape and then wound this, spiral-like, around a cylinder. The cylinder's surface was divided into 16 parts, based on the atomic weight of oxygen. Chancourtois noted that certain triads came together down the cylinder, such as the alkali metals, lithium, sodium and potassium, whose atomic weights are 7, 23 (7 + 16) and 39 (23 + 16). This coincidence was also true of the tetrad oxygen–sulfur–selenium–tellurium. He called his model the *Vis Tellurique* (telluric screw) and published it in 1862, but parts of it made no sense at all. For example, one of his groups consisted of boron, aluminium, nickel, arsenic, lanthanum and palladium.

What Chancourtois had discovered was the *periodic* nature of the elements. In other words, elements with the same properties occurred at regular intervals, in this case when the atomic weights differed by 16, as with the alkali metals, or if they were multiples of 16, or nearly so, such as we find with oxygen = 16, sulfur = 32, selenium = 79 ($5 \times 16 = 80$) and tellurium = 128 (8×16). Clearly there appeared to be an underlying numerical rhythm to the elements, even though we now know that it was purely by chance that the elements of the oxygen–tellurium tetrad are multiples of 16.

Another attempt to classify the elements was made by an Englishman, 27 year old John Alexander Reina Newlands (1837–98). In 1864, he read a paper entitled 'The Law of Octaves' to a meeting of the London Chemical Society; in it he pointed out that there was a periodic similarity of elements at intervals of 8. The title of his paper was chosen by analogy with octaves in music. This was an ill-judged choice, and it is said that one member of his audience sarcastically asked Newlands whether he had ever thought of arranging the elements alphabetically instead. The society's journal refused to publish his talk as a paper,[215] but he wrote an account for *Chemical News* in 1865, so we know what he was proposing. He had arranged 56 elements into groups and noted that there seemed to be a periodic repetition of properties with every eighth element.

Although Newlands was moving in the right direction – we can see some of the groups in his table in a modern periodic table – he was clearly not aware of any underlying imperative for what he had done, nor did he build on it. For example, he noticed that silicon and tin should be part of a triad, but he left no vacant place in the table for the missing element between them. Nor had he any qualms about putting two elements together in some of the boxes in his table.

William Odling, successor to Michael Faraday at the Royal Institution in London, was another chemist to speculate about relationships among the elements. He

[215] Nevertheless, the Royal Society of London awarded him its prestigious Davy Medal in 1887 in belated recognition of his achievements.

published a paper entitled: 'On the proportional numbers of the elements' in the first volume of the *Quarterly Journal of Science* in 1864. His arrangement of the elements came surprisingly close to that of Mendeleyev's, and he left gaps where there were missing elements. In fact he even left gaps that were later to be filled by the noble gases helium and neon, although he thought these would be lighter elements of a pentad that expanded the triad of zinc–cadmium–mercury. Despite his realization that the table could have predictive power, it was little more than a convenient way of classifying the elements.

Another scientist to come near to discovering the periodic table was the German, Julius Lothar Meyer (1830–95). He published a table of 49 elements with their 'valencies'[216] in 1864. Meyer drew a graph in which he plotted the atomic volumes of the elements against their atomic weights in 1868; the graph shows a periodic rise and fall in atomic volume, and from this Meyer deduced a periodic table. He passed this paper to a colleague, Professor Remelé, for his comments. Unfortunately these were slow in coming, and before he could submit it for publication, Mendeleyev's definitive paper was published. Meyer's paper was eventually published in 1870, and there then followed a heated debate between Meyer and Mendeleyev as to who had priority in the discovery.

COMETH THE HOUR, COMETH THE MAN: DIMITRI MENDELEYEV

It still strikes one as odd that, in 1869, a relatively unknown Russian Professor of Chemistry, Dimitri Ivanovich Mendeleyev, produced the first truly successful periodic table of the elements. Mendeleyev was born in Tobolsk in western Siberia on 8 February 1834, the fourteenth child of a local schoolmaster. His mother, Maria, raised the children after her husband became blind. She came from a family with interests in glass-works and paper mills and these she ran while struggling to educate her favourite son, Dimitri, whom she recognized as highly intelligent. She was determined that he should have a university education and when he was 18 she took him to Moscow, but failed to secure him a place at the university there.

Nothing daunted, they travelled on to St Petersburg, where Mendeleyev was allowed to enrol at the Central Pedagogic Institute to study physics and mathematics. He graduated with a gold medal for excellence in scholarship, and in 1859 went to study for his doctorate, first to Paris, working under Henri-Victor Regnault, and then to Heidelberg, where he spent time with the great German chemist, Robert Bunsen.

In 1861, he returned to St Petersburg, that great city on the Baltic, which was at the leading edge of reform in Russia. There, science and the arts were encouraged in an

[216] Valency is a term usually reserved for non-metals, so that if an element forms a compound of formula MCl_x we would say it was *x*-valent; for example, in PCl_3 phosphorus has a valency of 3 and in PCl_5 it has a valency of 5. When it comes to talking about metals the preferred term is oxidation state rather than valency; oxidation state is shown by adding a Roman number suffix to the name of the element, so we would call $PtCl_2$ platinum(II) chloride and $PtCl_4$ platinum(IV) chloride. Nevertheless, valency was an essential part of chemical theory until relatively recently, and it is still a term that is used.

atmosphere that was well ahead of its time; for example, women were encouraged to become educated and take up professions. In this atmosphere of enlightenment, chemistry flourished.

Mendeleyev became a Professor of Chemistry in 1865, and began to write his textbook *The Principles of Chemistry*, which ran to many editions and was translated into French, German and English. He wondered how best to deal with the many elements with their diverse properties, and in the end he decided to group them according to their valencies. What followed was to transform a large part of chemistry from a disorganized jumble of facts into a disciplined science.

Mendeleyev's movements on the day of his great discovery are well documented. For him it was 17 February but for the rest of the Western world it was 1 March, because reform in Russia had not yet changed the yearly dating system from the Julian calendar of the Roman Empire to the Gregorian calendar that we use today. Mendeleyev had planned to visit a cheese factory that morning, but the weather was bad so he stayed at home and worked on the manuscript for his book.

He had written the name of each element on separate pieces of card, on which he also wrote its atomic weight, a few physical properties and the formulae of any hydrides and oxides it formed, which revealed its preferred valency. These cards he began to arrange like those in a game of patience (solitaire), with rows of elements having the same valency in descending order of their atomic weights. Suddenly things began to fall into place. He produced one arrangement that particularly impressed him and wrote it down on to the back of an envelope. This can be considered his first periodic table, and the envelope still exists: it is kept in a museum in St. Petersburg, Russia.

			Ti=50	Zr=90	?=180.
			V=51	Nb=94	Ta=182.
			Cr=52	Mo=96	W=186.
			Mn=55	Rh=104,4	Pt=197,4
			Fe=56	Ru=104,4	Ir=198.
			Ni=Co=59	Pl=106,6,	Os=199.
H=1			Cu=63,4	Ag=108	Hg=200.
	Be=9,4	Mg=24	Zn=65,2	Cd=112	
	B=11	Al=27,4	?=68	Ur=116	Au=197?
	C=12	Si=28	?=70	Sn=118	
	N=14	P=31	As=75	Sb=122	Bi=210
	O=16	S=32	Se=79,4	Te=128?	
	F=19	Cl=35,5	Br=80	I=127	
Li=7	Na=23	K=39	Rb=85,4	Cs=133	Tl=204
		Ca=40	Sr=87,6	Ba=137	Pb=207.
		?=45	Ce=92		
		?Er=56	La=94		
		?Yt=60	Di=95		
		?In=75,6	Th=118?		

Fig. 2 Mendeleyev's first periodic table.

After his midday meal, Mendeleyev took a nap. On waking he decided that a vertical arrangement of groups was a better way of depicting the periodic table. This event was to give rise to the romantic notion that the periodic table came to him in a dream. In any case he redrew his table and this has remained the standard format to this day. What makes Mendeleyev's table so important, was that he realized that he had stumbled on an underlying order to the elements and this gave him the confidence to make a few predictions, most of which were soon to be proved right.

Mendeleyev realized that if his table of the elements meant anything, then he had to leave gaps where elements, yet unknown, should come. Below boron, aluminium and silicon were such vacant spaces. These elements he named eka-boron, eka-aluminium and eka-silicon. He predicted their atomic weights, melting points and densities, and said what the chemical composition of their oxides and other compounds would be.

The most convincing demonstration of the correctness of Mendeleyev's periodic table was the discovery of these three missing elements. Eka-aluminium was discovered by Paul-Émile Lecoq de Boisbaudran in 1875 at Paris; he called it gallium after the Latin name for his native France. He measured its properties, including the density, which he said was 4.7 grams per cubic centimetre. He was informed that the newly discovered element was Mendeleyev's eka-aluminium and that most of its properties had already been foretold. However, Boisbaudran was aware of the discrepancy between the density he had calculated and that predicted by Mendeleyev, which was 5.9 grams per cubic centimetre, so he checked his measurements, discovered he had made an error, and reported a corrected value of 5.956 grams per cubic centimetre.

In 1879 Lars Fredrik Nilson discovered eka-boron at Uppsala, Sweden; this element too was named after the region of its discovery, scandium, from the Latin for Scandinavia. It too had the properties Mendeleyev predicted. For example, its atomic weight was 44 (predicted 44) and density 3.86 grams per cubic centimetre (predicted 3.5 grams per cubic centimetre).

Finally, in 1886, Clemens Alexander Winkler discovered eka-silicon at Freiberg in Germany; he also named it after the Latin name for his native land. Germanium was almost exactly as Mendeleyev had predicted, right down to the density of its oxide – which he said would be around 4.7 grams per cubic centimetre and turned out to be 4.703 grams per cubic centimetre – and the boiling point of its chloride – which he said would be a few degrees below 100°C, and which turned out to be 86°C.

Mendeleyev had also noted that his group IV, the titanium group, was lacking a member and urged that this be sought among titanium ores. It only came to light in 1923 when D. Coster and George von Hevsey found it in a zirconium ore and named it hafnium. Mendeleyev had predicted its atomic weight would be 180 – it was measured as 178.5.

Sometimes Mendeleyev got it wrong. For example, he placed mercury in a group with copper and silver, even though this meant placing it in the wrong order of atomic weights.[217] His difficulty with tellurium and iodine, whose atomic weights were in the wrong order, is noted on p. 429, yet despite this, he knew he must put them in the

[217] These elements share a common valency of 1, but this is deceptive as a way of classifying them because mercury's dominant valency is 2.

correct group of his table, so sure was he that the periodic table was correct. And it was correct. What surprises us today is that Mendeleyev was able to deduce it long before chemists were even aware of the fundamental entities that shape it: the protons in an atom's nucleus and the electrons surrounding it. Mendeleyev based his table on the two properties, atomic weight and valency, which are reasonable approximations: atomic weight closely mirrors the number of protons, while valency is linked to the accessibility of electrons.

What distinguished Mendeleyev from the others was his ability to grasp that the known elements fitted into a scheme that was predetermined. The elements were not being arranged to make *a* periodic table, but to fit *the* periodic table. The difference may seem trivial to us today, but for its time the mental jump was truly one of genius. The certainty that he was right gave Mendeleyev the confidence not only to forecast the discovery of 'new' elements but even to predict that a whole row of his periodic table awaited discovery because of the gap in atomic weights between cerium (140) and tantalum (182). These missing elements we now recognize as the lanthanides, only two of which, erbium and terbium, were correctly known in Mendeleyev's day.

THE DEVELOPMENT OF THE PERIODIC TABLE

Considering his isolation from the main centres of chemistry, Mendeleyev's discovery might easily have gone unnoticed. However, he was determined that it should be publicized and immediately wrote a paper, entitled 'The relationship of properties and the atomic weights of the elements', which was published in the very first edition of a new journal, *The Russian Journal of General Chemistry*. When copies of this reached Germany later that year, Mendeleyev's article was summarized in a German journal, *Zeitschrift für Chemie*, thereby ensuring it a wide circulation among the chemical community. The importance of the discovery was immediately recognized, and a German translation of Mendeleyev's paper was published that same year. The periodic table that this contained (see Figure 3) remained the standard version for almost a century.

Mendeleyev's table had eight columns labelled with the Roman numerals I to VIII, corresponding to the chemical valencies (oxidation states) of the elements. This property is revealed by the chemical formula of the highest oxide. Yet, it brought together elements that may be quite dissimilar in other ways, such as metals and non-metals. For example, in group V we find vanadium and phosphorus, which have almost no chemistry in common. Mendeleyev consequently split the columns of his periodic table into two sub-groups, labelled A and B. Vanadium was in VA and phosphorus in VB. The same pattern was repeated in the other columns with the exception of group VIII, which contained metals which were very similar and which occurred in three sets of three. These were iron–cobalt–nickel, ruthenium–rhodium–palladium and osmium–iridium–platinum.

Within 2 years of Mendeleyev's table appearing, other chemists produced alternative versions. These have continued to appear year-by-year ever since, until today there are well in excess of 600 variants, although many are very similar to one another. Most tables are two-dimensional representations, although there have been several three-

Reihen	Gruppe I. — R^2O	Gruppe II. — RO	Gruppe III. — R^2O^3	Gruppe IV. RH^4 RO^2	Gruppe V. RH^3 R^2O^5	Gruppe VI. RH^2 RO^3	Gruppe VII. RH R^2O^7	Gruppe VIII. — RO^4
1	H = 1							
2	Li = 7	Be = 9,4	B = 11	C = 12	N = 14	O = 16	F = 19	
3	Na = 23	Mg = 24	Al = 27,3	Si = 28	P = 31	S = 32	Cl = 35,5	
4	K = 39	Ca = 40	— = 44	Ti = 48	V = 51	Cr = 52	Mn = 55	Fe = 56, Co = 59, Ni = 59, Cu = 63.
5	(Cu = 63)	Zn = 65	— = 68	— = 72	As = 75	Se = 78	Br = 80	
6	Rb = 85	Sr = 87	?Yt = 88	Zr = 90	Nb = 94	Mo = 96	— = 100	Ru = 104, Rh = 104, Pd = 106, Ag = 108.
7	(Ag = 108)	Cd = 112	In = 113	Sn = 118	Sb = 122	Te = 125	J = 127	
8	Cs = 133	Ba = 137	?Di = 138	?Ce = 140	—	—	—	— — — —
9	(—)	—	—	—	—	—	—	
10	—	—	?Er = 178	?La = 180	Ta = 182	W = 184	—	Os = 195, Ir = 197, Pt = 198, Au = 199.
11	(Au = 199)	Hg = 200	Tl = 204	Pb = 207	Bi = 208	—	—	
12	—	—	—	Th = 231	—	U = 240	—	— — — —

Fig. 3 Mendeleyev's periodic table as published in Germany in 1870.

dimensional versions in the shapes of cylinders, pyramids, spirals and even trees, which make excellent displays for museums and exhibitions.

Basically there have been two approaches to devising a periodic table: the first lists all the elements in a continuous line, rather like the numbers on a tape measure, and this is then looped in such a way that like elements come together. The second version chops the tape into segments and stacks these in rows, again bringing like elements together. The former approach is what Chancourtois used in 1862, and what many others have done since. The others are direct descendants of Mendeleyev's table.

Circular versions of the continuous type of table have also been proposed, but despite their elegance and the tantalizing analogy with electrons in shells around a nucleus (see Figure 4), they all suffer the drawback of being difficult to read and interpret, because they tend to crowd together the more important elements at their centre while giving the less important elements more room at the periphery.

The preferred way of arranging the elements is still the matrix format that Mendeleyev used, but now there is more purpose to it. The modern table denotes each row as the filling of an electron orbit, which finally results in a noble gas element when the orbit is full. These noble gases have a special place in the development of chemical theory and the layout of the periodic table.

NOBLE GASES

Although Mendeleyev did not realize it, there was a group missing from his periodic table. These were the noble gases, and when they were discovered 30 years later, they were to exert an influence on the way the table was perceived.

The lightest noble gas, helium, had in fact been reported the year before Mendeleyev produced his table, by two astronomers, Pierre Janssen and Norman Lockyer, who

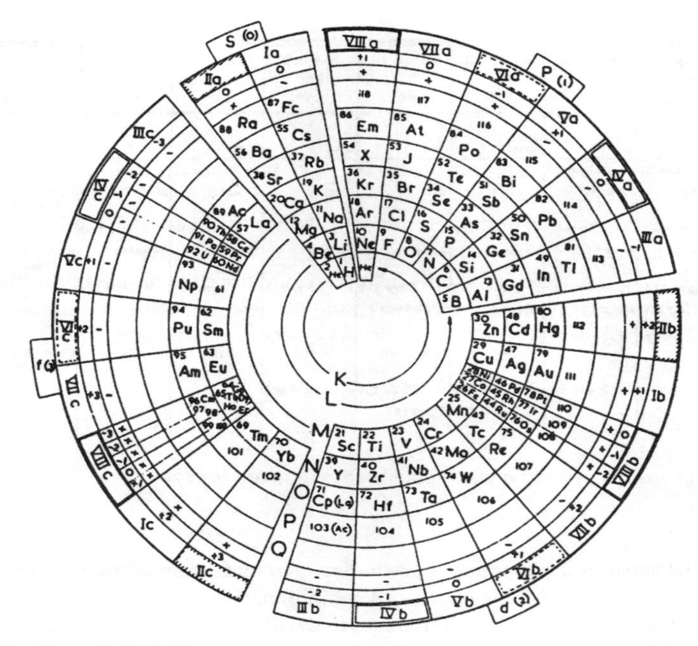

Fig. 4 A circular periodic table.

detected it in the spectrum of light from the Sun in 1868. The newly discovered element was called helium (from the Greek word *helios*) but it was generally assumed that it would not be found on Earth.

Then, in 1894, Lord Rayleigh and William Ramsay separated a new gas, argon, from air. A terrestrial source of helium was discovered a few months later (see p. 176). Having measured the atomic weight of helium as 4 and that of argon as 40, Ramsay was able to see that there was a missing group in the periodic table, in which case there must be another gaseous element of atomic number about 20, and others heavier than argon with atomic weights around 82 and 130. These would fill gaps in the list of atomic weights between bromine (80) and rubidium (85), and between iodine (127) and caesium (133).

By 1899 Rayleigh and Travers had extracted the three missing gases: neon (atomic weight 20), krypton (84) and xenon (131). The heaviest member of the group is radioactive radon, whose longest-lived isotope is radon-222; F. E. Dorn at Halle in Germany discovered this inside sealed ampoules of radium in 1900.

We now see the noble gases as *completing* the rows of the periodic table, but chemists were first inclined to place them as group 0 at the start of the table because they had no tendency to form compounds, in other words their valency was 0. Today we put them at the end of the rows of the periodic table since they represent the culmination of adding electrons to a particular electron shell, when their outer shells are full.

THE STRUCTURE OF ATOMS AND THE PERIODIC TABLE

The noble gases completed the format of the periodic table as we know it (see p. 539). In the long form there are seven rows, or periods, containing 2, 8, 8, 18, 18, 32 and 32 elements, respectively. This pattern can be understood in terms of the underlying electronic structure of atoms; indeed the pattern is really that of electron orbits, which hold 2, 6, 10 and 14 electrons. The innermost shell holds two electrons, the next one six, etc. Combinations of these numbers give rise to the numbers 8 (= 2 + 6), 18 (= 2 + 6 + 10) and 32 (= 2 + 6 + 10 + 14).

This underlying structure could only be understood once the nature of the atoms was revealed. This puzzle began to unravel with the discovery of the electron, in 1897, by J. J. Thomson (1856-1940). In 1904, he said that elements with similar electron configurations had similar properties, and he suggested that the electron configuration was the basis for the periodicity of the elements. A few years after this, in 1909, Ernest Rutherford (1871-1937) bombarded thin gold foil with α-particles, and discovered that atoms must consist of a tiny, positively-charged nucleus in which almost all the mass was concentrated.

In 1913, Henry G. H. Moseley (1887-1915), using X-ray spectroscopy, showed that the periodic table should be set up in order of the atomic *numbers* of the elements. He came to this idea after having heard of the hypothesis of Anton van den Broek, a Dutch physicist, who had said earlier that year that the positive charge of the nucleus may be the key to the periodic table. Moseley showed that the sequence of elements in the table was really the order of their atomic numbers. Again in that remarkable year, Niels Bohr (1885-1962) linked the form of the periodic table to the atomic structure of atoms, using quantum theory to explain the arrangement of the atom's electrons in orbits around the nucleus. He published a periodic table based on electron energy levels in 1922.

If the elements are arranged in rows of increasing atomic number, and in columns having the same electron outer shell, then we arrive at the long form of the periodic table. Across a row of the periodic table we are adding electrons to a particular shell until that shell is full, when we arrive at one of the noble gases. Consequently these represent a natural break in the table.

Despite these advances in atomic theory, Mendeleyev's eight-column periodic table remained the common type for almost a hundred years. Then, the extended or long form slowly displaced it as the synthesized elements were announced and interest centred on the final row. The long form had earlier been championed by the great Swiss inorganic chemist Alfred Werner (1866–1919) in 1905.

To understand the periodic table, it is necessary to realize that atoms have arrays of electron orbits around the nucleus. We can imagine these being filled one-by-one with electrons, starting at the levels of lowest energy nearest the nucleus, and continuing until the number of negative electrons balances the number of positive protons in the atom's nucleus. This came to be understood with the development of quantum mechanics in the 1920s. Studies of the spectra of atoms showed there to be lines corresponding to the energies of electromagnetic radiation (ultraviolet, visible and infrared) that represented electrons jumping between these various levels.

The primary orbits are numbered from 1 to 7 from the nucleus outwards. Secondary orbits are given the letters s, p, d and f (derived respectively from the words 'sharp', 'principal', 'diffuse' and 'fundamental', used to describe the lines in the spectrum corresponding to electrons moving in and out of the orbits). The orbit nearest the nucleus is just a single orbit and is labelled 1s, the next is a pair of orbits labelled 2s and 2p, the next is a trio of orbits, 3s, 3p and 3d, and so on. However, as these fan out from the nucleus they begin to overlap with sub-orbits even further out so that the next one, 4s, is in fact occupied before the 3d sub-orbit.

Orbits can hold increasing numbers of electrons the further from the nucleus they are. The s ones can hold 2 electrons, the p ones 6, the d ones 10 and the f ones 14. These are the basis of the various blocks of the periodic table. The s-block elements consist of two groups numbered 1 (the alkali metals) and 2 (the alkaline earths), the p-block elements consist of six groups (numbered 13 to 18), the d-block elements have 10 groups (numbered 3 to 12) and the f-block elements consist of two rows, 4f and 5f, each of 14 elements, which are not given group numbers.

The changeover from the eight-column periodic table to the modern form was not without its difficulties. When Mendeleyev's periodic table of eight groups was turned into the long form of 18 groups, the Europeans numbered the groups in the left-hand half IA to VIII, and those in the right-hand half IB to VIIB, with the noble gases being group 0. The Americans, on the other hand, kept the IA/IB system of Mendeleyev, so their long form of the table was numbered IA, IIA, etc. for the s- and p-block elements, and IB, IIB, etc. for the d-block. Both systems numbered the alkali metals groups IA, and the alkaline earth metals IIA, but after that they diverged.

Various suggestions were made to resolve the difficulty, but finally the International Union of Pure and Applied Chemists (IUPAC) suggested that the groups simply be numbered 1 through to 18 across the periodic table. After much heart-searching the American Chemical Society (ACS) finally agreed. The f-block does not fit into the numbering system but this poses no problem since the 4f and 5f periods of elements are best dealt with as rows of the periodic table rather than groups.

PROBLEM ELEMENTS

There are four elements that pose a problem for anyone devising a periodic table: hydrogen (atomic number 1), helium (2), lanthanum (57) and actinium (89).

Hydrogen has an electron configuration 1s, which should place it in the s-block of the table above lithium (2s) and this is where it is to be found in some tables. But hydrogen is not a metal like lithium. Indeed, by the same logic, the noble gas helium should be placed above beryllium, but no form of the table places it so because it is clearly a noble gas and must go above neon even though this has the electron configuration of a filled p-shell.

Some tables place hydrogen by itself, or with helium, in the very centre of the table, floating free above the other elements. Others place hydrogen above fluorine, although it shares little in common with the halogen gases. Some tables give it double billing and place it above both lithium and fluorine. The periodic table at the end of this book has hydrogen next to helium and above fluorine, but it has both hydrogen and helium in a separate block, labelled 1s, as a mark of their uniqueness.

The other problem concerns the f-block of elements, which comes between the s-block and d-block at the bottom of the table. For reasons of economy of space, this block is generally written below the d-block.

Which elements belong to the f-block? Traditionally, and mainly for historical reasons, lanthanum and actinium were placed in group 3, below scandium and yttrium. After lanthanum come the other so-called rare-earth elements (the lanthanides), which are in fact the upper row of the f-block, and this begins with cerium. The weight of evidence is that lanthanum and actinium are the first members of their respective rows of the f-block. These rows must contain 14 elements and so the final ones in each are ytterbium and nobelium respectively.[218]

As long as chemistry is studied, there will be a periodic table. Even if some day we communicate with another part of the Universe, we can be sure that one thing both cultures will have in common is an ordered system of the elements that will be instantly recognizable by both intelligent life forms. Meanwhile the periodic table has inspired artists to construct periodic tables in which the elements are symbolized by flowers and foods. Perhaps the most dramatic periodic table is that by the Glasgow artist, Murray Robertson, which is entitled 'Visual Elements' and for which he had created a stunning computer graphic for each element. This can be accessed, via the Internet, on the Royal Society of Chemistry's website.

[218] In 1982 William B. Jensen argued cogently for this change to be made to periodic tables in an article in the *Journal of Chemical Education* (volume 59, page 634).

Appendix: The discovery of the elements in chronological order

Elements known to the ancient world

Date	Element	Discoverer	Place
Prehistory	carbon	not known	not known
Prehistory	sulfur	not known	not known
Prehistory	copper	not known	not known
c.5000 BC	silver	not known	not known
c.5000 BC	gold	not known	not known
c.2500 BC	iron	not known	not known
c.2100 BC	tin	not known	not known
c.1600 BC	antimony	not known	not known
c.1500 BC	mercury	not known	not known
c.1000 BC	lead	not known	not known

Elements discovered in the Middle Ages

Date	Element	Discoverer	Place
c.1200	zinc	not known	India
c.1250	arsenic	Magnus	Germany
c.1500	bismuth	not known	Central Europe
1669	phosphorus	Brandt	Hamburg
Pre-1700	platinum	not known	Central America

Elements discovered in the eighteenth century

Year	Element	Discoverer(s)	Place
1735	cobalt	Brandt	Stockholm
1751	nickel	Cronstedt	Stockholm
1755	magnesium	Black	Edinburgh
1766	hydrogen	Cavendish	London
1772	nitrogen	Rutherford	Edinburgh
1772/4	oxygen	Scheele/Priestley	Uppsala/Wiltshire
1774	chlorine	Scheele	Uppsala
1774	manganese	Gahn	Stockholm

Elements discovered in the eighteenth century *(continued)*

Year	Element	Discoverer(s)	Place
1780	chromium	Vauquelin	Paris
1781	molybdenum	Hjelm	Uppsala
1783	tellurium	Müller von Reichenstein	Sibiu, Romania
1783	tungsten	Elhuiyar and Elhuiyar	Vergara, Spain
1789	zirconium	Klaproth	Berlin
1789	uranium	Klaproth	Berlin
1791	titanium	Gregor	Creed, Cornwall
		Klaproth	Berlin
1794	yttrium	Gadolin	Turku, Finland
1797	beryllium	Vauquelin	Paris

Elements discovered in the nineteenth century*

Year	Element	Discoverer(s)	Place
1801	vanadium	del Rio	Mexico
1801	niobium	Hatchett	London
1802	tantalum	Ekeberg	Uppsala
1803	rhodium	Wollaston	London
1803	palladium	Wollaston	London
1803	osmium	Tennant	London
1803	iridium	Tennant	London
1803	cerium	Berzelius and Hisinger	Vestmanland, Sweden
1807	potassium	Davy	London
1807	sodium	Davy	London
1808	boron	Gay-Lussac and Thenard/Davy	Paris/London
1808	calcium	Davy	London
1808	strontium	Crawford	Edinburgh
1808	ruthenium	Sniadecki	Vilno, Poland
1808	barium	Stromeyer	Göttingen
1811	iodine	Courtois	Paris
1815	thorium	Berzelius	Stockholm
1817	lithium	Arfvedson	Stockholm
1817	selenium	Berzelius	Stockholm
1817	cadmium	Stromeyer	Göttingen
1824	silicon	Berzelius	Stockholm
1825	aluminium	Oersted	Copenhagen
1825/6	bromine	Löwig/Balard	Heidelberg/Montpellier

1839	lanthanum	Mosander	Stockholm
1842	erbium	Mosander	Stockholm
1843	terbium	Mosander	Stockholm
1860	caesium	Bunsen and Kirchhoff	Heidelberg
1861	rubidium	Bunsen and Kirchhoff	Heidelberg
1861	thallium	Crookes	London
1863	indium	Reich and Richter	Freiberg
[1868	helium	Janssen/Lockyer	India/England]
1875	gallium	de Boisbaudran	Paris
1878	holmium	Cleve/Delafontaine and Soret	Uppsala/Geneva
1878	ytterbium	de Marignac	Geneva
1879	scandium	Nilson	Uppsala
1879	samarium	de Boisbaudran	Paris
1879	thulium	Cleve	Uppsala
1880	gadolinium	de Marignac	Geneva
1885	praseodymium	von Welsbach	Vienna
1885	neodymium	von Welsbach	Vienna
1886	germanium	Winkler	Freiberg
1886	fluorine	Moissan	Paris
1886	dysprosium	de Boisbaudran	Paris
1894	argon	Rayleigh and Ramsay	London
1895	helium	Ramsay	London
1898	krypton	Ramsay and Travers	London
1898	neon	Ramsay and Travers	London
1898	xenon	Ramsay and Travers	London
1898	polonium	M. Curie	Paris
1898	radium	M. Curie and P. Curie	Paris
1899	actinium	Debierne	Paris

*Elements in square brackets were discovered but not confirmed at the time.

Elements discovered and synthesized in the twentieth century*

Year	Element	Discoverer(s)	Place
1900	radon	Dorn	Halle, Germany
1901	europium	Demarçay	Paris
1907	lutetium	Urbain	Paris
		James (with Auer, see p. 241)	New Hampshire, USA
1917	protactinium	Hahn and Meitner	Berlin
		Fajans	Karlsruhe
		Soddy, Cranston and Fleck	Glasgow

Elements discovered and synthesized in the twentieth century *(continued)*

Year	Element	Discoverer(s)	Place
1923	hafnium	Coster and Hevesey	Copenhagen
1925	rhenium	Noddack, Tacke and Berg	Berlin
[1925	technetium, masurium	Noddack, Tacke and Berg	Berlin]

* Elements in square brackets were discovered but not confirmed at the time.

Year	Element	Synthesized by	Place
1937	technetium	Perrier and Segré	Palermo
1938	promethium, cyclonium	Law, Pool, Kurbatov, Quill	Ohio
1939	francium	Perey	Paris
1940	neptunium	McMillan and Abelson	Berkeley, California
1940	astatine	Corson, Mackenzie and Segré	Berkeley, California
1940	plutonium	Seaborg and colleagues	Berkeley, California
1944	curium	Seaborg and colleagues	Berkeley, California
1944	americium	Seaborg and colleagues	Chicago
1945	promethium	Marinsky, Glendenin and Coryell	Oak Ridge, USA
1949	berkelium	Ghiorso and Seaborg	Berkeley, California
1950	californium	Ghiorso, Seaborg and colleagues	Berkeley, California
1952	einsteinium	Ghiorso and colleagues	Berkeley, California
1952	fermium	Ghiorso and colleagues	Berkeley, California
1955	mendelevium	Ghiorso and colleagues	Berkeley, California
1956	nobelium	JINR team	Dubna, USSR
1961	lawrencium	Ghiorso and colleagues	Berkeley, California
1964	rutherfordium	Flerov and colleagues/ Ghiorso and colleagues	Dubna, USSR/ Berkeley, California
1967	dubnium	Flerov and colleagues/ Ghiorso and colleagues	Dubna, USSR/ Berkeley, California
1974	seaborgium	Ghiorso and colleagues	Berkeley, California
1981	bohrium	Armbruster, Münzenberg and colleagues	Darmstadt, Germany
1982	meitnerium	Armbruster, Münzenberg and colleagues	Darmstadt, Germany
1984	hassium	Armbruster, Münzenberg and colleagues	Darmstadt, Germany
1994	darmstadtium	Armbruster, Münzenberg and colleagues	Darmstadt, Germany
1994	element 111	Armbruster, Münzenberg and colleagues	Darmstadt, Germany
1996	element 112	Armbruster, Hofmann and colleagues	Darmstadt, Germany
1998	element 114	Oganessian, Utyonkov and colleagues	Dubna, Russia
2000	element 116	Oganessian and colleagues	Dubna, Russia

Bibliography

In compiling this reference work I have drawn on more than 20 years' accumulation of hundreds of magazine and newspaper articles culled from a wide range of sources. Many of these were taken from *Chemical and Engineering News*, published by the American Chemical Society, *Chemistry in Britain*, published by the Royal Society of Chemistry, *Chemistry and Industry*, published by the Society for Chemical Industry, London, and *New Scientist*. More recently I have drawn upon newsletters and websites published by industry associations, which share a common interest in a particular element or metal. These are not only highly informative but carry details of current usages and production figures.

Despite all this wealth of information, I have found that books are still the most comprehensive sources of information; the following ones have proved invaluable (listed in alphabetical order of first-named author):

Belitz, H.-D. and Grosch, W. (1987). *Food Chemistry.* Springer-Verlag, Heidelberg.

Bowen, H. J. M. (1979). *Environmental Chemistry of the Elements.* Academic Press, London.

Brown, S. S. and Kodama, Y. (eds) (1987). *Toxicology of Metals.* Ellis Horwood, Chichester.

Büchner, W., Schliebs, R., Winter, G. and Büchel, K. H. (1989). *Industrial Inorganic Chemistry.* VCH, Weinheim.

Budavari, S. (ed.) (1989). *The Merck Index,* 11th edn. Merck & Co. Inc., Rahway, NJ.

Cardarelli, F. (2000). *Materials Handbook: a Concise Desktop Reference.* Springer, London.

Chown, M. (1999). *The Magic Furnace.* Jonathan Cape, London.

Cotton, F. A. and Wilkinson, G. *Advanced Inorganic Chemistry,* 5th edn., John Wiley & Sons, New York.

Cox, P. A. (1989). *The Elements: Their Origin, Abundance and Distribution.* Oxford University Press, Oxford.

Cox, P. A. (1997). *The Elements on Earth.* Oxford University Press, Oxford.

Crystal, D. (ed.) (1992). *The Cambridge Encyclopedia.* Cambridge University Press, Cambridge.

Daintith, J. and Gjertsen, D. (eds) (1999). *Oxford Dictionary of Scientists.* Oxford University Press, Oxford.

Dean, J. A. (ed.) (1992). *Llange's Handbook of Chemistry,* 14th edn. McGraw-Hill Inc., New York.

Department of Health (1991). *Dietary Reference Values for Food Energy and Nutrients for the UK.* HMSO, London.

Diem, K. and Lentner, C. (eds) (1970). *Scientific Tables,* 7th edn. Documenta Giegy, Basel.

Emsley, J. (1998). *The Elements,* 3rd edn. Clarendon Press, Oxford.

Fergusson, J. E. (1990). *The Heavy Elements: Chemistry, Environmental Impact and Health Effects.* Pergamon Press, Oxford.

Foth, H. D. (1990). *Fundamentals of Soil Science,* 8th edn. John Wiley & Sons, New York.

Fricke, B. (1975). 'Superheavy elements' in *Structure and Bonding,* vol. 12.

Gray, H. B., Simon, J. D. and Trogler, W. C. (1995). *Braving the Elements.* University Science Books, Sausalito, California.

Greenwood, N. N. and Earnshaw, A. (1997). *The Chemistry of the Elements,* 2nd edn. Pergamon Press, Oxford.

Hampel, C. A. (ed.) (1968). *The Encyclopedia of the Chemical Elements.* Reinhold Book Corpn., New York.

Hawley, G. (ed.) (1981). *The Condensed Chemical Dictionary,* 10th edn. Van Nostrand Reinhold, New York.

Hubbard, E., Stephenson, M. and Waddington, D. (eds) (1999). *The Essential Chemical Industry,* 4th edn. Chemical Industry Education Centre, University of York.

Hunter, D. (1976). *The Diseases of Occupations,* 5th edn. Hodder and Stoughton, London.

Ihde, A. J. (1966). *The Development of Modern Chemistry.* Harper & Row, New York.

Isaacs, A., Daintith, J. and Martin, E. (eds) (1999). *Oxford Dictionary of Science,* 4th edn. Oxford University Press, Oxford and New York.

Jakubke, H-D. and Jeschkeit, H. (1993). *Concise Encyclopedia of Chemistry* (translated by M. Eagleston). Walter de Gruyter, Berlin.

James, A. M. and Lord, M. P. (1992). *Macmillan's Chemical and Physical Data.* Macmillan, London.

James, L. K. (ed.) (1993). *Noble Laureates in Chemistry, 1901-1992.* American Chemical Society and the Chemical Heritage Foundation, USA.

Kabata-Pendias, A. and Pendias, H. (1991). *CRC Trace Elements in Soils and Plants,* 2nd edn. CRC Press, Boca Raton, Florida.

Kaye, G. W. C. and Laby, T. H. (1993). *Tables of Physical and Chemical Constants,* 15th edn. Longman Scientific & Technical, London.

Kent, J. A. (ed.) (1992). *Riegel's Handbook of Industrial Chemistry,* 9th edn. Van Nostrand Reinhold, New York.

Kilbourn, B. T. (1993). *A Lanthanide Lanthology,* Vols A-L and M-Z. Molycorp Inc., Fairfield, New Jersey.

Kutsky, R. J. (1981). *Handbook of Vitamins, Minerals and Hormones,* 2nd edn. Van Nostrand Reinhold, New York.

Lehrer, T. (1999). *Too Many Songs.* Methuen, London.

Lide, D. R. (ed.) (1995). *CRC Handbook of Chemistry and Physics,* 75th edn. CRC Press, Boca Raton, Florida.

Luckey, T. D. and Venugopal, B. (1977). *Metal Toxicity in Mammals,* Vol. 1. Plenum Press, New York.

Marshall, P. (2001). *The Philosopher's Stone.* Macmillan, London.

Mason, S. F. (1991). *Chemical Evolution.* Clarendon Press, Oxford.

Merian, E. (ed.) (1991). *Metals and Their Compounds in the Environment.* VCH, Weinheim.

Morss, L. R. and Fuger, J. (eds) (1992). *Transuranium Elements: a Half Century.* American Chemical Society, Washington, DC.

Moses, A. J. (1978). *The Practising Scientist's Handbook.* Van Nostrand Reinhold, New York.

Muir, H. (ed.) (1994). *Larousse Dictionary of Scientists.* Larousse, Edinburgh.

Nechaev, I. and Jenkins, G. W. (1998). *The Chemical Elements.* Tarquin Publications, Diss, Norfolk.

Nickel, E. H. and Nichols, M. C. (1991). *Mineral Reference Manual.* Van Nostrand Reinhold, New York.

Pearce, J. (ed.) (1987). *Gardner's Chemical Synonyms and Trade Names,* 9th edn. Gower Technical Press, Aldershot.

Prasad, A. S. (ed.) (1976). *Trace Elements in Human Health and Disease*, Vol. II: Essential and Toxic Elements. Academic Press, New York.

Purves, D. (1985). *Trace Element Contamination of the Environment*, revised edn. Elsevier, Amsterdam.

Quadbeck-Seeger, H-J. (ed.) (1999). *World Records in Chemistry*. Wiley-VCH, Weinheim.

Roberts, W. L., Campbell, T. J.and Rapp, G. R., Jr (1990). *Encyclopedia of Minerals*. Van Nostrand Reinhold, New York.

Rossotti, H. (1998). *Diverse Atoms: Profiles of the Chemical Elements*. Oxford University Press, Oxford.

Sax, N. I. and Lewis, R. J., Sr (1989). *Dangerous Properties of Industrial Materials*, 7th edn. Van Nostrand Reinhold, New York.

Schwochau, K. (1984). 'Extraction of metals from sea water'. *Topics in Current Chemistry*, no. 124, p. 91. Springer Verlag, Berlin.

Seaborg, G. T. and Loveland, W. D. (1990). *The Elements Beyond Uranium*. Wiley Interscience, New York.

Selinger, B. (1998). *Chemistry in the Marketplace*, 5th edn. Harcourt Brace, Sydney.

Snyder, C. H. (1992). *The Extraordinary Chemistry of Ordinary Things*. John Wiley & Sons, Inc., New York.

Stwertka, A. (1996). *A Guide to the Elements*. Oxford University Press, New York.

Venetsky, S. I. (1981). *On Rare and Scattered Metals: Tales about Metals* (translated by N. G. Kittel). Mir Publishers, Moscow.

Wade, A. (ed.) (1977). *Martindale, the Extra Pharmacopoeia*. The Pharmaceutical Press, London.

Weeks, M. E. and Leicester, H. M. (1968). *Discovery of the Elements*, 7th edn. Chemical Education, Easton, Pennsylvania.

Williams, J. P. and Fraústo da Silva, J. J. R. (1996). *The Natural Selection of the Chemical Elements*. Clarendon Press, Oxford.

Williams, J. P. and Fraústo da Silva, J. J. R. (1991). *The Biological Chemistry of the Elements*. Clarendon Press, Oxford.

For some of the elements there are individual monographs devoted entirely to the chemistry of that element, and sometimes even to one aspect of its chemistry. These were of secondary importance in compiling this book, but those that proved helpful were (listed in alphabetical order of the element concerned):

Sprague, A. (1972). *The Ecological Significance of Boron*. US Borax, Valencia, California.

Travis, N. J. and Cocks, E. J. (1984). *The Tincal Trail: a History of Borax*. Harrap, London.

Kilbourn, B. T. (1992). *Cerium: a Guide to its Role in Chemical Technology*. Molycorp Inc., Fairfield, New Jersey.

Foster, R. P. (ed.) (1991). *Gold Metallogeny and Exploration*. Blackie, Glasgow.

Hetzel, S. (1989). *The Story of Iodine Deficiency*. Oxford University Press, Oxford.

Evans, C. H. (1990). *Biochemistry of the Lanthanides*. Plenum, New York.

Griffin, T. B. and Knelson, J. H. (1975). *Lead*. Georg Thieme Publishers, Stuttgart.

D'Itri, A. and D'Itri, F. M. (1977). *Mercury Contamination: a Human Tragedy*. Wiley Interscience, New York.

Mitra, S. (1986). *Mercury in the Ecosystem*. Tans Tech Publications, Aedermannsdorf, Switzerland.

Braithwaite E. R. and Haber, J. (eds) (1995). *Molybdenum: an Outline of its Chemistry and Uses*. Elsevier Science, Amsterdam.

Abelson, P. H. (1992). Discovery of neptunium. In *Transuranium Elements* (eds L.R. Morss and J. Fuger), pp. 50–55. American Chemical Society, Washington, DC.

Combs, G. F., Jr, and Combs, S. B. (1986). *The Role of Selenium in Nutrition.* Academic Press, Orlando, Florida.

Lewis, A. (1982). *Selenium, The Essential Element You Might Not be Getting Enough of.* Thorsons, Wellingborough, UK.

Shamberger, R. J. (1983). *Biochemistry of Selenium.* Plenum, New York.

Zingaro, R. A. and Cooper, W. C. (eds) (1974). *Selenium.* Van Nostrand Reinhold, New York.

Smith, I. C. and Carson, B. L. (1977). *Trace Metals in the Environment: Silver.* Ann Arbor Science, Ann Arbor, Michigan.

Smith, I. C. and Carson, B. L. (1977). *Trace Metals in the Environment: Thallium.* Ann Arbor Science, Ann Arbor, Michigan.

Harrion, P. G. (ed.) (1989). *Chemistry of Tin.* Blackie, Glasgow.

Horovitz, C. T. (1995). 'Two hundred years of research and development of yttrium'. *Trace Elements and Electrolytes*, vol. 12, p. 153.

For further reading on the periodic table consult:

Atkins, P. (1995). *The Periodic Kingdom.* Weidenfeld & Nicolson, London.

Bensaude-Vincent, B. (1984). 'La genèse du tableau de Mendeleyev'. *La Recherche*, vol. 15, p. 1206.

Brock, W. H. (1992). *The Fontana History of Chemistry.* Fontana Press, London.

Cassebaum, H. and Kauffman, G. B. (1971). The periodic system of the chemical elements: the search for its discoverer. *Isis*, vol. 62, p. 314.

Emsley, J. (1984). 'Mendeleyev's dream table'. *New Scientist*, 7 March, p. 32.

Emsley, J. (1987). The development of the periodic table of the chemical elements. *Interdisciplinary Science Reviews*, vol. 12, p. 23.

Gilreath, E. S. (1958). *Fundamental Concepts of Inorganic Chemistry.* McGraw-Hill, New York.

Holden, N. E. (1984). 'Mendeleyev and the periodic classification of the elements'. *Chemistry International*, no. 6, p. 18.

Mazurs, E.G. (1974). *Graphic Representations of the Periodic System During One Hundred Years.* University of Alabama Press, Tuscloosa, Alabama.

Rouvray, D. H. (1994). 'Turning the tables on Mendeleyev'. *Chemistry in Britain*, May issue, p. 373.

Sanderson, R. T. (1967). *Chemical Periodicity.* Reinhold Publishing Corp., New York.

Scerri, E. R. (1994). 'Plus ça change...' . *Chemistry in Britain*, May issue, p. 379.

Seaborg, G. T. (1990). *The Elements Beyond Uranium.* John Wiley & Sons Inc., New York.

Strathern, P. (2000). *Mendeleyev's Dream: the Quest for the Elements.* Hamish Hamilton, London.

Van Spronsen, J. W. (1969). *The Periodic System of Chemical Elements: a History of the First Hundred Years.* Elsevier, Amsterdam.

Venables, F. P. (1896). *The Development of the Periodic Law.* Chemical Publishing Co., Easton, Pennsylvania.

Weeks, M. E. and Leicester, H. M. (1968). *Discovery of the Elements*, 7th edn. Journal of Chemical Education, Easton, Pennsylvania.

The chemical elements in order of their atomic numbers, with chemical symbol

1	Hydrogen	H
2	Helium	He
3	Lithium	Li
4	Beryllium	Be
5	Boron	B
6	Carbon	C
7	Nitrogen	N
8	Oxygen	O
9	Fluorine	F
10	Neon	Ne
11	Sodium	Na
12	Magnesium	Mg
13	Aluminium	Al
14	Silicon	Si
15	Phosphorus	P
16	Sulfur	S
17	Chlorine	Cl
18	Argon	Ar
19	Potassium	K
20	Calcium	Ca
21	Scandium	Sc
22	Titanium	Ti
23	Vanadium	V
24	Chromium	Cr
25	Manganese	Mn
26	Iron	Fe
27	Cobalt	Co
28	Nickel	Ni
29	Copper	Cu
30	Zinc	Zn
31	Gallium	Ga
32	Germanium	Ge
33	Arsenic	As
34	Selenium	Se
35	Bromine	Br
36	Krypton	Kr
37	Rubidium	Rb
38	Strontium	Sr
39	Yttrium	Y
40	Zirconium	Zr
41	Niobium	Nb
42	Molybdenum	Mo
43	Technetium	Tc
44	Ruthenium	Ru
45	Rhodium	Rh
46	Palladium	Pd
47	Silver	Ag
48	Cadmium	Cd
49	Indium	In
50	Tin	Sn
51	Antimony	Sb
52	Tellurium	Te
53	Iodine	I
54	Xenon	Xe
55	Caesium	Cs
56	Barium	Ba
57	Lanthanum	La
58	Cerium	Ce
59	Praseodymium	Pr
60	Neodymium	Nd
61	Promethium	Pm
62	Samarium	Sm
63	Europium	Eu
64	Gadolinium	Gd
65	Terbium	Tb
66	Dysprosium	Dy
67	Holmium	Ho
68	Erbium	Er
69	Thulium	Tm
70	Ytterbium	Yb
71	Lutetium	Lu
72	Hafnium	Hf
73	Tantalum	Ta
74	Tungsten	W
75	Rhenium	Re
76	Osmium	Os
77	Iridium	Ir
78	Platinum	Pt
79	Gold	Au
80	Mercury	Hg
81	Thallium	Tl
82	Lead	Pb
83	Bismuth	Bi
84	Polonium	Po
85	Astatine	At
86	Radon	Rn
87	Francium	Fr
88	Radium	Ra
89	Actinium	Ac
90	Thorium	Th
91	Protactinium	Pa
92	Uranium	U
93	Neptunium	Np
94	Plutonium	Pu
95	Americium	Am
96	Curium	Cm
97	Berkelium	Bk
98	Californium	Cf
99	Einsteinium	Es
100	Fermium	Fm
101	Mendelevium	Md
102	Nobelium	No
103	Lawrencium	Lr
104	Rutherfordium	Rf
105	Dubnium	Db
106	Seaborgium	Sg
107	Bohrium	Bh
108	Hassium	Hs
109	Meitnerium	Mt
110	Darmstadtium	Ds
111	Unununium	Uuu
112	Ununbiium	Uub
114	Ununquadium	Uuq
116	Ununhexium	Uuh
118	Ununoctium	Uuo

The chemical elements in alphabetical order of their chemical symbol, with atomic numbers

Symbol	Element	Atomic number
Ac	Actinium	89
Ag	Silver	47
Al	Aluminium	13
Am	Americium	95
Ar	Argon	18
As	Arsenic	33
At	Astatine	85
Au	Gold	79
B	Boron	5
Ba	Barium	56
Be	Beryllium	4
Bh	Bohrium	107
Bi	Bismuth	83
Bk	Berkelium	97
Br	Bromine	35
C	Carbon	6
Ca	Calcium	20
Cd	Cadmium	48
Ce	Cerium	58
Cf	Californium	98
Cl	Chlorine	17
Cm	Curium	96
Co	Cobalt	27
Cr	Chromium	24
Cs	Caesium	55
Cu	Copper	29
Db	Dubnium	105
Ds	Darmstadtium	110
Dy	Dysprosium	66
Er	Erbium	68
Es	Einsteinium	99
Eu	Europium	63
F	Fluorine	9
Fe	Iron	26
Fm	Fermium	100
Fr	Francium	87
Ga	Gallium	31
Gd	Gadolinium	64
Ge	Germanium	32
H	Hydrogen	1
He	Helium	2
Hf	Hafnium	72
Hg	Mercury	80
Ho	Holmium	67
Hs	Hassium	108
I	Iodine	53
In	Indium	49
Ir	Iridium	77
K	Potassium	19
Kr	Krypton	36
La	Lanthanum	57
Li	Lithium	3
Lr	Lawrencium	103
Lu	Lutetium	71
Md	Mendelevium	101
Mg	Magnesium	12
Mn	Manganese	25
Mo	Molybdenum	42
Mt	Meitnerium	109
N	Nitrogen	7
Na	Sodium	11
Nb	Niobium	41
Nd	Neodymium	60
Ne	Neon	10
Ni	Nickel	28
No	Nobelium	102
Np	Neptunium	93
O	Oxygen	8
Os	Osmium	76
P	Phosphorus	15
Pa	Protactinium	91
Pb	Lead	82
Pd	Palladium	46
Pm	Promethium	61
Po	Polonium	84
Pr	Praseodymium	59
Pt	Platinum	78
Pu	Plutonium	94
Ra	Radium	88
Rb	Rubidium	37
Re	Rhenium	75
Rf	Rutherfordium	104
Rh	Rhodium	45
Rn	Radon	86
Ru	Ruthenium	44
S	Sulfur	16
Sb	Antimony	51
Sc	Scandium	21
Se	Selenium	34
Sg	Seaborgium	106
Si	Silicon	14
Sm	Samarium	62
Sn	Tin	50
Sr	Strontium	38
Ta	Tantalum	73
Tb	Terbium	65
Tc	Technetium	43
Te	Tellurium	52
Th	Thorium	90
Ti	Titanium	22
Tl	Thallium	81
Tm	Thulium	69
U	Uranium	92
Uub	Ununbiium	112
Uuh	Ununhexium	116
Uuo	Ununoctium	118
Uuq	Ununquadium	114
Uuu	Unununium	111
V	Vanadium	23
W	Tungsten	74
Xe	Xenon	54
Y	Yttrium	39
Yb	Ytterbium	70
Zn	Zinc	30
Zr	Zirconium	40

Periodic Table

	1 I (s)	2 II (s)	3 III (d)	4 IV	5 V	6 VI	7 VII	8 VIII	9 VIII	10 VIII	11 I	12 II	13 III (p)	14 IV	15 V	16 VI	17 VII	18 VIII
1s																	1 H 1.008	2 He 4.003
	3 Li 6.941	4 Be 9.012											5 B 10.811	6 C 12.011	7 N 14.007	8 O 15.999	9 F 18.998	10 Ne 20.180
	11 Na 22.990	12 Mg 24.305											13 Al 26.982	14 Si 28.086	15 P 30.974	16 S 32.066	17 Cl 35.453	18 Ar 39.948
	19 K 39.098	20 Ca 40.078	21 Sc 44.956	22 Ti 47.867	23 V 50.942	24 Cr 51.996	25 Mn 54.938	26 Fe 55.845	27 Co 58.933	28 Ni 58.693	29 Cu 63.546	30 Zn 65.39	31 Ga 69.723	32 Ge 72.61	33 As 74.922	34 Se 78.96	35 Br 79.904	36 Kr 83.80
	37 Rb 85.468	38 Sr 87.62	39 Y 88.906	40 Zr 91.224	41 Nb 92.906	42 Mo 95.94	43 Tc 98.906	44 Ru 101.07	45 Rh 102.906	46 Pd 106.42	47 Ag 107.868	48 Cd 112.411	49 In 114.818	50 Sn 118.710	51 Sb 121.760	52 Te 127.60	53 I 126.904	54 Xe 131.29
	55 Cs 132.905	56 Ba 137.327	71 Lu 174.967	72 Hf 178.49	73 Ta 180.948	74 W 183.84	75 Re 186.207	76 Os 190.23	77 Ir 192.217	78 Pt 195.08	79 Au 196.967	80 Hg 200.59	81 Tl 204.383	82 Pb 207.2	83 Bi 208.980	84 Po (209)	85 At (210)	86 Rn (222)
	87 Fr (223)	88 Ra 226.025	103 Lr (262)	104 Rf (261)	105 Db (262)	106 Sg (266)	107 Bh (267)	108 Hs (277)	109 Mt (268)	110 Ds (281)	111 Uuu (272)	112 Uub (285)	113 ?	114 Uuq (289)	115 ?	116 Uuh (292)	117 ?	118 ?
	119 ?	120 ?																

f

57 La 138.906	58 Ce 140.115	59 Pr 140.908	60 Nd 144.24	61 Pm 144.913	62 Sm 150.36	63 Eu 151.965	64 Gd 157.25	65 Tb 158.925	66 Dy 162.50	67 Ho 164.50	68 Er 167.26	69 Tm 168.934	70 Yb 173.04
89 Ac (227)	90 Th 232.038	91 Pa 231.036	92 U 236.029	93 Np 237.048	94 Pu 244.064	95 Am 243.061	96 Cm 247.070	97 Bk 247.070	98 Cf 251.080	99 Es (254)	100 Fm (257)	101 Md (258)	102 No (259)

III